D0076765

Concepts of

Biology

Senior Contributing Authors
Samantha Fowler, Clayton State University
Rebecca Roush, Sandhills Community College
James Wise, Hampton University

Table of Contents

PREFACE

Welcome to *Concepts of Biology,* an OpenStax resource. This textbook has been created with several goals in mind: accessibility, customization, and student engagement—all while encouraging students toward high levels of academic scholarship. Instructors and students alike will find that this textbook offers a strong introduction to biology in an accessible format.

About OpenStax

OpenStax is a non-profit organization committed to improving student access to quality learning materials. Our free textbooks are developed and peer-reviewed by educators to ensure they are readable, accurate, and meet the scope and sequence requirements of today's college courses. Unlike traditional textbooks, OpenStax resources live online and are owned by the community of educators using them. Through our partnerships with companies and foundations committed to reducing costs for students, OpenStax is working to improve access to higher education for all. OpenStax is an initiative of Rice University and is made possible through the generous support of several philanthropic foundations.

About OpenStax's Resources

OpenStax resources provide quality academic instruction. Three key features set our materials apart from others: they can be customized by instructors for each class, they are a "living" resource that grows online through contributions from science educators, and they are available free or for minimal cost.

Customization

OpenStax learning resources are designed to be customized for each course. Our textbooks provide a solid foundation on which instructors can build, and our resources are conceived and written with flexibility in mind. Instructors can select the sections most relevant to their curricula and create a textbook that speaks directly to the needs of their classes and student body. Teachers are encouraged to expand on existing examples by adding unique context via geographically localized applications and topical connections.

Concepts of Biology can be easily customized using our online platform. Simply select the content most relevant to your syllabus and create a textbook that speaks directly to the needs of your class. *Concepts of Biology* is organized as a collection of sections that can be rearranged, modified, and enhanced through localized examples or to incorporate a specific theme of your course. This customization feature will help bring biology to life for your students and will ensure that your textbook truly reflects the goals of your course.

Curation

To broaden access and encourage community curation, *Concepts of Biology* is "open source" licensed under a Creative Commons Attribution (CC-BY) license. The scientific community is invited to submit examples, emerging research, and other feedback to enhance and strengthen the material and keep it current and relevant for today's students. You can submit your suggestions to info@openstaxcollege.org.

Cost

Our textbooks are available for free online, and in low-cost print and e-book editions.

About *Concepts of Biology*

Concepts of Biology is designed for the single-semester introduction to biology course for non-science majors, which for many students is their only college-level science course. As such, this course represents an important opportunity for students to develop the necessary knowledge, tools, and skills to make informed decisions as they continue with their lives. Rather than being mired down with facts and vocabulary, the typical non-science major student needs information presented in a way that is easy to read and understand. Even more importantly, the content should be meaningful. Students do much better when they understand why biology is relevant to their everyday lives. For these reasons, *Concepts of Biology* is grounded on an evolutionary basis and includes exciting features that highlight careers in the biological sciences and everyday applications of the concepts at hand. We also strive to show the interconnectedness of topics within this extremely broad discipline. In order to meet the needs of today's instructors and students, we maintain the overall organization and coverage found in most syllabi for this course. A strength of *Concepts of Biology* is that instructors can customize the book, adapting it to the approach that works best in their classroom. Concepts of Biology also includes an innovative art program that incorporates critical thinking and clicker questions to help students understand—and apply—key concepts.

Coverage and Scope

Our *Concepts of Biology* textbook adheres to the scope and sequence of most one-semester non-majors courses nationwide. We also strive to make biology, as a discipline, interesting and accessible to students. In addition to a comprehensive coverage of core concepts and foundational research, we have incorporated features that draw learners into the discipline in meaningful ways. Our scope of content was developed after surveying over a hundred biology professors and listening to their coverage needs. We provide a thorough treatment of biology's fundamental concepts with a scope that is manageable for instructors and students alike.

Unit 1: **The Cellular Foundation of Life**. Our opening unit introduces students to the sciences, including the process of science and the underlying concepts from the physical sciences that provide a framework within which learners comprehend biological processes. Additionally, students will gain solid understanding of the structures, functions, and processes of the most basic unit of life: the cell.

Unit 2: **Cell Division and Genetics**. Our genetics unit takes learners from the foundations of cellular reproduction to the experiments that revealed the basis of genetics and laws of inheritance.

Unit 3: **Molecular Biology and Biotechnology**. Students will learn the intricacies of DNA, protein synthesis, and gene regulation and current applications of biotechnology and genomics.

Unit 4: **Evolution and the Diversity of Life**. The core concepts of evolution are discussed in this unit with examples illustrating evolutionary processes. Additionally, the evolutionary basis of biology reappears throughout the textbook in general discussion and is reinforced through special call-out features highlighting specific evolution-based topics. The diversity of life is explored with detailed study of various organisms and discussion of emerging phylogenetic relationships between and among bacteria, protist kingdoms, fungi, plants, and animals.

Unit 5: **Animal Structure and Function**. An introduction to the form and function of the animal body is followed by chapters on the immune system and animal development. This unit touches on the biology of all organisms while maintaining an engaging focus on human anatomy and physiology that helps students connect to the topics.

Unit 6: **Ecology**. Ecological concepts are broadly covered in this unit, with features highlighting localized, real-world issues of conservation and biodiversity.

Pedagogical Foundation and Features

Because of the impact science has on students and society, an important goal of science education is to achieve a scientifically literate population that consistently makes informed decisions. Scientific literacy transcends a basic understanding of scientific principles and processes to include the ability to make sense of the myriad instances where people encounter science in day-to-day life. Thus, a scientifically literate person is one who uses science content knowledge to make informed decisions, either personally or socially, about topics or issues that have a connection with science. Concepts of Biology is grounded on a solid scientific base and designed to promote scientific literacy. Throughout the text, you will find features that engage the students in scientific inquiry by taking selected topics a step further.

Evolution in Action features uphold the importance of evolution to all biological study through discussions like "Global Decline of Coral Reefs" and "The Red Queen Hypothesis."

Career in Action features present information on a variety of careers in the biological sciences, introducing students to the educational requirements and day-to-day work life of a variety of professions, such as forensic scientists, registered dietitians, and biogeographers.

Biology in Action features tie biological concepts to emerging issues and discuss science in terms of everyday life. Topics include "Invasive Species" and "Photosynthesis at the Grocery Store."

Art and Animations that Engage

Our art program takes a straightforward approach designed to help students learn the concepts of biology through simple, effective illustrations, photos, and micrographs. Concepts of Biology also incorporates links to relevant animations and interactive exercises that help bring biology to life for students.

Art Connection features call out core figures in each chapter for student attention. Questions about key figures, including clicker questions that can be used in the classroom, engage students' critical thinking and analytical abilities to ensure their genuine understanding of the concept at hand.

Concepts in Action features direct students to online interactive exercises and animations to add a fuller context and examples to core content.

About Our Team

Concepts of Biology would not be possible if not for the tremendous contributions of the authors and community reviewing team

Senior Contributing Authors

Samantha Fowler   Clayton State University

Rebecca Roush   Sandhills Community College

James Wise   Hampton University

Contributing Authors and Reviewers

Mark Belk   Brigham Young University

Lisa Boggs   Southwestern Oklahoma State University

Sherryl Broverman   Duke University

David Byres   Florida State College at Jacksonville

Aaron Cassill   The University of Texas at San Antonio

Karen Champ   College of Central Florida

Sue Chaplin   University of St. Thomas

Diane Day   Clayton State University

Jean DeSaix   University of North Carolina at Chapel Hill

David Hunnicutt   St. Norbert College

Barbara Kuehner   Hawaii Community College

Brenda Leady   University of Toledo

Bernie Marcus   Genesee Community College

Flora Mhlanga   Lipscomb University

Madeline Mignone   Dominican College

Elizabeth Nash   Long Beach City College

Mark Newton   San Jose City College

Diana Oliveras   University of Colorado Boulder

Ann Paterson   Williams Baptist College

Joel Piperberg   Millersville University

Nick Reeves   Mt. San Jacinto College

Ann Reisenauer   San Jose State University

Lynn Rumfelt   Gordon College

Michael Rutledge   Middle Tennessee State University

Edward Saiff   Ramapo College of New Jersey

Brian Shmaefsky   Kingwood College

Gary Shultz   Marshall University

Donald Slish   SUNY Plattsburgh

Anh-Hue Tu   Georgia Southwestern State University

Elena Zoubina   Bridgewater State University

Learning Resources

Wiley Plus for Biology-Fall 2013 Pilot

WileyPLUS provides an engaging online environment for effective teaching and learning. WileyPLUS builds students' confidence because it takes the guesswork out of studying by providing a clear roadmap; what to do, how to do it, and if they did it right. With WileyPLUS, students take more initiative. Therefore, the course has a greater impact on their learning experience. Adaptive tools provide students with a personal, adaptive learning experience so they can build their proficiency on topics and use their study time most effectively. Please let us know if you would like to participate in a Fall 2013 Pilot.

Concepts of Biology Powerpoint Slides (faculty only)

The PowerPoint slides are based on the extensive illustrations from College Physics. They can be edited, incorporated into lecture notes, and you are free to share with anyone in the community. This is a restricted item requiring faculty registration. NOTE: This file is very large and may take some time to download.

SimBio (Laboratory)

SimBio's interactive modules (virtual labs and interactive tutorials and chapters) provide engaging, discovery-based learning tools that complement many of the chapters of Concepts of Biology. SimBio is best known for their EcoBeaker® and EvoBeaker® suites of simulated ecology and evolution laboratories that guide students through the "discovery" of important concepts via a mix of structured and open-ended experimentation on simulated systems. In response to popular demand, SimBio has begun applying the same powerful approaches to topics in cell biology, genetics, and neurobiology. All of SimBio's modules include instant-feedback questions that enhance student comprehension and auto-graded questions that facilitate implementation.

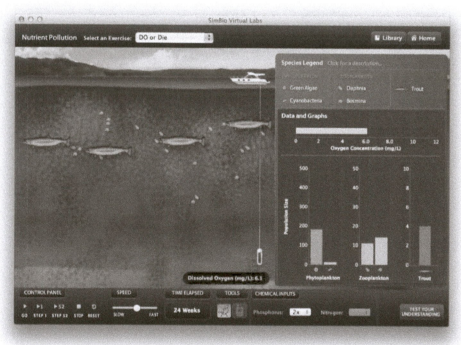

1 | INTRODUCTION TO BIOLOGY

Figure 1.1 This NASA image is a composite of several satellite-based views of Earth. To make the whole-Earth image, NASA scientists combine observations of different parts of the planet. (credit: modification of work by NASA)

Chapter Outline
1.1: Themes and Concepts of Biology
1.2: The Process of Science

Introduction

Viewed from space, Earth (Figure 1.1) offers few clues about the diversity of life forms that reside there. The first forms of life on Earth are thought to have been microorganisms that existed for billions of years before plants and animals appeared. The mammals, birds, and flowers so familiar to us are all relatively recent, originating 130 to 200 million years ago. Humans have inhabited this planet for only the last 2.5 million years, and only in the last 200,000 years have humans started looking like we do today.

1.1 | Themes and Concepts of Biology

By the end of this section, you will be able to:

- Identify and describe the properties of life
- Describe the levels of organization among living things
- List examples of different sub disciplines in biology

Biology is the science that studies life. What exactly is life? This may sound like a silly question with an obvious answer, but it is not easy to define life. For example, a branch of biology called virology studies viruses, which exhibit some of the characteristics of living entities but lack others. It turns out that although viruses can attack living organisms, cause diseases, and even reproduce, they do not meet the criteria that biologists use to define life.

From its earliest beginnings, biology has wrestled with four questions: What are the shared properties that make something "alive"? How do those various living things function? When faced with the remarkable diversity of life, how do we organize the different kinds of organisms so that we can better understand them? And, finally—what biologists ultimately seek to understand—how did this diversity arise and how is it continuing? As new organisms are discovered every day, biologists continue to seek answers to these and other questions.

Properties of Life

All groups of living organisms share several key characteristics or functions: order, sensitivity or response to stimuli, reproduction, adaptation, growth and development, regulation, homeostasis, and energy processing. When viewed together, these eight characteristics serve to define life.

Order

Organisms are highly organized structures that consist of one or more cells. Even very simple, single-celled organisms are remarkably complex. Inside each cell, atoms make up molecules. These in turn make up cell components or organelles. Multicellular organisms, which may consist of millions of individual cells, have an advantage over single-celled organisms in that their cells can be specialized to perform specific functions, and even sacrificed in certain situations for the good of the organism as a whole. How these specialized cells come together to form organs such as the heart, lung, or skin in organisms like the toad shown in Figure 1.2 will be discussed later.

Figure 1.2 A toad represents a highly organized structure consisting of cells, tissues, organs, and organ systems. (credit: "Ivengo(RUS)"/Wikimedia Commons)

Sensitivity or Response to Stimuli

Organisms respond to diverse stimuli. For example, plants can bend toward a source of light or respond to touch (Figure 1.3). Even tiny bacteria can move toward or away from chemicals (a process called chemotaxis) or light (phototaxis). Movement toward a stimulus is considered a positive response, while movement away from a stimulus is considered a negative response.

Figure 1.3 The leaves of this sensitive plant (*Mimosa pudica*) will instantly droop and fold when touched. After a few minutes, the plant returns to its normal state. (credit: Alex Lomas)

Watch this video (http://openstaxcollege.org/l/thigmonasty) to see how the sensitive plant responds to a touch stimulus.

Reproduction

Single-celled organisms reproduce by first duplicating their DNA, which is the genetic material, and then dividing it equally as the cell prepares to divide to form two new cells. Many multicellular organisms (those made up of more than one cell) produce specialized reproductive cells that will form new individuals. When reproduction occurs, DNA containing genes is passed along to an organism's offspring. These genes are the reason that the offspring will belong to the same species and will have characteristics similar to the parent, such as fur color and blood type.

Adaptation

All living organisms exhibit a "fit" to their environment. Biologists refer to this fit as adaptation and it is a consequence of evolution by natural selection, which operates in every lineage of reproducing organisms. Examples of adaptations are diverse and unique, from heat-resistant Archaea that live in boiling hot springs to the tongue length of a nectar-feeding moth that matches the size of the flower from which it feeds. All adaptations enhance the reproductive potential of the individual exhibiting them, including their ability to survive to reproduce. Adaptations are not constant. As an environment changes, natural selection causes the characteristics of the individuals in a population to track those changes.

Growth and Development

Organisms grow and develop according to specific instructions coded for by their genes. These genes provide instructions that will direct cellular growth and development, ensuring that a species' young (Figure 1.4) will grow up to exhibit many of the same characteristics as its parents.

Figure 1.4 Although no two look alike, these kittens have inherited genes from both parents and share many of the same characteristics. (credit: Pieter & Renée Lanser)

Regulation

Even the smallest organisms are complex and require multiple regulatory mechanisms to coordinate internal functions, such as the transport of nutrients, response to stimuli, and coping with environmental stresses. For example, organ systems such as the digestive or circulatory systems perform specific functions like carrying oxygen throughout the body, removing wastes, delivering nutrients to every cell, and cooling the body.

Homeostasis

To function properly, cells require appropriate conditions such as proper temperature, pH, and concentrations of diverse chemicals. These conditions may, however, change from one moment to the next. Organisms are able to maintain internal conditions within a narrow range almost constantly, despite environmental changes, through a process called **homeostasis** or "steady state"—the ability of an organism to maintain constant internal conditions. For example, many organisms regulate their body temperature in a process known as thermoregulation. Organisms that live in cold climates, such as the polar bear (Figure 1.5), have body structures that help them withstand low temperatures and conserve body heat. In hot climates, organisms have methods (such as perspiration in humans or panting in dogs) that help them to shed excess body heat.

Figure 1.5 Polar bears and other mammals living in ice-covered regions maintain their body temperature by generating heat and reducing heat loss through thick fur and a dense layer of fat under their skin. (credit: "longhorndave"/Flickr)

Energy Processing

All organisms (such as the California condor shown in Figure 1.6) use a source of energy for their metabolic activities. Some organisms capture energy from the Sun and convert it into chemical energy in food; others use chemical energy from molecules they take in.

Download for free at https://openstax.org/details/books/concepts-biology

Figure 1.6 A lot of energy is required for a California condor to fly. Chemical energy derived from food is used to power flight. California condors are an endangered species; scientists have strived to place a wing tag on each bird to help them identify and locate each individual bird. (credit: Pacific Southwest Region U.S. Fish and Wildlife)

Levels of Organization of Living Things

Living things are highly organized and structured, following a hierarchy on a scale from small to large. The **atom** is the smallest and most fundamental unit of matter. It consists of a nucleus surrounded by electrons. Atoms form molecules. A **molecule** is a chemical structure consisting of at least two atoms held together by a chemical bond. Many molecules that are biologically important are **macromolecules**, large molecules that are typically formed by combining smaller units called monomers. An example of a macromolecule is deoxyribonucleic acid (DNA) (Figure 1.7), which contains the instructions for the functioning of the organism that contains it.

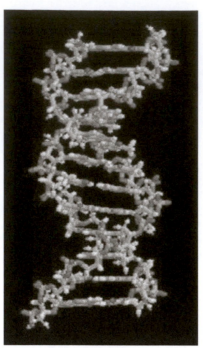

Figure 1.7 A molecule, like this large DNA molecule, is composed of atoms. (credit: "Brian0918"/Wikimedia Commons)

To see an animation of this DNA molecule, click here (http://openstaxcollege.org/l/rotating_DNA2) .

Some cells contain aggregates of macromolecules surrounded by membranes; these are called **organelles**. Organelles are small structures that exist within cells and perform specialized functions. All living things are made of cells; the **cell** itself is the smallest fundamental unit of structure and function in living organisms. (This requirement is why viruses are not considered living: they are not made of cells. To make new viruses, they have to invade and hijack a living cell; only then can they obtain the materials they need to reproduce.) Some organisms consist of a single cell and others are multicellular. Cells are classified as prokaryotic or eukaryotic. **Prokaryotes** are single-celled organisms that lack organelles surrounded by a membrane and do not have nuclei surrounded by nuclear membranes; in contrast, the cells of **eukaryotes** do have membrane-bound organelles and nuclei.

In most multicellular organisms, cells combine to make **tissues**, which are groups of similar cells carrying out the same function. **Organs** are collections of tissues grouped together based on a common function. Organs are present not only in animals but also in plants. An **organ system** is a higher level of organization that consists of functionally related organs. For example vertebrate animals have many organ systems, such as the circulatory system that transports blood throughout the body and to and from the lungs; it includes organs such as the heart and blood vessels. **Organisms** are individual living entities. For example, each tree in a forest is an organism. Single-celled prokaryotes and single-celled eukaryotes are also considered organisms and are typically referred to as microorganisms.

a r t CONNECTION

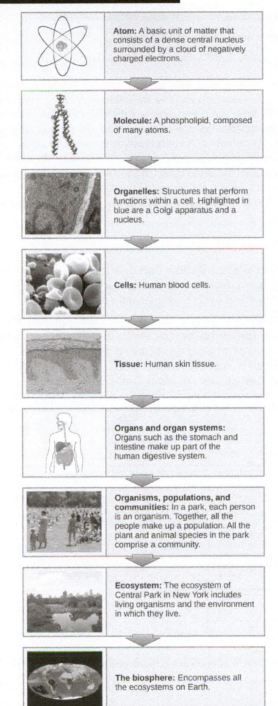

Atom: A basic unit of matter that consists of a dense central nucleus surrounded by a cloud of negatively charged electrons.

Molecule: A phospholipid, composed of many atoms.

Organelles: Structures that perform functions within a cell. Highlighted in blue are a Golgi apparatus and a nucleus.

Cells: Human blood cells.

Tissue: Human skin tissue.

Organs and organ systems: Organs such as the stomach and intestine make up part of the human digestive system.

Organisms, populations, and communities: In a park, each person is an organism. Together, all the people make up a population. All the plant and animal species in the park comprise a community.

Ecosystem: The ecosystem of Central Park in New York includes living organisms and the environment in which they live.

The biosphere: Encompasses all the ecosystems on Earth.

Figure 1.8 From an atom to the entire Earth, biology examines all aspects of life. (credit "molecule": modification of work by Jane Whitney; credit "organelles": modification of work by Louisa Howard; credit "cells": modification of work by Bruce Wetzel, Harry Schaefer, National Cancer Institute; credit "tissue": modification of work by "Kilbad"/Wikimedia Commons; credit "organs": modification of work by Mariana Ruiz Villareal, Joaquim Alves Gaspar; credit "organisms": modification of work by Peter Dutton; credit "ecosystem": modification of work by "gigi4791"/Flickr; credit "biosphere": modification of work by NASA)

Which of the following statements is false?

a. Tissues exist within organs which exist within organ systems.
b. Communities exist within populations which exist within ecosystems.
c. Organelles exist within cells which exist within tissues.
d. Communities exist within ecosystems which exist in the biosphere.

All the individuals of a species living within a specific area are collectively called a **population**. For example, a forest may include many white pine trees. All of these pine trees represent the population of white pine trees in this forest. Different populations may live in the same specific area. For example, the forest with the pine trees includes populations of flowering plants and also insects and microbial populations. A **community** is the set of populations inhabiting a particular area. For instance, all of the trees, flowers, insects, and other populations in a forest form the forest's community. The forest itself is an ecosystem. An **ecosystem** consists of all the living things in a particular area together with the abiotic, or non-living, parts of that environment such as nitrogen in the soil or rainwater. At the highest level of organization (Figure 1.8), the **biosphere** is the collection of all ecosystems, and it represents the zones of life on Earth. It includes land, water, and portions of the atmosphere.

The Diversity of Life

The science of biology is very broad in scope because there is a tremendous diversity of life on Earth. The source of this diversity is **evolution**, the process of gradual change during which new species arise from older species. Evolutionary biologists study the evolution of living things in everything from the microscopic world to ecosystems.

In the 18th century, a scientist named Carl Linnaeus first proposed organizing the known species of organisms into a hierarchical taxonomy. In this system, species that are most similar to each other are put together within a grouping known as a genus. Furthermore, similar genera (the plural of genus) are put together within a family. This grouping continues until all organisms are collected together into groups at the highest level. The current taxonomic system now has eight levels in its hierarchy, from lowest to highest, they are: species, genus, family, order, class, phylum, kingdom, domain. Thus species are grouped within genera, genera are grouped within families, families are grouped within orders, and so on (Figure 1.9).

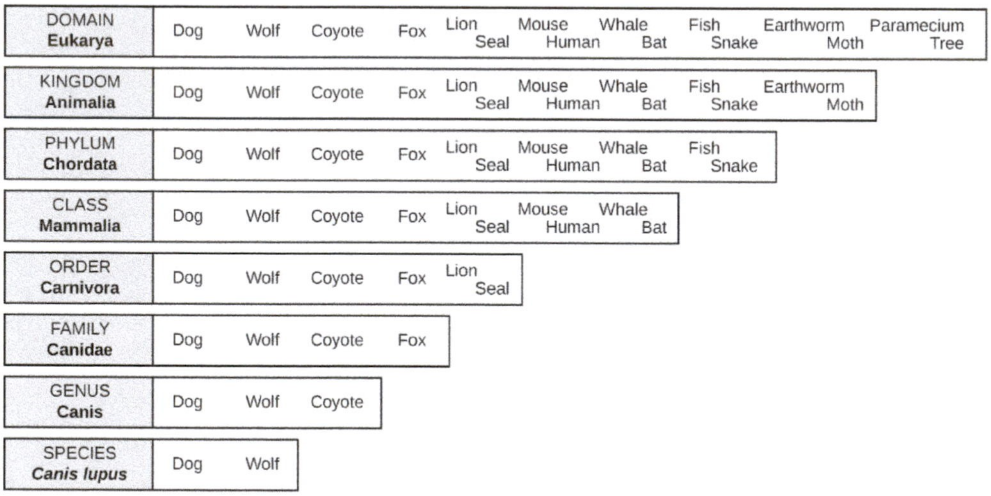

DOMAIN **Eukarya**	Dog	Wolf	Coyote	Fox	Lion Seal	Mouse Human	Whale Bat	Fish Snake	Earthworm Moth	Paramecium Tree
KINGDOM **Animalia**	Dog	Wolf	Coyote	Fox	Lion Seal	Mouse Human	Whale Bat	Fish Snake	Earthworm Moth	
PHYLUM **Chordata**	Dog	Wolf	Coyote	Fox	Lion Seal	Mouse Human	Whale Bat	Fish Snake		
CLASS **Mammalia**	Dog	Wolf	Coyote	Fox	Lion Seal	Mouse Human	Whale Bat			
ORDER **Carnivora**	Dog	Wolf	Coyote	Fox	Lion Seal					
FAMILY **Canidae**	Dog	Wolf	Coyote	Fox						
GENUS **Canis**	Dog	Wolf	Coyote							
SPECIES **Canis lupus**	Dog	Wolf								

Figure 1.9 This diagram shows the levels of taxonomic hierarchy for a dog, from the broadest category—domain—to the most specific—species.

The highest level, domain, is a relatively new addition to the system since the 1990s. Scientists now recognize three domains of life, the Eukarya, the Archaea, and the Bacteria. The domain Eukarya contains organisms that have cells with nuclei. It includes the kingdoms of fungi, plants, animals, and several kingdoms of protists. The Archaea, are single-celled organisms without nuclei and include many extremophiles that live in harsh environments like hot springs. The Bacteria are another quite different group of single-celled organisms without nuclei (Figure 1.10). Both the Archaea and the Bacteria are prokaryotes, an informal name for cells without nuclei. The recognition in the 1990s that certain "bacteria," now known as the Archaea, were as different genetically and biochemically from other bacterial cells as they were from eukaryotes, motivated the recommendation to divide life into three domains. This dramatic change in our knowledge of the tree of life demonstrates that classifications are not permanent and will change when new information becomes available.

In addition to the hierarchical taxonomic system, Linnaeus was the first to name organisms using two unique names, now called the binomial naming system. Before Linnaeus, the use of common names to refer to organisms caused confusion because there were regional differences in these common names. Binomial names consist of the genus name (which is capitalized) and the species name (all lower-case). Both names are set in italics when they are printed. Every species is given a unique binomial which is recognized the world over, so that a scientist in any location can know which organism is being referred to. For example, the North American blue jay is known uniquely as *Cyanocitta cristata*. Our own species is *Homo sapiens*.

(a)   (b)   (c)   (d)

Figure 1.10 These images represent different domains. The scanning electron micrograph shows (a) bacterial cells belong to the domain Bacteria, while the (b) extremophiles, seen all together as colored mats in this hot spring, belong to domain Archaea. Both the (c) sunflower and (d) lion are part of domain Eukarya. (credit a: modification of work by Rocky Mountain Laboratories, NIAID, NIH; credit b: modification of work by Steve Jurvetson; credit c: modification of work by Michael Arrighi; credit d: modification of work by Frank Vassen)

Carl Woese and the Phylogenetic Tree

The evolutionary relationships of various life forms on Earth can be summarized in a phylogenetic tree. A **phylogenetic tree** is a diagram showing the evolutionary relationships among biological species based on similarities and differences in genetic or physical traits or both. A phylogenetic tree is composed of branch points, or nodes, and branches. The internal nodes represent ancestors and are points in evolution when, based on scientific evidence, an ancestor is thought to have diverged to form two new species. The length of each branch can be considered as estimates of relative time.

In the past, biologists grouped living organisms into five kingdoms: animals, plants, fungi, protists, and bacteria. The pioneering work of American microbiologist Carl Woese in the early 1970s has shown, however, that life on Earth has evolved along three lineages, now called domains—Bacteria, Archaea, and Eukarya. Woese proposed the domain as a new taxonomic level and Archaea as a new domain, to reflect the new phylogenetic tree (Figure 1.11). Many organisms belonging to the Archaea domain live under extreme conditions and are called extremophiles. To construct his tree, Woese used genetic relationships rather than similarities based on morphology (shape). Various genes were used in phylogenetic studies. Woese's tree was constructed from comparative sequencing of the genes that are universally distributed, found in some slightly altered form in every organism, conserved (meaning that these genes have remained only slightly changed throughout evolution), and of an appropriate length.

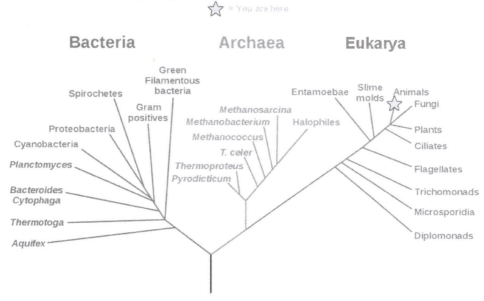

Figure 1.11 This phylogenetic tree was constructed by microbiologist Carl Woese using genetic relationships. The tree shows the separation of living organisms into three domains: Bacteria, Archaea, and Eukarya. Bacteria and Archaea are organisms without a nucleus or other organelles surrounded by a membrane and, therefore, are prokaryotes. (credit: modification of work by Eric Gaba)

Branches of Biological Study

The scope of biology is broad and therefore contains many branches and sub disciplines. Biologists may pursue one of those sub disciplines and work in a more focused field. For instance, molecular biology studies biological processes at the molecular level, including interactions among molecules such as DNA, RNA, and proteins, as well as the way they are regulated. Microbiology is the study of the structure and function of microorganisms. It is quite a broad branch itself, and depending on the subject of study, there are also microbial physiologists, ecologists, and geneticists, among others.

Another field of biological study, neurobiology, studies the biology of the nervous system, and although it is considered a branch of biology, it is also recognized as an interdisciplinary field of study known as neuroscience. Because of its interdisciplinary nature, this sub discipline studies different functions of the nervous system using molecular, cellular, developmental, medical, and computational approaches.

Figure 1.12 Researchers work on excavating dinosaur fossils at a site in Castellón, Spain. (credit: Mario Modesto)

Paleontology, another branch of biology, uses fossils to study life's history (Figure 1.12). Zoology and botany are the study of animals and plants, respectively. Biologists can also specialize as biotechnologists, ecologists, or physiologists, to name just a few areas. Biotechnologists apply the knowledge of biology to create useful products. Ecologists study the interactions of organisms in their environments. Physiologists study the workings of cells, tissues and organs. This is just a small sample of the many fields that biologists can pursue. From our own bodies to the world we live in, discoveries in biology can affect us in very direct and important ways. We depend on these discoveries for our health, our food sources, and the benefits provided by our ecosystem. Because of this, knowledge of biology can benefit us in making decisions in our day-to-day lives.

The development of technology in the twentieth century that continues today, particularly the technology to describe and manipulate the genetic material, DNA, has transformed biology. This transformation will allow biologists to continue to understand the history of life in greater detail, how the human body works, our human origins, and how humans can survive as a species on this planet despite the stresses caused by our increasing numbers. Biologists continue to decipher huge mysteries about life suggesting that we have only begun to understand life on the planet, its history, and our relationship to it. For this and other reasons, the knowledge of biology gained through this textbook and other printed and electronic media should be a benefit in whichever field you enter.

careers IN ACTION

Forensic Scientist

Forensic science is the application of science to answer questions related to the law. Biologists as well as chemists and biochemists can be forensic scientists. Forensic scientists provide scientific evidence for use in courts, and their job involves examining trace material associated with crimes. Interest in forensic science has increased in the last few years, possibly because of popular television shows that feature forensic scientists on the job. Also, the development of molecular techniques and the establishment of DNA databases have updated the types of work that forensic scientists can do. Their job activities are primarily related to crimes against people such as murder, rape, and assault. Their work involves analyzing samples such as hair, blood, and other body fluids and also processing DNA (Figure 1.13) found in many different environments and materials. Forensic scientists also analyze other biological evidence left at crime scenes, such as insect parts or pollen grains. Students who want to pursue careers in forensic science will most likely be required to take chemistry and biology courses as well as some intensive math courses.

Figure 1.13 This forensic scientist works in a DNA extraction room at the U.S. Army Criminal Investigation Laboratory. (credit: U.S. Army CID Command Public Affairs)

1.2 | The Process of Science

By the end of this section, you will be able to:

- Identify the shared characteristics of the natural sciences
- Understand the process of scientific inquiry
- Compare inductive reasoning with deductive reasoning
- Describe the goals of basic science and applied science

(a)   (b)

Figure 1.14 Formerly called blue-green algae, the (a) cyanobacteria seen through a light microscope are some of Earth's oldest life forms. These (b) stromatolites along the shores of Lake Thetis in Western Australia are ancient structures formed by the layering of cyanobacteria in shallow waters. (credit a: modification of work by NASA; scale-bar data from Matt Russell; credit b: modification of work by Ruth Ellison)

Like geology, physics, and chemistry, biology is a science that gathers knowledge about the natural world. Specifically, biology is the study of life. The discoveries of biology are made by a community of researchers who work individually and together using agreed-on methods. In this sense, biology, like all sciences is a social enterprise like politics or the arts. The methods of science include careful observation, record keeping, logical and mathematical reasoning, experimentation, and submitting conclusions to the scrutiny of others. Science also requires considerable imagination and creativity; a well-designed experiment is commonly described as elegant, or beautiful. Like politics, science has considerable practical implications and some science is dedicated to practical applications, such as the prevention of disease (see Figure 1.15). Other science proceeds largely motivated by curiosity. Whatever its goal, there is no doubt that science, including biology, has transformed human existence and will continue to do so.

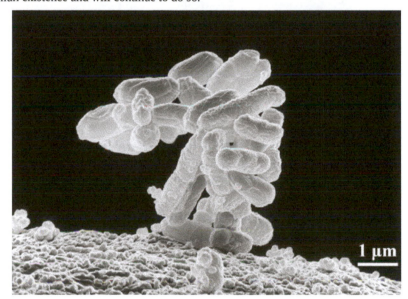

Figure 1.15 Biologists may choose to study *Escherichia coli* (*E. coli*), a bacterium that is a normal resident of our digestive tracts but which is also sometimes responsible for disease outbreaks. In this micrograph, the bacterium is visualized using a scanning electron microscope and digital colorization. (credit: Eric Erbe; digital colorization by Christopher Pooley, USDA-ARS)

The Nature of Science

Biology is a science, but what exactly is science? What does the study of biology share with other scientific disciplines? **Science** (from the Latin *scientia*, meaning "knowledge") can be defined as knowledge about the natural world.

Science is a very specific way of learning, or knowing, about the world. The history of the past 500 years demonstrates that science is a very powerful way of knowing about the world; it is largely responsible for the technological revolutions that have taken place during this time. There are however, areas of knowledge and human experience that the methods of science cannot be applied to. These include such things as answering purely moral questions, aesthetic questions, or what can be generally categorized as spiritual questions. Science has cannot investigate these areas because they are outside the realm of material phenomena, the phenomena of matter and energy, and cannot be observed and measured.

The **scientific method** is a method of research with defined steps that include experiments and careful observation. The steps of the scientific method will be examined in detail later, but one of the most important aspects of this method is the testing of hypotheses. A **hypothesis** is a suggested explanation for an event, which can be tested. Hypotheses, or tentative explanations, are generally produced within the context of a **scientific theory**. A scientific theory is a generally accepted, thoroughly tested and confirmed explanation for a set of observations or phenomena. Scientific theory is the foundation of scientific knowledge. In addition, in many scientific disciplines (less so in biology) there are **scientific laws**, often expressed in mathematical formulas, which describe how elements of nature will behave under certain specific conditions. There is not an evolution of hypotheses through theories to laws as if they represented some increase in certainty about the world. Hypotheses are the day-to-day material that scientists work with and they are developed within the context of theories. Laws are concise descriptions of parts of the world that are amenable to formulaic or mathematical description.

Natural Sciences

What would you expect to see in a museum of natural sciences? Frogs? Plants? Dinosaur skeletons? Exhibits about how the brain functions? A planetarium? Gems and minerals? Or maybe all of the above? Science includes such diverse fields as astronomy, biology, computer sciences, geology, logic, physics, chemistry, and mathematics (Figure 1.16). However, those fields of science related to the physical world and its phenomena and processes are considered **natural sciences**. Thus, a museum of natural sciences might contain any of the items listed above.

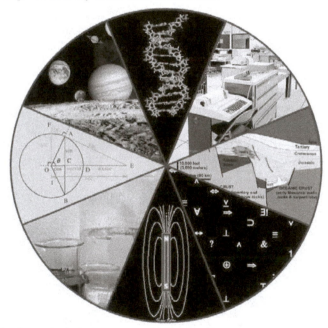

Figure 1.16 Some fields of science include astronomy, biology, computer science, geology, logic, physics, chemistry, and mathematics. (credit: "Image Editor"/Flickr)

There is no complete agreement when it comes to defining what the natural sciences include. For some experts, the natural sciences are astronomy, biology, chemistry, earth science, and physics. Other scholars choose to divide natural sciences into **life sciences**, which study living things and include biology, and **physical sciences**, which study nonliving matter and include astronomy, physics, and chemistry. Some disciplines such as biophysics and biochemistry build on two sciences and are interdisciplinary.

Scientific Inquiry

One thing is common to all forms of science: an ultimate goal "to know." Curiosity and inquiry are the driving forces for the development of science. Scientists seek to understand the world and the way it operates. Two methods of logical thinking are used: inductive reasoning and deductive reasoning.

Inductive reasoning is a form of logical thinking that uses related observations to arrive at a general conclusion. This type of reasoning is common in descriptive science. A life scientist such as a biologist makes observations and records them.

These data can be qualitative (descriptive) or quantitative (consisting of numbers), and the raw data can be supplemented with drawings, pictures, photos, or videos. From many observations, the scientist can infer conclusions (inductions) based on evidence. Inductive reasoning involves formulating generalizations inferred from careful observation and the analysis of a large amount of data. Brain studies often work this way. Many brains are observed while people are doing a task. The part of the brain that lights up, indicating activity, is then demonstrated to be the part controlling the response to that task.

Deductive reasoning or deduction is the type of logic used in hypothesis-based science. In deductive reasoning, the pattern of thinking moves in the opposite direction as compared to inductive reasoning. **Deductive reasoning** is a form of logical thinking that uses a general principle or law to forecast specific results. From those general principles, a scientist can extrapolate and predict the specific results that would be valid as long as the general principles are valid. For example, a prediction would be that if the climate is becoming warmer in a region, the distribution of plants and animals should change. Comparisons have been made between distributions in the past and the present, and the many changes that have been found are consistent with a warming climate. Finding the change in distribution is evidence that the climate change conclusion is a valid one.

Both types of logical thinking are related to the two main pathways of scientific study: descriptive science and hypothesis-based science. **Descriptive** (or discovery) **science** aims to observe, explore, and discover, while **hypothesis-based science** begins with a specific question or problem and a potential answer or solution that can be tested. The boundary between these two forms of study is often blurred, because most scientific endeavors combine both approaches. Observations lead to questions, questions lead to forming a hypothesis as a possible answer to those questions, and then the hypothesis is tested. Thus, descriptive science and hypothesis-based science are in continuous dialogue.

Hypothesis Testing

Biologists study the living world by posing questions about it and seeking science-based responses. This approach is common to other sciences as well and is often referred to as the scientific method. The scientific method was used even in ancient times, but it was first documented by England's Sir Francis Bacon (1561–1626) (Figure 1.17), who set up inductive methods for scientific inquiry. The scientific method is not exclusively used by biologists but can be applied to almost anything as a logical problem-solving method.

Figure 1.17 Sir Francis Bacon is credited with being the first to document the scientific method.

The scientific process typically starts with an observation (often a problem to be solved) that leads to a question. Let's think about a simple problem that starts with an observation and apply the scientific method to solve the problem. One Monday morning, a student arrives at class and quickly discovers that the classroom is too warm. That is an observation that also describes a problem: the classroom is too warm. The student then asks a question: "Why is the classroom so warm?"

Recall that a hypothesis is a suggested explanation that can be tested. To solve a problem, several hypotheses may be proposed. For example, one hypothesis might be, "The classroom is warm because no one turned on the air conditioning."

But there could be other responses to the question, and therefore other hypotheses may be proposed. A second hypothesis might be, "The classroom is warm because there is a power failure, and so the air conditioning doesn't work."

Once a hypothesis has been selected, a prediction may be made. A prediction is similar to a hypothesis but it typically has the format "If . . . then . . . ." For example, the prediction for the first hypothesis might be, "*If* the student turns on the air conditioning, *then* the classroom will no longer be too warm."

A hypothesis must be testable to ensure that it is valid. For example, a hypothesis that depends on what a bear thinks is not testable, because it can never be known what a bear thinks. It should also be **falsifiable**, meaning that it can be disproven by experimental results. An example of an unfalsifiable hypothesis is "Botticelli's *Birth of Venus* is beautiful." There is no experiment that might show this statement to be false. To test a hypothesis, a researcher will conduct one or more experiments designed to eliminate one or more of the hypotheses. This is important. A hypothesis can be disproven, or eliminated, but it can never be proven. Science does not deal in proofs like mathematics. If an experiment fails to disprove a hypothesis, then we find support for that explanation, but this is not to say that down the road a better explanation will not be found, or a more carefully designed experiment will be found to falsify the hypothesis.

Each experiment will have one or more variables and one or more controls. A **variable** is any part of the experiment that can vary or change during the experiment. A **control** is a part of the experiment that does not change. Look for the variables and controls in the example that follows. As a simple example, an experiment might be conducted to test the hypothesis that phosphate limits the growth of algae in freshwater ponds. A series of artificial ponds are filled with water and half of them are treated by adding phosphate each week, while the other half are treated by adding a salt that is known not to be used by algae. The variable here is the phosphate (or lack of phosphate), the experimental or treatment cases are the ponds with added phosphate and the control ponds are those with something inert added, such as the salt. Just adding something is also a control against the possibility that adding extra matter to the pond has an effect. If the treated ponds show lesser growth of algae, then we have found support for our hypothesis. If they do not, then we reject our hypothesis. Be aware that rejecting one hypothesis does not determine whether or not the other hypotheses can be accepted; it simply eliminates one hypothesis that is not valid (Figure 1.18). Using the scientific method, the hypotheses that are inconsistent with experimental data are rejected.

art CONNECTION

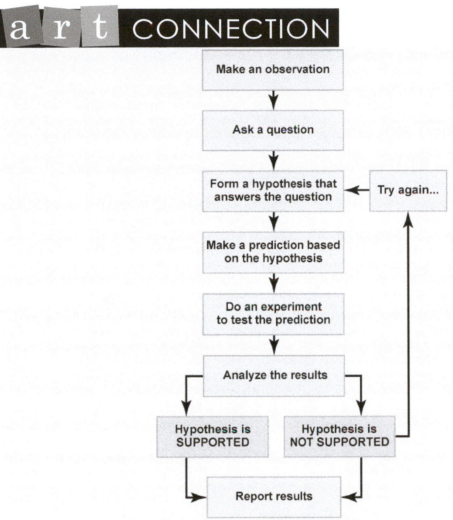

Figure 1.18 The scientific method is a series of defined steps that include experiments and careful observation. If a hypothesis is not supported by data, a new hypothesis can be proposed.

In the example below, the scientific method is used to solve an everyday problem. Which part in the example below is the hypothesis? Which is the prediction? Based on the results of the experiment, is the hypothesis supported? If it is not supported, propose some alternative hypotheses.

1. My toaster doesn't toast my bread.

2. Why doesn't my toaster work?

3. There is something wrong with the electrical outlet.

4. If something is wrong with the outlet, my coffeemaker also won't work when plugged into it.

5. I plug my coffeemaker into the outlet.

6. My coffeemaker works.

In practice, the scientific method is not as rigid and structured as it might at first appear. Sometimes an experiment leads to conclusions that favor a change in approach; often, an experiment brings entirely new scientific questions to the puzzle. Many times, science does not operate in a linear fashion; instead, scientists continually draw inferences and make generalizations, finding patterns as their research proceeds. Scientific reasoning is more complex than the scientific method alone suggests.

Basic and Applied Science

The scientific community has been debating for the last few decades about the value of different types of science. Is it valuable to pursue science for the sake of simply gaining knowledge, or does scientific knowledge only have worth if we can apply it to solving a specific problem or bettering our lives? This question focuses on the differences between two types of science: basic science and applied science.

Basic science or "pure" science seeks to expand knowledge regardless of the short-term application of that knowledge. It is not focused on developing a product or a service of immediate public or commercial value. The immediate goal of basic science is knowledge for knowledge's sake, though this does not mean that in the end it may not result in an application.

In contrast, **applied science** or "technology," aims to use science to solve real-world problems, making it possible, for example, to improve a crop yield, find a cure for a particular disease, or save animals threatened by a natural disaster. In applied science, the problem is usually defined for the researcher.

Some individuals may perceive applied science as "useful" and basic science as "useless." A question these people might pose to a scientist advocating knowledge acquisition would be, "What for?" A careful look at the history of science, however, reveals that basic knowledge has resulted in many remarkable applications of great value. Many scientists think that a basic understanding of science is necessary before an application is developed; therefore, applied science relies on the results generated through basic science. Other scientists think that it is time to move on from basic science and instead to find solutions to actual problems. Both approaches are valid. It is true that there are problems that demand immediate attention; however, few solutions would be found without the help of the knowledge generated through basic science.

One example of how basic and applied science can work together to solve practical problems occurred after the discovery of DNA structure led to an understanding of the molecular mechanisms governing DNA replication. Strands of DNA, unique in every human, are found in our cells, where they provide the instructions necessary for life. During DNA replication, new copies of DNA are made, shortly before a cell divides to form new cells. Understanding the mechanisms of DNA replication enabled scientists to develop laboratory techniques that are now used to identify genetic diseases, pinpoint individuals who were at a crime scene, and determine paternity. Without basic science, it is unlikely that applied science would exist.

Another example of the link between basic and applied research is the Human Genome Project, a study in which each human chromosome was analyzed and mapped to determine the precise sequence of DNA subunits and the exact location of each gene. (The gene is the basic unit of heredity; an individual's complete collection of genes is his or her genome.) Other organisms have also been studied as part of this project to gain a better understanding of human chromosomes. The Human Genome Project (Figure 1.19) relied on basic research carried out with non-human organisms and, later, with the human genome. An important end goal eventually became using the data for applied research seeking cures for genetically related diseases.

Figure 1.19 The Human Genome Project was a 13-year collaborative effort among researchers working in several different fields of science. The project was completed in 2003. (credit: the U.S. Department of Energy Genome Programs)

While research efforts in both basic science and applied science are usually carefully planned, it is important to note that some discoveries are made by serendipity, that is, by means of a fortunate accident or a lucky surprise. Penicillin was discovered when biologist Alexander Fleming accidentally left a petri dish of *Staphylococcus* bacteria open. An unwanted mold grew, killing the bacteria. The mold turned out to be *Penicillium*, and a new antibiotic was discovered. Even in the highly organized world of science, luck—when combined with an observant, curious mind—can lead to unexpected breakthroughs.

Reporting Scientific Work

Whether scientific research is basic science or applied science, scientists must share their findings for other researchers to expand and build upon their discoveries. Communication and collaboration within and between sub disciplines of science are key to the advancement of knowledge in science. For this reason, an important aspect of a scientist's work is disseminating results and communicating with peers. Scientists can share results by presenting them at a scientific meeting or conference, but this approach can reach only the limited few who are present. Instead, most scientists present their results in peer-reviewed articles that are published in scientific journals. **Peer-reviewed articles** are scientific papers that are reviewed, usually anonymously by a scientist's colleagues, or peers. These colleagues are qualified individuals, often experts in the same research area, who judge whether or not the scientist's work is suitable for publication. The process of peer review helps to ensure that the research described in a scientific paper or grant proposal is original, significant, logical, and thorough. Grant proposals, which are requests for research funding, are also subject to peer review. Scientists publish their work so other scientists can reproduce their experiments under similar or different conditions to expand on the findings. The experimental results must be consistent with the findings of other scientists.

There are many journals and the popular press that do not use a peer-review system. A large number of online open-access journals, journals with articles available without cost, are now available many of which use rigorous peer-review systems, but some of which do not. Results of any studies published in these forums without peer review are not reliable and should not form the basis for other scientific work. In one exception, journals may allow a researcher to cite a personal communication from another researcher about unpublished results with the cited author's permission.

KEY TERMS

applied science a form of science that solves real-world problems

atom a basic unit of matter that cannot be broken down by normal chemical reactions

basic science science that seeks to expand knowledge regardless of the short-term application of that knowledge

biology the study of living organisms and their interactions with one another and their environments

biosphere a collection of all ecosystems on Earth

cell the smallest fundamental unit of structure and function in living things

community a set of populations inhabiting a particular area

control a part of an experiment that does not change during the experiment

deductive reasoning a form of logical thinking that uses a general statement to forecast specific results

descriptive science a form of science that aims to observe, explore, and find things out

ecosystem all living things in a particular area together with the abiotic, nonliving parts of that environment

eukaryote an organism with cells that have nuclei and membrane-bound organelles

evolution the process of gradual change in a population that can also lead to new species arising from older species

falsifiable able to be disproven by experimental results

homeostasis the ability of an organism to maintain constant internal conditions

hypothesis a suggested explanation for an event, which can be tested

hypothesis-based science a form of science that begins with a specific explanation that is then tested

inductive reasoning a form of logical thinking that uses related observations to arrive at a general conclusion

life science a field of science, such as biology, that studies living things

macromolecule a large molecule typically formed by the joining of smaller molecules

molecule a chemical structure consisting of at least two atoms held together by a chemical bond

natural science a field of science that studies the physical world, its phenomena, and processes

organ a structure formed of tissues operating together to perform a common function

organ system the higher level of organization that consists of functionally related organs

organelle a membrane-bound compartment or sac within a cell

organism an individual living entity

peer-reviewed article a scientific report that is reviewed by a scientist's colleagues before publication

phylogenetic tree a diagram showing the evolutionary relationships among biological species based on similarities and differences in genetic or physical traits or both

physical science a field of science, such as astronomy, physics, and chemistry, that studies nonliving matter

population all individuals within a species living within a specific area

prokaryote a unicellular organism that lacks a nucleus or any other membrane-bound organelle

science knowledge that covers general truths or the operation of general laws, especially when acquired and tested by the scientific method

scientific law a description, often in the form of a mathematical formula, for the behavior of some aspect of nature under certain specific conditions

scientific method a method of research with defined steps that include experiments and careful observation

scientific theory a thoroughly tested and confirmed explanation for observations or phenomena

tissue a group of similar cells carrying out the same function

variable a part of an experiment that can vary or change

CHAPTER SUMMARY

1.1 Themes and Concepts of Biology

Biology is the science of life. All living organisms share several key properties such as order, sensitivity or response to stimuli, reproduction, adaptation, growth and development, regulation, homeostasis, and energy processing. Living things are highly organized following a hierarchy that includes atoms, molecules, organelles, cells, tissues, organs, and organ systems. Organisms, in turn, are grouped as populations, communities, ecosystems, and the biosphere. Evolution is the source of the tremendous biological diversity on Earth today. A diagram called a phylogenetic tree can be used to show evolutionary relationships among organisms. Biology is very broad and includes many branches and sub disciplines. Examples include molecular biology, microbiology, neurobiology, zoology, and botany, among others.

1.2 The Process of Science

Biology is the science that studies living organisms and their interactions with one another and their environments. Science attempts to describe and understand the nature of the universe in whole or in part. Science has many fields; those fields related to the physical world and its phenomena are considered natural sciences.

A hypothesis is a tentative explanation for an observation. A scientific theory is a well-tested and consistently verified explanation for a set of observations or phenomena. A scientific law is a description, often in the form of a mathematical formula, of the behavior of an aspect of nature under certain circumstances. Two types of logical reasoning are used in science. Inductive reasoning uses results to produce general scientific principles. Deductive reasoning is a form of logical thinking that predicts results by applying general principles. The common thread throughout scientific research is the use of the scientific method. Scientists present their results in peer-reviewed scientific papers published in scientific journals.

Science can be basic or applied. The main goal of basic science is to expand knowledge without any expectation of short-term practical application of that knowledge. The primary goal of applied research, however, is to solve practical problems.

ART CONNECTION QUESTIONS

1. Figure 1.8 Which of the following statements is false?

   A. Tissues exist within organs which exist within organ systems.
   B. Communities exist within populations which exist within ecosystems.
   C. Organelles exist within cells which exist within tissues.
   D. Communities exist within ecosystems which exist in the biosphere.

2. Figure 1.18 In the example below, the scientific method is used to solve an everyday problem. Which part in the example below is the hypothesis? Which is the prediction?

Based on the results of the experiment, is the hypothesis supported? If it is not supported, propose some alternative hypotheses.

   1. My toaster doesn't toast my bread.
   2. Why doesn't my toaster work?
   3. There is something wrong with the electrical outlet.
   4. If something is wrong with the outlet, my coffeemaker also won't work when plugged into it.
   5. I plug my coffeemaker into the outlet.
   6. My coffeemaker works.

REVIEW QUESTIONS

3. The smallest unit of biological structure that meets the functional requirements of "living" is the _____.

    a. organ
    b. organelle
    c. cell
    d. macromolecule

4. Which of the following sequences represents the hierarchy of biological organization from the most complex to the least complex level?

    a. organelle, tissue, biosphere, ecosystem, population
    b. organ, organism, tissue, organelle, molecule
    c. organism, community, biosphere, molecule, tissue, organ

    d. biosphere, ecosystem, community, population, organism

5. A suggested and testable explanation for an event is called a _____.

    a. hypothesis
    b. variable
    c. theory
    d. control

6. The type of logical thinking that uses related observations to arrive at a general conclusion is called _____.

    a. deductive reasoning
    b. the scientific method
    c. hypothesis-based science
    d. inductive reasoning

CRITICAL THINKING QUESTIONS

7. Using examples, explain how biology can be studied from a microscopic approach to a global approach.

8. Give an example of how applied science has had a direct effect on your daily life.

2 | CHEMISTRY OF LIFE

Figure 2.1 Foods such as bread, fruit, and cheese are rich sources of biological macromolecules. (credit: modification of work by Bengt Nyman)

Chapter Outline

2.1: The Building Blocks of Molecules

2.2: Water

2.3: Biological Molecules

Introduction

The elements carbon, hydrogen, nitrogen, oxygen, sulfur, and phosphorus are the key building blocks of the chemicals found in living things. They form the carbohydrates, nucleic acids, proteins, and lipids (all of which will be defined later in this chapter) that are the fundamental molecular components of all organisms. In this chapter, we will discuss these important building blocks and learn how the unique properties of the atoms of different elements affect their interactions with other atoms to form the molecules of life.

Food provides an organism with nutrients—the matter it needs to survive. Many of these critical nutrients come in the form of biological macromolecules, or large molecules necessary for life. These macromolecules are built from different combinations of smaller organic molecules. What specific types of biological macromolecules do living things require? How are these molecules formed? What functions do they serve? In this chapter, we will explore these questions.

2.1 | The Building Blocks of Molecules

By the end of this section, you will be able to:

- Describe matter and elements
- Describe the interrelationship between protons, neutrons, and electrons, and the ways in which electrons can be donated or shared between atoms

At its most fundamental level, life is made up of matter. **Matter** occupies space and has mass. All matter is composed of **elements**, substances that cannot be broken down or transformed chemically into other substances. Each element is made of atoms, each with a constant number of protons and unique properties. A total of 118 elements have been defined; however, only 92 occur naturally, and fewer than 30 are found in living cells. The remaining 26 elements are unstable and, therefore, do not exist for very long or are theoretical and have yet to be detected.

Each element is designated by its chemical symbol (such as H, N, O, C, and Na), and possesses unique properties. These unique properties allow elements to combine and to bond with each other in specific ways.

Atoms

An atom is the smallest component of an element that retains all of the chemical properties of that element. For example, one hydrogen atom has all of the properties of the element hydrogen, such as it exists as a gas at room temperature, and it bonds with oxygen to create a water molecule. Hydrogen atoms cannot be broken down into anything smaller while still retaining the properties of hydrogen. If a hydrogen atom were broken down into subatomic particles, it would no longer have the properties of hydrogen.

At the most basic level, all organisms are made of a combination of elements. They contain atoms that combine together to form molecules. In multicellular organisms, such as animals, molecules can interact to form cells that combine to form tissues, which make up organs. These combinations continue until entire multicellular organisms are formed.

All atoms contain protons, electrons, and neutrons (Figure 2.2). The only exception is hydrogen (H), which is made of one proton and one electron. A **proton** is a positively charged particle that resides in the **nucleus** (the core of the atom) of an atom and has a mass of 1 and a charge of +1. An **electron** is a negatively charged particle that travels in the space around the nucleus. In other words, it resides outside of the nucleus. It has a negligible mass and has a charge of −1.

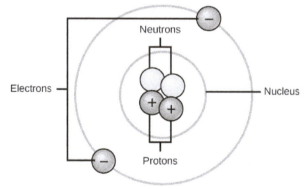

Figure 2.2 Atoms are made up of protons and neutrons located within the nucleus, and electrons surrounding the nucleus.

Neutrons, like protons, reside in the nucleus of an atom. They have a mass of 1 and no charge. The positive (protons) and negative (electrons) charges balance each other in a neutral atom, which has a net zero charge.

Because protons and neutrons each have a mass of 1, the mass of an atom is equal to the number of protons and neutrons of that atom. The number of electrons does not factor into the overall mass, because their mass is so small.

As stated earlier, each element has its own unique properties. Each contains a different number of protons and neutrons, giving it its own atomic number and mass number. The **atomic number** of an element is equal to the number of protons that element contains. The **mass number**, or atomic mass, is the number of protons plus the number of neutrons of that element. Therefore, it is possible to determine the number of neutrons by subtracting the atomic number from the mass number.

Download for free at https://openstax.org/details/books/concepts-biology

These numbers provide information about the elements and how they will react when combined. Different elements have different melting and boiling points, and are in different states (liquid, solid, or gas) at room temperature. They also combine in different ways. Some form specific types of bonds, whereas others do not. How they combine is based on the number of electrons present. Because of these characteristics, the elements are arranged into the **periodic table of elements**, a chart of the elements that includes the atomic number and relative atomic mass of each element. The periodic table also provides key information about the properties of elements (Figure 2.2)—often indicated by color-coding. The arrangement of the table also shows how the electrons in each element are organized and provides important details about how atoms will react with each other to form molecules.

Isotopes are different forms of the same element that have the same number of protons, but a different number of neutrons. Some elements, such as carbon, potassium, and uranium, have naturally occurring isotopes. Carbon-12, the most common isotope of carbon, contains six protons and six neutrons. Therefore, it has a mass number of 12 (six protons and six neutrons) and an atomic number of 6 (which makes it carbon). Carbon-14 contains six protons and eight neutrons. Therefore, it has a mass number of 14 (six protons and eight neutrons) and an atomic number of 6, meaning it is still the element carbon. These two alternate forms of carbon are isotopes. Some isotopes are unstable and will lose protons, other subatomic particles, or energy to form more stable elements. These are called **radioactive isotopes** or radioisotopes.

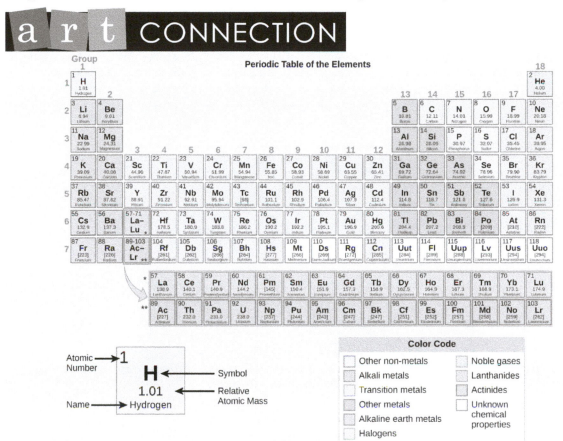

Figure 2.3 Arranged in columns and rows based on the characteristics of the elements, the periodic table provides key information about the elements and how they might interact with each other to form molecules. Most periodic tables provide a key or legend to the information they contain.

How many neutrons do (K) potassium-39 and potassium-40 have, respectively?

e✓olution IN ACTION

Carbon Dating

Carbon-14 (^{14}C) is a naturally occurring radioisotope that is created in the atmosphere by cosmic rays. This is a continuous process, so more ^{14}C is always being created. As a living organism develops, the relative level of ^{14}C in its body is equal to the concentration of ^{14}C in the atmosphere. When an organism dies, it is no longer ingesting ^{14}C, so the ratio will decline. ^{14}C decays to ^{14}N by a process called beta decay; it gives off energy in this slow process.

After approximately 5,730 years, only one-half of the starting concentration of ^{14}C will have been converted to ^{14}N. The time it takes for half of the original concentration of an isotope to decay to its more stable form is called its half-life. Because the half-life of ^{14}C is long, it is used to age formerly living objects, such as fossils. Using the ratio of the ^{14}C concentration found in an object to the amount of ^{14}C detected in the atmosphere, the amount of the isotope that has not yet decayed can be determined. Based on this amount, the age of the fossil can be calculated to about 50,000 years (Figure 2.4). Isotopes with longer half-lives, such as potassium-40, are used to calculate the ages of older fossils. Through the use of carbon dating, scientists can reconstruct the ecology and biogeography of organisms living within the past 50,000 years.

Figure 2.4 The age of remains that contain carbon and are less than about 50,000 years old, such as this pygmy mammoth, can be determined using carbon dating. (credit: Bill Faulkner/NPS)

To learn more about atoms and isotopes, and how you can tell one isotope from another, visit this site (http://openstaxcollege.org/l/isotopes) and run the simulation.

Chemical Bonds

How elements interact with one another depends on how their electrons are arranged and how many openings for electrons exist at the outermost region where electrons are present in an atom. Electrons exist at energy levels that form shells around the nucleus. The closest shell can hold up to two electrons. The closest shell to the nucleus is always filled first, before any other shell can be filled. Hydrogen has one electron; therefore, it has only one spot occupied within the lowest shell. Helium has two electrons; therefore, it can completely fill the lowest shell with its two electrons. If you look at the periodic table, you will see that hydrogen and helium are the only two elements in the first row. This is because they only have electrons in their first shell. Hydrogen and helium are the only two elements that have the lowest shell and no other shells.

The second and third energy levels can hold up to eight electrons. The eight electrons are arranged in four pairs and one position in each pair is filled with an electron before any pairs are completed.

Looking at the periodic table again (Figure 2.3), you will notice that there are seven rows. These rows correspond to the number of shells that the elements within that row have. The elements within a particular row have increasing numbers of electrons as the columns proceed from left to right. Although each element has the same number of shells, not all of the shells are completely filled with electrons. If you look at the second row of the periodic table, you will find lithium (Li), beryllium (Be), boron (B), carbon (C), nitrogen (N), oxygen (O), fluorine (F), and neon (Ne). These all have electrons that occupy only the first and second shells. Lithium has only one electron in its outermost shell, beryllium has two electrons, boron has three, and so on, until the entire shell is filled with eight electrons, as is the case with neon.

Not all elements have enough electrons to fill their outermost shells, but an atom is at its most stable when all of the electron positions in the outermost shell are filled. Because of these vacancies in the outermost shells, we see the formation of **chemical bonds**, or interactions between two or more of the same or different elements that result in the formation of molecules. To achieve greater stability, atoms will tend to completely fill their outer shells and will bond with other elements to accomplish this goal by sharing electrons, accepting electrons from another atom, or donating electrons to another atom. Because the outermost shells of the elements with low atomic numbers (up to calcium, with atomic number 20) can hold eight electrons, this is referred to as the **octet rule**. An element can donate, accept, or share electrons with other elements to fill its outer shell and satisfy the octet rule.

When an atom does not contain equal numbers of protons and electrons, it is called an **ion**. Because the number of electrons does not equal the number of protons, each ion has a net charge. Positive ions are formed by losing electrons and are called **cations**. Negative ions are formed by gaining electrons and are called **anions**.

For example, sodium only has one electron in its outermost shell. It takes less energy for sodium to donate that one electron than it does to accept seven more electrons to fill the outer shell. If sodium loses an electron, it now has 11 protons and only 10 electrons, leaving it with an overall charge of +1. It is now called a sodium ion.

The chlorine atom has seven electrons in its outer shell. Again, it is more energy-efficient for chlorine to gain one electron than to lose seven. Therefore, it tends to gain an electron to create an ion with 17 protons and 18 electrons, giving it a net negative (–1) charge. It is now called a chloride ion. This movement of electrons from one element to another is referred to as **electron transfer**. As Figure 2.5 illustrates, a sodium atom (Na) only has one electron in its outermost shell, whereas a chlorine atom (Cl) has seven electrons in its outermost shell. A sodium atom will donate its one electron to empty its shell, and a chlorine atom will accept that electron to fill its shell, becoming chloride. Both ions now satisfy the octet rule and have complete outermost shells. Because the number of electrons is no longer equal to the number of protons, each is now an ion and has a +1 (sodium) or –1 (chloride) charge.

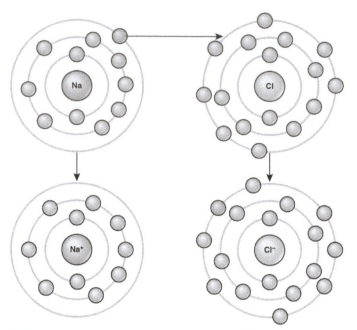

Figure 2.5 Elements tend to fill their outermost shells with electrons. To do this, they can either donate or accept electrons from other elements.

Ionic Bonds

There are four types of bonds or interactions: ionic, covalent, hydrogen bonds, and van der Waals interactions. Ionic and covalent bonds are strong interactions that require a larger energy input to break apart. When an element donates an electron from its outer shell, as in the sodium atom example above, a positive ion is formed. The element accepting the electron is now negatively charged. Because positive and negative charges attract, these ions stay together and form an **ionic bond**, or a bond between ions. The elements bond together with the electron from one element staying predominantly with the other element. When Na^+ and Cl^- ions combine to produce NaCl, an electron from a sodium atom stays with the other seven from the chlorine atom, and the sodium and chloride ions attract each other in a lattice of ions with a net zero charge.

Covalent Bonds

Another type of strong chemical bond between two or more atoms is a **covalent bond**. These bonds form when an electron is shared between two elements and are the strongest and most common form of chemical bond in living organisms. Covalent bonds form between the elements that make up the biological molecules in our cells. Unlike ionic bonds, covalent bonds do not dissociate in water.

The hydrogen and oxygen atoms that combine to form water molecules are bound together by covalent bonds. The electron from the hydrogen atom divides its time between the outer shell of the hydrogen atom and the incomplete outer shell of the oxygen atom. To completely fill the outer shell of an oxygen atom, two electrons from two hydrogen atoms are needed, hence the subscript "2" in H_2O. The electrons are shared between the atoms, dividing their time between them to "fill" the outer shell of each. This sharing is a lower energy state for all of the atoms involved than if they existed without their outer shells filled.

There are two types of covalent bonds: polar and nonpolar. **Nonpolar covalent bonds** form between two atoms of the same element or between different elements that share the electrons equally. For example, an oxygen atom can bond with another oxygen atom to fill their outer shells. This association is nonpolar because the electrons will be equally distributed between each oxygen atom. Two covalent bonds form between the two oxygen atoms because oxygen requires two shared electrons to fill its outermost shell. Nitrogen atoms will form three covalent bonds (also called triple covalent) between two atoms of nitrogen because each nitrogen atom needs three electrons to fill its outermost shell. Another example of a nonpolar covalent bond is found in the methane (CH_4) molecule. The carbon atom has four electrons in its outermost shell and needs four more to fill it. It gets these four from four hydrogen atoms, each atom providing one. These elements all share the electrons equally, creating four nonpolar covalent bonds (Figure 2.6).

In a **polar covalent bond**, the electrons shared by the atoms spend more time closer to one nucleus than to the other nucleus. Because of the unequal distribution of electrons between the different nuclei, a slightly positive (δ+) or slightly negative (δ−) charge develops. The covalent bonds between hydrogen and oxygen atoms in water are polar covalent bonds. The shared electrons spend more time near the oxygen nucleus, giving it a small negative charge, than they spend near the hydrogen nuclei, giving these molecules a small positive charge.

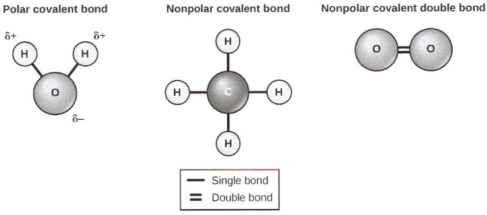

Figure 2.6 The water molecule (left) depicts a polar bond with a slightly positive charge on the hydrogen atoms and a slightly negative charge on the oxygen. Examples of nonpolar bonds include methane (middle) and oxygen (right).

Hydrogen Bonds

Ionic and covalent bonds are strong bonds that require considerable energy to break. However, not all bonds between elements are ionic or covalent bonds. Weaker bonds can also form. These are attractions that occur between positive and negative charges that do not require much energy to break. Two weak bonds that occur frequently are hydrogen bonds and van der Waals interactions. These bonds give rise to the unique properties of water and the unique structures of DNA and proteins.

When polar covalent bonds containing a hydrogen atom form, the hydrogen atom in that bond has a slightly positive charge. This is because the shared electron is pulled more strongly toward the other element and away from the hydrogen nucleus. Because the hydrogen atom is slightly positive (δ+), it will be attracted to neighboring negative partial charges (δ−). When this happens, a weak interaction occurs between the δ+ charge of the hydrogen atom of one molecule and the δ− charge of the other molecule. This interaction is called a **hydrogen bond**. This type of bond is common; for example, the liquid nature of water is caused by the hydrogen bonds between water molecules (Figure 2.7). Hydrogen bonds give water the unique properties that sustain life. If it were not for hydrogen bonding, water would be a gas rather than a liquid at room temperature.

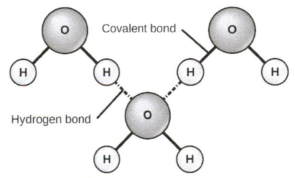

Figure 2.7 Hydrogen bonds form between slightly positive (δ+) and slightly negative (δ−) charges of polar covalent molecules, such as water.

Hydrogen bonds can form between different molecules and they do not always have to include a water molecule. Hydrogen atoms in polar bonds within any molecule can form bonds with other adjacent molecules. For example, hydrogen bonds hold together two long strands of DNA to give the DNA molecule its characteristic double-stranded structure. Hydrogen bonds are also responsible for some of the three-dimensional structure of proteins.

van der Waals Interactions

Like hydrogen bonds, **van der Waals interactions** are weak attractions or interactions between molecules. They occur between polar, covalently bound, atoms in different molecules. Some of these weak attractions are caused by temporary partial charges formed when electrons move around a nucleus. These weak interactions between molecules are important in biological systems.

careers IN ACTION

Radiography Technician

Have you or anyone you know ever had a magnetic resonance imaging (MRI) scan, a mammogram, or an X-ray? These tests produce images of your soft tissues and organs (as with an MRI or mammogram) or your bones (as happens in an X-ray) by using either radiowaves or special isotopes (radiolabeled or fluorescently labeled) that are ingested or injected into the body. These tests provide data for disease diagnoses by creating images of your organs or skeletal system.

MRI imaging works by subjecting hydrogen nuclei, which are abundant in the water in soft tissues, to fluctuating magnetic fields, which cause them to emit their own magnetic field. This signal is then read by sensors in the machine and interpreted by a computer to form a detailed image.

Some radiography technologists and technicians specialize in computed tomography, MRI, and mammography. They produce films or images of the body that help medical professionals examine and diagnose. Radiologists work directly with patients, explaining machinery, preparing them for exams, and ensuring that their body or body parts are positioned correctly to produce the needed images. Physicians or radiologists then analyze the test results.

Radiography technicians can work in hospitals, doctors' offices, or specialized imaging centers. Training to become a radiography technician happens at hospitals, colleges, and universities that offer certificates, associate's degrees, or bachelor's degrees in radiography.

2.2 | Water

By the end of this section, you will be able to:

* Describe the properties of water that are critical to maintaining life

Do you ever wonder why scientists spend time looking for water on other planets? It is because water is essential to life; even minute traces of it on another planet can indicate that life could or did exist on that planet. Water is one of the more abundant molecules in living cells and the one most critical to life as we know it. Approximately 60–70 percent of your body is made up of water. Without it, life simply would not exist.

Water Is Polar

The hydrogen and oxygen atoms within water molecules form polar covalent bonds. The shared electrons spend more time associated with the oxygen atom than they do with hydrogen atoms. There is no overall charge to a water molecule, but there is a slight positive charge on each hydrogen atom and a slight negative charge on the oxygen atom. Because of these charges, the slightly positive hydrogen atoms repel each other and form the unique shape seen in Figure 2.7. Each water molecule attracts other water molecules because of the positive and negative charges in the different parts of the molecule. Water also attracts other polar molecules (such as sugars), forming hydrogen bonds. When a substance readily forms hydrogen bonds with water, it can dissolve in water and is referred to as **hydrophilic** ("water-loving"). Hydrogen bonds are not readily formed with nonpolar substances like oils and fats (Figure 2.8). These nonpolar compounds are **hydrophobic** ("water-fearing") and will not dissolve in water.

Figure 2.8 As this macroscopic image of oil and water show, oil is a nonpolar compound and, hence, will not dissolve in water. Oil and water do not mix. (credit: Gautam Dogra)

Water Stabilizes Temperature

The hydrogen bonds in water allow it to absorb and release heat energy more slowly than many other substances. **Temperature** is a measure of the motion (kinetic energy) of molecules. As the motion increases, energy is higher and thus temperature is higher. Water absorbs a great deal of energy before its temperature rises. Increased energy disrupts the hydrogen bonds between water molecules. Because these bonds can be created and disrupted rapidly, water absorbs an increase in energy and temperature changes only minimally. This means that water moderates temperature changes within organisms and in their environments. As energy input continues, the balance between hydrogen-bond formation and destruction swings toward the destruction side. More bonds are broken than are formed. This process results in the release of individual water molecules at the surface of the liquid (such as a body of water, the leaves of a plant, or the skin of an organism) in a process called **evaporation**. Evaporation of sweat, which is 90 percent water, allows for cooling of an organism, because breaking hydrogen bonds requires an input of energy and takes heat away from the body.

Conversely, as molecular motion decreases and temperatures drop, less energy is present to break the hydrogen bonds between water molecules. These bonds remain intact and begin to form a rigid, lattice-like structure (e.g., ice) (Figure 2.9a). When frozen, ice is less dense than liquid water (the molecules are farther apart). This means that ice floats on the surface of a body of water (Figure 2.9b). In lakes, ponds, and oceans, ice will form on the surface of the water, creating an insulating barrier to protect the animal and plant life beneath from freezing in the water. If this did not happen, plants and animals living in water would freeze in a block of ice and could not move freely, making life in cold temperatures difficult or impossible.

(a)   (b)

Figure 2.9 (a) The lattice structure of ice makes it less dense than the freely flowing molecules of liquid water. Ice's lower density enables it to (b) float on water. (credit a: modification of work by Jane Whitney; credit b: modification of work by Carlos Ponte)

Click here (http://openstaxcollege.org/l/ice_lattice) to see a 3-D animation of the structure of an ice lattice. (credit: image created by Jane Whitney using Visual Molecular Dynamics (VMD) software[1])

Water Is an Excellent Solvent

Because water is polar, with slight positive and negative charges, ionic compounds and polar molecules can readily dissolve in it. Water is, therefore, what is referred to as a **solvent**—a substance capable of dissolving another substance. The charged particles will form hydrogen bonds with a surrounding layer of water molecules. This is referred to as a sphere of hydration and serves to keep the particles separated or dispersed in the water. In the case of table salt (NaCl) mixed in water (Figure 2.10), the sodium and chloride ions separate, or dissociate, in the water, and spheres of hydration are formed around the ions. A positively charged sodium ion is surrounded by the partially negative charges of oxygen atoms in water molecules. A negatively charged chloride ion is surrounded by the partially positive charges of hydrogen atoms in water molecules. These spheres of hydration are also referred to as hydration shells. The polarity of the water molecule makes it an effective solvent and is important in its many roles in living systems.

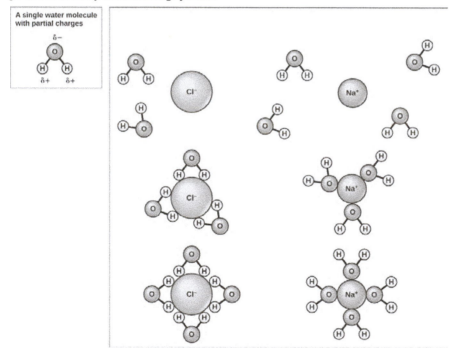

Figure 2.10 When table salt (NaCl) is mixed in water, spheres of hydration form around the ions.

Water Is Cohesive

Have you ever filled up a glass of water to the very top and then slowly added a few more drops? Before it overflows, the water actually forms a dome-like shape above the rim of the glass. This water can stay above the glass because of the property of **cohesion**. In cohesion, water molecules are attracted to each other (because of hydrogen bonding), keeping the molecules together at the liquid-air (gas) interface, although there is no more room in the glass. Cohesion gives rise to **surface tension**, the capacity of a substance to withstand rupture when placed under tension or stress. When you drop

1. Humphrey, W., Dalke, A. and Schulten, K., "VMD—Visual Molecular Dynamics", *J. Molec. Graphics*, 1996, vol. 14, pp. 33-38. http://www.ks.uiuc.edu/Research/vmd/

a small scrap of paper onto a droplet of water, the paper floats on top of the water droplet, although the object is denser (heavier) than the water. This occurs because of the surface tension that is created by the water molecules. Cohesion and surface tension keep the water molecules intact and the item floating on the top. It is even possible to "float" a steel needle on top of a glass of water if you place it gently, without breaking the surface tension (Figure 2.11).

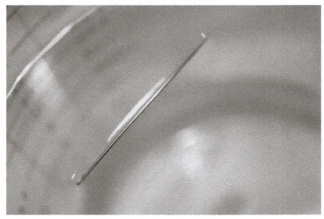

Figure 2.11 The weight of a needle on top of water pulls the surface tension downward; at the same time, the surface tension of the water is pulling it up, suspending the needle on the surface of the water and keeping it from sinking. Notice the indentation in the water around the needle. (credit: Cory Zanker)

These cohesive forces are also related to the water's property of **adhesion,** or the attraction between water molecules and other molecules. This is observed when water "climbs" up a straw placed in a glass of water. You will notice that the water appears to be higher on the sides of the straw than in the middle. This is because the water molecules are attracted to the straw and therefore adhere to it.

Cohesive and adhesive forces are important for sustaining life. For example, because of these forces, water can flow up from the roots to the tops of plants to feed the plant.

To learn more about water, visit the U.S. Geological Survey Water Science for Schools: All About Water! website. (http://openstaxcollege.org/l/about_water)

Buffers, pH, Acids, and Bases

The pH of a solution is a measure of its acidity or alkalinity. You have probably used **litmus paper**, paper that has been treated with a natural water-soluble dye so it can be used as a pH indicator, to test how much acid or base (alkalinity) exists in a solution. You might have even used some to make sure the water in an outdoor swimming pool is properly treated. In both cases, this pH test measures the amount of hydrogen ions that exists in a given solution. High concentrations of hydrogen ions yield a low pH, whereas low levels of hydrogen ions result in a high pH. The overall concentration of hydrogen ions is inversely related to its pH and can be measured on the **pH scale** (Figure 2.12). Therefore, the more hydrogen ions present, the lower the pH; conversely, the fewer hydrogen ions, the higher the pH.

The pH scale ranges from 0 to 14. A change of one unit on the pH scale represents a change in the concentration of hydrogen ions by a factor of 10, a change in two units represents a change in the concentration of hydrogen ions by a factor of 100. Thus, small changes in pH represent large changes in the concentrations of hydrogen ions. Pure water is neutral. It is neither acidic nor basic, and has a pH of 7.0. Anything below 7.0 (ranging from 0.0 to 6.9) is acidic, and anything above 7.0 (from 7.1 to 14.0) is alkaline. The blood in your veins is slightly alkaline (pH = 7.4). The environment in your stomach is highly acidic (pH = 1 to 2). Orange juice is mildly acidic (pH = approximately 3.5), whereas baking soda is basic (pH = 9.0).

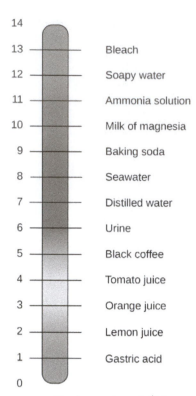

14 — Bleach (at 13)
13 — Bleach
12 — Soapy water
11 — Ammonia solution
10 — Milk of magnesia
9 — Baking soda
8 — Seawater
7 — Distilled water
6 — Urine
5 — Black coffee
4 — Tomato juice
3 — Orange juice
2 — Lemon juice
1 — Gastric acid
0

Figure 2.12 The pH scale measures the amount of hydrogen ions (H^+) in a substance. (credit: modification of work by Edward Stevens)

Acids are substances that provide hydrogen ions (H^+) and lower pH, whereas **bases** provide hydroxide ions (OH^-) and raise pH. The stronger the acid, the more readily it donates H^+. For example, hydrochloric acid and lemon juice are very acidic and readily give up H^+ when added to water. Conversely, bases are those substances that readily donate OH^-. The OH^- ions combine with H^+ to produce water, which raises a substance's pH. Sodium hydroxide and many household cleaners are very alkaline and give up OH^- rapidly when placed in water, thereby raising the pH.

Most cells in our bodies operate within a very narrow window of the pH scale, typically ranging only from 7.2 to 7.6. If the pH of the body is outside of this range, the respiratory system malfunctions, as do other organs in the body. Cells no longer function properly, and proteins will break down. Deviation outside of the pH range can induce coma or even cause death.

So how is it that we can ingest or inhale acidic or basic substances and not die? Buffers are the key. **Buffers** readily absorb excess H^+ or OH^-, keeping the pH of the body carefully maintained in the aforementioned narrow range. Carbon dioxide is part of a prominent buffer system in the human body; it keeps the pH within the proper range. This buffer system involves carbonic acid (H_2CO_3) and bicarbonate (HCO_3^-) anion. If too much H^+ enters the body, bicarbonate will combine with the H^+ to create carbonic acid and limit the decrease in pH. Likewise, if too much OH^- is introduced into the system, carbonic acid will rapidly dissociate into bicarbonate and H^+ ions. The H^+ ions can combine with the OH^- ions, limiting the increase in pH. While carbonic acid is an important product in this reaction, its presence is fleeting because the carbonic acid is released from the body as carbon dioxide gas each time we breathe. Without this buffer system, the pH in our bodies would fluctuate too much and we would fail to survive.

2.3 | Biological Molecules

By the end of this section, you will be able to:

- Describe the ways in which carbon is critical to life
- Explain the impact of slight changes in amino acids on organisms
- Describe the four major types of biological molecules
- Understand the functions of the four major types of molecules

The large molecules necessary for life that are built from smaller organic molecules are called biological **macromolecules**. There are four major classes of biological macromolecules (carbohydrates, lipids, proteins, and nucleic acids), and each is an important component of the cell and performs a wide array of functions. Combined, these molecules make up the majority of a cell's mass. Biological macromolecules are organic, meaning that they contain carbon. In addition, they may contain hydrogen, oxygen, nitrogen, phosphorus, sulfur, and additional minor elements.

Carbon

It is often said that life is "carbon-based." This means that carbon atoms, bonded to other carbon atoms or other elements, form the fundamental components of many, if not most, of the molecules found uniquely in living things. Other elements play important roles in biological molecules, but carbon certainly qualifies as the "foundation" element for molecules in living things. It is the bonding properties of carbon atoms that are responsible for its important role.

Carbon Bonding

Carbon contains four electrons in its outer shell. Therefore, it can form four covalent bonds with other atoms or molecules. The simplest organic carbon molecule is methane (CH_4), in which four hydrogen atoms bind to a carbon atom (Figure 2.13).

Methane

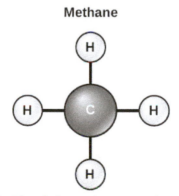

Figure 2.13 Carbon can form four covalent bonds to create an organic molecule. The simplest carbon molecule is methane (CH_4), depicted here.

However, structures that are more complex are made using carbon. Any of the hydrogen atoms can be replaced with another carbon atom covalently bonded to the first carbon atom. In this way, long and branching chains of carbon compounds can be made (Figure 2.14a). The carbon atoms may bond with atoms of other elements, such as nitrogen, oxygen, and phosphorus (Figure 2.14b). The molecules may also form rings, which themselves can link with other rings (Figure 2.14c). This diversity of molecular forms accounts for the diversity of functions of the biological macromolecules and is based to a large degree on the ability of carbon to form multiple bonds with itself and other atoms.

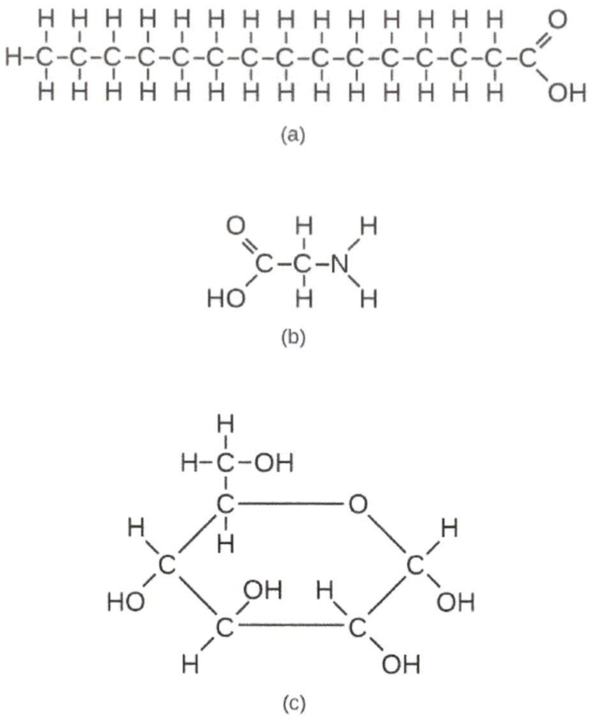

(a)

(b)

(c)

Figure 2.14 These examples show three molecules (found in living organisms) that contain carbon atoms bonded in various ways to other carbon atoms and the atoms of other elements. (a) This molecule of stearic acid has a long chain of carbon atoms. (b) Glycine, a component of proteins, contains carbon, nitrogen, oxygen, and hydrogen atoms. (c) Glucose, a sugar, has a ring of carbon atoms and one oxygen atom.

Carbohydrates

Carbohydrates are macromolecules with which most consumers are somewhat familiar. To lose weight, some individuals adhere to "low-carb" diets. Athletes, in contrast, often "carb-load" before important competitions to ensure that they have sufficient energy to compete at a high level. Carbohydrates are, in fact, an essential part of our diet; grains, fruits, and vegetables are all natural sources of carbohydrates. Carbohydrates provide energy to the body, particularly through glucose, a simple sugar. Carbohydrates also have other important functions in humans, animals, and plants.

Carbohydrates can be represented by the formula $(CH_2O)_n$, where n is the number of carbon atoms in the molecule. In other words, the ratio of carbon to hydrogen to oxygen is 1:2:1 in carbohydrate molecules. Carbohydrates are classified into three subtypes: monosaccharides, disaccharides, and polysaccharides.

Monosaccharides (mono- = "one"; sacchar- = "sweet") are simple sugars, the most common of which is glucose. In monosaccharides, the number of carbon atoms usually ranges from three to six. Most monosaccharide names end with the suffix -ose. Depending on the number of carbon atoms in the sugar, they may be known as trioses (three carbon atoms), pentoses (five carbon atoms), and hexoses (six carbon atoms).

Monosaccharides may exist as a linear chain or as ring-shaped molecules; in aqueous solutions, they are usually found in the ring form.

The chemical formula for glucose is $C_6H_{12}O_6$. In most living species, glucose is an important source of energy. During cellular respiration, energy is released from glucose, and that energy is used to help make adenosine triphosphate (ATP). Plants synthesize glucose using carbon dioxide and water by the process of photosynthesis, and the glucose, in turn, is used for the energy requirements of the plant. The excess synthesized glucose is often stored as starch that is broken down by other organisms that feed on plants.

Galactose (part of lactose, or milk sugar) and fructose (found in fruit) are other common monosaccharides. Although glucose, galactose, and fructose all have the same chemical formula ($C_6H_{12}O_6$), they differ structurally and chemically (and are known as isomers) because of differing arrangements of atoms in the carbon chain (Figure 2.15).

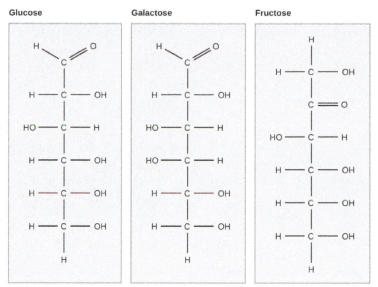

Figure 2.15 Glucose, galactose, and fructose are isomeric monosaccharides, meaning that they have the same chemical formula but slightly different structures.

Disaccharides (di- = "two") form when two monosaccharides undergo a dehydration reaction (a reaction in which the removal of a water molecule occurs). During this process, the hydroxyl group (–OH) of one monosaccharide combines with a hydrogen atom of another monosaccharide, releasing a molecule of water (H_2O) and forming a covalent bond between atoms in the two sugar molecules.

Common disaccharides include lactose, maltose, and sucrose. Lactose is a disaccharide consisting of the monomers glucose and galactose. It is found naturally in milk. Maltose, or malt sugar, is a disaccharide formed from a dehydration reaction between two glucose molecules. The most common disaccharide is sucrose, or table sugar, which is composed of the monomers glucose and fructose.

A long chain of monosaccharides linked by covalent bonds is known as a **polysaccharide** (poly- = "many"). The chain may be branched or unbranched, and it may contain different types of monosaccharides. Polysaccharides may be very large molecules. Starch, glycogen, cellulose, and chitin are examples of polysaccharides.

Starch is the stored form of sugars in plants and is made up of amylose and amylopectin (both polymers of glucose). Plants are able to synthesize glucose, and the excess glucose is stored as starch in different plant parts, including roots and seeds. The starch that is consumed by animals is broken down into smaller molecules, such as glucose. The cells can then absorb the glucose.

Glycogen is the storage form of glucose in humans and other vertebrates, and is made up of monomers of glucose. Glycogen is the animal equivalent of starch and is a highly branched molecule usually stored in liver and muscle cells. Whenever glucose levels decrease, glycogen is broken down to release glucose.

Cellulose is one of the most abundant natural biopolymers. The cell walls of plants are mostly made of cellulose, which provides structural support to the cell. Wood and paper are mostly cellulosic in nature. Cellulose is made up of glucose monomers that are linked by bonds between particular carbon atoms in the glucose molecule.

Every other glucose monomer in cellulose is flipped over and packed tightly as extended long chains. This gives cellulose its rigidity and high tensile strength—which is so important to plant cells. Cellulose passing through our digestive system is called dietary fiber. While the glucose-glucose bonds in cellulose cannot be broken down by human digestive enzymes, herbivores such as cows, buffalos, and horses are able to digest grass that is rich in cellulose and use it as a food source. In these animals, certain species of bacteria reside in the rumen (part of the digestive system of herbivores) and secrete the enzyme cellulase. The appendix also contains bacteria that break down cellulose, giving it an important role in the digestive systems of ruminants. Cellulases can break down cellulose into glucose monomers that can be used as an energy source by the animal.

Carbohydrates serve other functions in different animals. Arthropods, such as insects, spiders, and crabs, have an outer skeleton, called the exoskeleton, which protects their internal body parts. This exoskeleton is made of the biological macromolecule **chitin**, which is a nitrogenous carbohydrate. It is made of repeating units of a modified sugar containing nitrogen.

Thus, through differences in molecular structure, carbohydrates are able to serve the very different functions of energy storage (starch and glycogen) and structural support and protection (cellulose and chitin) (Figure 2.16).

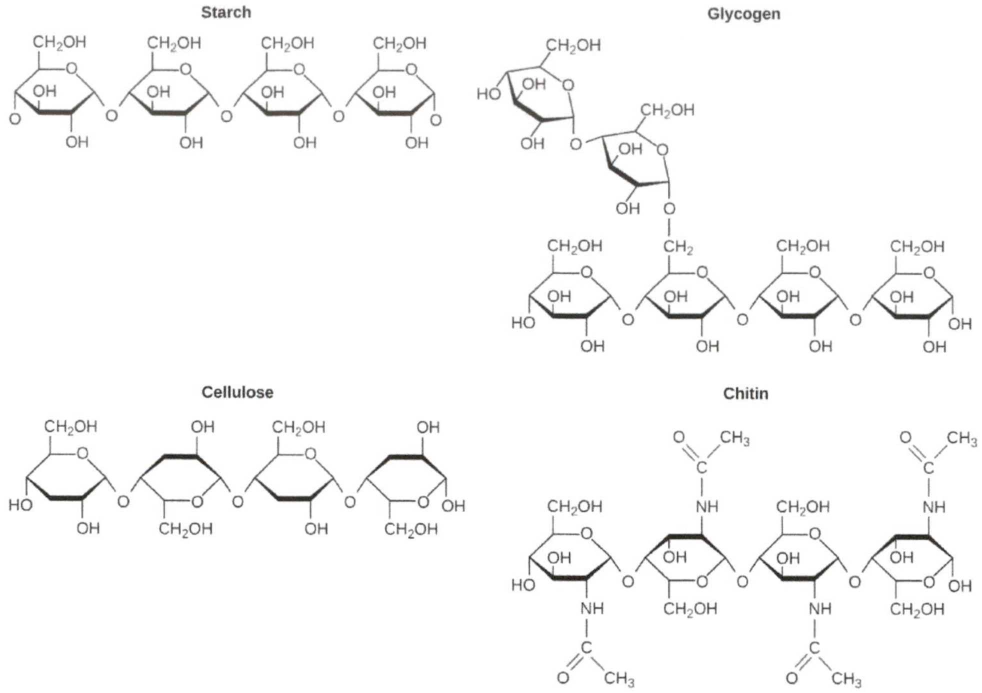

Figure 2.16 Although their structures and functions differ, all polysaccharide carbohydrates are made up of monosaccharides and have the chemical formula $(CH_2O)n$.

careers IN ACTION

Registered Dietitian

Obesity is a worldwide health concern, and many diseases, such as diabetes and heart disease, are becoming more prevalent because of obesity. This is one of the reasons why registered dietitians are increasingly sought after for advice. Registered dietitians help plan food and nutrition programs for individuals in various settings. They often work with patients in health-care facilities, designing nutrition plans to prevent and treat diseases. For example, dietitians may teach a patient with diabetes how to manage blood-sugar levels by eating the correct types and amounts of carbohydrates. Dietitians may also work in nursing homes, schools, and private practices.

To become a registered dietitian, one needs to earn at least a bachelor's degree in dietetics, nutrition, food technology, or a related field. In addition, registered dietitians must complete a supervised internship program and pass a national exam. Those who pursue careers in dietetics take courses in nutrition, chemistry, biochemistry, biology, microbiology, and human physiology. Dietitians must become experts in the chemistry and functions of food (proteins, carbohydrates, and fats).

Lipids

Lipids include a diverse group of compounds that are united by a common feature. **Lipids** are hydrophobic ("water-fearing"), or insoluble in water, because they are nonpolar molecules. This is because they are hydrocarbons that include only nonpolar carbon-carbon or carbon-hydrogen bonds. Lipids perform many different functions in a cell. Cells store energy for long-term use in the form of lipids called fats. Lipids also provide insulation from the environment for plants and animals (Figure 2.17). For example, they help keep aquatic birds and mammals dry because of their water-repelling nature. Lipids are also the building blocks of many hormones and are an important constituent of the plasma membrane. Lipids include fats, oils, waxes, phospholipids, and steroids.

Figure 2.17 Hydrophobic lipids in the fur of aquatic mammals, such as this river otter, protect them from the elements. (credit: Ken Bosma)

A **fat** molecule, such as a triglyceride, consists of two main components—glycerol and fatty acids. Glycerol is an organic compound with three carbon atoms, five hydrogen atoms, and three hydroxyl (–OH) groups. Fatty acids have a long chain of hydrocarbons to which an acidic carboxyl group is attached, hence the name "fatty acid." The number of carbons in the fatty acid may range from 4 to 36; most common are those containing 12–18 carbons. In a fat molecule, a fatty acid is attached to each of the three oxygen atoms in the –OH groups of the glycerol molecule with a covalent bond (Figure 2.18).

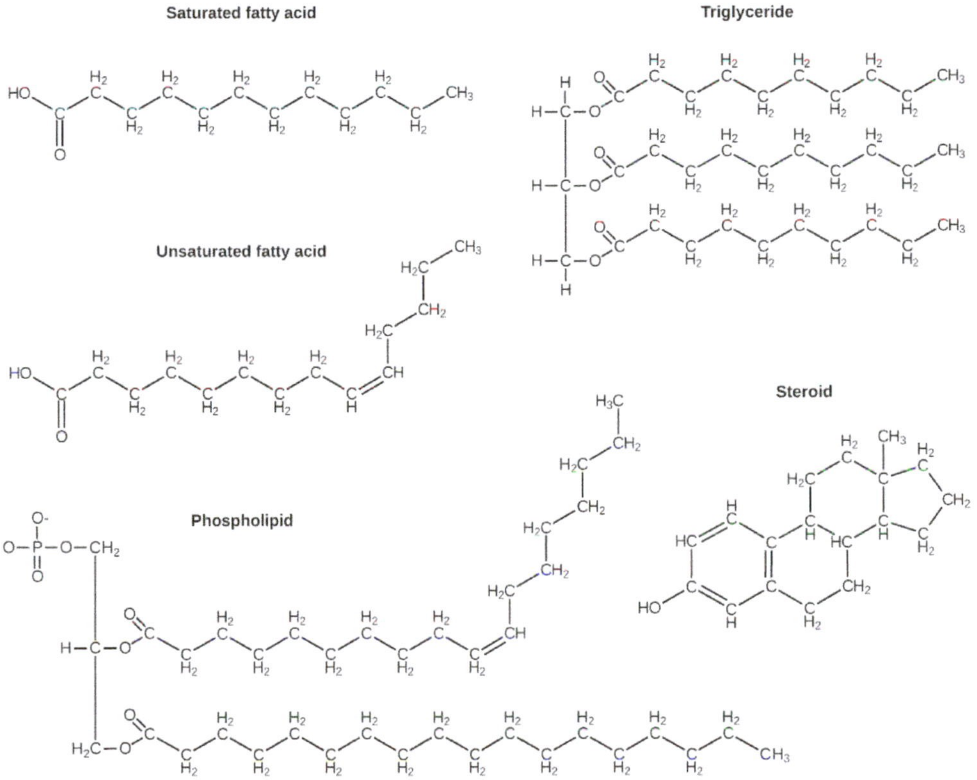

Figure 2.18 Lipids include fats, such as triglycerides, which are made up of fatty acids and glycerol, phospholipids, and steroids.

During this covalent bond formation, three water molecules are released. The three fatty acids in the fat may be similar or dissimilar. These fats are also called **triglycerides** because they have three fatty acids. Some fatty acids have common names that specify their origin. For example, palmitic acid, a saturated fatty acid, is derived from the palm tree. Arachidic acid is derived from *Arachis hypogaea*, the scientific name for peanuts.

Fatty acids may be saturated or unsaturated. In a fatty acid chain, if there are only single bonds between neighboring carbons in the hydrocarbon chain, the fatty acid is saturated. **Saturated fatty acids** are saturated with hydrogen; in other words, the number of hydrogen atoms attached to the carbon skeleton is maximized.

When the hydrocarbon chain contains a double bond, the fatty acid is an **unsaturated fatty acid**.

Most unsaturated fats are liquid at room temperature and are called **oils**. If there is one double bond in the molecule, then it is known as a monounsaturated fat (e.g., olive oil), and if there is more than one double bond, then it is known as a polyunsaturated fat (e.g., canola oil).

Saturated fats tend to get packed tightly and are solid at room temperature. Animal fats with stearic acid and palmitic acid contained in meat, and the fat with butyric acid contained in butter, are examples of saturated fats. Mammals store fats in specialized cells called adipocytes, where globules of fat occupy most of the cell. In plants, fat or oil is stored in seeds and is used as a source of energy during embryonic development.

Unsaturated fats or oils are usually of plant origin and contain unsaturated fatty acids. The double bond causes a bend or a "kink" that prevents the fatty acids from packing tightly, keeping them liquid at room temperature. Olive oil, corn oil, canola oil, and cod liver oil are examples of unsaturated fats. Unsaturated fats help to improve blood cholesterol levels, whereas saturated fats contribute to plaque formation in the arteries, which increases the risk of a heart attack.

In the food industry, oils are artificially hydrogenated to make them semi-solid, leading to less spoilage and increased shelf life. Simply speaking, hydrogen gas is bubbled through oils to solidify them. During this hydrogenation process, double bonds of the *cis*-conformation in the hydrocarbon chain may be converted to double bonds in the *trans*-conformation. This forms a ***trans*-fat** from a *cis*-fat. The orientation of the double bonds affects the chemical properties of the fat (Figure 2.19).

cis-fat molecule

trans-fat molecule

Figure 2.19 During the hydrogenation process, the orientation around the double bonds is changed, making a *trans*-fat from a *cis*-fat. This changes the chemical properties of the molecule.

Margarine, some types of peanut butter, and shortening are examples of artificially hydrogenated *trans*-fats. Recent studies have shown that an increase in *trans*-fats in the human diet may lead to an increase in levels of low-density lipoprotein (LDL), or "bad" cholesterol, which, in turn, may lead to plaque deposition in the arteries, resulting in heart disease. Many fast food restaurants have recently eliminated the use of *trans*-fats, and U.S. food labels are now required to list their *trans*-fat content.

Essential fatty acids are fatty acids that are required but not synthesized by the human body. Consequently, they must be supplemented through the diet. Omega-3 fatty acids fall into this category and are one of only two known essential fatty acids for humans (the other being omega-6 fatty acids). They are a type of polyunsaturated fat and are called omega-3 fatty acids because the third carbon from the end of the fatty acid participates in a double bond.

Salmon, trout, and tuna are good sources of omega-3 fatty acids. Omega-3 fatty acids are important in brain function and normal growth and development. They may also prevent heart disease and reduce the risk of cancer.

Like carbohydrates, fats have received a lot of bad publicity. It is true that eating an excess of fried foods and other "fatty" foods leads to weight gain. However, fats do have important functions. Fats serve as long-term energy storage. They also provide insulation for the body. Therefore, "healthy" unsaturated fats in moderate amounts should be consumed on a regular basis.

Phospholipids are the major constituent of the plasma membrane. Like fats, they are composed of fatty acid chains attached to a glycerol or similar backbone. Instead of three fatty acids attached, however, there are two fatty acids and the third carbon of the glycerol backbone is bound to a phosphate group. The phosphate group is modified by the addition of an alcohol.

A phospholipid has both hydrophobic and hydrophilic regions. The fatty acid chains are hydrophobic and exclude themselves from water, whereas the phosphate is hydrophilic and interacts with water.

Cells are surrounded by a membrane, which has a bilayer of phospholipids. The fatty acids of phospholipids face inside, away from water, whereas the phosphate group can face either the outside environment or the inside of the cell, which are both aqueous.

Steroids and Waxes

Unlike the phospholipids and fats discussed earlier, **steroids** have a ring structure. Although they do not resemble other lipids, they are grouped with them because they are also hydrophobic. All steroids have four, linked carbon rings and several of them, like cholesterol, have a short tail.

Cholesterol is a steroid. Cholesterol is mainly synthesized in the liver and is the precursor of many steroid hormones, such as testosterone and estradiol. It is also the precursor of vitamins E and K. Cholesterol is the precursor of bile salts, which help in the breakdown of fats and their subsequent absorption by cells. Although cholesterol is often spoken of in negative terms, it is necessary for the proper functioning of the body. It is a key component of the plasma membranes of animal cells.

Waxes are made up of a hydrocarbon chain with an alcohol (–OH) group and a fatty acid. Examples of animal waxes include beeswax and lanolin. Plants also have waxes, such as the coating on their leaves, that helps prevent them from drying out.

For an additional perspective on lipids, explore "Biomolecules: The Lipids" through this interactive animation (http://openstaxcollege.org/l/lipids) .

Proteins

Proteins are one of the most abundant organic molecules in living systems and have the most diverse range of functions of all macromolecules. Proteins may be structural, regulatory, contractile, or protective; they may serve in transport, storage, or membranes; or they may be toxins or enzymes. Each cell in a living system may contain thousands of different proteins, each with a unique function. Their structures, like their functions, vary greatly. They are all, however, polymers of amino acids, arranged in a linear sequence.

The functions of proteins are very diverse because there are 20 different chemically distinct amino acids that form long chains, and the amino acids can be in any order. For example, proteins can function as enzymes or hormones. **Enzymes**, which are produced by living cells, are catalysts in biochemical reactions (like digestion) and are usually proteins. Each enzyme is specific for the substrate (a reactant that binds to an enzyme) upon which it acts. Enzymes can function to break molecular bonds, to rearrange bonds, or to form new bonds. An example of an enzyme is salivary amylase, which breaks down amylose, a component of starch.

Hormones are chemical signaling molecules, usually proteins or steroids, secreted by an endocrine gland or group of endocrine cells that act to control or regulate specific physiological processes, including growth, development, metabolism, and reproduction. For example, insulin is a protein hormone that maintains blood glucose levels.

Proteins have different shapes and molecular weights; some proteins are globular in shape whereas others are fibrous in nature. For example, hemoglobin is a globular protein, but collagen, found in our skin, is a fibrous protein. Protein shape is critical to its function. Changes in temperature, pH, and exposure to chemicals may lead to permanent changes in the shape of the protein, leading to a loss of function or **denaturation** (to be discussed in more detail later). All proteins are made up of different arrangements of the same 20 kinds of amino acids.

Amino acids are the monomers that make up proteins. Each amino acid has the same fundamental structure, which consists of a central carbon atom bonded to an amino group ($-NH_2$), a carboxyl group ($-COOH$), and a hydrogen atom. Every amino acid also has another variable atom or group of atoms bonded to the central carbon atom known as the R group. The R group is the only difference in structure between the 20 amino acids; otherwise, the amino acids are identical (Figure 2.20).

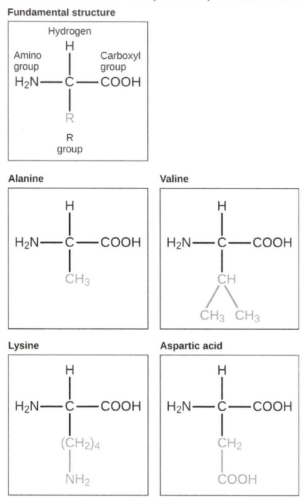

Figure 2.20 Amino acids are made up of a central carbon bonded to an amino group ($-NH_2$), a carboxyl group ($-COOH$), and a hydrogen atom. The central carbon's fourth bond varies among the different amino acids, as seen in these examples of alanine, valine, lysine, and aspartic acid.

The chemical nature of the R group determines the chemical nature of the amino acid within its protein (that is, whether it is acidic, basic, polar, or nonpolar).

The sequence and number of amino acids ultimately determine a protein's shape, size, and function. Each amino acid is attached to another amino acid by a covalent bond, known as a peptide bond, which is formed by a dehydration reaction. The carboxyl group of one amino acid and the amino group of a second amino acid combine, releasing a water molecule. The resulting bond is the peptide bond.

The products formed by such a linkage are called polypeptides. While the terms polypeptide and protein are sometimes used interchangeably, a **polypeptide** is technically a polymer of amino acids, whereas the term protein is used for a polypeptide or polypeptides that have combined together, have a distinct shape, and have a unique function.

The Evolutionary Significance of Cytochrome c

Cytochrome c is an important component of the molecular machinery that harvests energy from glucose. Because this protein's role in producing cellular energy is crucial, it has changed very little over millions of years. Protein sequencing has shown that there is a considerable amount of sequence similarity among cytochrome c molecules of different species; evolutionary relationships can be assessed by measuring the similarities or differences among various species' protein sequences.

For example, scientists have determined that human cytochrome c contains 104 amino acids. For each cytochrome c molecule that has been sequenced to date from different organisms, 37 of these amino acids appear in the same position in each cytochrome c. This indicates that all of these organisms are descended from a common ancestor. On comparing the human and chimpanzee protein sequences, no sequence difference was found. When human and rhesus monkey sequences were compared, a single difference was found in one amino acid. In contrast, human-to-yeast comparisons show a difference in 44 amino acids, suggesting that humans and chimpanzees have a more recent common ancestor than humans and the rhesus monkey, or humans and yeast.

Protein Structure

As discussed earlier, the shape of a protein is critical to its function. To understand how the protein gets its final shape or conformation, we need to understand the four levels of protein structure: primary, secondary, tertiary, and quaternary (Figure 2.21).

The unique sequence and number of amino acids in a polypeptide chain is its primary structure. The unique sequence for every protein is ultimately determined by the gene that encodes the protein. Any change in the gene sequence may lead to a different amino acid being added to the polypeptide chain, causing a change in protein structure and function. In sickle cell anemia, the hemoglobin β chain has a single amino acid substitution, causing a change in both the structure and function of the protein. What is most remarkable to consider is that a hemoglobin molecule is made up of two alpha chains and two beta chains that each consist of about 150 amino acids. The molecule, therefore, has about 600 amino acids. The structural difference between a normal hemoglobin molecule and a sickle cell molecule—that dramatically decreases life expectancy in the affected individuals—is a single amino acid of the 600.

Because of this change of one amino acid in the chain, the normally biconcave, or disc-shaped, red blood cells assume a crescent or "sickle" shape, which clogs arteries. This can lead to a myriad of serious health problems, such as breathlessness, dizziness, headaches, and abdominal pain for those who have this disease.

Folding patterns resulting from interactions between the non-R group portions of amino acids give rise to the secondary structure of the protein. The most common are the alpha (α)-helix and beta (β)-pleated sheet structures. Both structures are held in shape by hydrogen bonds. In the alpha helix, the bonds form between every fourth amino acid and cause a twist in the amino acid chain.

In the β-pleated sheet, the "pleats" are formed by hydrogen bonding between atoms on the backbone of the polypeptide chain. The R groups are attached to the carbons, and extend above and below the folds of the pleat. The pleated segments align parallel to each other, and hydrogen bonds form between the same pairs of atoms on each of the aligned amino acids. The α-helix and β-pleated sheet structures are found in many globular and fibrous proteins.

The unique three-dimensional structure of a polypeptide is known as its tertiary structure. This structure is caused by chemical interactions between various amino acids and regions of the polypeptide. Primarily, the interactions among R groups create the complex three-dimensional tertiary structure of a protein. There may be ionic bonds formed between R groups on different amino acids, or hydrogen bonding beyond that involved in the secondary structure. When protein folding takes place, the hydrophobic R groups of nonpolar amino acids lay in the interior of the protein, whereas the hydrophilic R groups lay on the outside. The former types of interactions are also known as hydrophobic interactions.

In nature, some proteins are formed from several polypeptides, also known as subunits, and the interaction of these subunits forms the quaternary structure. Weak interactions between the subunits help to stabilize the overall structure. For example, hemoglobin is a combination of four polypeptide subunits.

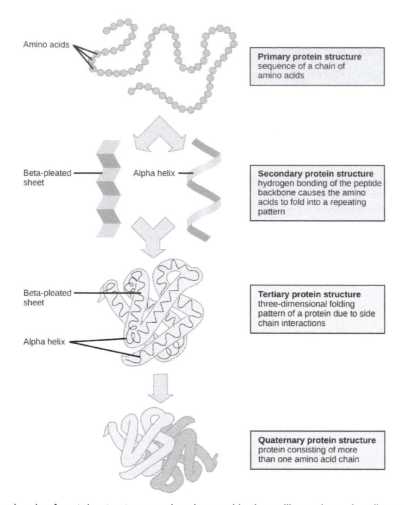

Figure 2.21 The four levels of protein structure can be observed in these illustrations. (credit: modification of work by National Human Genome Research Institute)

Each protein has its own unique sequence and shape held together by chemical interactions. If the protein is subject to changes in temperature, pH, or exposure to chemicals, the protein structure may change, losing its shape in what is known as denaturation as discussed earlier. Denaturation is often reversible because the primary structure is preserved if the denaturing agent is removed, allowing the protein to resume its function. Sometimes denaturation is irreversible, leading to a loss of function. One example of protein denaturation can be seen when an egg is fried or boiled. The albumin protein in the liquid egg white is denatured when placed in a hot pan, changing from a clear substance to an opaque white substance. Not all proteins are denatured at high temperatures; for instance, bacteria that survive in hot springs have proteins that are adapted to function at those temperatures.

For an additional perspective on proteins, explore "Biomolecules: The Proteins" through this interactive animation (http://openstaxcollege.org/l/proteins) .

Nucleic Acids

Nucleic acids are key macromolecules in the continuity of life. They carry the genetic blueprint of a cell and carry instructions for the functioning of the cell.

The two main types of **nucleic acids** are **deoxyribonucleic acid (DNA)** and **ribonucleic acid (RNA)**. DNA is the genetic material found in all living organisms, ranging from single-celled bacteria to multicellular mammals.

The other type of nucleic acid, RNA, is mostly involved in protein synthesis. The DNA molecules never leave the nucleus, but instead use an RNA intermediary to communicate with the rest of the cell. Other types of RNA are also involved in protein synthesis and its regulation.

DNA and RNA are made up of monomers known as **nucleotides**. The nucleotides combine with each other to form a polynucleotide, DNA or RNA. Each nucleotide is made up of three components: a nitrogenous base, a pentose (five-carbon) sugar, and a phosphate group (Figure 2.22). Each nitrogenous base in a nucleotide is attached to a sugar molecule, which is attached to a phosphate group.

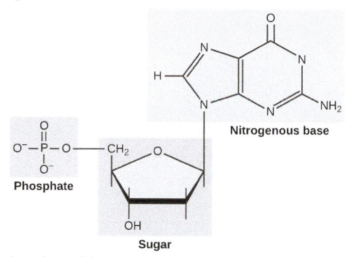

Figure 2.22 A nucleotide is made up of three components: a nitrogenous base, a pentose sugar, and a phosphate group.

DNA Double-Helical Structure

DNA has a double-helical structure (Figure 2.23). It is composed of two strands, or polymers, of nucleotides. The strands are formed with bonds between phosphate and sugar groups of adjacent nucleotides. The strands are bonded to each other at their bases with hydrogen bonds, and the strands coil about each other along their length, hence the "double helix" description, which means a double spiral.

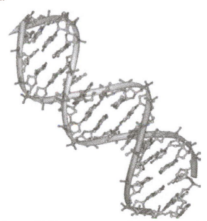

Figure 2.23 The double-helix model shows DNA as two parallel strands of intertwining molecules. (credit: Jerome Walker, Dennis Myts)

The alternating sugar and phosphate groups lie on the outside of each strand, forming the backbone of the DNA. The nitrogenous bases are stacked in the interior, like the steps of a staircase, and these bases pair; the pairs are bound to each other by hydrogen bonds. The bases pair in such a way that the distance between the backbones of the two strands is the same all along the molecule.

KEY TERMS

acid a substance that donates hydrogen ions and therefore lowers pH

adhesion the attraction between water molecules and molecules of a different substance

amino acid a monomer of a protein

anion a negative ion formed by gaining electrons

atomic number the number of protons in an atom

base a substance that absorbs hydrogen ions and therefore raises pH

buffer a solution that resists a change in pH by absorbing or releasing hydrogen or hydroxide ions

carbohydrate a biological macromolecule in which the ratio of carbon to hydrogen to oxygen is 1:2:1; carbohydrates serve as energy sources and structural support in cells

cation a positive ion formed by losing electrons

cellulose a polysaccharide that makes up the cell walls of plants and provides structural support to the cell

chemical bond an interaction between two or more of the same or different elements that results in the formation of molecules

chitin a type of carbohydrate that forms the outer skeleton of arthropods, such as insects and crustaceans, and the cell walls of fungi

cohesion the intermolecular forces between water molecules caused by the polar nature of water; creates surface tension

covalent bond a type of strong bond between two or more of the same or different elements; forms when electrons are shared between elements

denaturation the loss of shape in a protein as a result of changes in temperature, pH, or exposure to chemicals

deoxyribonucleic acid (DNA) a double-stranded polymer of nucleotides that carries the hereditary information of the cell

disaccharide two sugar monomers that are linked together by a peptide bond

electron a negatively charged particle that resides outside of the nucleus in the electron orbital; lacks functional mass and has a charge of −1

electron transfer the movement of electrons from one element to another

element one of 118 unique substances that cannot be broken down into smaller substances and retain the characteristic of that substance; each element has a specified number of protons and unique properties

enzyme a catalyst in a biochemical reaction that is usually a complex or conjugated protein

evaporation the release of water molecules from liquid water to form water vapor

fat a lipid molecule composed of three fatty acids and a glycerol (triglyceride) that typically exists in a solid form at room temperature

glycogen a storage carbohydrate in animals

hormone a chemical signaling molecule, usually a protein or steroid, secreted by an endocrine gland or group of endocrine cells; acts to control or regulate specific physiological processes

hydrogen bond a weak bond between partially positively charged hydrogen atoms and partially negatively charged elements or molecules

hydrophilic describes a substance that dissolves in water; water-loving

hydrophobic describes a substance that does not dissolve in water; water-fearing

ion an atom or compound that does not contain equal numbers of protons and electrons, and therefore has a net charge

ionic bond a chemical bond that forms between ions of opposite charges

isotope one or more forms of an element that have different numbers of neutrons

lipids a class of macromolecules that are nonpolar and insoluble in water

litmus paper filter paper that has been treated with a natural water-soluble dye so it can be used as a pH indicator

macromolecule a large molecule, often formed by polymerization of smaller monomers

mass number the number of protons plus neutrons in an atom

matter anything that has mass and occupies space

monosaccharide a single unit or monomer of carbohydrates

neutron a particle with no charge that resides in the nucleus of an atom; has a mass of 1

nonpolar covalent bond a type of covalent bond that forms between atoms when electrons are shared equally between atoms, resulting in no regions with partial charges as in polar covalent bonds

nucleic acid a biological macromolecule that carries the genetic information of a cell and carries instructions for the functioning of the cell

nucleotide a monomer of nucleic acids; contains a pentose sugar, a phosphate group, and a nitrogenous base

nucleus (chemistry) the dense center of an atom made up of protons and (except in the case of a hydrogen atom) neutrons

octet rule states that the outermost shell of an element with a low atomic number can hold eight electrons

oil an unsaturated fat that is a liquid at room temperature

periodic table of elements an organizational chart of elements, indicating the atomic number and mass number of each element; also provides key information about the properties of elements

pH scale a scale ranging from 0 to 14 that measures the approximate concentration of hydrogen ions of a substance

phospholipid a major constituent of the membranes of cells; composed of two fatty acids and a phosphate group attached to the glycerol backbone

polar covalent bond a type of covalent bond in which electrons are pulled toward one atom and away from another, resulting in slightly positive and slightly negative charged regions of the molecule

polypeptide a long chain of amino acids linked by peptide bonds

polysaccharide a long chain of monosaccharides; may be branched or unbranched

protein a biological macromolecule composed of one or more chains of amino acids

proton a positively charged particle that resides in the nucleus of an atom; has a mass of 1 and a charge of +1

radioactive isotope an isotope that spontaneously emits particles or energy to form a more stable element

ribonucleic acid (RNA) a single-stranded polymer of nucleotides that is involved in protein synthesis

saturated fatty acid a long-chain hydrocarbon with single covalent bonds in the carbon chain; the number of hydrogen atoms attached to the carbon skeleton is maximized

solvent a substance capable of dissolving another substance

starch a storage carbohydrate in plants

steroid a type of lipid composed of four fused hydrocarbon rings

surface tension the cohesive force at the surface of a body of liquid that prevents the molecules from separating

temperature a measure of molecular motion

***trans*-fat** a form of unsaturated fat with the hydrogen atoms neighboring the double bond across from each other rather than on the same side of the double bond

triglyceride a fat molecule; consists of three fatty acids linked to a glycerol molecule

unsaturated fatty acid a long-chain hydrocarbon that has one or more than one double bonds in the hydrocarbon chain

van der Waals interaction a weak attraction or interaction between molecules caused by slightly positively charged or slightly negatively charged atoms

CHAPTER SUMMARY

2.1 The Building Blocks of Molecules

Matter is anything that occupies space and has mass. It is made up of atoms of different elements. All of the 92 elements that occur naturally have unique qualities that allow them to combine in various ways to create compounds or molecules. Atoms, which consist of protons, neutrons, and electrons, are the smallest units of an element that retain all of the properties of that element. Electrons can be donated or shared between atoms to create bonds, including ionic, covalent, and hydrogen bonds, as well as van der Waals interactions.

2.2 Water

Water has many properties that are critical to maintaining life. It is polar, allowing for the formation of hydrogen bonds, which allow ions and other polar molecules to dissolve in water. Therefore, water is an excellent solvent. The hydrogen bonds between water molecules give water the ability to hold heat better than many other substances. As the temperature rises, the hydrogen bonds between water continually break and reform, allowing for the overall temperature to remain stable, although increased energy is added to the system. Water's cohesive forces allow for the property of surface tension. All of these unique properties of water are important in the chemistry of living organisms.

The pH of a solution is a measure of the concentration of hydrogen ions in the solution. A solution with a high number of hydrogen ions is acidic and has a low pH value. A solution with a high number of hydroxide ions is basic and has a high pH value. The pH scale ranges from 0 to 14, with a pH of 7 being neutral. Buffers are solutions that moderate pH changes when an acid or base is added to the buffer system. Buffers are important in biological systems because of their ability to maintain constant pH conditions.

2.3 Biological Molecules

Living things are carbon-based because carbon plays such a prominent role in the chemistry of living things. The four covalent bonding positions of the carbon atom can give rise to a wide diversity of compounds with many functions, accounting for the importance of carbon in living things. Carbohydrates are a group of macromolecules that are a vital energy source for the cell, provide structural support to many organisms, and can be found on the surface of the cell as receptors or for cell recognition. Carbohydrates are classified as monosaccharides, disaccharides, and polysaccharides, depending on the number of monomers in the molecule.

Lipids are a class of macromolecules that are nonpolar and hydrophobic in nature. Major types include fats and oils, waxes, phospholipids, and steroids. Fats and oils are a stored form of energy and can include triglycerides. Fats and oils are usually made up of fatty acids and glycerol.

Proteins are a class of macromolecules that can perform a diverse range of functions for the cell. They help in metabolism by providing structural support and by acting as enzymes, carriers or as hormones. The building blocks of proteins are amino acids. Proteins are organized at four levels: primary, secondary, tertiary, and quaternary. Protein shape and function are intricately linked; any change in shape caused by changes in temperature, pH, or chemical exposure may lead to protein denaturation and a loss of function.

Nucleic acids are molecules made up of repeating units of nucleotides that direct cellular activities such as cell division and protein synthesis. Each nucleotide is made up of a pentose sugar, a nitrogenous base, and a phosphate group. There are two types of nucleic acids: DNA and RNA.

ART CONNECTION QUESTIONS

1. Figure 2.3 How many neutrons do (K) potassium-39 and potassium-40 have, respectively?

REVIEW QUESTIONS

2. Magnesium has an atomic number of 12. Which of the following statements is true of a neutral magnesium atom?

    a. It has 12 protons, 12 electrons, and 12 neutrons.
    b. It has 12 protons, 12 electrons, and six neutrons.
    c. It has six protons, six electrons, and no neutrons.
    d. It has six protons, six electrons, and six neutrons.

3. Which type of bond represents a weak chemical bond?

    a. hydrogen bond
    b. ionic bond
    c. covalent bond
    d. polar covalent bond

4. An isotope of sodium (Na) has a mass number of 22. How many neutrons does it have?

    a. 11
    b. 12
    c. 22
    d. 44

5. Which of the following statements is not true?

    a. Water is polar.
    b. Water stabilizes temperature.
    c. Water is essential for life.
    d. Water is the most abundant atom in Earth's atmosphere.

6. Using a pH meter, you find the pH of an unknown solution to be 8.0. How would you describe this solution?

    a. weakly acidic
    b. strongly acidic
    c. weakly basic
    d. strongly basic

7. The pH of lemon juice is about 2.0, whereas tomato juice's pH is about 4.0. Approximately how much of an increase in hydrogen ion concentration is there between tomato juice and lemon juice?

    a. 2 times
    b. 10 times
    c. 100 times
    d. 1000 times

8. An example of a monosaccharide is _____.

    a. fructose
    b. glucose
    c. galactose
    d. all of the above

9. Cellulose and starch are examples of _____.

    a. monosaccharides
    b. disaccharides
    c. lipids
    d. polysaccharides

10. Phospholipids are important components of _____.

    a. the plasma membrane of cells
    b. the ring structure of steroids
    c. the waxy covering on leaves
    d. the double bond in hydrocarbon chains

11. The monomers that make up proteins are called _____.

    a. nucleotides
    b. disaccharides
    c. amino acids
    d. chaperones

CRITICAL THINKING QUESTIONS

12. Why are hydrogen bonds and van der Waals interactions necessary for cells?

13. Why can some insects walk on water?

14. Explain why water is an excellent solvent.

15. Explain at least three functions that lipids serve in plants and/or animals.

16. Explain what happens if even one amino acid is substituted for another in a polypeptide chain. Provide a specific example.

3 | CELL STRUCTURE AND FUNCTION

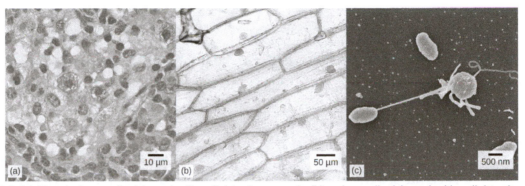

Figure 3.1 (a) Nasal sinus cells (viewed with a light microscope), (b) onion cells (viewed with a light microscope), and (c) *Vibrio tasmaniensis* bacterial cells (viewed using a scanning electron microscope) are from very different organisms, yet all share certain characteristics of basic cell structure. (credit a: modification of work by Ed Uthman, MD; credit b: modification of work by Umberto Salvagnin; credit c: modification of work by Anthony D'Onofrio; scale-bar data from Matt Russell)

Chapter Outline

3.1: How Cells Are Studied

3.2: Comparing Prokaryotic and Eukaryotic Cells

3.3: Eukaryotic Cells

3.4: The Cell Membrane

3.5: Passive Transport

3.6: Active Transport

Introduction

Close your eyes and picture a brick wall. What is the basic building block of that wall? It is a single brick, of course. Like a brick wall, your body is composed of basic building blocks, and the building blocks of your body are cells.

Your body has many kinds of cells, each specialized for a specific purpose. Just as a home is made from a variety of building materials, the human body is constructed from many cell types. For example, epithelial cells protect the surface of the body and cover the organs and body cavities within. Bone cells help to support and protect the body. Cells of the immune system fight invading bacteria. Additionally, red blood cells carry oxygen throughout the body. Each of these cell types plays a vital role during the growth, development, and day-to-day maintenance of the body. In spite of their enormous variety, however, all cells share certain fundamental characteristics.

3.1 | How Cells Are Studied

By the end of this section, you will be able to:

- Describe the roles of cells in organisms
- Compare and contrast light microscopy and electron microscopy
- Summarize the cell theory

A cell is the smallest unit of a living thing. A living thing, like you, is called an organism. Thus, cells are the basic building blocks of all organisms.

In multicellular organisms, several cells of one particular kind interconnect with each other and perform shared functions to form tissues (for example, muscle tissue, connective tissue, and nervous tissue), several tissues combine to form an organ (for example, stomach, heart, or brain), and several organs make up an organ system (such as the digestive system, circulatory system, or nervous system). Several systems functioning together form an organism (such as an elephant, for example).

There are many types of cells, and all are grouped into one of two broad categories: prokaryotic and eukaryotic. Animal cells, plant cells, fungal cells, and protist cells are classified as eukaryotic, whereas bacteria and archaea cells are classified as prokaryotic. Before discussing the criteria for determining whether a cell is prokaryotic or eukaryotic, let us first examine how biologists study cells.

Microscopy

Cells vary in size. With few exceptions, individual cells are too small to be seen with the naked eye, so scientists use microscopes to study them. A **microscope** is an instrument that magnifies an object. Most images of cells are taken with a microscope and are called micrographs.

Light Microscopes

To give you a sense of the size of a cell, a typical human red blood cell is about eight millionths of a meter or eight micrometers (abbreviated as μm) in diameter; the head of a pin is about two thousandths of a meter (millimeters, or mm) in diameter. That means that approximately 250 red blood cells could fit on the head of a pin.

The optics of the lenses of a light microscope changes the orientation of the image. A specimen that is right-side up and facing right on the microscope slide will appear upside-down and facing left when viewed through a microscope, and vice versa. Similarly, if the slide is moved left while looking through the microscope, it will appear to move right, and if moved down, it will seem to move up. This occurs because microscopes use two sets of lenses to magnify the image. Due to the manner in which light travels through the lenses, this system of lenses produces an inverted image (binoculars and a dissecting microscope work in a similar manner, but include an additional magnification system that makes the final image appear to be upright).

Most student microscopes are classified as light microscopes (Figure 3.2**a**). Visible light both passes through and is bent by the lens system to enable the user to see the specimen. Light microscopes are advantageous for viewing living organisms, but since individual cells are generally transparent, their components are not distinguishable unless they are colored with special stains. Staining, however, usually kills the cells.

Light microscopes commonly used in the undergraduate college laboratory magnify up to approximately 400 times. Two parameters that are important in microscopy are magnification and resolving power. Magnification is the degree of enlargement of an object. Resolving power is the ability of a microscope to allow the eye to distinguish two adjacent structures as separate; the higher the resolution, the closer those two objects can be, and the better the clarity and detail of the image. When oil immersion lenses are used, magnification is usually increased to 1,000 times for the study of smaller cells, like most prokaryotic cells. Because light entering a specimen from below is focused onto the eye of an observer, the specimen can be viewed using light microscopy. For this reason, for light to pass through a specimen, the sample must be thin or translucent.

For another perspective on cell size, try the HowBig (http://openstaxcollege.org/l/cell_sizes2) interactive.

A second type of microscope used in laboratories is the dissecting microscope (Figure 3.2**b**). These microscopes have a lower magnification (20 to 80 times the object size) than light microscopes and can provide a three-dimensional view of the specimen. Thick objects can be examined with many components in focus at the same time. These microscopes are designed to give a magnified and clear view of tissue structure as well as the anatomy of the whole organism. Like light microscopes, most modern dissecting microscopes are also binocular, meaning that they have two separate lens systems, one for each eye. The lens systems are separated by a certain distance, and therefore provide a sense of depth in the view of their subject to make manipulations by hand easier. Dissecting microscopes also have optics that correct the image so that it appears as if being seen by the naked eye and not as an inverted image. The light illuminating a sample under a dissecting microscope typically comes from above the sample, but may also be directed from below.

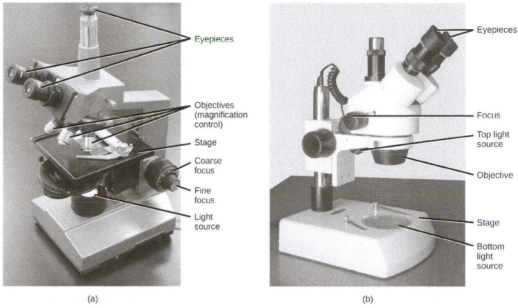

(a) (b)

Figure 3.2 (a) Most light microscopes used in a college biology lab can magnify cells up to approximately 400 times. (b) Dissecting microscopes have a lower magnification than light microscopes and are used to examine larger objects, such as tissues.

Electron Microscopes

In contrast to light microscopes, electron microscopes use a beam of electrons instead of a beam of light. Not only does this allow for higher magnification and, thus, more detail (Figure 3.3), it also provides higher resolving power. Preparation of a specimen for viewing under an electron microscope will kill it; therefore, live cells cannot be viewed using this type of microscopy. In addition, the electron beam moves best in a vacuum, making it impossible to view living materials.

In a scanning electron microscope, a beam of electrons moves back and forth across a cell's surface, rendering the details of cell surface characteristics by reflection. Cells and other structures are usually coated with a metal like gold. In a transmission electron microscope, the electron beam is transmitted through the cell and provides details of a cell's internal structures. As you might imagine, electron microscopes are significantly more bulky and expensive than are light microscopes.

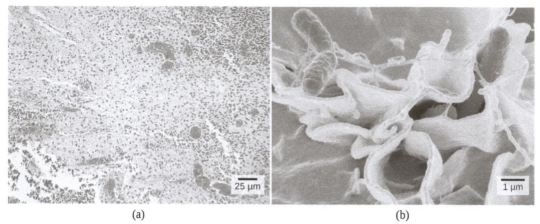

(a)   (b)

Figure 3.3 (a) *Salmonella* bacteria are viewed with a light microscope. (b) This scanning electron micrograph shows *Salmonella* bacteria (in red) invading human cells. (credit a: modification of work by CDC, Armed Forces Institute of Pathology, Charles N. Farmer; credit b: modification of work by Rocky Mountain Laboratories, NIAID, NIH; scale-bar data from Matt Russell)

careers IN ACTION

Cytotechnologist

Have you ever heard of a medical test called a Pap smear (Figure 3.4)? In this test, a doctor takes a small sample of cells from the uterine cervix of a patient and sends it to a medical lab where a cytotechnologist stains the cells and examines them for any changes that could indicate cervical cancer or a microbial infection.

Cytotechnologists (*cyto-* = cell) are professionals who study cells through microscopic examinations and other laboratory tests. They are trained to determine which cellular changes are within normal limits or are abnormal. Their focus is not limited to cervical cells; they study cellular specimens that come from all organs. When they notice abnormalities, they consult a pathologist, who is a medical doctor who can make a clinical diagnosis.

Cytotechnologists play vital roles in saving people's lives. When abnormalities are discovered early, a patient's treatment can begin sooner, which usually increases the chances of successful treatment.

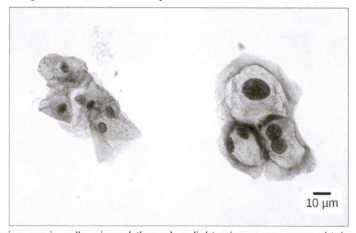

Figure 3.4 These uterine cervix cells, viewed through a light microscope, were obtained from a Pap smear. Normal cells are on the left. The cells on the right are infected with human papillomavirus. (credit: modification of work by Ed Uthman; scale-bar data from Matt Russell)

Cell Theory

The microscopes we use today are far more complex than those used in the 1600s by Antony van Leeuwenhoek, a Dutch shopkeeper who had great skill in crafting lenses. Despite the limitations of his now-ancient lenses, van Leeuwenhoek observed the movements of protists (a type of single-celled organism) and sperm, which he collectively termed "animalcules."

In a 1665 publication called *Micrographia*, experimental scientist Robert Hooke coined the term "cell" (from the Latin *cella*, meaning "small room") for the box-like structures he observed when viewing cork tissue through a lens. In the 1670s, van Leeuwenhoek discovered bacteria and protozoa. Later advances in lenses and microscope construction enabled other scientists to see different components inside cells.

By the late 1830s, botanist Matthias Schleiden and zoologist Theodor Schwann were studying tissues and proposed the **unified cell theory**, which states that all living things are composed of one or more cells, that the cell is the basic unit of life, and that all new cells arise from existing cells. These principles still stand today.

3.2 | Comparing Prokaryotic and Eukaryotic Cells

By the end of this section, you will be able to:

- Name examples of prokaryotic and eukaryotic organisms
- Compare and contrast prokaryotic cells and eukaryotic cells
- Describe the relative sizes of different kinds of cells

Cells fall into one of two broad categories: prokaryotic and eukaryotic. The predominantly single-celled organisms of the domains Bacteria and Archaea are classified as prokaryotes (*pro-* = before; *-karyon-* = nucleus). Animal cells, plant cells, fungi, and protists are eukaryotes (*eu-* = true).

Components of Prokaryotic Cells

All cells share four common components: 1) a plasma membrane, an outer covering that separates the cell's interior from its surrounding environment; 2) cytoplasm, consisting of a jelly-like region within the cell in which other cellular components are found; 3) DNA, the genetic material of the cell; and 4) ribosomes, particles that synthesize proteins. However, prokaryotes differ from eukaryotic cells in several ways.

A **prokaryotic cell** is a simple, single-celled (unicellular) organism that lacks a nucleus, or any other membrane-bound organelle. We will shortly come to see that this is significantly different in eukaryotes. Prokaryotic DNA is found in the central part of the cell: a darkened region called the nucleoid (Figure 3.5).

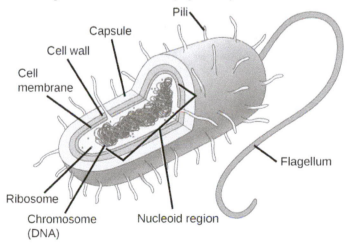

Figure 3.5 This figure shows the generalized structure of a prokaryotic cell.

Unlike Archaea and eukaryotes, bacteria have a cell wall made of peptidoglycan, comprised of sugars and amino acids, and many have a polysaccharide capsule (Figure 3.5). The cell wall acts as an extra layer of protection, helps the cell maintain its shape, and prevents dehydration. The capsule enables the cell to attach to surfaces in its environment. Some prokaryotes have flagella, pili, or fimbriae. Flagella are used for locomotion, while most pili are used to exchange genetic material during a type of reproduction called conjugation.

Eukaryotic Cells

In nature, the relationship between form and function is apparent at all levels, including the level of the cell, and this will become clear as we explore eukaryotic cells. The principle "form follows function" is found in many contexts. For example, birds and fish have streamlined bodies that allow them to move quickly through the medium in which they live, be it air or water. It means that, in general, one can deduce the function of a structure by looking at its form, because the two are matched.

A **eukaryotic cell** is a cell that has a membrane-bound nucleus and other membrane-bound compartments or sacs, called **organelles**, which have specialized functions. The word eukaryotic means "true kernel" or "true nucleus," alluding to the presence of the membrane-bound nucleus in these cells. The word "organelle" means "little organ," and, as already mentioned, organelles have specialized cellular functions, just as the organs of your body have specialized functions.

Cell Size

At 0.1–5.0 µm in diameter, prokaryotic cells are significantly smaller than eukaryotic cells, which have diameters ranging from 10–100 µm (Figure 3.6). The small size of prokaryotes allows ions and organic molecules that enter them to quickly spread to other parts of the cell. Similarly, any wastes produced within a prokaryotic cell can quickly move out. However, larger eukaryotic cells have evolved different structural adaptations to enhance cellular transport. Indeed, the large size of these cells would not be possible without these adaptations. In general, cell size is limited because volume increases much more quickly than does cell surface area. As a cell becomes larger, it becomes more and more difficult for the cell to acquire sufficient materials to support the processes inside the cell, because the relative size of the surface area across which materials must be transported declines.

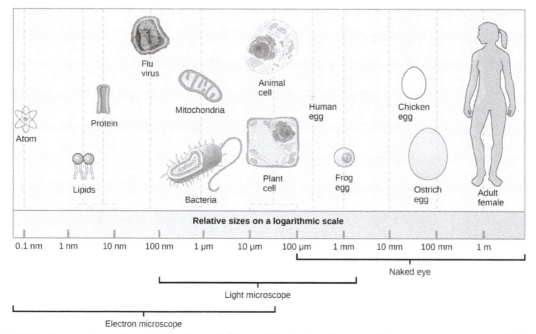

Figure 3.6 This figure shows the relative sizes of different kinds of cells and cellular components. An adult human is shown for comparison.

3.3 | Eukaryotic Cells

By the end of this section, you will be able to:

- Describe the structure of eukaryotic plant and animal cells
- State the role of the plasma membrane
- Summarize the functions of the major cell organelles
- Describe the cytoskeleton and extracellular matrix

At this point, it should be clear that eukaryotic cells have a more complex structure than do prokaryotic cells. Organelles allow for various functions to occur in the cell at the same time. Before discussing the functions of organelles within a eukaryotic cell, let us first examine two important components of the cell: the plasma membrane and the cytoplasm.

art CONNECTION

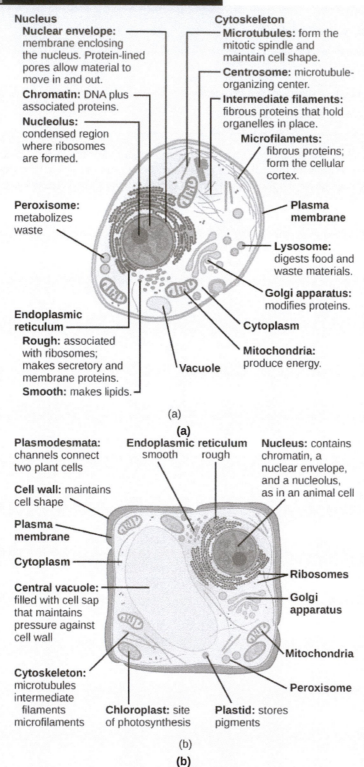

Nucleus
  Nuclear envelope: membrane enclosing the nucleus. Protein-lined pores allow material to move in and out.
  Chromatin: DNA plus associated proteins.
  Nucleolus: condensed region where ribosomes are formed.

Peroxisome: metabolizes waste

Endoplasmic reticulum
  Rough: associated with ribosomes; makes secretory and membrane proteins.
  Smooth: makes lipids.

Cytoskeleton
  Microtubules: form the mitotic spindle and maintain cell shape.
  Centrosome: microtubule-organizing center.
  Intermediate filaments: fibrous proteins that hold organelles in place.
  Microfilaments: fibrous proteins; form the cellular cortex.

Plasma membrane

Lysosome: digests food and waste materials.

Golgi apparatus: modifies proteins.

Cytoplasm

Mitochondria: produce energy.

Vacuole

(a)

(a)

Plasmodesmata: channels connect two plant cells

Cell wall: maintains cell shape

Plasma membrane

Cytoplasm

Central vacuole: filled with cell sap that maintains pressure against cell wall

Cytoskeleton: microtubules intermediate filaments microfilaments

Endoplasmic reticulum
smooth   rough

Chloroplast: site of photosynthesis

Nucleus: contains chromatin, a nuclear envelope, and a nucleolus, as in an animal cell

Ribosomes

Golgi apparatus

Mitochondria

Peroxisome

Plastid: stores pigments

(b)

(b)

Figure 3.7 This figure shows (a) a typical animal cell and (b) a typical plant cell.

What structures does a plant cell have that an animal cell does not have? What structures does an animal cell have that a plant cell does not have?

The Plasma Membrane

Like prokaryotes, eukaryotic cells have a **plasma membrane** (Figure 3.8) made up of a phospholipid bilayer with embedded proteins that separates the internal contents of the cell from its surrounding environment. A phospholipid is a lipid molecule composed of two fatty acid chains, a glycerol backbone, and a phosphate group. The plasma membrane regulates the passage of some substances, such as organic molecules, ions, and water, preventing the passage of some to maintain internal conditions, while actively bringing in or removing others. Other compounds move passively across the membrane.

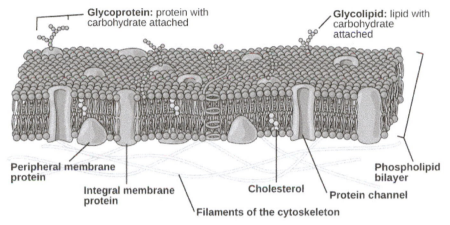

Figure 3.8 The plasma membrane is a phospholipid bilayer with embedded proteins. There are other components, such as cholesterol and carbohydrates, which can be found in the membrane in addition to phospholipids and protein.

The plasma membranes of cells that specialize in absorption are folded into fingerlike projections called microvilli (singular = microvillus). This folding increases the surface area of the plasma membrane. Such cells are typically found lining the small intestine, the organ that absorbs nutrients from digested food. This is an excellent example of form matching the function of a structure.

People with celiac disease have an immune response to gluten, which is a protein found in wheat, barley, and rye. The immune response damages microvilli, and thus, afflicted individuals cannot absorb nutrients. This leads to malnutrition, cramping, and diarrhea. Patients suffering from celiac disease must follow a gluten-free diet.

The Cytoplasm

The **cytoplasm** comprises the contents of a cell between the plasma membrane and the nuclear envelope (a structure to be discussed shortly). It is made up of organelles suspended in the gel-like **cytosol**, the cytoskeleton, and various chemicals (Figure 3.7). Even though the cytoplasm consists of 70 to 80 percent water, it has a semi-solid consistency, which comes from the proteins within it. However, proteins are not the only organic molecules found in the cytoplasm. Glucose and other simple sugars, polysaccharides, amino acids, nucleic acids, fatty acids, and derivatives of glycerol are found there too. Ions of sodium, potassium, calcium, and many other elements are also dissolved in the cytoplasm. Many metabolic reactions, including protein synthesis, take place in the cytoplasm.

The Cytoskeleton

If you were to remove all the organelles from a cell, would the plasma membrane and the cytoplasm be the only components left? No. Within the cytoplasm, there would still be ions and organic molecules, plus a network of protein fibers that helps to maintain the shape of the cell, secures certain organelles in specific positions, allows cytoplasm and vesicles to move within the cell, and enables unicellular organisms to move independently. Collectively, this network of protein fibers is known as the **cytoskeleton**. There are three types of fibers within the cytoskeleton: microfilaments, also known as actin filaments, intermediate filaments, and microtubules (Figure 3.9).

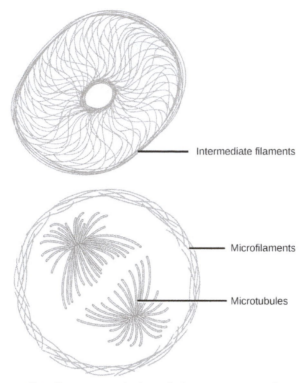

Figure 3.9 Microfilaments, intermediate filaments, and microtubules compose a cell's cytoskeleton.

Microfilaments are the thinnest of the cytoskeletal fibers and function in moving cellular components, for example, during cell division. They also maintain the structure of microvilli, the extensive folding of the plasma membrane found in cells dedicated to absorption. These components are also common in muscle cells and are responsible for muscle cell contraction. Intermediate filaments are of intermediate diameter and have structural functions, such as maintaining the shape of the cell and anchoring organelles. Keratin, the compound that strengthens hair and nails, forms one type of intermediate filament. Microtubules are the thickest of the cytoskeletal fibers. These are hollow tubes that can dissolve and reform quickly. Microtubules guide organelle movement and are the structures that pull chromosomes to their poles during cell division. They are also the structural components of flagella and cilia. In cilia and flagella, the microtubules are organized as a circle of nine double microtubules on the outside and two microtubules in the center.

The centrosome is a region near the nucleus of animal cells that functions as a microtubule-organizing center. It contains a pair of centrioles, two structures that lie perpendicular to each other. Each centriole is a cylinder of nine triplets of microtubules.

The centrosome replicates itself before a cell divides, and the centrioles play a role in pulling the duplicated chromosomes to opposite ends of the dividing cell. However, the exact function of the centrioles in cell division is not clear, since cells that have the centrioles removed can still divide, and plant cells, which lack centrioles, are capable of cell division.

Flagella and Cilia

Flagella (singular = flagellum) are long, hair-like structures that extend from the plasma membrane and are used to move an entire cell, (for example, sperm, *Euglena*). When present, the cell has just one flagellum or a few flagella. When **cilia** (singular = cilium) are present, however, they are many in number and extend along the entire surface of the plasma membrane. They are short, hair-like structures that are used to move entire cells (such as paramecium) or move substances along the outer surface of the cell (for example, the cilia of cells lining the fallopian tubes that move the ovum toward the uterus, or cilia lining the cells of the respiratory tract that move particulate matter toward the throat that mucus has trapped).

The Endomembrane System

The **endomembrane system** (*endo* = within) is a group of membranes and organelles (Figure 3.13) in eukaryotic cells that work together to modify, package, and transport lipids and proteins. It includes the nuclear envelope, lysosomes, and vesicles, the endoplasmic reticulum and Golgi apparatus, which we will cover shortly. Although not technically *within* the cell, the plasma membrane is included in the endomembrane system because, as you will see, it interacts with the other endomembranous organelles.

The Nucleus

Typically, the nucleus is the most prominent organelle in a cell (Figure 3.7). The **nucleus** (plural = nuclei) houses the cell's DNA in the form of chromatin and directs the synthesis of ribosomes and proteins. Let us look at it in more detail (Figure 3.10).

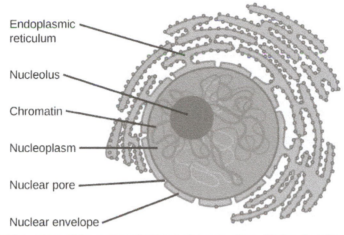

Endoplasmic reticulum

Nucleolus

Chromatin

Nucleoplasm

Nuclear pore

Nuclear envelope

Figure 3.10 The outermost boundary of the nucleus is the nuclear envelope. Notice that the nuclear envelope consists of two phospholipid bilayers (membranes)—an outer membrane and an inner membrane—in contrast to the plasma membrane (Figure 3.8), which consists of only one phospholipid bilayer. (credit: modification of work by NIGMS, NIH)

The **nuclear envelope** is a double-membrane structure that constitutes the outermost portion of the nucleus (Figure 3.10). Both the inner and outer membranes of the nuclear envelope are phospholipid bilayers.

The nuclear envelope is punctuated with pores that control the passage of ions, molecules, and RNA between the nucleoplasm and the cytoplasm.

To understand chromatin, it is helpful to first consider chromosomes. Chromosomes are structures within the nucleus that are made up of DNA, the hereditary material, and proteins. This combination of DNA and proteins is called chromatin. In eukaryotes, chromosomes are linear structures. Every species has a specific number of chromosomes in the nucleus of its body cells. For example, in humans, the chromosome number is 46, whereas in fruit flies, the chromosome number is eight.

Chromosomes are only visible and distinguishable from one another when the cell is getting ready to divide. When the cell is in the growth and maintenance phases of its life cycle, the chromosomes resemble an unwound, jumbled bunch of threads.

We already know that the nucleus directs the synthesis of ribosomes, but how does it do this? Some chromosomes have sections of DNA that encode ribosomal RNA. A darkly staining area within the nucleus, called the **nucleolus** (plural = nucleoli), aggregates the ribosomal RNA with associated proteins to assemble the ribosomal subunits that are then transported through the nuclear pores into the cytoplasm.

The Endoplasmic Reticulum

The **endoplasmic reticulum (ER)** (Figure 3.13) is a series of interconnected membranous tubules that collectively modify proteins and synthesize lipids. However, these two functions are performed in separate areas of the endoplasmic reticulum: the rough endoplasmic reticulum and the smooth endoplasmic reticulum, respectively.

The hollow portion of the ER tubules is called the lumen or cisternal space. The membrane of the ER, which is a phospholipid bilayer embedded with proteins, is continuous with the nuclear envelope.

The **rough endoplasmic reticulum (RER)** is so named because the ribosomes attached to its cytoplasmic surface give it a studded appearance when viewed through an electron microscope.

The ribosomes synthesize proteins while attached to the ER, resulting in transfer of their newly synthesized proteins into the lumen of the RER where they undergo modifications such as folding or addition of sugars. The RER also makes phospholipids for cell membranes.

If the phospholipids or modified proteins are not destined to stay in the RER, they will be packaged within vesicles and transported from the RER by budding from the membrane (Figure 3.13). Since the RER is engaged in modifying proteins that will be secreted from the cell, it is abundant in cells that secrete proteins, such as the liver.

The **smooth endoplasmic reticulum (SER)** is continuous with the RER but has few or no ribosomes on its cytoplasmic surface (see Figure 3.7). The SER's functions include synthesis of carbohydrates, lipids (including phospholipids), and steroid hormones; detoxification of medications and poisons; alcohol metabolism; and storage of calcium ions.

The Golgi Apparatus

We have already mentioned that vesicles can bud from the ER, but where do the vesicles go? Before reaching their final destination, the lipids or proteins within the transport vesicles need to be sorted, packaged, and tagged so that they wind up in the right place. The sorting, tagging, packaging, and distribution of lipids and proteins take place in the **Golgi apparatus** (also called the Golgi body), a series of flattened membranous sacs (Figure 3.11).

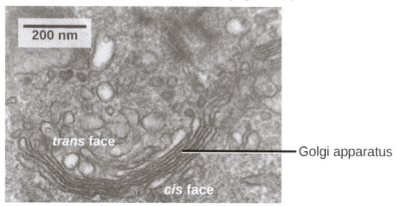

Figure 3.11 The Golgi apparatus in this transmission electron micrograph of a white blood cell is visible as a stack of semicircular flattened rings in the lower portion of this image. Several vesicles can be seen near the Golgi apparatus. (credit: modification of work by Louisa Howard; scale-bar data from Matt Russell)

The Golgi apparatus has a receiving face near the endoplasmic reticulum and a releasing face on the side away from the ER, toward the cell membrane. The transport vesicles that form from the ER travel to the receiving face, fuse with it, and empty their contents into the lumen of the Golgi apparatus. As the proteins and lipids travel through the Golgi, they undergo further modifications. The most frequent modification is the addition of short chains of sugar molecules. The newly modified proteins and lipids are then tagged with small molecular groups to enable them to be routed to their proper destinations.

Finally, the modified and tagged proteins are packaged into vesicles that bud from the opposite face of the Golgi. While some of these vesicles, transport vesicles, deposit their contents into other parts of the cell where they will be used, others, secretory vesicles, fuse with the plasma membrane and release their contents outside the cell.

The amount of Golgi in different cell types again illustrates that form follows function within cells. Cells that engage in a great deal of secretory activity (such as cells of the salivary glands that secrete digestive enzymes or cells of the immune system that secrete antibodies) have an abundant number of Golgi.

In plant cells, the Golgi has an additional role of synthesizing polysaccharides, some of which are incorporated into the cell wall and some of which are used in other parts of the cell.

Lysosomes

In animal cells, the **lysosomes** are the cell's "garbage disposal." Digestive enzymes within the lysosomes aid the breakdown of proteins, polysaccharides, lipids, nucleic acids, and even worn-out organelles. In single-celled eukaryotes, lysosomes are important for digestion of the food they ingest and the recycling of organelles. These enzymes are active at a much lower pH (more acidic) than those located in the cytoplasm. Many reactions that take place in the cytoplasm could not occur at a low pH, thus the advantage of compartmentalizing the eukaryotic cell into organelles is apparent.

Lysosomes also use their hydrolytic enzymes to destroy disease-causing organisms that might enter the cell. A good example of this occurs in a group of white blood cells called macrophages, which are part of your body's immune system. In a process known as phagocytosis, a section of the plasma membrane of the macrophage invaginates (folds in) and engulfs a pathogen. The invaginated section, with the pathogen inside, then pinches itself off from the plasma membrane and becomes a vesicle. The vesicle fuses with a lysosome. The lysosome's hydrolytic enzymes then destroy the pathogen (Figure 3.12).

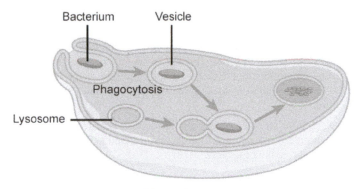

Figure 3.12 A macrophage has phagocytized a potentially pathogenic bacterium into a vesicle, which then fuses with a lysosome within the cell so that the pathogen can be destroyed. Other organelles are present in the cell, but for simplicity, are not shown.

Vesicles and Vacuoles

Vesicles and **vacuoles** are membrane-bound sacs that function in storage and transport. Vacuoles are somewhat larger than vesicles, and the membrane of a vacuole does not fuse with the membranes of other cellular components. Vesicles can fuse with other membranes within the cell system. Additionally, enzymes within plant vacuoles can break down macromolecules.

a r t CONNECTION

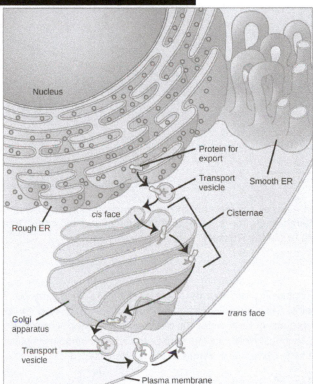

Figure 3.13 The endomembrane system works to modify, package, and transport lipids and proteins. (credit: modification of work by Magnus Manske)

Why does the *cis* face of the Golgi not face the plasma membrane?

Ribosomes

Ribosomes are the cellular structures responsible for protein synthesis. When viewed through an electron microscope, free ribosomes appear as either clusters or single tiny dots floating freely in the cytoplasm. Ribosomes may be attached to either the cytoplasmic side of the plasma membrane or the cytoplasmic side of the endoplasmic reticulum (Figure 3.7). Electron microscopy has shown that ribosomes consist of large and small subunits. Ribosomes are enzyme complexes that are responsible for protein synthesis.

Because protein synthesis is essential for all cells, ribosomes are found in practically every cell, although they are smaller in prokaryotic cells. They are particularly abundant in immature red blood cells for the synthesis of hemoglobin, which functions in the transport of oxygen throughout the body.

Mitochondria

Mitochondria (singular = mitochondrion) are often called the "powerhouses" or "energy factories" of a cell because they are responsible for making adenosine triphosphate (ATP), the cell's main energy-carrying molecule. The formation of ATP from the breakdown of glucose is known as cellular respiration. Mitochondria are oval-shaped, double-membrane organelles (Figure 3.14) that have their own ribosomes and DNA. Each membrane is a phospholipid bilayer embedded with proteins. The inner layer has folds called cristae, which increase the surface area of the inner membrane. The area surrounded by the folds is called the mitochondrial matrix. The cristae and the matrix have different roles in cellular respiration.

In keeping with our theme of form following function, it is important to point out that muscle cells have a very high concentration of mitochondria because muscle cells need a lot of energy to contract.

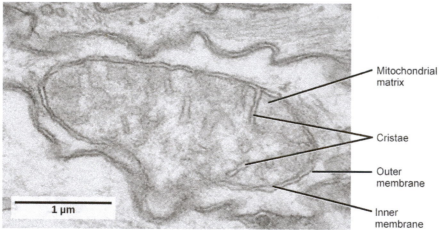

Figure 3.14 This transmission electron micrograph shows a mitochondrion as viewed with an electron microscope. Notice the inner and outer membranes, the cristae, and the mitochondrial matrix. (credit: modification of work by Matthew Britton; scale-bar data from Matt Russell)

Peroxisomes

Peroxisomes are small, round organelles enclosed by single membranes. They carry out oxidation reactions that break down fatty acids and amino acids. They also detoxify many poisons that may enter the body. Alcohol is detoxified by peroxisomes in liver cells. A byproduct of these oxidation reactions is hydrogen peroxide, H_2O_2, which is contained within the peroxisomes to prevent the chemical from causing damage to cellular components outside of the organelle. Hydrogen peroxide is safely broken down by peroxisomal enzymes into water and oxygen.

Animal Cells versus Plant Cells

Despite their fundamental similarities, there are some striking differences between animal and plant cells (see Table 3.1). Animal cells have centrioles, centrosomes (discussed under the cytoskeleton), and lysosomes, whereas plant cells do not. Plant cells have a cell wall, chloroplasts, plasmodesmata, and plastids used for storage, and a large central vacuole, whereas animal cells do not.

The Cell Wall

In Figure 3.7b, the diagram of a plant cell, you see a structure external to the plasma membrane called the cell wall. The **cell wall** is a rigid covering that protects the cell, provides structural support, and gives shape to the cell. Fungal and protist cells also have cell walls.

While the chief component of prokaryotic cell walls is peptidoglycan, the major organic molecule in the plant cell wall is cellulose, a polysaccharide made up of long, straight chains of glucose units. When nutritional information refers to dietary fiber, it is referring to the cellulose content of food.

Chloroplasts

Like mitochondria, chloroplasts also have their own DNA and ribosomes. **Chloroplasts** function in photosynthesis and can be found in eukaryotic cells such as plants and algae. In photosynthesis, carbon dioxide, water, and light energy are used to make glucose and oxygen. This is the major difference between plants and animals: Plants (autotrophs) are able to make their own food, like glucose, whereas animals (heterotrophs) must rely on other organisms for their organic compounds or food source.

Like mitochondria, chloroplasts have outer and inner membranes, but within the space enclosed by a chloroplast's inner membrane is a set of interconnected and stacked, fluid-filled membrane sacs called thylakoids (Figure 3.15). Each stack of thylakoids is called a granum (plural = grana). The fluid enclosed by the inner membrane and surrounding the grana is called the stroma.

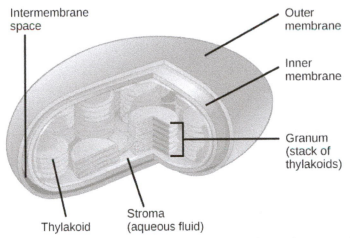

Figure 3.15 This simplified diagram of a chloroplast shows the outer membrane, inner membrane, thylakoids, grana, and stroma.

The chloroplasts contain a green pigment called chlorophyll, which captures the energy of sunlight for photosynthesis. Like plant cells, photosynthetic protists also have chloroplasts. Some bacteria also perform photosynthesis, but they do not have chloroplasts. Their photosynthetic pigments are located in the thylakoid membrane within the cell itself.

Endosymbiosis

We have mentioned that both mitochondria and chloroplasts contain DNA and ribosomes. Have you wondered why? Strong evidence points to endosymbiosis as the explanation.

Symbiosis is a relationship in which organisms from two separate species live in close association and typically exhibit specific adaptations to each other. Endosymbiosis (*endo-*= within) is a relationship in which one organism lives inside the other. Endosymbiotic relationships abound in nature. Microbes that produce vitamin K live inside the human gut. This relationship is beneficial for us because we are unable to synthesize vitamin K. It is also beneficial for the microbes because they are protected from other organisms and are provided a stable habitat and abundant food by living within the large intestine.

Scientists have long noticed that bacteria, mitochondria, and chloroplasts are similar in size. We also know that mitochondria and chloroplasts have DNA and ribosomes, just as bacteria do. Scientists believe that host cells and bacteria formed a mutually beneficial endosymbiotic relationship when the host cells ingested aerobic bacteria and cyanobacteria but did not destroy them. Through evolution, these ingested bacteria became more specialized in their functions, with the aerobic bacteria becoming mitochondria and the photosynthetic bacteria becoming chloroplasts.

The Central Vacuole

Previously, we mentioned vacuoles as essential components of plant cells. If you look at Figure 3.7, you will see that plant cells each have a large, central vacuole that occupies most of the cell. The **central vacuole** plays a key role in regulating the cell's concentration of water in changing environmental conditions. In plant cells, the liquid inside the central vacuole provides turgor pressure, which is the outward pressure caused by the fluid inside the cell. Have you ever noticed that if you forget to water a plant for a few days, it wilts? That is because as the water concentration in the soil becomes lower than the water concentration in the plant, water moves out of the central vacuoles and cytoplasm and into the soil. As the central vacuole shrinks, it leaves the cell wall unsupported. This loss of support to the cell walls of a plant results in the wilted appearance. Additionally, this fluid has a very bitter taste, which discourages consumption by insects and animals. The central vacuole also functions to store proteins in developing seed cells.

Extracellular Matrix of Animal Cells

Most animal cells release materials into the extracellular space. The primary components of these materials are glycoproteins and the protein collagen. Collectively, these materials are called the **extracellular matrix** (Figure 3.16). Not only does the extracellular matrix hold the cells together to form a tissue, but it also allows the cells within the tissue to communicate with each other.

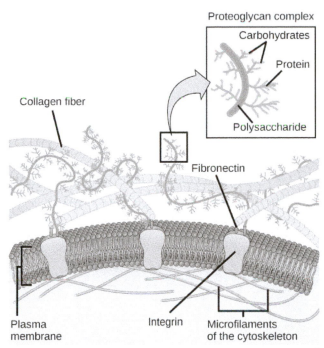

Figure 3.16 The extracellular matrix consists of a network of substances secreted by cells.

Blood clotting provides an example of the role of the extracellular matrix in cell communication. When the cells lining a blood vessel are damaged, they display a protein receptor called tissue factor. When tissue factor binds with another factor in the extracellular matrix, it causes platelets to adhere to the wall of the damaged blood vessel, stimulates adjacent smooth muscle cells in the blood vessel to contract (thus constricting the blood vessel), and initiates a series of steps that stimulate the platelets to produce clotting factors.

Intercellular Junctions

Cells can also communicate with each other by direct contact, referred to as intercellular junctions. There are some differences in the ways that plant and animal cells do this. **Plasmodesmata** (singular = plasmodesma) are junctions between plant cells, whereas animal cell contacts include tight and gap junctions, and desmosomes.

In general, long stretches of the plasma membranes of neighboring plant cells cannot touch one another because they are separated by the cell walls surrounding each cell. Plasmodesmata are numerous channels that pass between the cell walls of adjacent plant cells, connecting their cytoplasm and enabling signal molecules and nutrients to be transported from cell to cell (Figure 3.17a).

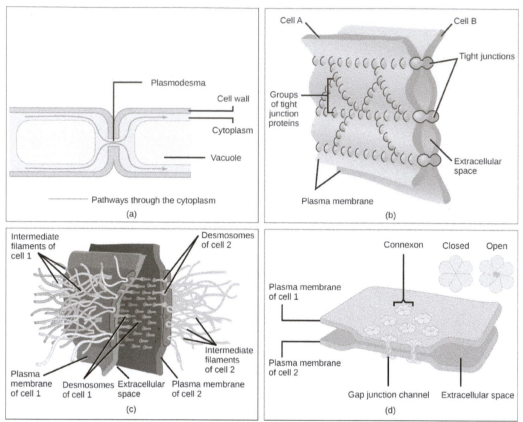

Figure 3.17 There are four kinds of connections between cells. (a) A plasmodesma is a channel between the cell walls of two adjacent plant cells. (b) Tight junctions join adjacent animal cells. (c) Desmosomes join two animal cells together. (d) Gap junctions act as channels between animal cells. (credit b, c, d: modification of work by Mariana Ruiz Villareal)

A **tight junction** is a watertight seal between two adjacent animal cells (Figure 3.17b). Proteins hold the cells tightly against each other. This tight adhesion prevents materials from leaking between the cells. Tight junctions are typically found in the epithelial tissue that lines internal organs and cavities, and composes most of the skin. For example, the tight junctions of the epithelial cells lining the urinary bladder prevent urine from leaking into the extracellular space.

Also found only in animal cells are **desmosomes**, which act like spot welds between adjacent epithelial cells (Figure 3.17c). They keep cells together in a sheet-like formation in organs and tissues that stretch, like the skin, heart, and muscles.

Gap junctions in animal cells are like plasmodesmata in plant cells in that they are channels between adjacent cells that allow for the transport of ions, nutrients, and other substances that enable cells to communicate (Figure 3.17d). Structurally, however, gap junctions and plasmodesmata differ.

Components of Prokaryotic and Eukaryotic Cells and Their Functions

Cell Component	Function	Present in Prokaryotes?	Present in Animal Cells?	Present in Plant Cells?
Plasma membrane	Separates cell from external environment; controls passage of organic molecules, ions, water, oxygen, and wastes into and out of the cell	Yes	Yes	Yes

Table 3.1

Components of Prokaryotic and Eukaryotic Cells and Their Functions

Cell Component	Function	Present in Prokaryotes?	Present in Animal Cells?	Present in Plant Cells?
Cytoplasm	Provides structure to cell; site of many metabolic reactions; medium in which organelles are found	Yes	Yes	Yes
Nucleoid	Location of DNA	Yes	No	No
Nucleus	Cell organelle that houses DNA and directs synthesis of ribosomes and proteins	No	Yes	Yes
Ribosomes	Protein synthesis	Yes	Yes	Yes
Mitochondria	ATP production/cellular respiration	No	Yes	Yes
Peroxisomes	Oxidizes and breaks down fatty acids and amino acids, and detoxifies poisons	No	Yes	Yes
Vesicles and vacuoles	Storage and transport; digestive function in plant cells	No	Yes	Yes
Centrosome	Unspecified role in cell division in animal cells; organizing center of microtubules in animal cells	No	Yes	No
Lysosomes	Digestion of macromolecules; recycling of worn-out organelles	No	Yes	No
Cell wall	Protection, structural support and maintenance of cell shape	Yes, primarily peptidoglycan in bacteria but not Archaea	No	Yes, primarily cellulose
Chloroplasts	Photosynthesis	No	No	Yes
Endoplasmic reticulum	Modifies proteins and synthesizes lipids	No	Yes	Yes
Golgi apparatus	Modifies, sorts, tags, packages, and distributes lipids and proteins	No	Yes	Yes
Cytoskeleton	Maintains cell's shape, secures organelles in specific positions, allows cytoplasm and vesicles to move within the cell, and enables unicellular organisms to move independently	Yes	Yes	Yes
Flagella	Cellular locomotion	Some	Some	No, except for some plant sperm.
Cilia	Cellular locomotion, movement of particles along extracellular surface of plasma membrane, and filtration	No	Some	No

Table 3.1

This table provides the components of prokaryotic and eukaryotic cells and their respective functions.

3.4 | The Cell Membrane

By the end of this section, you will be able to:

- Understand the fluid mosaic model of membranes
- Describe the functions of phospholipids, proteins, and carbohydrates in membranes

A cell's plasma membrane defines the boundary of the cell and determines the nature of its contact with the environment. Cells exclude some substances, take in others, and excrete still others, all in controlled quantities. Plasma membranes enclose the borders of cells, but rather than being a static bag, they are dynamic and constantly in flux. The plasma membrane must be sufficiently flexible to allow certain cells, such as red blood cells and white blood cells, to change shape as they pass through narrow capillaries. These are the more obvious functions of a plasma membrane. In addition, the surface of the plasma membrane carries markers that allow cells to recognize one another, which is vital as tissues and organs form during early development, and which later plays a role in the "self" versus "non-self" distinction of the immune response.

The plasma membrane also carries receptors, which are attachment sites for specific substances that interact with the cell. Each receptor is structured to bind with a specific substance. For example, surface receptors of the membrane create changes in the interior, such as changes in enzymes of metabolic pathways. These metabolic pathways might be vital for providing the cell with energy, making specific substances for the cell, or breaking down cellular waste or toxins for disposal. Receptors on the plasma membrane's exterior surface interact with hormones or neurotransmitters, and allow their messages to be transmitted into the cell. Some recognition sites are used by viruses as attachment points. Although they are highly specific, pathogens like viruses may evolve to exploit receptors to gain entry to a cell by mimicking the specific substance that the receptor is meant to bind. This specificity helps to explain why human immunodeficiency virus (HIV) or any of the five types of hepatitis viruses invade only specific cells.

Fluid Mosaic Model

In 1972, S. J. Singer and Garth L. Nicolson proposed a new model of the plasma membrane that, compared to earlier understanding, better explained both microscopic observations and the function of the plasma membrane. This was called the **fluid mosaic model**. The model has evolved somewhat over time, but still best accounts for the structure and functions of the plasma membrane as we now understand them. The fluid mosaic model describes the structure of the plasma membrane as a mosaic of components—including phospholipids, cholesterol, proteins, and carbohydrates—in which the components are able to flow and change position, while maintaining the basic integrity of the membrane. Both phospholipid molecules and embedded proteins are able to diffuse rapidly and laterally in the membrane. The fluidity of the plasma membrane is necessary for the activities of certain enzymes and transport molecules within the membrane. Plasma membranes range from 5–10 nm thick. As a comparison, human red blood cells, visible via light microscopy, are approximately 8 μm thick, or approximately 1,000 times thicker than a plasma membrane. (Figure 3.18)

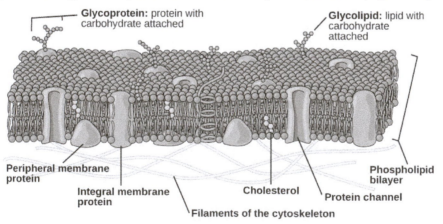

Figure 3.18 The fluid mosaic model of the plasma membrane structure describes the plasma membrane as a fluid combination of phospholipids, cholesterol, proteins, and carbohydrates.

The plasma membrane is made up primarily of a bilayer of phospholipids with embedded proteins, carbohydrates, glycolipids, and glycoproteins, and, in animal cells, cholesterol. The amount of cholesterol in animal plasma membranes regulates the fluidity of the membrane and changes based on the temperature of the cell's environment. In other words, cholesterol acts as antifreeze in the cell membrane and is more abundant in animals that live in cold climates.

The main fabric of the membrane is composed of two layers of phospholipid molecules, and the polar ends of these molecules (which look like a collection of balls in an artist's rendition of the model) (Figure 3.18) are in contact with aqueous fluid both inside and outside the cell. Thus, both surfaces of the plasma membrane are hydrophilic. In contrast, the interior of the membrane, between its two surfaces, is a hydrophobic or nonpolar region because of the fatty acid tails. This region has no attraction for water or other polar molecules.

Proteins make up the second major chemical component of plasma membranes. Integral proteins are embedded in the plasma membrane and may span all or part of the membrane. Integral proteins may serve as channels or pumps to move materials into or out of the cell. Peripheral proteins are found on the exterior or interior surfaces of membranes, attached either to integral proteins or to phospholipid molecules. Both integral and peripheral proteins may serve as enzymes, as structural attachments for the fibers of the cytoskeleton, or as part of the cell's recognition sites.

Carbohydrates are the third major component of plasma membranes. They are always found on the exterior surface of cells and are bound either to proteins (forming glycoproteins) or to lipids (forming glycolipids). These carbohydrate chains may consist of 2–60 monosaccharide units and may be either straight or branched. Along with peripheral proteins, carbohydrates form specialized sites on the cell surface that allow cells to recognize each other.

evolution IN ACTION

How Viruses Infect Specific Organs

Specific glycoprotein molecules exposed on the surface of the cell membranes of host cells are exploited by many viruses to infect specific organs. For example, HIV is able to penetrate the plasma membranes of specific kinds of white blood cells called T-helper cells and monocytes, as well as some cells of the central nervous system. The hepatitis virus attacks only liver cells.

These viruses are able to invade these cells, because the cells have binding sites on their surfaces that the viruses have exploited with equally specific glycoproteins in their coats. (Figure 3.19). The cell is tricked by the mimicry of the virus coat molecules, and the virus is able to enter the cell. Other recognition sites on the virus's surface interact with the human immune system, prompting the body to produce antibodies. Antibodies are made in response to the antigens (or proteins associated with invasive pathogens). These same sites serve as places for antibodies to attach, and either destroy or inhibit the activity of the virus. Unfortunately, these sites on HIV are encoded by genes that change quickly, making the production of an effective vaccine against the virus very difficult. The virus population within an infected individual quickly evolves through mutation into different populations, or variants, distinguished by differences in these recognition sites. This rapid change of viral surface markers decreases the effectiveness of the person's immune system in attacking the virus, because the antibodies will not recognize the new variations of the surface patterns.

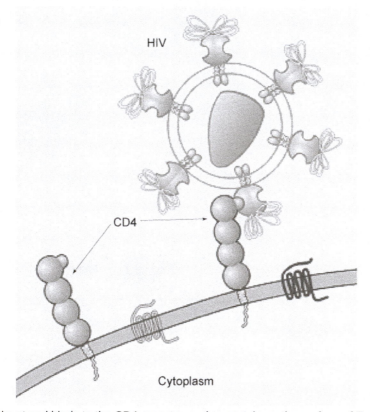

Figure 3.19 HIV docks at and binds to the CD4 receptor, a glycoprotein on the surface of T cells, before entering, or infecting, the cell. (credit: modification of work by US National Institutes of Health/National Institute of Allergy and Infectious Diseases)

3.5 | Passive Transport

By the end of this section, you will be able to:

- Explain why and how passive transport occurs
- Understand the processes of osmosis and diffusion
- Define tonicity and describe its relevance to passive transport

Plasma membranes must allow certain substances to enter and leave a cell, while preventing harmful material from entering and essential material from leaving. In other words, plasma membranes are **selectively permeable**—they allow some substances through but not others. If they were to lose this selectivity, the cell would no longer be able to sustain itself, and it would be destroyed. Some cells require larger amounts of specific substances than do other cells; they must have a way of obtaining these materials from the extracellular fluids. This may happen passively, as certain materials move back and forth, or the cell may have special mechanisms that ensure transport. Most cells expend most of their energy, in the form of adenosine triphosphate (ATP), to create and maintain an uneven distribution of ions on the opposite sides of their membranes. The structure of the plasma membrane contributes to these functions, but it also presents some problems.

The most direct forms of membrane transport are passive. **Passive transport** is a naturally occurring phenomenon and does not require the cell to expend energy to accomplish the movement. In passive transport, substances move from an area of higher concentration to an area of lower concentration in a process called diffusion. A physical space in which there is a different concentration of a single substance is said to have a **concentration gradient**.

Selective Permeability

Plasma membranes are asymmetric, meaning that despite the mirror image formed by the phospholipids, the interior of the membrane is not identical to the exterior of the membrane. Integral proteins that act as channels or pumps work in one direction. Carbohydrates, attached to lipids or proteins, are also found on the exterior surface of the plasma membrane. These carbohydrate complexes help the cell bind substances that the cell needs in the extracellular fluid. This adds considerably to the selective nature of plasma membranes.

Recall that plasma membranes have hydrophilic and hydrophobic regions. This characteristic helps the movement of certain materials through the membrane and hinders the movement of others. Lipid-soluble material can easily slip through the hydrophobic lipid core of the membrane. Substances such as the fat-soluble vitamins A, D, E, and K readily pass through the plasma membranes in the digestive tract and other tissues. Fat-soluble drugs also gain easy entry into cells and are readily transported into the body's tissues and organs. Molecules of oxygen and carbon dioxide have no charge and pass through by simple diffusion.

Polar substances, with the exception of water, present problems for the membrane. While some polar molecules connect easily with the outside of a cell, they cannot readily pass through the lipid core of the plasma membrane. Additionally, whereas small ions could easily slip through the spaces in the mosaic of the membrane, their charge prevents them from doing so. Ions such as sodium, potassium, calcium, and chloride must have a special means of penetrating plasma membranes. Simple sugars and amino acids also need help with transport across plasma membranes.

Diffusion

Diffusion is a passive process of transport. A single substance tends to move from an area of high concentration to an area of low concentration until the concentration is equal across the space. You are familiar with diffusion of substances through the air. For example, think about someone opening a bottle of perfume in a room filled with people. The perfume is at its highest concentration in the bottle and is at its lowest at the edges of the room. The perfume vapor will diffuse, or spread away, from the bottle, and gradually, more and more people will smell the perfume as it spreads. Materials move within the cell's cytosol by diffusion, and certain materials move through the plasma membrane by diffusion (Figure 3.20). Diffusion expends no energy. Rather the different concentrations of materials in different areas are a form of potential energy, and diffusion is the dissipation of that potential energy as materials move down their concentration gradients, from high to low.

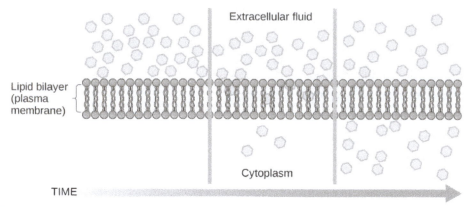

Figure 3.20 Diffusion through a permeable membrane follows the concentration gradient of a substance, moving the substance from an area of high concentration to one of low concentration. (credit: modification of work by Mariana Ruiz Villarreal)

Each separate substance in a medium, such as the extracellular fluid, has its own concentration gradient, independent of the concentration gradients of other materials. Additionally, each substance will diffuse according to that gradient.

Several factors affect the rate of diffusion.

- Extent of the concentration gradient: The greater the difference in concentration, the more rapid the diffusion. The closer the distribution of the material gets to equilibrium, the slower the rate of diffusion becomes.

- Mass of the molecules diffusing: More massive molecules move more slowly, because it is more difficult for them to move between the molecules of the substance they are moving through; therefore, they diffuse more slowly.

- Temperature: Higher temperatures increase the energy and therefore the movement of the molecules, increasing the rate of diffusion.

- Solvent density: As the density of the solvent increases, the rate of diffusion decreases. The molecules slow down because they have a more difficult time getting through the denser medium.

For an animation of the diffusion process in action, view **this short video (http://openstaxcollege.org/l/passive_trnsprt)** on cell membrane transport.

Facilitated transport

In **facilitated transport**, also called facilitated diffusion, material moves across the plasma membrane with the assistance of transmembrane proteins down a concentration gradient (from high to low concentration) without the expenditure of cellular energy. However, the substances that undergo facilitated transport would otherwise not diffuse easily or quickly across the plasma membrane. The solution to moving polar substances and other substances across the plasma membrane rests in the proteins that span its surface. The material being transported is first attached to protein or glycoprotein receptors on the exterior surface of the plasma membrane. This allows the material that is needed by the cell to be removed from the extracellular fluid. The substances are then passed to specific integral proteins that facilitate their passage, because they form channels or pores that allow certain substances to pass through the membrane. The integral proteins involved in facilitated transport are collectively referred to as transport proteins, and they function as either channels for the material or carriers.

Osmosis

Osmosis is the diffusion of water through a semipermeable membrane according to the concentration gradient of water across the membrane. Whereas diffusion transports material across membranes and within cells, osmosis transports *only water* across a membrane and the membrane limits the diffusion of solutes in the water. Osmosis is a special case of diffusion. Water, like other substances, moves from an area of higher concentration to one of lower concentration. Imagine a beaker with a semipermeable membrane, separating the two sides or halves (Figure 3.21). On both sides of the membrane, the water level is the same, but there are different concentrations on each side of a dissolved substance, or **solute**, that cannot cross the membrane. If the volume of the water is the same, but the concentrations of solute are different, then there are also different concentrations of water, the solvent, on either side of the membrane.

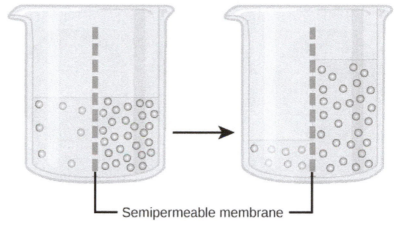

— Semipermeable membrane —

Figure 3.21 In osmosis, water always moves from an area of higher concentration (of water) to one of lower concentration (of water). In this system, the solute cannot pass through the selectively permeable membrane.

A principle of diffusion is that the molecules move around and will spread evenly throughout the medium if they can. However, only the material capable of getting through the membrane will diffuse through it. In this example, the solute cannot diffuse through the membrane, but the water can. Water has a concentration gradient in this system. Therefore, water will diffuse down its concentration gradient, crossing the membrane to the side where it is less concentrated. This diffusion of water through the membrane—osmosis—will continue until the concentration gradient of water goes to zero. Osmosis proceeds constantly in living systems.

Watch this video (http://openstaxcollege.org/l/passive_trnsprt) that illustrates diffusion in hot versus cold solutions.

Tonicity

Tonicity describes the amount of solute in a solution. The measure of the tonicity of a solution, or the total amount of solutes dissolved in a specific amount of solution, is called its **osmolarity**. Three terms—hypotonic, isotonic, and hypertonic—are used to relate the osmolarity of a cell to the osmolarity of the extracellular fluid that contains the cells. In a **hypotonic** solution, such as tap water, the extracellular fluid has a lower concentration of solutes than the fluid inside the cell, and water enters the cell. (In living systems, the point of reference is always the cytoplasm, so the prefix *hypo-* means that the extracellular fluid has a lower concentration of solutes, or a lower osmolarity, than the cell cytoplasm.) It also means that the extracellular fluid has a higher concentration of water than does the cell. In this situation, water will follow its concentration gradient and enter the cell. This may cause an animal cell to burst, or lyse.

In a **hypertonic** solution (the prefix *hyper-* refers to the extracellular fluid having a higher concentration of solutes than the cell's cytoplasm), the fluid contains less water than the cell does, such as seawater. Because the cell has a lower concentration of solutes, the water will leave the cell. In effect, the solute is drawing the water out of the cell. This may cause an animal cell to shrivel, or crenate.

In an **isotonic** solution, the extracellular fluid has the same osmolarity as the cell. If the concentration of solutes of the cell matches that of the extracellular fluid, there will be no net movement of water into or out of the cell. Blood cells in hypertonic, isotonic, and hypotonic solutions take on characteristic appearances (Figure 3.22).

art CONNECTION

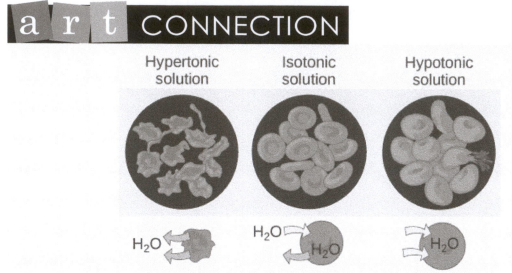

Figure 3.22 Osmotic pressure changes the shape of red blood cells in hypertonic, isotonic, and hypotonic solutions. (credit: modification of work by Mariana Ruiz Villarreal)

A doctor injects a patient with what the doctor thinks is isotonic saline solution. The patient dies, and autopsy reveals that many red blood cells have been destroyed. Do you think the solution the doctor injected was really isotonic?

Some organisms, such as plants, fungi, bacteria, and some protists, have cell walls that surround the plasma membrane and prevent cell lysis. The plasma membrane can only expand to the limit of the cell wall, so the cell will not lyse. In fact, the cytoplasm in plants is always slightly hypertonic compared to the cellular environment, and water will always enter a cell if water is available. This influx of water produces turgor pressure, which stiffens the cell walls of the plant (Figure 3.23). In nonwoody plants, turgor pressure supports the plant. If the plant cells become hypertonic, as occurs in drought or if a plant is not watered adequately, water will leave the cell. Plants lose turgor pressure in this condition and wilt.

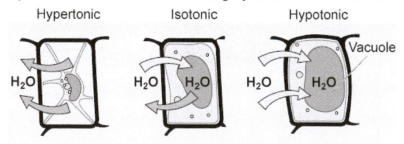

Figure 3.23 The turgor pressure within a plant cell depends on the tonicity of the solution that it is bathed in. (credit: modification of work by Mariana Ruiz Villarreal)

Download for free at https://openstax.org/details/books/concepts-biology

3.6 | Active Transport

By the end of this section, you will be able to:

- Understand how electrochemical gradients affect ions
- Describe endocytosis, including phagocytosis, pinocytosis, and receptor-mediated endocytosis
- Understand the process of exocytosis

Active transport mechanisms require the use of the cell's energy, usually in the form of adenosine triphosphate (ATP). If a substance must move into the cell against its concentration gradient, that is, if the concentration of the substance inside the cell must be greater than its concentration in the extracellular fluid, the cell must use energy to move the substance. Some active transport mechanisms move small-molecular weight material, such as ions, through the membrane.

In addition to moving small ions and molecules through the membrane, cells also need to remove and take in larger molecules and particles. Some cells are even capable of engulfing entire unicellular microorganisms. You might have correctly hypothesized that the uptake and release of large particles by the cell requires energy. A large particle, however, cannot pass through the membrane, even with energy supplied by the cell.

Electrochemical Gradient

We have discussed simple concentration gradients—differential concentrations of a substance across a space or a membrane—but in living systems, gradients are more complex. Because cells contain proteins, most of which are negatively charged, and because ions move into and out of cells, there is an electrical gradient, a difference of charge, across the plasma membrane. The interior of living cells is electrically negative with respect to the extracellular fluid in which they are bathed; at the same time, cells have higher concentrations of potassium (K^+) and lower concentrations of sodium (Na^+) than does the extracellular fluid. Thus, in a living cell, the concentration gradient and electrical gradient of Na^+ promotes diffusion of the ion into the cell, and the electrical gradient of Na^+ (a positive ion) tends to drive it inward to the negatively charged interior. The situation is more complex, however, for other elements such as potassium. The electrical gradient of K^+ promotes diffusion of the ion *into* the cell, but the concentration gradient of K^+ promotes diffusion *out* of the cell (Figure 3.24). The combined gradient that affects an ion is called its **electrochemical gradient**, and it is especially important to muscle and nerve cells.

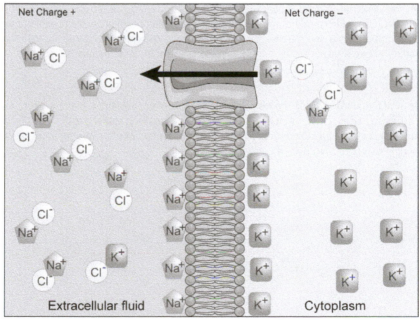

Figure 3.24 Electrochemical gradients arise from the combined effects of concentration gradients and electrical gradients. (credit: modification of work by "Synaptitude"/Wikimedia Commons)

Moving Against a Gradient

To move substances against a concentration or an electrochemical gradient, the cell must use energy. This energy is harvested from ATP that is generated through cellular metabolism. Active transport mechanisms, collectively called pumps or carrier proteins, work against electrochemical gradients. With the exception of ions, small substances constantly pass through plasma membranes. Active transport maintains concentrations of ions and other substances needed by living cells in the face of these passive changes. Much of a cell's supply of metabolic energy may be spent maintaining these processes. Because active transport mechanisms depend on cellular metabolism for energy, they are sensitive to many metabolic poisons that interfere with the supply of ATP.

Two mechanisms exist for the transport of small-molecular weight material and macromolecules. Primary active transport moves ions across a membrane and creates a difference in charge across that membrane. The primary active transport system uses ATP to move a substance, such as an ion, into the cell, and often at the same time, a second substance is moved out of the cell. The sodium-potassium pump, an important pump in animal cells, expends energy to move potassium ions into the cell and a different number of sodium ions out of the cell (Figure 3.25). The action of this pump results in a concentration and charge difference across the membrane.

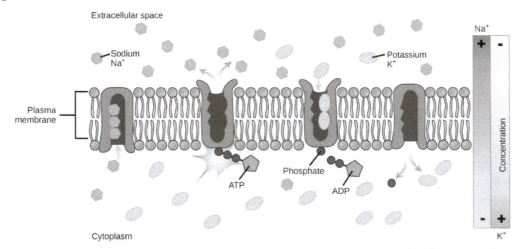

Figure 3.25 The sodium-potassium pump move potassium and sodium ions across the plasma membrane. (credit: modification of work by Mariana Ruiz Villarreal)

Secondary active transport describes the movement of material using the energy of the electrochemical gradient established by primary active transport. Using the energy of the electrochemical gradient created by the primary active transport system, other substances such as amino acids and glucose can be brought into the cell through membrane channels. ATP itself is formed through secondary active transport using a hydrogen ion gradient in the mitochondrion.

Endocytosis

Endocytosis is a type of active transport that moves particles, such as large molecules, parts of cells, and even whole cells, into a cell. There are different variations of endocytosis, but all share a common characteristic: The plasma membrane of the cell invaginates, forming a pocket around the target particle. The pocket pinches off, resulting in the particle being contained in a newly created vacuole that is formed from the plasma membrane.

Download for free at https://openstax.org/details/books/concepts-biology

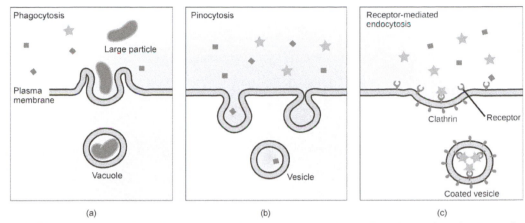

Figure 3.26 Three variations of endocytosis are shown. (a) In one form of endocytosis, phagocytosis, the cell membrane surrounds the particle and pinches off to form an intracellular vacuole. (b) In another type of endocytosis, pinocytosis, the cell membrane surrounds a small volume of fluid and pinches off, forming a vesicle. (c) In receptor-mediated endocytosis, uptake of substances by the cell is targeted to a single type of substance that binds at the receptor on the external cell membrane. (credit: modification of work by Mariana Ruiz Villarreal)

Phagocytosis is the process by which large particles, such as cells, are taken in by a cell. For example, when microorganisms invade the human body, a type of white blood cell called a neutrophil removes the invader through this process, surrounding and engulfing the microorganism, which is then destroyed by the neutrophil (Figure 3.26).

A variation of endocytosis is called **pinocytosis**. This literally means "cell drinking" and was named at a time when the assumption was that the cell was purposefully taking in extracellular fluid. In reality, this process takes in solutes that the cell needs from the extracellular fluid (Figure 3.26).

A targeted variation of endocytosis employs binding proteins in the plasma membrane that are specific for certain substances (Figure 3.26). The particles bind to the proteins and the plasma membrane invaginates, bringing the substance and the proteins into the cell. If passage across the membrane of the target of **receptor-mediated endocytosis** is ineffective, it will not be removed from the tissue fluids or blood. Instead, it will stay in those fluids and increase in concentration. Some human diseases are caused by a failure of receptor-mediated endocytosis. For example, the form of cholesterol termed low-density lipoprotein or LDL (also referred to as "bad" cholesterol) is removed from the blood by receptor-mediated endocytosis. In the human genetic disease familial hypercholesterolemia, the LDL receptors are defective or missing entirely. People with this condition have life-threatening levels of cholesterol in their blood, because their cells cannot clear the chemical from their blood.

See receptor-mediated endocytosis in action and click on different parts for a focused animation (http://openstaxcollege.org/l/endocytosis2) to learn more.

Exocytosis

In contrast to these methods of moving material into a cell is the process of exocytosis. **Exocytosis** is the opposite of the processes discussed above in that its purpose is to expel material from the cell into the extracellular fluid. A particle enveloped in membrane fuses with the interior of the plasma membrane. This fusion opens the membranous envelope to the exterior of the cell, and the particle is expelled into the extracellular space (Figure 3.27).

Exocytosis

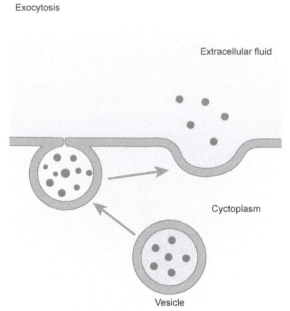

Figure 3.27 In exocytosis, a vesicle migrates to the plasma membrane, binds, and releases its contents to the outside of the cell. (credit: modification of work by Mariana Ruiz Villarreal)

KEY TERMS

active transport the method of transporting material that requires energy

cell wall a rigid cell covering made of cellulose in plants, peptidoglycan in bacteria, non-peptidoglycan compounds in Archaea, and chitin in fungi that protects the cell, provides structural support, and gives shape to the cell

central vacuole a large plant cell organelle that acts as a storage compartment, water reservoir, and site of macromolecule degradation

chloroplast a plant cell organelle that carries out photosynthesis

cilium (plural: cilia) a short, hair-like structure that extends from the plasma membrane in large numbers and is used to move an entire cell or move substances along the outer surface of the cell

concentration gradient an area of high concentration across from an area of low concentration

cytoplasm the entire region between the plasma membrane and the nuclear envelope, consisting of organelles suspended in the gel-like cytosol, the cytoskeleton, and various chemicals

cytoskeleton the network of protein fibers that collectively maintains the shape of the cell, secures some organelles in specific positions, allows cytoplasm and vesicles to move within the cell, and enables unicellular organisms to move

cytosol the gel-like material of the cytoplasm in which cell structures are suspended

desmosome a linkage between adjacent epithelial cells that forms when cadherins in the plasma membrane attach to intermediate filaments

diffusion a passive process of transport of low-molecular weight material down its concentration gradient

electrochemical gradient a gradient produced by the combined forces of the electrical gradient and the chemical gradient

endocytosis a type of active transport that moves substances, including fluids and particles, into a cell

endomembrane system the group of organelles and membranes in eukaryotic cells that work together to modify, package, and transport lipids and proteins

endoplasmic reticulum (ER) a series of interconnected membranous structures within eukaryotic cells that collectively modify proteins and synthesize lipids

eukaryotic cell a cell that has a membrane-bound nucleus and several other membrane-bound compartments or sacs

exocytosis a process of passing material out of a cell

extracellular matrix the material, primarily collagen, glycoproteins, and proteoglycans, secreted from animal cells that holds cells together as a tissue, allows cells to communicate with each other, and provides mechanical protection and anchoring for cells in the tissue

facilitated transport a process by which material moves down a concentration gradient (from high to low concentration) using integral membrane proteins

flagellum (plural: flagella) the long, hair-like structure that extends from the plasma membrane and is used to move the cell

fluid mosaic model a model of the structure of the plasma membrane as a mosaic of components, including phospholipids, cholesterol, proteins, and glycolipids, resulting in a fluid rather than static character

gap junction a channel between two adjacent animal cells that allows ions, nutrients, and other low-molecular weight substances to pass between the cells, enabling the cells to communicate

Golgi apparatus a eukaryotic organelle made up of a series of stacked membranes that sorts, tags, and packages lipids and proteins for distribution

hypertonic describes a solution in which extracellular fluid has higher osmolarity than the fluid inside the cell

hypotonic describes a solution in which extracellular fluid has lower osmolarity than the fluid inside the cell

isotonic describes a solution in which the extracellular fluid has the same osmolarity as the fluid inside the cell

lysosome an organelle in an animal cell that functions as the cell's digestive component; it breaks down proteins, polysaccharides, lipids, nucleic acids, and even worn-out organelles

microscope the instrument that magnifies an object

mitochondria (singular: mitochondrion) the cellular organelles responsible for carrying out cellular respiration, resulting in the production of ATP, the cell's main energy-carrying molecule

nuclear envelope the double-membrane structure that constitutes the outermost portion of the nucleus

nucleolus the darkly staining body within the nucleus that is responsible for assembling ribosomal subunits

nucleus the cell organelle that houses the cell's DNA and directs the synthesis of ribosomes and proteins

organelle a membrane-bound compartment or sac within a cell

osmolarity the total amount of substances dissolved in a specific amount of solution

osmosis the transport of water through a semipermeable membrane from an area of high water concentration to an area of low water concentration across a membrane

passive transport a method of transporting material that does not require energy

peroxisome a small, round organelle that contains hydrogen peroxide, oxidizes fatty acids and amino acids, and detoxifies many poisons

phagocytosis a process that takes macromolecules that the cell needs from the extracellular fluid; a variation of endocytosis

pinocytosis a process that takes solutes that the cell needs from the extracellular fluid; a variation of endocytosis

plasma membrane a phospholipid bilayer with embedded (integral) or attached (peripheral) proteins that separates the internal contents of the cell from its surrounding environment

plasmodesma (plural: plasmodesmata) a channel that passes between the cell walls of adjacent plant cells, connects their cytoplasm, and allows materials to be transported from cell to cell

prokaryotic cell a unicellular organism that lacks a nucleus or any other membrane-bound organelle

receptor-mediated endocytosis a variant of endocytosis that involves the use of specific binding proteins in the plasma membrane for specific molecules or particles

ribosome a cellular structure that carries out protein synthesis

rough endoplasmic reticulum (RER) the region of the endoplasmic reticulum that is studded with ribosomes and engages in protein modification

selectively permeable the characteristic of a membrane that allows some substances through but not others

smooth endoplasmic reticulum (SER) the region of the endoplasmic reticulum that has few or no ribosomes on its cytoplasmic surface and synthesizes carbohydrates, lipids, and steroid hormones; detoxifies chemicals like pesticides, preservatives, medications, and environmental pollutants, and stores calcium ions

solute a substance dissolved in another to form a solution

tight junction a firm seal between two adjacent animal cells created by protein adherence

tonicity the amount of solute in a solution.

unified cell theory the biological concept that states that all organisms are composed of one or more cells, the cell is the basic unit of life, and new cells arise from existing cells

vacuole a membrane-bound sac, somewhat larger than a vesicle, that functions in cellular storage and transport

vesicle a small, membrane-bound sac that functions in cellular storage and transport; its membrane is capable of fusing with the plasma membrane and the membranes of the endoplasmic reticulum and Golgi apparatus

CHAPTER SUMMARY

3.1 How Cells Are Studied

A cell is the smallest unit of life. Most cells are so small that they cannot be viewed with the naked eye. Therefore, scientists must use microscopes to study cells. Electron microscopes provide higher magnification, higher resolution, and more detail than light microscopes. The unified cell theory states that all organisms are composed of one or more cells, the cell is the basic unit of life, and new cells arise from existing cells.

3.2 Comparing Prokaryotic and Eukaryotic Cells

Prokaryotes are predominantly single-celled organisms of the domains Bacteria and Archaea. All prokaryotes have plasma membranes, cytoplasm, ribosomes, a cell wall, DNA, and lack membrane-bound organelles. Many also have polysaccharide capsules. Prokaryotic cells range in diameter from 0.1–5.0 μm.

Like a prokaryotic cell, a eukaryotic cell has a plasma membrane, cytoplasm, and ribosomes, but a eukaryotic cell is typically larger than a prokaryotic cell, has a true nucleus (meaning its DNA is surrounded by a membrane), and has other membrane-bound organelles that allow for compartmentalization of functions. Eukaryotic cells tend to be 10 to 100 times the size of prokaryotic cells.

3.3 Eukaryotic Cells

Like a prokaryotic cell, a eukaryotic cell has a plasma membrane, cytoplasm, and ribosomes, but a eukaryotic cell is typically larger than a prokaryotic cell, has a true nucleus (meaning its DNA is surrounded by a membrane), and has other membrane-bound organelles that allow for compartmentalization of functions. The plasma membrane is a phospholipid bilayer embedded with proteins. The nucleolus within the nucleus is the site for ribosome assembly. Ribosomes are found in the cytoplasm or are attached to the cytoplasmic side of the plasma membrane or endoplasmic reticulum. They perform protein synthesis. Mitochondria perform cellular respiration and produce ATP. Peroxisomes break down fatty acids, amino acids, and some toxins. Vesicles and vacuoles are storage and transport compartments. In plant cells, vacuoles also help break down macromolecules.

Animal cells also have a centrosome and lysosomes. The centrosome has two bodies, the centrioles, with an unknown role in cell division. Lysosomes are the digestive organelles of animal cells.

Plant cells have a cell wall, chloroplasts, and a central vacuole. The plant cell wall, whose primary component is cellulose, protects the cell, provides structural support, and gives shape to the cell. Photosynthesis takes place in chloroplasts. The central vacuole expands, enlarging the cell without the need to produce more cytoplasm.

The endomembrane system includes the nuclear envelope, the endoplasmic reticulum, Golgi apparatus, lysosomes, vesicles, as well as the plasma membrane. These cellular components work together to modify, package, tag, and transport membrane lipids and proteins.

The cytoskeleton has three different types of protein elements. Microfilaments provide rigidity and shape to the cell, and facilitate cellular movements. Intermediate filaments bear tension and anchor the nucleus and other organelles in place. Microtubules help the cell resist compression, serve as tracks for motor proteins that move vesicles through the cell, and pull replicated chromosomes to opposite ends of a dividing cell. They are also the structural elements of centrioles, flagella, and cilia.

Animal cells communicate through their extracellular matrices and are connected to each other by tight junctions, desmosomes, and gap junctions. Plant cells are connected and communicate with each other by plasmodesmata.

3.4 The Cell Membrane

The modern understanding of the plasma membrane is referred to as the fluid mosaic model. The plasma membrane is composed of a bilayer of phospholipids, with their hydrophobic, fatty acid tails in contact with each other. The landscape of the membrane is studded with proteins, some of which span the membrane. Some of these proteins serve to transport materials into or out of the cell. Carbohydrates are attached to some of the proteins and lipids on the outward-facing surface of the membrane. These form complexes that function to identify the cell to other cells. The fluid nature of the membrane owes itself to the configuration of the fatty acid tails, the presence of cholesterol embedded in the membrane (in animal cells), and the mosaic nature of the proteins and protein-carbohydrate complexes, which are not firmly fixed in place. Plasma membranes enclose the borders of cells, but rather than being a static bag, they are dynamic and constantly in flux.

3.5 Passive Transport

The passive forms of transport, diffusion and osmosis, move material of small molecular weight. Substances diffuse from areas of high concentration to areas of low concentration, and this process continues until the substance is evenly distributed in a system. In solutions of more than one substance, each type of molecule diffuses according to its own concentration gradient. Many factors can affect the rate of diffusion, including concentration gradient, the sizes of the particles that are diffusing, and the temperature of the system.

In living systems, diffusion of substances into and out of cells is mediated by the plasma membrane. Some materials diffuse readily through the membrane, but others are hindered, and their passage is only made possible by protein channels and carriers. The chemistry of living things occurs in aqueous solutions, and balancing the concentrations of those solutions is an ongoing problem. In living systems, diffusion of some substances would be slow or difficult without membrane proteins.

3.6 Active Transport

The combined gradient that affects an ion includes its concentration gradient and its electrical gradient. Living cells need certain substances in concentrations greater than they exist in the extracellular space. Moving substances up their electrochemical gradients requires energy from the cell. Active transport uses energy stored in ATP to fuel the transport. Active transport of small molecular-size material uses integral proteins in the cell membrane to move the material—these proteins are analogous to pumps. Some pumps, which carry out primary active transport, couple directly with ATP to drive their action. In secondary transport, energy from primary transport can be used to move another substance into the cell and up its concentration gradient.

Endocytosis methods require the direct use of ATP to fuel the transport of large particles such as macromolecules; parts of cells or whole cells can be engulfed by other cells in a process called phagocytosis. In phagocytosis, a portion of the membrane invaginates and flows around the particle, eventually pinching off and leaving the particle wholly enclosed by an envelope of plasma membrane. Vacuoles are broken down by the cell, with the particles used as food or dispatched in some other way. Pinocytosis is a similar process on a smaller scale. The cell expels waste and other particles through the reverse process, exocytosis. Wastes are moved outside the cell, pushing a membranous vesicle to the plasma membrane, allowing the vesicle to fuse with the membrane and incorporating itself into the membrane structure, releasing its contents to the exterior of the cell.

ART CONNECTION QUESTIONS

1. Figure 3.7 What structures does a plant cell have that an animal cell does not have? What structures does an animal cell have that a plant cell does not have?

2. Figure 3.13 Why does the *cis* face of the Golgi not face the plasma membrane?

3. Figure 3.22 A doctor injects a patient with what he thinks is isotonic saline solution. The patient dies, and autopsy reveals that many red blood cells have been destroyed. Do you think the solution the doctor injected was really isotonic?

REVIEW QUESTIONS

4. When viewing a specimen through a light microscope, scientists use _____ to distinguish the individual components of cells.
   a.   a beam of electrons
   b.   radioactive isotopes

   c.   special stains
   d.   high temperatures

5. The _____ is the basic unit of life.
   a.   organism

b. cell
c. tissue
d. organ

6. Which of these do all prokaryotes and eukaryotes share?
   a. nuclear envelope
   b. cell walls
   c. organelles
   d. plasma membrane

7. A typical prokaryotic cell _____ compared to a eukaryotic cell.
   a. is smaller in size by a factor of 100
   b. is similar in size
   c. is smaller in size by a factor of one million
   d. is larger in size by a factor of 10

8. Which of the following is found both in eukaryotic and prokaryotic cells?
   a. nucleus
   b. mitochondrion
   c. vacuole
   d. ribosome

9. Which of the following is not a component of the endomembrane system?
   a. mitochondrion
   b. Golgi apparatus
   c. endoplasmic reticulum
   d. lysosome

10. Which plasma membrane component can be either found on its surface or embedded in the membrane structure?
   a. protein

b. cholesterol
c. carbohydrate
d. phospholipid

11. The tails of the phospholipids of the plasma membrane are composed of _____ and are _____?
   a. phosphate groups; hydrophobic
   b. fatty acid groups; hydrophilic
   c. phosphate groups; hydrophilic
   d. fatty acid groups; hydrophobic

12. Water moves via osmosis _____.
   a. throughout the cytoplasm
   b. from an area with a high concentration of other solutes to a lower one
   c. from an area with a low concentration of solutes to an area with a higher one
   d. from an area with a low concentration of water to one of higher concentration

13. The principal force driving movement in diffusion is _____.
   a. temperature
   b. particle size
   c. concentration gradient
   d. membrane surface area

14. Active transport must function continuously because _____.
   a. plasma membranes wear out
   b. cells must be in constant motion
   c. facilitated transport opposes active transport
   d. diffusion is constantly moving the solutes in the other direction

CRITICAL THINKING QUESTIONS

15. What are the advantages and disadvantages of light, transmission, and scanning electron microscopes?

16. Describe the structures that are characteristic of a prokaryote cell.

17. In the context of cell biology, what do we mean by form follows function? What are at least two examples of this concept?

18. Why is it advantageous for the cell membrane to be fluid in nature?

19. Why does osmosis occur?

20. Where does the cell get energy for active transport processes?

4 | HOW CELLS OBTAIN ENERGY

Figure 4.1 A hummingbird needs energy to maintain prolonged flight. The bird obtains its energy from taking in food and transforming the energy contained in food molecules into forms of energy to power its flight through a series of biochemical reactions. (credit: modification of work by Cory Zanker)

Chapter Outline
4.1: Energy and Metabolism
4.2: Glycolysis
4.3: Citric Acid Cycle and Oxidative Phosphorylation
4.4: Fermentation
4.5: Connections to Other Metabolic Pathways

Introduction

Virtually every task performed by living organisms requires energy. Energy is needed to perform heavy labor and exercise, but humans also use energy while thinking, and even during sleep. In fact, the living cells of every organism constantly use energy. Nutrients and other molecules are imported into the cell, metabolized (broken down) and possibly synthesized into new molecules, modified if needed, transported around the cell, and possibly distributed to the entire organism. For example, the large proteins that make up muscles are built from smaller molecules imported from dietary amino acids. Complex carbohydrates are broken down into simple sugars that the cell uses for energy. Just as energy is required to both build and demolish a building, energy is required for the synthesis and breakdown of molecules as well as the transport of molecules into and out of cells. In addition, processes such as ingesting and breaking down pathogenic bacteria and viruses, exporting wastes and toxins, and movement of the cell require energy. From where, and in what form, does this energy come? How do living cells obtain energy, and how do they use it? This chapter will discuss different forms of energy and the physical laws that govern energy transfer. This chapter will also describe how cells use energy and replenish it, and how chemical reactions in the cell are performed with great efficiency.

4.1 | Energy and Metabolism

By the end of this section, you will be able to:

- Explain what metabolic pathways are
- State the first and second laws of thermodynamics
- Explain the difference between kinetic and potential energy
- Describe endergonic and exergonic reactions
- Discuss how enzymes function as molecular catalysts

Scientists use the term **bioenergetics** to describe the concept of energy flow (Figure 4.2) through living systems, such as cells. Cellular processes such as the building and breaking down of complex molecules occur through stepwise chemical reactions. Some of these chemical reactions are spontaneous and release energy, whereas others require energy to proceed. Just as living things must continually consume food to replenish their energy supplies, cells must continually produce more energy to replenish that used by the many energy-requiring chemical reactions that constantly take place. Together, all of the chemical reactions that take place inside cells, including those that consume or generate energy, are referred to as the cell's **metabolism**.

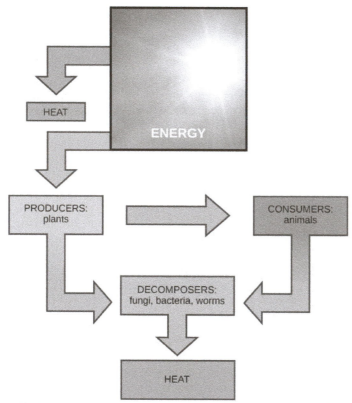

Figure 4.2 Ultimately, most life forms get their energy from the sun. Plants use photosynthesis to capture sunlight, and herbivores eat the plants to obtain energy. Carnivores eat the herbivores, and eventual decomposition of plant and animal material contributes to the nutrient pool.

Metabolic Pathways

Consider the metabolism of sugar. This is a classic example of one of the many cellular processes that use and produce energy. Living things consume sugars as a major energy source, because sugar molecules have a great deal of energy stored within their bonds. For the most part, photosynthesizing organisms like plants produce these sugars. During photosynthesis, plants use energy (originally from sunlight) to convert carbon dioxide gas (CO_2) into sugar molecules (like glucose: $C_6H_{12}O_6$). They consume carbon dioxide and produce oxygen as a waste product. This reaction is summarized as:

$$6CO_2 + 6H_2O \text{--}> C_6H_{12}O_6 + 6O_2$$

Because this process involves synthesizing an energy-storing molecule, it requires energy input to proceed. During the light reactions of photosynthesis, energy is provided by a molecule called adenosine triphosphate (ATP), which is the primary energy currency of all cells. Just as the dollar is used as currency to buy goods, cells use molecules of ATP as energy currency to perform immediate work. In contrast, energy-storage molecules such as glucose are consumed only to be broken down to use their energy. The reaction that harvests the energy of a sugar molecule in cells requiring oxygen to survive can be summarized by the reverse reaction to photosynthesis. In this reaction, oxygen is consumed and carbon dioxide is released as a waste product. The reaction is summarized as:

$$C_6H_{12}O_6 + 6O_2 \text{--}> 6H_2O + 6CO_2$$

Both of these reactions involve many steps.

The processes of making and breaking down sugar molecules illustrate two examples of metabolic pathways. A metabolic pathway is a series of chemical reactions that takes a starting molecule and modifies it, step-by-step, through a series of metabolic intermediates, eventually yielding a final product. In the example of sugar metabolism, the first metabolic pathway synthesized sugar from smaller molecules, and the other pathway broke sugar down into smaller molecules. These two opposite processes—the first requiring energy and the second producing energy—are referred to as **anabolic** pathways (building polymers) and **catabolic** pathways (breaking down polymers into their monomers), respectively. Consequently, metabolism is composed of synthesis (anabolism) and degradation (catabolism) (Figure 4.3).

It is important to know that the chemical reactions of metabolic pathways do not take place on their own. Each reaction step is facilitated, or catalyzed, by a protein called an enzyme. Enzymes are important for catalyzing all types of biological reactions—those that require energy as well as those that release energy.

Metabolic pathways

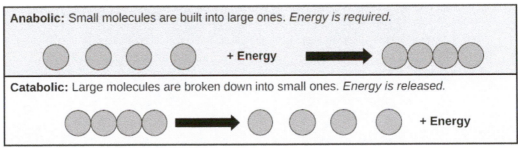

Figure 4.3 Catabolic pathways are those that generate energy by breaking down larger molecules. Anabolic pathways are those that require energy to synthesize larger molecules. Both types of pathways are required for maintaining the cell's energy balance.

Energy

Thermodynamics refers to the study of energy and energy transfer involving physical matter. The matter relevant to a particular case of energy transfer is called a system, and everything outside of that matter is called the surroundings. For instance, when heating a pot of water on the stove, the system includes the stove, the pot, and the water. Energy is transferred within the system (between the stove, pot, and water). There are two types of systems: open and closed. In an open system, energy can be exchanged with its surroundings. The stovetop system is open because heat can be lost to the air. A closed system cannot exchange energy with its surroundings.

Biological organisms are open systems. Energy is exchanged between them and their surroundings as they use energy from the sun to perform photosynthesis or consume energy-storing molecules and release energy to the environment by doing work and releasing heat. Like all things in the physical world, energy is subject to physical laws. The laws of thermodynamics govern the transfer of energy in and among all systems in the universe.

In general, energy is defined as the ability to do work, or to create some kind of change. Energy exists in different forms. For example, electrical energy, light energy, and heat energy are all different types of energy. To appreciate the way energy flows into and out of biological systems, it is important to understand two of the physical laws that govern energy.

Thermodynamics

The first law of thermodynamics states that the total amount of energy in the universe is constant and conserved. In other words, there has always been, and always will be, exactly the same amount of energy in the universe. Energy exists in many different forms. According to the first law of thermodynamics, energy may be transferred from place to place or transformed into different forms, but it cannot be created or destroyed. The transfers and transformations of energy take place around us all the time. Light bulbs transform electrical energy into light and heat energy. Gas stoves transform chemical energy from natural gas into heat energy. Plants perform one of the most biologically useful energy transformations on earth: that of converting the energy of sunlight to chemical energy stored within organic molecules (Figure 4.2). Some examples of energy transformations are shown in Figure 4.4.

The challenge for all living organisms is to obtain energy from their surroundings in forms that they can transfer or transform into usable energy to do work. Living cells have evolved to meet this challenge. Chemical energy stored within organic molecules such as sugars and fats is transferred and transformed through a series of cellular chemical reactions into energy within molecules of ATP. Energy in ATP molecules is easily accessible to do work. Examples of the types of work that cells need to do include building complex molecules, transporting materials, powering the motion of cilia or flagella, and contracting muscle fibers to create movement.

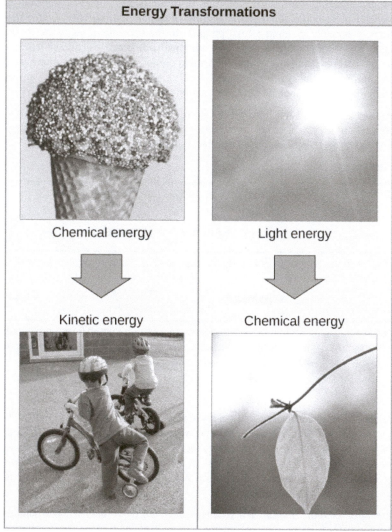

Figure 4.4 Shown are some examples of energy transferred and transformed from one system to another and from one form to another. The food we consume provides our cells with the energy required to carry out bodily functions, just as light energy provides plants with the means to create the chemical energy they need. (credit "ice cream": modification of work by D. Sharon Pruitt; credit "kids": modification of work by Max from Providence; credit "leaf": modification of work by Cory Zanker)

A living cell's primary tasks of obtaining, transforming, and using energy to do work may seem simple. However, the second law of thermodynamics explains why these tasks are harder than they appear. All energy transfers and transformations are never completely efficient. In every energy transfer, some amount of energy is lost in a form that is unusable. In most cases, this form is heat energy. Thermodynamically, **heat energy** is defined as the energy transferred from one system to another that is not work. For example, when a light bulb is turned on, some of the energy being converted from electrical energy into light energy is lost as heat energy. Likewise, some energy is lost as heat energy during cellular metabolic reactions.

An important concept in physical systems is that of order and disorder. The more energy that is lost by a system to its surroundings, the less ordered and more random the system is. Scientists refer to the measure of randomness or disorder within a system as entropy. High entropy means high disorder and low energy. Molecules and chemical reactions have varying entropy as well. For example, entropy increases as molecules at a high concentration in one place diffuse and spread out. The second law of thermodynamics says that energy will always be lost as heat in energy transfers or transformations.

Living things are highly ordered, requiring constant energy input to be maintained in a state of low entropy.

Potential and Kinetic Energy

When an object is in motion, there is energy associated with that object. Think of a wrecking ball. Even a slow-moving wrecking ball can do a great deal of damage to other objects. Energy associated with objects in motion is called **kinetic energy** (Figure 4.5). A speeding bullet, a walking person, and the rapid movement of molecules in the air (which produces heat) all have kinetic energy.

Now what if that same motionless wrecking ball is lifted two stories above ground with a crane? If the suspended wrecking ball is unmoving, is there energy associated with it? The answer is yes. The energy that was required to lift the wrecking ball did not disappear, but is now stored in the wrecking ball by virtue of its position and the force of gravity acting on it. This type of energy is called **potential energy** (Figure 4.5). If the ball were to fall, the potential energy would be transformed into kinetic energy until all of the potential energy was exhausted when the ball rested on the ground. Wrecking balls also swing like a pendulum; through the swing, there is a constant change of potential energy (highest at the top of the swing) to kinetic energy (highest at the bottom of the swing). Other examples of potential energy include the energy of water held behind a dam or a person about to skydive out of an airplane.

Figure 4.5 Still water has potential energy; moving water, such as in a waterfall or a rapidly flowing river, has kinetic energy. (credit "dam": modification of work by "Pascal"/Flickr; credit "waterfall": modification of work by Frank Gualtieri)

Potential energy is not only associated with the location of matter, but also with the structure of matter. Even a spring on the ground has potential energy if it is compressed; so does a rubber band that is pulled taut. On a molecular level, the bonds that hold the atoms of molecules together exist in a particular structure that has potential energy. Remember that anabolic cellular pathways require energy to synthesize complex molecules from simpler ones and catabolic pathways release energy when complex molecules are broken down. The fact that energy can be released by the breakdown of certain chemical bonds implies that those bonds have potential energy. In fact, there is potential energy stored within the bonds of all the food molecules we eat, which is eventually harnessed for use. This is because these bonds can release energy when broken. The type of potential energy that exists within chemical bonds, and is released when those bonds are broken, is called chemical energy. Chemical energy is responsible for providing living cells with energy from food. The release of energy occurs when the molecular bonds within food molecules are broken.

Visit the site (http://openstaxcollege.org/l/simple_pendulu2) and select "Pendulum" from the "Work and Energy" menu to see the shifting kinetic and potential energy of a pendulum in motion.

Free and Activation Energy

After learning that chemical reactions release energy when energy-storing bonds are broken, an important next question is the following: How is the energy associated with these chemical reactions quantified and expressed? How can the energy released from one reaction be compared to that of another reaction? A measurement of free energy is used to quantify these energy transfers. Recall that according to the second law of thermodynamics, all energy transfers involve the loss of some amount of energy in an unusable form such as heat. Free energy specifically refers to the energy associated with a chemical reaction that is available after the losses are accounted for. In other words, free energy is usable energy, or energy that is available to do work.

If energy is released during a chemical reaction, then the change in free energy, signified as ΔG (delta G) will be a negative number. A negative change in free energy also means that the products of the reaction have less free energy than the reactants, because they release some free energy during the reaction. Reactions that have a negative change in free energy and consequently release free energy are called **exergonic reactions**. Think: *ex*ergonic means energy is *ex*iting the system. These reactions are also referred to as spontaneous reactions, and their products have less stored energy than the reactants. An important distinction must be drawn between the term spontaneous and the idea of a chemical reaction occurring immediately. Contrary to the everyday use of the term, a spontaneous reaction is not one that suddenly or quickly occurs. The rusting of iron is an example of a spontaneous reaction that occurs slowly, little by little, over time.

If a chemical reaction absorbs energy rather than releases energy on balance, then the ΔG for that reaction will be a positive value. In this case, the products have more free energy than the reactants. Thus, the products of these reactions can be thought of as energy-storing molecules. These chemical reactions are called **endergonic reactions** and they are non-spontaneous. An endergonic reaction will not take place on its own without the addition of free energy.

art CONNECTION

(a) (b)

(c) (d)

Figure 4.6 Shown are some examples of endergonic processes (ones that require energy) and exergonic processes (ones that release energy). (credit a: modification of work by Natalie Maynor; credit b: modification of work by USDA; credit c: modification of work by Cory Zanker; credit d: modification of work by Harry Malsch)

Look at each of the processes shown and decide if it is endergonic or exergonic.

There is another important concept that must be considered regarding endergonic and exergonic reactions. Exergonic reactions require a small amount of energy input to get going, before they can proceed with their energy-releasing steps. These reactions have a net release of energy, but still require some energy input in the beginning. This small amount of energy input necessary for all chemical reactions to occur is called the **activation energy**.

Watch an animation (http://openstaxcollege.org/l/energy_reactio2) of the move from free energy to transition state of the reaction.

Enzymes

A substance that helps a chemical reaction to occur is called a catalyst, and the molecules that catalyze biochemical reactions are called **enzymes**. Most enzymes are proteins and perform the critical task of lowering the activation energies of chemical reactions inside the cell. Most of the reactions critical to a living cell happen too slowly at normal temperatures to be of any use to the cell. Without enzymes to speed up these reactions, life could not persist. Enzymes do this by binding to the reactant molecules and holding them in such a way as to make the chemical bond-breaking and -forming processes take place more easily. It is important to remember that enzymes do not change whether a reaction is exergonic (spontaneous) or endergonic. This is because they do not change the free energy of the reactants or products. They only reduce the activation energy required for the reaction to go forward (Figure 4.7). In addition, an enzyme itself is unchanged by the reaction it catalyzes. Once one reaction has been catalyzed, the enzyme is able to participate in other reactions.

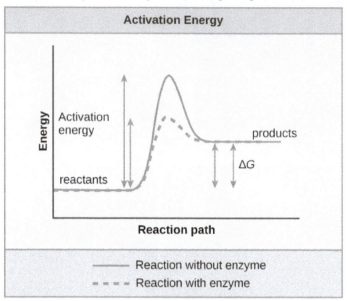

Figure 4.7 Enzymes lower the activation energy of the reaction but do not change the free energy of the reaction.

The chemical reactants to which an enzyme binds are called the enzyme's **substrates**. There may be one or more substrates, depending on the particular chemical reaction. In some reactions, a single reactant substrate is broken down into multiple products. In others, two substrates may come together to create one larger molecule. Two reactants might also enter a reaction and both become modified, but they leave the reaction as two products. The location within the enzyme where the substrate binds is called the enzyme's **active site**. The active site is where the "action" happens. Since enzymes are proteins, there is a unique combination of amino acid side chains within the active site. Each side chain is characterized by different properties. They can be large or small, weakly acidic or basic, hydrophilic or hydrophobic, positively or negatively charged, or neutral. The unique combination of side chains creates a very specific chemical environment within the active site. This specific environment is suited to bind to one specific chemical substrate (or substrates).

Active sites are subject to influences of the local environment. Increasing the environmental temperature generally increases reaction rates, enzyme-catalyzed or otherwise. However, temperatures outside of an optimal range reduce the rate at which an enzyme catalyzes a reaction. Hot temperatures will eventually cause enzymes to denature, an irreversible change in the three-dimensional shape and therefore the function of the enzyme. Enzymes are also suited to function best within a certain pH and salt concentration range, and, as with temperature, extreme pH, and salt concentrations can cause enzymes to denature.

For many years, scientists thought that enzyme-substrate binding took place in a simple "lock and key" fashion. This model asserted that the enzyme and substrate fit together perfectly in one instantaneous step. However, current research supports a model called induced fit (Figure 4.8). The induced-fit model expands on the lock-and-key model by describing a more dynamic binding between enzyme and substrate. As the enzyme and substrate come together, their interaction causes a mild shift in the enzyme's structure that forms an ideal binding arrangement between enzyme and substrate.

View an animation (http://openstaxcollege.org/l/hexokinase2) of induced fit.

When an enzyme binds its substrate, an enzyme-substrate complex is formed. This complex lowers the activation energy of the reaction and promotes its rapid progression in one of multiple possible ways. On a basic level, enzymes promote chemical reactions that involve more than one substrate by bringing the substrates together in an optimal orientation for reaction. Another way in which enzymes promote the reaction of their substrates is by creating an optimal environment within the active site for the reaction to occur. The chemical properties that emerge from the particular arrangement of amino acid R groups within an active site create the perfect environment for an enzyme's specific substrates to react.

The enzyme-substrate complex can also lower activation energy by compromising the bond structure so that it is easier to break. Finally, enzymes can also lower activation energies by taking part in the chemical reaction itself. In these cases, it is important to remember that the enzyme will always return to its original state by the completion of the reaction. One of the hallmark properties of enzymes is that they remain ultimately unchanged by the reactions they catalyze. After an enzyme has catalyzed a reaction, it releases its product(s) and can catalyze a new reaction.

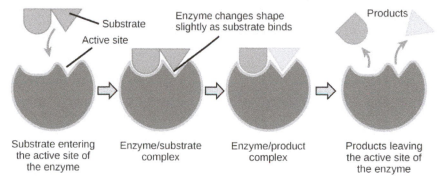

Figure 4.8 The induced-fit model is an adjustment to the lock-and-key model and explains how enzymes and substrates undergo dynamic modifications during the transition state to increase the affinity of the substrate for the active site.

It would seem ideal to have a scenario in which all of an organism's enzymes existed in abundant supply and functioned optimally under all cellular conditions, in all cells, at all times. However, a variety of mechanisms ensures that this does not happen. Cellular needs and conditions constantly vary from cell to cell, and change within individual cells over time. The required enzymes of stomach cells differ from those of fat storage cells, skin cells, blood cells, and nerve cells. Furthermore, a digestive organ cell works much harder to process and break down nutrients during the time that closely follows a meal compared with many hours after a meal. As these cellular demands and conditions vary, so must the amounts and functionality of different enzymes.

Since the rates of biochemical reactions are controlled by activation energy, and enzymes lower and determine activation energies for chemical reactions, the relative amounts and functioning of the variety of enzymes within a cell ultimately determine which reactions will proceed and at what rates. This determination is tightly controlled in cells. In certain cellular environments, enzyme activity is partly controlled by environmental factors like pH, temperature, salt concentration, and, in some cases, cofactors or coenzymes.

Enzymes can also be regulated in ways that either promote or reduce enzyme activity. There are many kinds of molecules that inhibit or promote enzyme function, and various mechanisms by which they do so. In some cases of enzyme inhibition, an inhibitor molecule is similar enough to a substrate that it can bind to the active site and simply block the substrate from binding. When this happens, the enzyme is inhibited through **competitive inhibition**, because an inhibitor molecule competes with the substrate for binding to the active site.

On the other hand, in **noncompetitive inhibition**, an inhibitor molecule binds to the enzyme in a location other than the active site, called an allosteric site, but still manages to block substrate binding to the active site. Some inhibitor molecules bind to enzymes in a location where their binding induces a conformational change that reduces the affinity of the enzyme for its substrate. This type of inhibition is called **allosteric inhibition** (Figure 4.9). Most allosterically regulated enzymes are made up of more than one polypeptide, meaning that they have more than one protein subunit. When an allosteric inhibitor binds to a region on an enzyme, all active sites on the protein subunits are changed slightly such that they bind their substrates with less efficiency. There are allosteric activators as well as inhibitors. Allosteric activators bind to locations on an enzyme away from the active site, inducing a conformational change that increases the affinity of the enzyme's active site(s) for its substrate(s) (Figure 4.9).

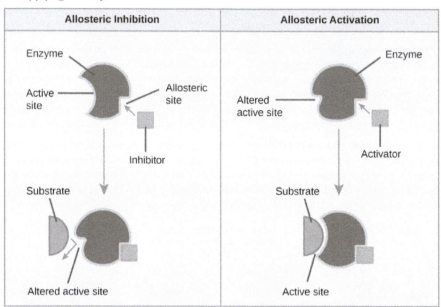

Figure 4.9 Allosteric inhibition works by indirectly inducing a conformational change to the active site such that the substrate no longer fits. In contrast, in allosteric activation, the activator molecule modifies the shape of the active site to allow a better fit of the substrate.

Pharmaceutical Drug Developer

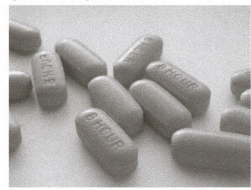

Figure 4.10 Have you ever wondered how pharmaceutical drugs are developed? (credit: Deborah Austin)

Enzymes are key components of metabolic pathways. Understanding how enzymes work and how they can be regulated are key principles behind the development of many of the pharmaceutical drugs on the market today. Biologists working in this field collaborate with other scientists to design drugs (Figure 4.10).

Consider statins for example—statins is the name given to one class of drugs that can reduce cholesterol levels. These compounds are inhibitors of the enzyme HMG-CoA reductase, which is the enzyme that synthesizes cholesterol from lipids in the body. By inhibiting this enzyme, the level of cholesterol synthesized in the body can be reduced. Similarly, acetaminophen, popularly marketed under the brand name Tylenol, is an inhibitor of the enzyme cyclooxygenase. While it is used to provide relief from fever and inflammation (pain), its mechanism of action is still not completely understood.

How are drugs discovered? One of the biggest challenges in drug discovery is identifying a drug target. A drug target is a molecule that is literally the target of the drug. In the case of statins, HMG-CoA reductase is the drug target. Drug targets are identified through painstaking research in the laboratory. Identifying the target alone is not enough; scientists also need to know how the target acts inside the cell and which reactions go awry in the case of disease. Once the target and the pathway are identified, then the actual process of drug design begins. In this stage, chemists and biologists work together to design and synthesize molecules that can block or activate a particular reaction. However, this is only the beginning: If and when a drug prototype is successful in performing its function, then it is subjected to many tests from in vitro experiments to clinical trials before it can get approval from the U.S. Food and Drug Administration to be on the market.

Many enzymes do not work optimally, or even at all, unless bound to other specific non-protein helper molecules. They may bond either temporarily through ionic or hydrogen bonds, or permanently through stronger covalent bonds. Binding to these molecules promotes optimal shape and function of their respective enzymes. Two examples of these types of helper molecules are cofactors and coenzymes. Cofactors are inorganic ions such as ions of iron and magnesium. Coenzymes are organic helper molecules, those with a basic atomic structure made up of carbon and hydrogen. Like enzymes, these molecules participate in reactions without being changed themselves and are ultimately recycled and reused. Vitamins are the source of coenzymes. Some vitamins are the precursors of coenzymes and others act directly as coenzymes. Vitamin C is a direct coenzyme for multiple enzymes that take part in building the important connective tissue, collagen. Therefore, enzyme function is, in part, regulated by the abundance of various cofactors and coenzymes, which may be supplied by an organism's diet or, in some cases, produced by the organism.

Feedback Inhibition in Metabolic Pathways

Molecules can regulate enzyme function in many ways. The major question remains, however: What are these molecules and where do they come from? Some are cofactors and coenzymes, as you have learned. What other molecules in the cell provide enzymatic regulation such as allosteric modulation, and competitive and non-competitive inhibition? Perhaps the most relevant sources of regulatory molecules, with respect to enzymatic cellular metabolism, are the products of the cellular metabolic reactions themselves. In a most efficient and elegant way, cells have evolved to use the products of their own reactions for feedback inhibition of enzyme activity. **Feedback inhibition** involves the use of a reaction product to regulate its own further production (Figure 4.11). The cell responds to an abundance of the products by slowing down production during anabolic or catabolic reactions. Such reaction products may inhibit the enzymes that catalyzed their production through the mechanisms described above.

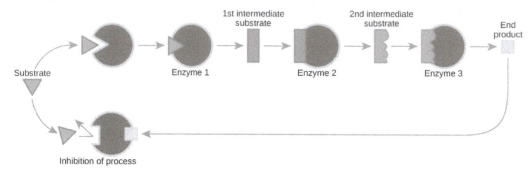

Figure 4.11 Metabolic pathways are a series of reactions catalyzed by multiple enzymes. Feedback inhibition, where the end product of the pathway inhibits an upstream process, is an important regulatory mechanism in cells.

The production of both amino acids and nucleotides is controlled through feedback inhibition. Additionally, ATP is an allosteric regulator of some of the enzymes involved in the catabolic breakdown of sugar, the process that creates ATP. In this way, when ATP is in abundant supply, the cell can prevent the production of ATP. On the other hand, ADP serves as a positive allosteric regulator (an allosteric activator) for some of the same enzymes that are inhibited by ATP. Thus, when relative levels of ADP are high compared to ATP, the cell is triggered to produce more ATP through sugar catabolism.

4.2 | Glycolysis

By the end of this section, you will be able to:

- Explain how ATP is used by the cell as an energy source
- Describe the overall result in terms of molecules produced of the breakdown of glucose by glycolysis

Even exergonic, energy-releasing reactions require a small amount of activation energy to proceed. However, consider endergonic reactions, which require much more energy input because their products have more free energy than their reactants. Within the cell, where does energy to power such reactions come from? The answer lies with an energy-supplying molecule called adenosine triphosphate, or **ATP**. ATP is a small, relatively simple molecule, but within its bonds contains the potential for a quick burst of energy that can be harnessed to perform cellular work. This molecule can be thought of as the primary energy currency of cells in the same way that money is the currency that people exchange for things they need. ATP is used to power the majority of energy-requiring cellular reactions.

ATP in Living Systems

A living cell cannot store significant amounts of free energy. Excess free energy would result in an increase of heat in the cell, which would denature enzymes and other proteins, and thus destroy the cell. Rather, a cell must be able to store energy safely and release it for use only as needed. Living cells accomplish this using ATP, which can be used to fill any energy need of the cell. How? It functions as a rechargeable battery.

When ATP is broken down, usually by the removal of its terminal phosphate group, energy is released. This energy is used to do work by the cell, usually by the binding of the released phosphate to another molecule, thus activating it. For example, in the mechanical work of muscle contraction, ATP supplies energy to move the contractile muscle proteins.

ATP Structure and Function

At the heart of ATP is a molecule of adenosine monophosphate (AMP), which is composed of an adenine molecule bonded to both a ribose molecule and a single phosphate group (Figure 4.12). Ribose is a five-carbon sugar found in RNA and AMP is one of the nucleotides in RNA. The addition of a second phosphate group to this core molecule results in adenosine diphosphate (ADP); the addition of a third phosphate group forms adenosine triphosphate (ATP).

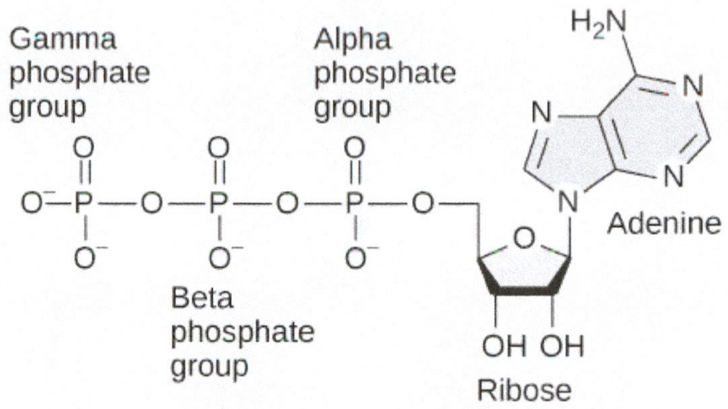

Figure 4.12 The structure of ATP shows the basic components of a two-ring adenine, five-carbon ribose, and three phosphate groups.

The addition of a phosphate group to a molecule requires a high amount of energy and results in a high-energy bond. Phosphate groups are negatively charged and thus repel one another when they are arranged in series, as they are in ADP and ATP. This repulsion makes the ADP and ATP molecules inherently unstable. The release of one or two phosphate groups from ATP, a process called hydrolysis, releases energy.

Glycolysis

You have read that nearly all of the energy used by living things comes to them in the bonds of the sugar, glucose. **Glycolysis** is the first step in the breakdown of glucose to extract energy for cell metabolism. Many living organisms carry out glycolysis as part of their metabolism. Glycolysis takes place in the cytoplasm of most prokaryotic and all eukaryotic cells.

Glycolysis begins with the six-carbon, ring-shaped structure of a single glucose molecule and ends with two molecules of a three-carbon sugar called pyruvate. Glycolysis consists of two distinct phases. In the first part of the glycolysis pathway, energy is used to make adjustments so that the six-carbon sugar molecule can be split evenly into two three-carbon pyruvate molecules. In the second part of glycolysis, ATP and nicotinamide-adenine dinucleotide (NADH) are produced (Figure 4.13).

If the cell cannot catabolize the pyruvate molecules further, it will harvest only two ATP molecules from one molecule of glucose. For example, mature mammalian red blood cells are only capable of glycolysis, which is their sole source of ATP. If glycolysis is interrupted, these cells would eventually die.

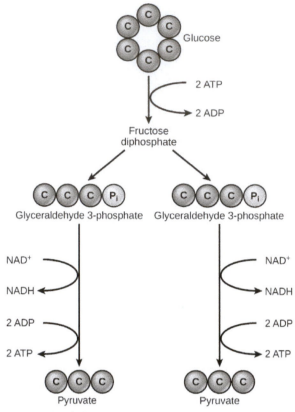

Figure 4.13 In glycolysis, a glucose molecule is converted into two pyruvate molecules.

4.3 | Citric Acid Cycle and Oxidative Phosphorylation

By the end of this section, you will be able to:

- Describe the location of the citric acid cycle and oxidative phosphorylation in the cell
- Describe the overall outcome of the citric acid cycle and oxidative phosphorylation in terms of the products of each
- Describe the relationships of glycolysis, the citric acid cycle, and oxidative phosphorylation in terms of their inputs and outputs.

The Citric Acid Cycle

In eukaryotic cells, the pyruvate molecules produced at the end of glycolysis are transported into mitochondria, which are sites of cellular respiration. If oxygen is available, aerobic respiration will go forward. In mitochondria, pyruvate will be transformed into a two-carbon acetyl group (by removing a molecule of carbon dioxide) that will be picked up by a carrier compound called coenzyme A (CoA), which is made from vitamin B5. The resulting compound is called **acetyl CoA**. (Figure 4.14). Acetyl CoA can be used in a variety of ways by the cell, but its major function is to deliver the acetyl group derived from pyruvate to the next pathway in glucose catabolism.

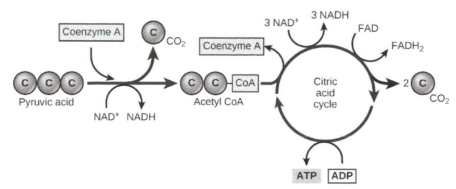

Figure 4.14 Pyruvate is converted into acetyl-CoA before entering the citric acid cycle.

Like the conversion of pyruvate to acetyl CoA, the **citric acid cycle** in eukaryotic cells takes place in the matrix of the mitochondria. Unlike glycolysis, the citric acid cycle is a closed loop: The last part of the pathway regenerates the compound used in the first step. The eight steps of the cycle are a series of chemical reactions that produces two carbon dioxide molecules, one ATP molecule (or an equivalent), and reduced forms (NADH and $FADH_2$) of NAD^+ and FAD^+, important coenzymes in the cell. Part of this is considered an aerobic pathway (oxygen-requiring) because the NADH and $FADH_2$ produced must transfer their electrons to the next pathway in the system, which will use oxygen. If oxygen is not present, this transfer does not occur.

Two carbon atoms come into the citric acid cycle from each acetyl group. Two carbon dioxide molecules are released on each turn of the cycle; however, these do not contain the same carbon atoms contributed by the acetyl group on that turn of the pathway. The two acetyl-carbon atoms will eventually be released on later turns of the cycle; in this way, all six carbon atoms from the original glucose molecule will be eventually released as carbon dioxide. It takes two turns of the cycle to process the equivalent of one glucose molecule. Each turn of the cycle forms three high-energy NADH molecules and one high-energy $FADH_2$ molecule. These high-energy carriers will connect with the last portion of aerobic respiration to produce ATP molecules. One ATP (or an equivalent) is also made in each cycle. Several of the intermediate compounds in the citric acid cycle can be used in synthesizing non-essential amino acids; therefore, the cycle is both anabolic and catabolic.

Oxidative Phosphorylation

You have just read about two pathways in glucose catabolism—glycolysis and the citric acid cycle—that generate ATP. Most of the ATP generated during the aerobic catabolism of glucose, however, is not generated directly from these pathways. Rather, it derives from a process that begins with passing electrons through a series of chemical reactions to a final electron acceptor, oxygen. These reactions take place in specialized protein complexes located in the inner membrane of the mitochondria of eukaryotic organisms and on the inner part of the cell membrane of prokaryotic organisms. The energy of the electrons is harvested and used to generate a electrochemical gradient across the inner mitochondrial membrane. The potential energy of this gradient is used to generate ATP. The entirety of this process is called **oxidative phosphorylation**.

The electron transport chain (Figure 4.15a) is the last component of aerobic respiration and is the only part of metabolism that uses atmospheric oxygen. Oxygen continuously diffuses into plants for this purpose. In animals, oxygen enters the body through the respiratory system. Electron transport is a series of chemical reactions that resembles a bucket brigade in that electrons are passed rapidly from one component to the next, to the endpoint of the chain where oxygen is the final electron acceptor and water is produced. There are four complexes composed of proteins, labeled I through IV in Figure 4.15c, and the aggregation of these four complexes, together with associated mobile, accessory electron carriers, is called the **electron transport chain**. The electron transport chain is present in multiple copies in the inner mitochondrial membrane of eukaryotes and in the plasma membrane of prokaryotes. In each transfer of an electron through the electron transport chain, the electron loses energy, but with some transfers, the energy is stored as potential energy by using it to pump hydrogen ions across the inner mitochondrial membrane into the intermembrane space, creating an electrochemical gradient.

art CONNECTION

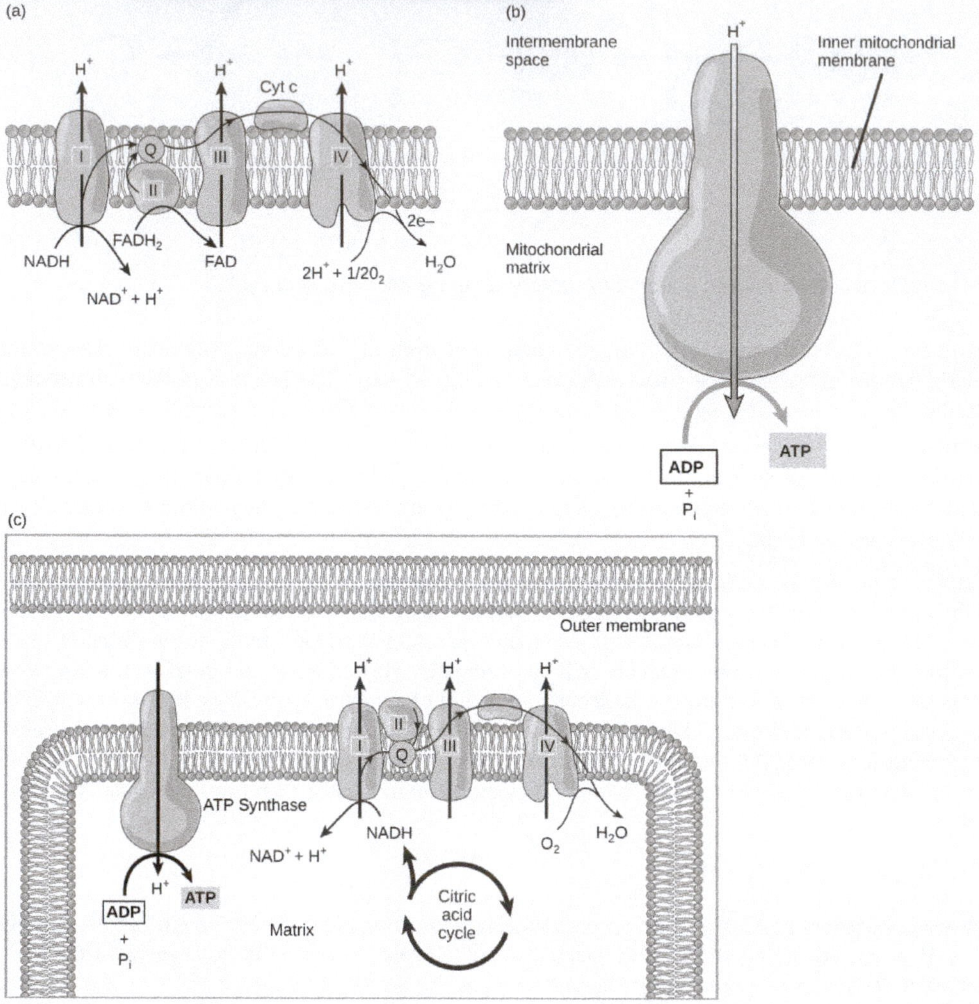

Figure 4.15 (a) The electron transport chain is a set of molecules that supports a series of oxidation-reduction reactions. (b) ATP synthase is a complex, molecular machine that uses an H^+ gradient to regenerate ATP from ADP. (c) Chemiosmosis relies on the potential energy provided by the H^+ gradient across the membrane.

Cyanide inhibits cytochrome c oxidase, a component of the electron transport chain. If cyanide poisoning occurs, would you expect the pH of the intermembrane space to increase or decrease? What affect would cyanide have on ATP synthesis?

Electrons from NADH and $FADH_2$ are passed to protein complexes in the electron transport chain. As they are passed from one complex to another (there are a total of four), the electrons lose energy, and some of that energy is used to pump hydrogen ions from the mitochondrial matrix into the intermembrane space. In the fourth protein complex, the electrons are accepted by oxygen, the terminal acceptor. The oxygen with its extra electrons then combines with two hydrogen ions, further enhancing the electrochemical gradient, to form water. If there were no oxygen present in the mitochondrion, the electrons could not be removed from the system, and the entire electron transport chain would back up and stop. The mitochondria would be unable to generate new ATP in this way, and the cell would ultimately die from lack of energy. This is the reason we must breathe to draw in new oxygen.

In the electron transport chain, the free energy from the series of reactions just described is used to pump hydrogen ions across the membrane. The uneven distribution of H^+ ions across the membrane establishes an electrochemical gradient, owing to the H^+ ions' positive charge and their higher concentration on one side of the membrane.

Download for free at https://openstax.org/details/books/concepts-biology

Hydrogen ions diffuse through the inner membrane through an integral membrane protein called **ATP synthase** (Figure 4.15b). This complex protein acts as a tiny generator, turned by the force of the hydrogen ions diffusing through it, down their electrochemical gradient from the intermembrane space, where there are many mutually repelling hydrogen ions to the matrix, where there are few. The turning of the parts of this molecular machine regenerate ATP from ADP. This flow of hydrogen ions across the membrane through ATP synthase is called **chemiosmosis**.

Chemiosmosis (Figure 4.15c) is used to generate 90 percent of the ATP made during aerobic glucose catabolism. The result of the reactions is the production of ATP from the energy of the electrons removed from hydrogen atoms. These atoms were originally part of a glucose molecule. At the end of the electron transport system, the electrons are used to reduce an oxygen molecule to oxygen ions. The extra electrons on the oxygen ions attract hydrogen ions (protons) from the surrounding medium, and water is formed. The electron transport chain and the production of ATP through chemiosmosis are collectively called oxidative phosphorylation.

ATP Yield

The number of ATP molecules generated from the catabolism of glucose varies. For example, the number of hydrogen ions that the electron transport chain complexes can pump through the membrane varies between species. Another source of variance stems from the shuttle of electrons across the mitochondrial membrane. The NADH generated from glycolysis cannot easily enter mitochondria. Thus, electrons are picked up on the inside of the mitochondria by either NAD^+ or FAD^+. Fewer ATP molecules are generated when FAD^+ acts as a carrier. NAD^+ is used as the electron transporter in the liver and FAD^+ in the brain, so ATP yield depends on the tissue being considered.

Another factor that affects the yield of ATP molecules generated from glucose is that intermediate compounds in these pathways are used for other purposes. Glucose catabolism connects with the pathways that build or break down all other biochemical compounds in cells, and the result is somewhat messier than the ideal situations described thus far. For example, sugars other than glucose are fed into the glycolytic pathway for energy extraction. Other molecules that would otherwise be used to harvest energy in glycolysis or the citric acid cycle may be removed to form nucleic acids, amino acids, lipids, or other compounds. Overall, in living systems, these pathways of glucose catabolism extract about 34 percent of the energy contained in glucose.

Mitochondrial Disease Physician

What happens when the critical reactions of cellular respiration do not proceed correctly? Mitochondrial diseases are genetic disorders of metabolism. Mitochondrial disorders can arise from mutations in nuclear or mitochondrial DNA, and they result in the production of less energy than is normal in body cells. Symptoms of mitochondrial diseases can include muscle weakness, lack of coordination, stroke-like episodes, and loss of vision and hearing. Most affected people are diagnosed in childhood, although there are some adult-onset diseases. Identifying and treating mitochondrial disorders is a specialized medical field. The educational preparation for this profession requires a college education, followed by medical school with a specialization in medical genetics. Medical geneticists can be board certified by the American Board of Medical Genetics and go on to become associated with professional organizations devoted to the study of mitochondrial disease, such as the Mitochondrial Medicine Society and the Society for Inherited Metabolic Disease.

4.4 | Fermentation

By the end of this section, you will be able to:

- Discuss the fundamental difference between anaerobic cellular respiration and fermentation
- Describe the type of fermentation that readily occurs in animal cells and the conditions that initiate that fermentation

In aerobic respiration, the final electron acceptor is an oxygen molecule, O_2. If aerobic respiration occurs, then ATP will be produced using the energy of the high-energy electrons carried by NADH or $FADH_2$ to the electron transport chain. If aerobic respiration does not occur, NADH must be reoxidized to NAD^+ for reuse as an electron carrier for glycolysis to continue. How is this done? Some living systems use an organic molecule as the final electron acceptor. Processes that use an organic molecule to regenerate NAD^+ from NADH are collectively referred to as **fermentation**. In contrast, some living systems use an inorganic molecule (other than oxygen) as a final electron acceptor to regenerate NAD^+; both methods are anaerobic (do not require oxygen) to achieve NAD^+ regeneration and enable organisms to convert energy for their use in the absence of oxygen.

Lactic Acid Fermentation

The fermentation method used by animals and some bacteria like those in yogurt is lactic acid fermentation (Figure 4.16). This occurs routinely in mammalian red blood cells and in skeletal muscle that has insufficient oxygen supply to allow aerobic respiration to continue (that is, in muscles used to the point of fatigue). In muscles, lactic acid produced by fermentation must be removed by the blood circulation and brought to the liver for further metabolism. The chemical reaction of lactic acid fermentation is the following:

$$\text{Pyruvic acid} + \text{NADH} \leftrightarrow \text{lactic acid} + \text{NAD}^+$$

The enzyme that catalyzes this reaction is lactate dehydrogenase. The reaction can proceed in either direction, but the left-to-right reaction is inhibited by acidic conditions. This lactic acid build-up causes muscle stiffness and fatigue. Once the lactic acid has been removed from the muscle and is circulated to the liver, it can be converted back to pyruvic acid and further catabolized for energy.

a r t CONNECTION

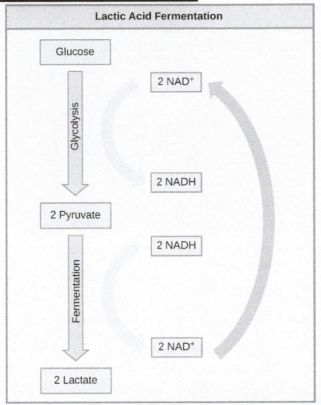

Figure 4.16 Lactic acid fermentation is common in muscles that have become exhausted by use.

Tremetol, a metabolic poison found in white snake root plant, prevents the metabolism of lactate. When cows eat this plant, Tremetol is concentrated in the milk. Humans who consume the milk become ill. Symptoms of this disease, which include vomiting, abdominal pain, and tremors, become worse after exercise. Why do you think this is the case?

Alcohol Fermentation

Another familiar fermentation process is alcohol fermentation (Figure 4.17), which produces ethanol, an alcohol. The alcohol fermentation reaction is the following:

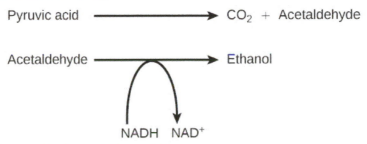

Figure 4.17 The reaction resulting in alcohol fermentation is shown.

In the first reaction, a carboxyl group is removed from pyruvic acid, releasing carbon dioxide as a gas. The loss of carbon dioxide reduces the molecule by one carbon atom, making acetaldehyde. The second reaction removes an electron from NADH, forming NAD^+ and producing ethanol from the acetaldehyde, which accepts the electron. The fermentation of pyruvic acid by yeast produces the ethanol found in alcoholic beverages (Figure 4.18). If the carbon dioxide produced by the reaction is not vented from the fermentation chamber, for example in beer and sparkling wines, it remains dissolved in the medium until the pressure is released. Ethanol above 12 percent is toxic to yeast, so natural levels of alcohol in wine occur at a maximum of 12 percent.

Figure 4.18 Fermentation of grape juice to make wine produces CO_2 as a byproduct. Fermentation tanks have valves so that pressure inside the tanks can be released.

Anaerobic Cellular Respiration

Certain prokaryotes, including some species of bacteria and Archaea, use anaerobic respiration. For example, the group of Archaea called methanogens reduces carbon dioxide to methane to oxidize NADH. These microorganisms are found in soil and in the digestive tracts of ruminants, such as cows and sheep. Similarly, sulfate-reducing bacteria and Archaea, most of which are anaerobic (Figure 4.19), reduce sulfate to hydrogen sulfide to regenerate NAD^+ from NADH.

Figure 4.19 The green color seen in these coastal waters is from an eruption of hydrogen sulfide. Anaerobic, sulfate-reducing bacteria release hydrogen sulfide gas as they decompose algae in the water. (credit: NASA image courtesy Jeff Schmaltz, MODIS Land Rapid Response Team at NASA GSFC)

Visit this site (http://openstaxcollege.org/l/fermentation2) to see anaerobic cellular respiration in action.

Other fermentation methods occur in bacteria. Many prokaryotes are facultatively anaerobic. This means that they can switch between aerobic respiration and fermentation, depending on the availability of oxygen. Certain prokaryotes, like *Clostridia* bacteria, are obligate anaerobes. Obligate anaerobes live and grow in the absence of molecular oxygen. Oxygen is a poison to these microorganisms and kills them upon exposure. It should be noted that all forms of fermentation, except lactic acid fermentation, produce gas. The production of particular types of gas is used as an indicator of the fermentation of specific carbohydrates, which plays a role in the laboratory identification of the bacteria. The various methods of fermentation are used by different organisms to ensure an adequate supply of NAD^+ for the sixth step in glycolysis. Without these pathways, that step would not occur, and no ATP would be harvested from the breakdown of glucose.

4.5 | Connections to Other Metabolic Pathways

By the end of this section, you will be able to:

- Discuss the way in which carbohydrate metabolic pathways, glycolysis, and the citric acid cycle interrelate with protein and lipid metabolic pathways
- Explain why metabolic pathways are not considered closed systems

You have learned about the catabolism of glucose, which provides energy to living cells. But living things consume more than just glucose for food. How does a turkey sandwich, which contains protein, provide energy to your cells? This happens because all of the catabolic pathways for carbohydrates, proteins, and lipids eventually connect into glycolysis and the citric acid cycle pathways (Figure 4.20). Metabolic pathways should be thought of as porous—that is, substances enter from other pathways, and other substances leave for other pathways. These pathways are not closed systems. Many of the products in a particular pathway are reactants in other pathways.

Connections of Other Sugars to Glucose Metabolism

Glycogen, a polymer of glucose, is a short-term energy storage molecule in animals. When there is adequate ATP present, excess glucose is converted into glycogen for storage. Glycogen is made and stored in the liver and muscle. Glycogen will be taken out of storage if blood sugar levels drop. The presence of glycogen in muscle cells as a source of glucose allows ATP to be produced for a longer time during exercise.

Sucrose is a disaccharide made from glucose and fructose bonded together. Sucrose is broken down in the small intestine, and the glucose and fructose are absorbed separately. Fructose is one of the three dietary monosaccharides, along with glucose and galactose (which is part of milk sugar, the disaccharide lactose), that are absorbed directly into the bloodstream during digestion. The catabolism of both fructose and galactose produces the same number of ATP molecules as glucose.

Connections of Proteins to Glucose Metabolism

Proteins are broken down by a variety of enzymes in cells. Most of the time, amino acids are recycled into new proteins. If there are excess amino acids, however, or if the body is in a state of famine, some amino acids will be shunted into pathways of glucose catabolism. Each amino acid must have its amino group removed prior to entry into these pathways. The amino group is converted into ammonia. In mammals, the liver synthesizes urea from two ammonia molecules and a carbon dioxide molecule. Thus, urea is the principal waste product in mammals from the nitrogen originating in amino acids, and it leaves the body in urine.

Connections of Lipids to Glucose Metabolism

The lipids that are connected to the glucose pathways are cholesterol and triglycerides. Cholesterol is a lipid that contributes to cell membrane flexibility and is a precursor of steroid hormones. The synthesis of cholesterol starts with acetyl CoA and proceeds in only one direction. The process cannot be reversed, and ATP is not produced.

Triglycerides are a form of long-term energy storage in animals. Triglycerides store about twice as much energy as carbohydrates. Triglycerides are made of glycerol and three fatty acids. Animals can make most of the fatty acids they need. Triglycerides can be both made and broken down through parts of the glucose catabolism pathways. Glycerol can be phosphorylated and proceeds through glycolysis. Fatty acids are broken into two-carbon units that enter the citric acid cycle.

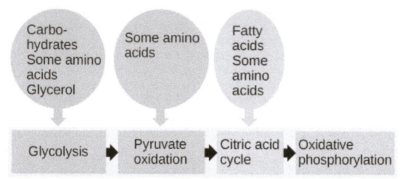

Figure 4.20 Glycogen from the liver and muscles, together with fats, can feed into the catabolic pathways for carbohydrates.

e)olution IN ACTION

Pathways of Photosynthesis and Cellular Metabolism

Photosynthesis and cellular metabolism consist of several very complex pathways. It is generally thought that the first cells arose in an aqueous environment—a "soup" of nutrients. If these cells reproduced successfully and their numbers climbed steadily, it follows that the cells would begin to deplete the nutrients from the medium in which they lived, as they shifted the nutrients into their own cells. This hypothetical situation would have resulted in natural selection favoring those organisms that could exist by using the nutrients that remained in their environment and by manipulating these nutrients into materials that they could use to survive. Additionally, selection would favor those organisms that could extract maximal value from the available nutrients.

An early form of photosynthesis developed that harnessed the sun's energy using compounds other than water as a source of hydrogen atoms, but this pathway did not produce free oxygen. It is thought that glycolysis developed prior to this time and could take advantage of simple sugars being produced, but these reactions were not able to fully extract the energy stored in the carbohydrates. A later form of photosynthesis used water as a source of hydrogen ions and generated free oxygen. Over time, the atmosphere became oxygenated. Living things adapted to exploit this new atmosphere and allowed respiration as we know it to evolve. When the full process of photosynthesis as we know it developed and the atmosphere became oxygenated, cells were finally able to use the oxygen expelled by photosynthesis to extract more energy from the sugar molecules using the citric acid cycle.

KEY TERMS

acetyl CoA the combination of an acetyl group derived from pyruvic acid and coenzyme A which is made from pantothenic acid (a B-group vitamin)

activation energy the amount of initial energy necessary for reactions to occur

active site a specific region on the enzyme where the substrate binds

allosteric inhibition the mechanism for inhibiting enzyme action in which a regulatory molecule binds to a second site (not the active site) and initiates a conformation change in the active site, preventing binding with the substrate

anabolic describes the pathway that requires a net energy input to synthesize complex molecules from simpler ones

anaerobic cellular respiration the use of an electron acceptor other than oxygen to complete metabolism using electron transport-based chemiosmosis

ATP (also, adenosine triphosphate) the cell's energy currency

ATP synthase a membrane-embedded protein complex that regenerates ATP from ADP with energy from protons diffusing through it

bioenergetics the concept of energy flow through living systems

catabolic describes the pathway in which complex molecules are broken down into simpler ones, yielding energy as an additional product of the reaction

chemiosmosis the movement of hydrogen ions down their electrochemical gradient across a membrane through ATP synthase to generate ATP

citric acid cycle a series of enzyme-catalyzed chemical reactions of central importance in all living cells that harvests the energy in carbon-carbon bonds of sugar molecules to generate ATP; the citric acid cycle is an aerobic metabolic pathway because it requires oxygen in later reactions to proceed

competitive inhibition a general mechanism of enzyme activity regulation in which a molecule other than the enzyme's substrate is able to bind the active site and prevent the substrate itself from binding, thus inhibiting the overall rate of reaction for the enzyme

electron transport chain a series of four large, multi-protein complexes embedded in the inner mitochondrial membrane that accepts electrons from donor compounds and harvests energy from a series of chemical reactions to generate a hydrogen ion gradient across the membrane

endergonic describes a chemical reaction that results in products that store more chemical potential energy than the reactants

enzyme a molecule that catalyzes a biochemical reaction

exergonic describes a chemical reaction that results in products with less chemical potential energy than the reactants, plus the release of free energy

feedback inhibition a mechanism of enzyme activity regulation in which the product of a reaction or the final product of a series of sequential reactions inhibits an enzyme for an earlier step in the reaction series

fermentation the steps that follow the partial oxidation of glucose via glycolysis to regenerate NAD^+; occurs in the absence of oxygen and uses an organic compound as the final electron acceptor

glycolysis the process of breaking glucose into two three-carbon molecules with the production of ATP and NADH

heat energy the energy transferred from one system to another that is not work

kinetic energy the type of energy associated with objects in motion

metabolism all the chemical reactions that take place inside cells, including those that use energy and those that release energy

noncompetitive inhibition a general mechanism of enzyme activity regulation in which a regulatory molecule binds to a site other than the active site and prevents the active site from binding the substrate; thus, the inhibitor molecule does not compete with the substrate for the active site; allosteric inhibition is a form of noncompetitive inhibition

oxidative phosphorylation the production of ATP by the transfer of electrons down the electron transport chain to create a proton gradient that is used by ATP synthase to add phosphate groups to ADP molecules

potential energy the type of energy that refers to the potential to do work

substrate a molecule on which the enzyme acts

thermodynamics the science of the relationships between heat, energy, and work

CHAPTER SUMMARY

4.1 Energy and Metabolism

Cells perform the functions of life through various chemical reactions. A cell's metabolism refers to the combination of chemical reactions that take place within it. Catabolic reactions break down complex chemicals into simpler ones and are associated with energy release. Anabolic processes build complex molecules out of simpler ones and require energy.

In studying energy, the term system refers to the matter and environment involved in energy transfers. Entropy is a measure of the disorder of a system. The physical laws that describe the transfer of energy are the laws of thermodynamics. The first law states that the total amount of energy in the universe is constant. The second law of thermodynamics states that every energy transfer involves some loss of energy in an unusable form, such as heat energy. Energy comes in different forms: kinetic, potential, and free. The change in free energy of a reaction can be negative (releases energy, exergonic) or positive (consumes energy, endergonic). All reactions require an initial input of energy to proceed, called the activation energy.

Enzymes are chemical catalysts that speed up chemical reactions by lowering their activation energy. Enzymes have an active site with a unique chemical environment that fits particular chemical reactants for that enzyme, called substrates. Enzymes and substrates are thought to bind according to an induced-fit model. Enzyme action is regulated to conserve resources and respond optimally to the environment.

4.2 Glycolysis

ATP functions as the energy currency for cells. It allows cells to store energy briefly and transport it within itself to support endergonic chemical reactions. The structure of ATP is that of an RNA nucleotide with three phosphate groups attached. As ATP is used for energy, a phosphate group is detached, and ADP is produced. Energy derived from glucose catabolism is used to recharge ADP into ATP.

Glycolysis is the first pathway used in the breakdown of glucose to extract energy. Because it is used by nearly all organisms on earth, it must have evolved early in the history of life. Glycolysis consists of two parts: The first part prepares the six-carbon ring of glucose for separation into two three-carbon sugars. Energy from ATP is invested into the molecule during this step to energize the separation. The second half of glycolysis extracts ATP and high-energy electrons from hydrogen atoms and attaches them to NAD^+. Two ATP molecules are invested in the first half and four ATP molecules are formed during the second half. This produces a net gain of two ATP molecules per molecule of glucose for the cell.

4.3 Citric Acid Cycle and Oxidative Phosphorylation

The citric acid cycle is a series of chemical reactions that removes high-energy electrons and uses them in the electron transport chain to generate ATP. One molecule of ATP (or an equivalent) is produced per each turn of the cycle.

The electron transport chain is the portion of aerobic respiration that uses free oxygen as the final electron acceptor for electrons removed from the intermediate compounds in glucose catabolism. The electrons are passed through a series of chemical reactions, with a small amount of free energy used at three points to transport hydrogen ions across the membrane. This contributes to the gradient used in chemiosmosis. As the electrons are passed from NADH or $FADH_2$ down the electron transport chain, they lose energy. The products of the electron transport chain are water and ATP. A number of intermediate compounds can be diverted into the anabolism of other biochemical molecules, such as nucleic acids, non-essential amino acids, sugars, and lipids. These same molecules, except nucleic acids, can serve as energy sources for the glucose pathway.

4.4 Fermentation

If NADH cannot be metabolized through aerobic respiration, another electron acceptor is used. Most organisms will use some form of fermentation to accomplish the regeneration of NAD^+, ensuring the continuation of glycolysis. The regeneration of NAD^+ in fermentation is not accompanied by ATP production; therefore, the potential for NADH to produce ATP using an electron transport chain is not utilized.

4.5 Connections to Other Metabolic Pathways

The breakdown and synthesis of carbohydrates, proteins, and lipids connect with the pathways of glucose catabolism. The carbohydrates that can also feed into glucose catabolism include galactose, fructose, and glycogen. These connect with glycolysis. The amino acids from proteins connect with glucose catabolism through pyruvate, acetyl CoA, and components of the citric acid cycle. Cholesterol synthesis starts with acetyl CoA, and the components of triglycerides are picked up by acetyl CoA and enter the citric acid cycle.

ART CONNECTION QUESTIONS

1. Figure 4.6 Look at each of the processes shown and decide if it is endergonic or exergonic.

2. Figure 4.15 Cyanide inhibits cytochrome c oxidase, a component of the electron transport chain. If cyanide poisoning occurs, would you expect the pH of the intermembrane space to increase or decrease? What affect would cyanide have on ATP synthesis?

3. Figure 4.16 Tremetol, a metabolic poison found in white snake root plant, prevents the metabolism of lactate. When cows eat this plant, Tremetol is concentrated in the milk. Humans who consume the milk become ill. Symptoms of this disease, which include vomiting, abdominal pain, and tremors, become worse after exercise. Why do you think this is the case?

REVIEW QUESTIONS

4. Which of the following is not an example of an energy transformation?
    a. Heating up dinner in a microwave
    b. Solar panels at work
    c. Formation of static electricity
    d. None of the above

5. Which of the following is not true about enzymes?

    a. They are consumed by the reactions they catalyze.
    b. They are usually made of amino acids.
    c. They lower the activation energy of chemical reactions.
    d. Each one is specific to the particular substrate(s) to which it binds.

6. Energy is stored long-term in the bonds of _____ and used short-term to perform work from a(n) _____ molecule.
    a. ATP : glucose
    b. an anabolic molecule : catabolic molecule
    c. glucose : ATP
    d. a catabolic molecule : anabolic molecule

7. The energy currency used by cells is _____.
    a. ATP
    b. ADP

    c. AMP
    d. adenosine

8. The glucose that enters the glycolysis pathway is split into two molecules of _____.
    a. ATP
    b. phosphate
    c. NADH
    d. pyruvate

9. What do the electrons added to NAD^+ do?
    a. They become part of a fermentation pathway.
    b. They go to another pathway for ATP production.
    c. They energize the entry of the acetyl group into the citric acid cycle.
    d. They are converted into NADP.

10. Chemiosmosis involves
    a. the movement of electrons across the cell membrane
    b. the movement of hydrogen atoms across a mitochondrial membrane
    c. the movement of hydrogen ions across a mitochondrial membrane
    d. the movement of glucose through the cell membrane

11. Which of the following fermentation methods can occur in animal skeletal muscles?

a.   lactic acid fermentation
b.   alcohol fermentation
c.   mixed acid fermentation
d.   propionic fermentation

12. The cholesterol synthesized by cells uses which component of the glycolytic pathway as a starting point?

a.   glucose

b.   acetyl CoA
c.   pyruvate
d.   carbon dioxide

13. Beta oxidation is _____.
a.   the breakdown of sugars
b.   the assembly of sugars
c.   the breakdown of fatty acids
d.   the removal of amino groups from amino acids

CRITICAL THINKING QUESTIONS

14. Does physical exercise to increase muscle mass involve anabolic and/or catabolic processes? Give evidence for your answer.

15. Explain in your own terms the difference between a spontaneous reaction and one that occurs instantaneously, and what causes this difference.

16. With regard to enzymes, why are vitamins and minerals necessary for good health? Give examples.

17. Both prokaryotic and eukaryotic organisms carry out some form of glycolysis. How does that fact support or not

support the assertion that glycolysis is one of the oldest metabolic pathways?

18. We inhale oxygen when we breathe and exhale carbon dioxide. What is the oxygen used for and where does the carbon dioxide come from?

19. When muscle cells run out of oxygen, what happens to the potential for energy extraction from sugars and what pathways do the cell use?

20. Would you describe metabolic pathways as inherently wasteful or inherently economical, and why?

5 | PHOTOSYNTHESIS

Figure 5.1 This sage thrasher's diet, like that of almost all organisms, depends on photosynthesis. (credit: modification of work by Dave Menke, U.S. Fish and Wildlife Service)

Chapter Outline
5.1: Overview of Photosynthesis
5.2: The Light-Dependent Reactions of Photosynthesis
5.3: The Calvin Cycle

Introduction

No matter how complex or advanced a machine, such as the latest cellular phone, the device cannot function without energy. Living things, similar to machines, have many complex components; they too cannot do anything without energy, which is why humans and all other organisms must "eat" in some form or another. That may be common knowledge, but how many people realize that every bite of every meal ingested depends on the process of photosynthesis?

5.1 | Overview of Photosynthesis

By the end of this section, you will be able to:

- Summarize the process of photosynthesis
- Explain the relevance of photosynthesis to other living things
- Identify the reactants and products of photosynthesis
- Describe the main structures involved in photosynthesis

All living organisms on earth consist of one or more cells. Each cell runs on the chemical energy found mainly in carbohydrate molecules (food), and the majority of these molecules are produced by one process: photosynthesis. Through photosynthesis, certain organisms convert solar energy (sunlight) into chemical energy, which is then used to build carbohydrate molecules. The energy used to hold these molecules together is released when an organism breaks down food. Cells then use this energy to perform work, such as cellular respiration.

The energy that is harnessed from photosynthesis enters the ecosystems of our planet continuously and is transferred from one organism to another. Therefore, directly or indirectly, the process of photosynthesis provides most of the energy required by living things on earth.

Photosynthesis also results in the release of oxygen into the atmosphere. In short, to eat and breathe, humans depend almost entirely on the organisms that carry out photosynthesis.

Click the following link (http://openstaxcollege.org/l/photosynthesis2) to learn more about photosynthesis.

Solar Dependence and Food Production

Some organisms can carry out photosynthesis, whereas others cannot. An **autotroph** is an organism that can produce its own food. The Greek roots of the word *autotroph* mean "self" (*auto*) "feeder" (*troph*). Plants are the best-known autotrophs, but others exist, including certain types of bacteria and algae (Figure 5.2). Oceanic algae contribute enormous quantities of food and oxygen to global food chains. Plants are also **photoautotrophs**, a type of autotroph that uses sunlight and carbon from carbon dioxide to synthesize chemical energy in the form of carbohydrates. All organisms carrying out photosynthesis require sunlight.

  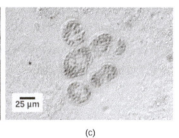

(a)   (b)   (c)

Figure 5.2 (a) Plants, (b) algae, and (c) certain bacteria, called cyanobacteria, are photoautotrophs that can carry out photosynthesis. Algae can grow over enormous areas in water, at times completely covering the surface. (credit a: Steve Hillebrand, U.S. Fish and Wildlife Service; credit b: "eutrophication&hypoxia"/Flickr; credit c: NASA; scale-bar data from Matt Russell)

Heterotrophs are organisms incapable of photosynthesis that must therefore obtain energy and carbon from food by consuming other organisms. The Greek roots of the word *heterotroph* mean "other" (*hetero*) "feeder" (*troph*), meaning that their food comes from other organisms. Even if the food organism is another animal, this food traces its origins back to autotrophs and the process of photosynthesis. Humans are heterotrophs, as are all animals. Heterotrophs depend on autotrophs, either directly or indirectly. Deer and wolves are heterotrophs. A deer obtains energy by eating plants. A wolf eating a deer obtains energy that originally came from the plants eaten by that deer. The energy in the plant came from photosynthesis, and therefore it is the only autotroph in this example (Figure 5.3). Using this reasoning, all food eaten by humans also links back to autotrophs that carry out photosynthesis.

Figure 5.3 The energy stored in carbohydrate molecules from photosynthesis passes through the food chain. The predator that eats these deer is getting energy that originated in the photosynthetic vegetation that the deer consumed. (credit: Steve VanRiper, U.S. Fish and Wildlife Service)

biology IN ACTION

Photosynthesis at the Grocery Store

Figure 5.4 Photosynthesis is the origin of the products that comprise the main elements of the human diet. (credit: Associação Brasileira de Supermercados)

Major grocery stores in the United States are organized into departments, such as dairy, meats, produce, bread, cereals, and so forth. Each aisle contains hundreds, if not thousands, of different products for customers to buy and consume (Figure 5.4).

Although there is a large variety, each item links back to photosynthesis. Meats and dairy products link to photosynthesis because the animals were fed plant-based foods. The breads, cereals, and pastas come largely from grains, which are the seeds of photosynthetic plants. What about desserts and drinks? All of these products contain sugar—the basic carbohydrate molecule produced directly from photosynthesis. The photosynthesis connection applies to every meal and every food a person consumes.

Main Structures and Summary of Photosynthesis

Photosynthesis requires sunlight, carbon dioxide, and water as starting reactants (Figure 5.5). After the process is complete, photosynthesis releases oxygen and produces carbohydrate molecules, most commonly glucose. These sugar molecules contain the energy that living things need to survive.

Figure 5.5 Photosynthesis uses solar energy, carbon dioxide, and water to release oxygen and to produce energy-storing sugar molecules.

The complex reactions of photosynthesis can be summarized by the chemical equation shown in Figure 5.6.

Photosynthesis Equation						
Carbon dioxide	+	Water	SUNLIGHT →	Sugar	+	Oxygen
$6CO_2$		$6H_2O$		$C_6H_{12}O_6$		$6O_2$

Figure 5.6 The process of photosynthesis can be represented by an equation, wherein carbon dioxide and water produce sugar and oxygen using energy from sunlight.

Although the equation looks simple, the many steps that take place during photosynthesis are actually quite complex, as in the way that the reaction summarizing cellular respiration represented many individual reactions. Before learning the details of how photoautotrophs turn sunlight into food, it is important to become familiar with the physical structures involved.

In plants, photosynthesis takes place primarily in leaves, which consist of many layers of cells and have differentiated top and bottom sides. The process of photosynthesis occurs not on the surface layers of the leaf, but rather in a middle layer called the **mesophyll** (Figure 5.7). The gas exchange of carbon dioxide and oxygen occurs through small, regulated openings called **stomata**.

In all autotrophic eukaryotes, photosynthesis takes place inside an organelle called a **chloroplast**. In plants, chloroplast-containing cells exist in the mesophyll. Chloroplasts have a double (inner and outer) membrane. Within the chloroplast is a third membrane that forms stacked, disc-shaped structures called **thylakoids**. Embedded in the thylakoid membrane are molecules of **chlorophyll**, a **pigment** (a molecule that absorbs light) through which the entire process of photosynthesis begins. Chlorophyll is responsible for the green color of plants. The thylakoid membrane encloses an internal space called the thylakoid space. Other types of pigments are also involved in photosynthesis, but chlorophyll is by far the most important. As shown in Figure 5.7, a stack of thylakoids is called a **granum**, and the space surrounding the granum is called **stroma** (not to be confused with stomata, the openings on the leaves).

art CONNECTION

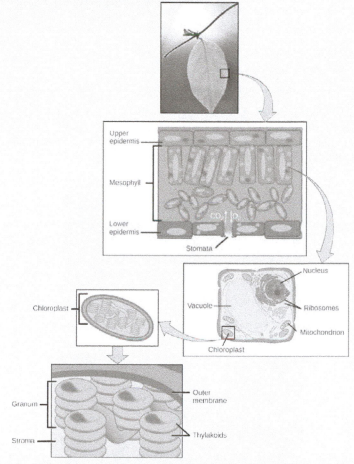

Figure 5.7 Not all cells of a leaf carry out photosynthesis. Cells within the middle layer of a leaf have chloroplasts, which contain the photosynthetic apparatus. (credit "leaf": modification of work by Cory Zanker)

On a hot, dry day, plants close their stomata to conserve water. What impact will this have on photosynthesis?

The Two Parts of Photosynthesis

Photosynthesis takes place in two stages: the light-dependent reactions and the Calvin cycle. In the **light-dependent reactions**, which take place at the thylakoid membrane, chlorophyll absorbs energy from sunlight and then converts it into chemical energy with the use of water. The light-dependent reactions release oxygen from the hydrolysis of water as a byproduct. In the Calvin cycle, which takes place in the stroma, the chemical energy derived from the light-dependent reactions drives both the capture of carbon in carbon dioxide molecules and the subsequent assembly of sugar molecules. The two reactions use carrier molecules to transport the energy from one to the other. The carriers that move energy from the light-dependent reactions to the Calvin cycle reactions can be thought of as "full" because they bring energy. After the energy is released, the "empty" energy carriers return to the light-dependent reactions to obtain more energy.

5.2 | The Light-Dependent Reactions of Photosynthesis

By the end of this section, you will be able to:

- Explain how plants absorb energy from sunlight
- Describe how the wavelength of light affects its energy and color
- Describe how and where photosynthesis takes place within a plant

How can light be used to make food? It is easy to think of light as something that exists and allows living organisms, such as humans, to see, but light is a form of energy. Like all energy, light can travel, change form, and be harnessed to do work. In the case of photosynthesis, light energy is transformed into chemical energy, which autotrophs use to build carbohydrate molecules. However, autotrophs only use a specific component of sunlight (Figure 5.8).

Figure 5.8 Autotrophs can capture light energy from the sun, converting it into chemical energy used to build food molecules. (credit: modification of work by Gerry Atwell, U.S. Fish and Wildlife Service)

Visit this site (http://openstaxcollege.org/l/light_reaction2) and click through the animation to view the process of photosynthesis within a leaf.

What Is Light Energy?

The sun emits an enormous amount of electromagnetic radiation (solar energy). Humans can see only a fraction of this energy, which is referred to as "visible light." The manner in which solar energy travels can be described and measured as

Download for free at https://openstax.org/details/books/concepts-biology

waves. Scientists can determine the amount of energy of a wave by measuring its **wavelength**, the distance between two consecutive, similar points in a series of waves, such as from crest to crest or trough to trough (Figure 5.9).

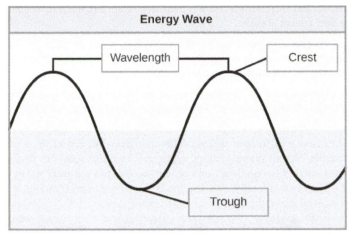

Figure 5.9 The wavelength of a single wave is the distance between two consecutive points along the wave.

Visible light constitutes only one of many types of electromagnetic radiation emitted from the sun. The **electromagnetic spectrum** is the range of all possible wavelengths of radiation (Figure 5.10). Each wavelength corresponds to a different amount of energy carried.

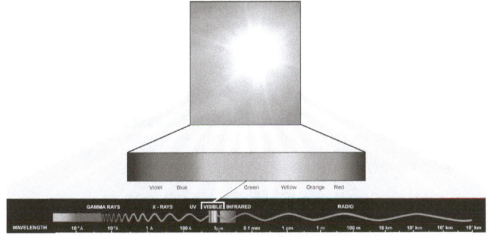

Figure 5.10 The sun emits energy in the form of electromagnetic radiation. This radiation exists in different wavelengths, each of which has its own characteristic energy. Visible light is one type of energy emitted from the sun.

Each type of electromagnetic radiation has a characteristic range of wavelengths. The longer the wavelength (or the more stretched out it appears), the less energy is carried. Short, tight waves carry the most energy. This may seem illogical, but think of it in terms of a piece of moving rope. It takes little effort by a person to move a rope in long, wide waves. To make a rope move in short, tight waves, a person would need to apply significantly more energy.

The sun emits (Figure 5.10) a broad range of electromagnetic radiation, including X-rays and ultraviolet (UV) rays. The higher-energy waves are dangerous to living things; for example, X-rays and UV rays can be harmful to humans.

Absorption of Light

Light energy enters the process of photosynthesis when pigments absorb the light. In plants, pigment molecules absorb only visible light for photosynthesis. The visible light seen by humans as white light actually exists in a rainbow of colors. Certain objects, such as a prism or a drop of water, disperse white light to reveal these colors to the human eye. The visible light portion of the electromagnetic spectrum is perceived by the human eye as a rainbow of colors, with violet and blue having shorter wavelengths and, therefore, higher energy. At the other end of the spectrum toward red, the wavelengths are longer and have lower energy.

Understanding Pigments

Different kinds of pigments exist, and each absorbs only certain wavelengths (colors) of visible light. Pigments reflect the color of the wavelengths that they cannot absorb.

All photosynthetic organisms contain a pigment called **chlorophyll *a***, which humans see as the common green color associated with plants. Chlorophyll *a* absorbs wavelengths from either end of the visible spectrum (blue and red), but not from green. Because green is reflected, chlorophyll appears green.

Other pigment types include **chlorophyll *b*** (which absorbs blue and red-orange light) and the carotenoids. Each type of pigment can be identified by the specific pattern of wavelengths it absorbs from visible light, which is its **absorption spectrum**.

Many photosynthetic organisms have a mixture of pigments; between them, the organism can absorb energy from a wider range of visible-light wavelengths. Not all photosynthetic organisms have full access to sunlight. Some organisms grow underwater where light intensity decreases with depth, and certain wavelengths are absorbed by the water. Other organisms grow in competition for light. Plants on the rainforest floor must be able to absorb any bit of light that comes through, because the taller trees block most of the sunlight (Figure 5.11).

Figure 5.11 Plants that commonly grow in the shade benefit from having a variety of light-absorbing pigments. Each pigment can absorb different wavelengths of light, which allows the plant to absorb any light that passes through the taller trees. (credit: Jason Hollinger)

How Light-Dependent Reactions Work

The overall purpose of the light-dependent reactions is to convert light energy into chemical energy. This chemical energy will be used by the Calvin cycle to fuel the assembly of sugar molecules.

The light-dependent reactions begin in a grouping of pigment molecules and proteins called a **photosystem**. Photosystems exist in the membranes of thylakoids. A pigment molecule in the photosystem absorbs one **photon**, a quantity or "packet" of light energy, at a time.

A photon of light energy travels until it reaches a molecule of chlorophyll. The photon causes an electron in the chlorophyll to become "excited." The energy given to the electron allows it to break free from an atom of the chlorophyll molecule. Chlorophyll is therefore said to "donate" an electron (Figure 5.12).

To replace the electron in the chlorophyll, a molecule of water is split. This splitting releases an electron and results in the formation of oxygen (O_2) and hydrogen ions (H^+) in the thylakoid space. Technically, each breaking of a water molecule releases a pair of electrons, and therefore can replace two donated electrons.

Download for free at https://openstax.org/details/books/concepts-biology

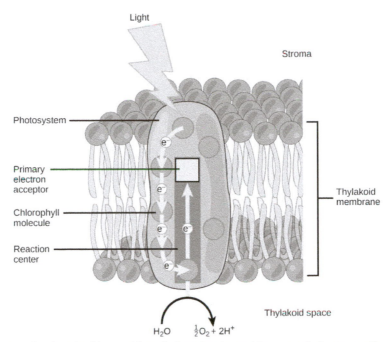

Figure 5.12 Light energy is absorbed by a chlorophyll molecule and is passed along a pathway to other chlorophyll molecules. The energy culminates in a molecule of chlorophyll found in the reaction center. The energy "excites" one of its electrons enough to leave the molecule and be transferred to a nearby primary electron acceptor. A molecule of water splits to release an electron, which is needed to replace the one donated. Oxygen and hydrogen ions are also formed from the splitting of water.

The replacing of the electron enables chlorophyll to respond to another photon. The oxygen molecules produced as byproducts find their way to the surrounding environment. The hydrogen ions play critical roles in the remainder of the light-dependent reactions.

Keep in mind that the purpose of the light-dependent reactions is to convert solar energy into chemical carriers that will be used in the Calvin cycle. In eukaryotes and some prokaryotes, two photosystems exist. The first is called photosystem II, which was named for the order of its discovery rather than for the order of the function.

After the photon hits, photosystem II transfers the free electron to the first in a series of proteins inside the thylakoid membrane called the electron transport chain. As the electron passes along these proteins, energy from the electron fuels membrane pumps that actively move hydrogen ions against their concentration gradient from the stroma into the thylakoid space. This is quite analogous to the process that occurs in the mitochondrion in which an electron transport chain pumps hydrogen ions from the mitochondrial stroma across the inner membrane and into the intermembrane space, creating an electrochemical gradient. After the energy is used, the electron is accepted by a pigment molecule in the next photosystem, which is called photosystem I (Figure 5.13).

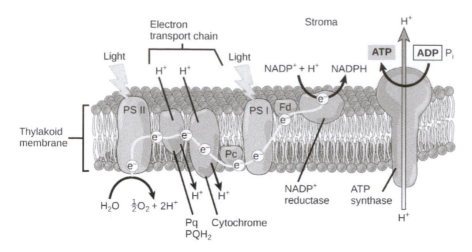

Figure 5.13 From photosystem II, the electron travels along a series of proteins. This electron transport system uses the energy from the electron to pump hydrogen ions into the interior of the thylakoid. A pigment molecule in photosystem I accepts the electron.

Generating an Energy Carrier: ATP

In the light-dependent reactions, energy absorbed by sunlight is stored by two types of energy-carrier molecules: ATP and NADPH. The energy that these molecules carry is stored in a bond that holds a single atom to the molecule. For ATP, it is a phosphate atom, and for NADPH, it is a hydrogen atom. Recall that NADH was a similar molecule that carried energy in the mitochondrion from the citric acid cycle to the electron transport chain. When these molecules release energy into the Calvin cycle, they each lose atoms to become the lower-energy molecules ADP and NADP$^+$.

The buildup of hydrogen ions in the thylakoid space forms an electrochemical gradient because of the difference in the concentration of protons (H$^+$) and the difference in the charge across the membrane that they create. This potential energy is harvested and stored as chemical energy in ATP through chemiosmosis, the movement of hydrogen ions down their electrochemical gradient through the transmembrane enzyme ATP synthase, just as in the mitochondrion.

The hydrogen ions are allowed to pass through the thylakoid membrane through an embedded protein complex called ATP synthase. This same protein generated ATP from ADP in the mitochondrion. The energy generated by the hydrogen ion stream allows ATP synthase to attach a third phosphate to ADP, which forms a molecule of ATP in a process called photophosphorylation. The flow of hydrogen ions through ATP synthase is called chemiosmosis, because the ions move from an area of high to low concentration through a semi-permeable structure.

Generating Another Energy Carrier: NADPH

The remaining function of the light-dependent reaction is to generate the other energy-carrier molecule, NADPH. As the electron from the electron transport chain arrives at photosystem I, it is re-energized with another photon captured by chlorophyll. The energy from this electron drives the formation of NADPH from NADP$^+$ and a hydrogen ion (H$^+$). Now that the solar energy is stored in energy carriers, it can be used to make a sugar molecule.

5.3 | The Calvin Cycle

By the end of this section, you will be able to:

- Describe the Calvin cycle
- Define carbon fixation
- Explain how photosynthesis works in the energy cycle of all living organisms

After the energy from the sun is converted and packaged into ATP and NADPH, the cell has the fuel needed to build food in the form of carbohydrate molecules. The carbohydrate molecules made will have a backbone of carbon atoms. Where does the carbon come from? The carbon atoms used to build carbohydrate molecules comes from carbon dioxide, the gas that animals exhale with each breath. The **Calvin cycle** is the term used for the reactions of photosynthesis that use the energy stored by the light-dependent reactions to form glucose and other carbohydrate molecules.

The Interworkings of the Calvin Cycle

In plants, carbon dioxide (CO_2) enters the chloroplast through the stomata and diffuses into the stroma of the chloroplast—the site of the Calvin cycle reactions where sugar is synthesized. The reactions are named after the scientist who discovered them, and reference the fact that the reactions function as a cycle. Others call it the Calvin-Benson cycle to include the name of another scientist involved in its discovery (Figure 5.14).

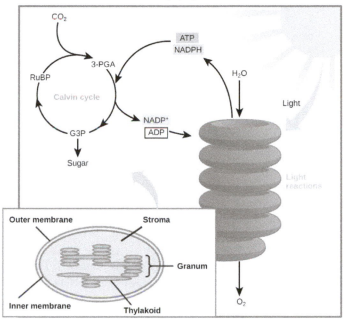

Figure 5.14 Light-dependent reactions harness energy from the sun to produce ATP and NADPH. These energy-carrying molecules travel into the stroma where the Calvin cycle reactions take place.

The Calvin cycle reactions (Figure 5.15) can be organized into three basic stages: fixation, reduction, and regeneration. In the stroma, in addition to CO_2, two other chemicals are present to initiate the Calvin cycle: an enzyme abbreviated RuBisCO, and the molecule ribulose bisphosphate (RuBP). RuBP has five atoms of carbon and a phosphate group on each end.

RuBisCO catalyzes a reaction between CO_2 and RuBP, which forms a six-carbon compound that is immediately converted into two three-carbon compounds. This process is called **carbon fixation**, because CO_2 is "fixed" from its inorganic form into organic molecules.

ATP and NADPH use their stored energy to convert the three-carbon compound, 3-PGA, into another three-carbon compound called G3P. This type of reaction is called a reduction reaction, because it involves the gain of electrons. A reduction is the gain of an electron by an atom or molecule. The molecules of ADP and NAD^+, resulting from the reduction reaction, return to the light-dependent reactions to be re-energized.

One of the G3P molecules leaves the Calvin cycle to contribute to the formation of the carbohydrate molecule, which is commonly glucose ($C_6H_{12}O_6$). Because the carbohydrate molecule has six carbon atoms, it takes six turns of the Calvin cycle to make one carbohydrate molecule (one for each carbon dioxide molecule fixed). The remaining G3P molecules regenerate RuBP, which enables the system to prepare for the carbon-fixation step. ATP is also used in the regeneration of RuBP.

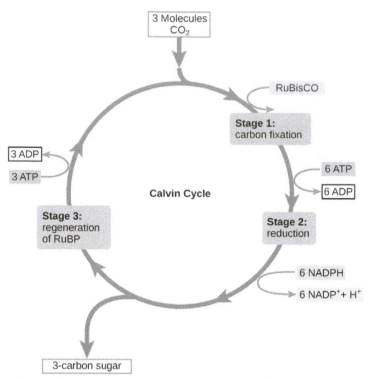

Figure 5.15 The Calvin cycle has three stages. In stage 1, the enzyme RuBisCO incorporates carbon dioxide into an organic molecule. In stage 2, the organic molecule is reduced. In stage 3, RuBP, the molecule that starts the cycle, is regenerated so that the cycle can continue.

In summary, it takes six turns of the Calvin cycle to fix six carbon atoms from CO_2. These six turns require energy input from 12 ATP molecules and 12 NADPH molecules in the reduction step and 6 ATP molecules in the regeneration step.

The following is a link (http://openstaxcollege.org/l/calvin_cycle2) to an animation of the Calvin cycle. Click Stage 1, Stage 2, and then Stage 3 to see G3P and ATP regenerate to form RuBP.

Photosynthesis

The shared evolutionary history of all photosynthetic organisms is conspicuous, as the basic process has changed little over eras of time. Even between the giant tropical leaves in the rainforest and tiny cyanobacteria, the process and components of photosynthesis that use water as an electron donor remain largely the same. Photosystems function to absorb light and use electron transport chains to convert energy. The Calvin cycle reactions assemble carbohydrate molecules with this energy.

However, as with all biochemical pathways, a variety of conditions leads to varied adaptations that affect the basic pattern. Photosynthesis in dry-climate plants (Figure 5.16) has evolved with adaptations that conserve water. In the harsh dry heat, every drop of water and precious energy must be used to survive. Two adaptations have evolved in such plants. In one form, a more efficient use of CO_2 allows plants to photosynthesize even when CO_2 is in short supply, as when the stomata are closed on hot days. The other adaptation performs preliminary reactions of the Calvin cycle at night, because opening the stomata at this time conserves water due to cooler temperatures. In addition, this adaptation has allowed plants to carry out low levels of photosynthesis without opening stomata at all, an extreme mechanism to face extremely dry periods.

Figure 5.16 Living in the harsh conditions of the desert has led plants like this cactus to evolve variations in reactions outside the Calvin cycle. These variations increase efficiency and help conserve water and energy. (credit: Piotr Wojtkowski)

Photosynthesis in Prokaryotes

The two parts of photosynthesis—the light-dependent reactions and the Calvin cycle—have been described, as they take place in chloroplasts. However, prokaryotes, such as cyanobacteria, lack membrane-bound organelles. Prokaryotic photosynthetic autotrophic organisms have infoldings of the plasma membrane for chlorophyll attachment and photosynthesis (Figure 5.17). It is here that organisms like cyanobacteria can carry out photosynthesis.

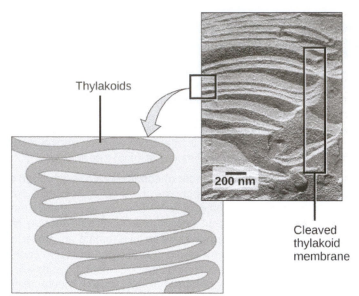

Figure 5.17 A photosynthetic prokaryote has infolded regions of the plasma membrane that function like thylakoids. Although these are not contained in an organelle, such as a chloroplast, all of the necessary components are present to carry out photosynthesis. (credit: scale-bar data from Matt Russell)

The Energy Cycle

Living things access energy by breaking down carbohydrate molecules. However, if plants make carbohydrate molecules, why would they need to break them down? Carbohydrates are storage molecules for energy in all living things. Although energy can be stored in molecules like ATP, carbohydrates are much more stable and efficient reservoirs for chemical energy. Photosynthetic organisms also carry out the reactions of respiration to harvest the energy that they have stored in carbohydrates, for example, plants have mitochondria in addition to chloroplasts.

You may have noticed that the overall reaction for photosynthesis:

$$6CO_2 + 6H_2O \rightarrow C_6H_{12}O_6 + 6O_2$$

is the reverse of the overall reaction for cellular respiration:

$$6O_2 + C_6H_{12}O_6 \rightarrow 6CO_2 + 6H_2O$$

Photosynthesis produces oxygen as a byproduct, and respiration produces carbon dioxide as a byproduct.

In nature, there is no such thing as waste. Every single atom of matter is conserved, recycling indefinitely. Substances change form or move from one type of molecule to another, but never disappear (Figure 5.18).

CO_2 is no more a form of waste produced by respiration than oxygen is a waste product of photosynthesis. Both are byproducts of reactions that move on to other reactions. Photosynthesis absorbs energy to build carbohydrates in chloroplasts, and aerobic cellular respiration releases energy by using oxygen to break down carbohydrates. Both organelles use electron transport chains to generate the energy necessary to drive other reactions. Photosynthesis and cellular respiration function in a biological cycle, allowing organisms to access life-sustaining energy that originates millions of miles away in a star.

Figure 5.18 In the carbon cycle, the reactions of photosynthesis and cellular respiration share reciprocal reactants and products. (credit: modification of work by Stuart Bassil)

KEY TERMS

absorption spectrum the specific pattern of absorption for a substance that absorbs electromagnetic radiation

autotroph an organism capable of producing its own food

Calvin cycle the reactions of photosynthesis that use the energy stored by the light-dependent reactions to form glucose and other carbohydrate molecules

carbon fixation the process of converting inorganic CO_2 gas into organic compounds

chlorophyll the green pigment that captures the light energy that drives the reactions of photosynthesis

chlorophyll *a* the form of chlorophyll that absorbs violet-blue and red light

chlorophyll *b* the form of chlorophyll that absorbs blue and red-orange light

chloroplast the organelle where photosynthesis takes place

electromagnetic spectrum the range of all possible frequencies of radiation

granum a stack of thylakoids located inside a chloroplast

heterotroph an organism that consumes other organisms for food

light-dependent reaction the first stage of photosynthesis where visible light is absorbed to form two energy-carrying molecules (ATP and NADPH)

mesophyll the middle layer of cells in a leaf

photoautotroph an organism capable of synthesizing its own food molecules (storing energy), using the energy of light

photon a distinct quantity or "packet" of light energy

photosystem a group of proteins, chlorophyll, and other pigments that are used in the light-dependent reactions of photosynthesis to absorb light energy and convert it into chemical energy

pigment a molecule that is capable of absorbing light energy

stoma the opening that regulates gas exchange and water regulation between leaves and the environment; plural: stomata

stroma the fluid-filled space surrounding the grana inside a chloroplast where the Calvin cycle reactions of photosynthesis take place

thylakoid a disc-shaped membranous structure inside a chloroplast where the light-dependent reactions of photosynthesis take place using chlorophyll embedded in the membranes

wavelength the distance between consecutive points of a wave

CHAPTER SUMMARY

5.1 Overview of Photosynthesis

The process of photosynthesis transformed life on earth. By harnessing energy from the sun, photosynthesis allowed living things to access enormous amounts of energy. Because of photosynthesis, living things gained access to sufficient energy, allowing them to evolve new structures and achieve the biodiversity that is evident today.

Only certain organisms, called autotrophs, can perform photosynthesis; they require the presence of chlorophyll, a specialized pigment that can absorb light and convert light energy into chemical energy. Photosynthesis uses carbon dioxide and water to assemble carbohydrate molecules (usually glucose) and releases oxygen into the air. Eukaryotic autotrophs, such as plants and algae, have organelles called chloroplasts in which photosynthesis takes place.

5.2 The Light-Dependent Reactions of Photosynthesis

In the first part of photosynthesis, the light-dependent reaction, pigment molecules absorb energy from sunlight. The most common and abundant pigment is chlorophyll *a*. A photon strikes photosystem II to initiate photosynthesis. Energy travels through the electron transport chain, which pumps hydrogen ions into the thylakoid space. This forms an electrochemical gradient. The ions flow through ATP synthase from the thylakoid space into the stroma in a process called chemiosmosis to form molecules of ATP, which are used for the formation of sugar molecules in the second stage of photosynthesis. Photosystem I absorbs a second photon, which results in the formation of an NADPH molecule, another energy carrier for the Calvin cycle reactions.

5.3 The Calvin Cycle

Using the energy carriers formed in the first stage of photosynthesis, the Calvin cycle reactions fix CO_2 from the environment to build carbohydrate molecules. An enzyme, RuBisCO, catalyzes the fixation reaction, by combining CO_2 with RuBP. The resulting six-carbon compound is broken down into two three-carbon compounds, and the energy in ATP and NADPH is used to convert these molecules into G3P. One of the three-carbon molecules of G3P leaves the cycle to become a part of a carbohydrate molecule. The remaining G3P molecules stay in the cycle to be formed back into RuBP, which is ready to react with more CO_2. Photosynthesis forms a balanced energy cycle with the process of cellular respiration. Plants are capable of both photosynthesis and cellular respiration, since they contain both chloroplasts and mitochondria.

ART CONNECTION QUESTIONS

1. Figure 5.7 On a hot, dry day, plants close their stomata to conserve water. What impact will this have on photosynthesis?

REVIEW QUESTIONS

2. What two products result from photosynthesis?

   a. water and carbon dioxide
   b. water and oxygen
   c. glucose and oxygen
   d. glucose and carbon dioxide

3. Which statement about thylakoids in eukaryotes is *not* correct?
   a. Thylakoids are assembled into stacks.
   b. Thylakoids exist as a maze of folded membranes.
   c. The space surrounding thylakoids is called stroma.
   d. Thylakoids contain chlorophyll.

4. From where does a heterotroph directly obtain its energy?
   a. the sun
   b. the sun and eating other organisms
   c. eating other organisms
   d. simple chemicals in the environment

5. What is the energy of a photon first used to do in photosynthesis?
   a. split a water molecule
   b. energize an electron
   c. produce ATP
   d. synthesize glucose

6. Which molecule absorbs the energy of a photon in photosynthesis?

   a. ATP
   b. glucose
   c. chlorophyll
   d. water

7. Plants produce oxygen when they photosynthesize. Where does the oxygen come from?
   a. splitting water molecules
   b. ATP synthesis
   c. the electron transport chain
   d. chlorophyll

8. Which color(s) of light does chlorophyll *a* reflect?

   a. red and blue
   b. green
   c. red
   d. blue

9. Where in plant cells does the Calvin cycle take place?

   a. thylakoid membrane
   b. thylakoid space
   c. stroma
   d. granum

10. Which statement correctly describes carbon fixation?

   a. the conversion of CO_2 to an organic compound
   b. the use of RUBISCO to form 3-PGA
   c. the production of carbohydrate molecules from G3P

d.   the formation of RuBP from G3P molecules
e.   the use of ATP and NADPH to reduce CO_2

11. What is the molecule that leaves the Calvin cycle to be converted into glucose?

a.   ADP
b.   G3P
c.   RuBP
d.   3-PGA

CRITICAL THINKING QUESTIONS

12. What is the overall purpose of the light reactions in photosynthesis?

13. Why are carnivores, such as lions, dependent on photosynthesis to survive?

14. Describe the pathway of energy in light-dependent reactions.

15. Which part of the Calvin cycle would be affected if a cell could not produce the enzyme RuBisCO?

16. Explain the reciprocal nature of the net chemical reactions for photosynthesis and respiration.

6 | REPRODUCTION AT THE CELLULAR LEVEL

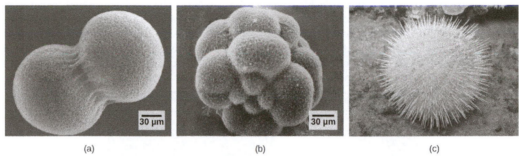

(a)   (b)   (c)

Figure 6.1 A sea urchin begins life as a single cell that (a) divides to form two cells, visible by scanning electron microscopy. After four rounds of cell division, (b) there are 16 cells, as seen in this SEM image. After many rounds of cell division, the individual develops into a complex, multicellular organism, as seen in this (c) mature sea urchin. (credit a: modification of work by Evelyn Spiegel, Louisa Howard; credit b: modification of work by Evelyn Spiegel, Louisa Howard; credit c: modification of work by Marco Busdraghi; scale-bar data from Matt Russell)

Chapter Outline

6.1: The Genome

6.2: The Cell Cycle

6.3: Cancer and the Cell Cycle

6.4: Prokaryotic Cell Division

Introduction

The individual sexually reproducing organism—including humans—begins life as a fertilized egg, or zygote. Trillions of cell divisions subsequently occur in a controlled manner to produce a complex, multicellular human. In other words, that original single cell was the ancestor of every other cell in the body. Once a human individual is fully grown, cell reproduction is still necessary to repair or regenerate tissues. For example, new blood and skin cells are constantly being produced. All multicellular organisms use cell division for growth, and in most cases, the maintenance and repair of cells and tissues. Single-celled organisms use cell division as their method of reproduction.

6.1 | The Genome

By the end of this section, you will be able to:

• Describe the prokaryotic and eukaryotic genome

• Distinguish between chromosomes, genes, and traits

The continuity of life from one cell to another has its foundation in the reproduction of cells by way of the cell cycle. The cell cycle is an orderly sequence of events in the life of a cell from the division of a single parent cell to produce two new daughter cells, to the subsequent division of those daughter cells. The mechanisms involved in the cell cycle are highly conserved across eukaryotes. Organisms as diverse as protists, plants, and animals employ similar steps.

Genomic DNA

Before discussing the steps a cell undertakes to replicate, a deeper understanding of the structure and function of a cell's genetic information is necessary. A cell's complete complement of DNA is called its **genome**. In prokaryotes, the genome is composed of a single, double-stranded DNA molecule in the form of a loop or circle. The region in the cell containing this genetic material is called a nucleoid. Some prokaryotes also have smaller loops of DNA called plasmids that are not essential for normal growth.

In eukaryotes, the genome comprises several double-stranded, linear DNA molecules (Figure 6.2) bound with proteins to form complexes called chromosomes. Each species of eukaryote has a characteristic number of chromosomes in the nuclei of its cells. Human body cells (somatic cells) have 46 chromosomes. A somatic cell contains two matched sets of chromosomes, a configuration known as **diploid**. The letter *n* is used to represent a single set of chromosomes; therefore a diploid organism is designated 2*n*. Human cells that contain one set of 23 chromosomes are called **gametes**, or sex cells; these eggs and sperm are designated *n*, or **haploid**.

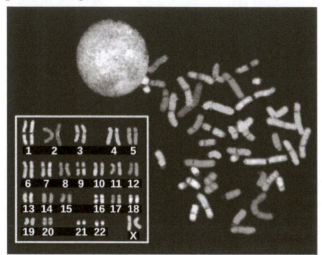

Figure 6.2 There are 23 pairs of homologous chromosomes in a female human somatic cell. These chromosomes are viewed within the nucleus (top), removed from a cell in mitosis (right), and arranged according to length (left) in an arrangement called a karyotype. In this image, the chromosomes were exposed to fluorescent stains to distinguish them. (credit: "718 Bot"/Wikimedia Commons, National Human Genome Research)

The matched pairs of chromosomes in a diploid organism are called **homologous chromosomes**. Homologous chromosomes are the same length and have specific nucleotide segments called **genes** in exactly the same location, or **locus**. Genes, the functional units of chromosomes, determine specific characteristics by coding for specific proteins. Traits are the different forms of a characteristic. For example, the shape of earlobes is a characteristic with traits of free or attached.

Each copy of the homologous pair of chromosomes originates from a different parent; therefore, the copies of each of the genes themselves may not be identical. The variation of individuals within a species is caused by the specific combination of the genes inherited from both parents. For example, there are three possible gene sequences on the human chromosome that codes for blood type: sequence A, sequence B, and sequence O. Because all diploid human cells have two copies of the chromosome that determines blood type, the blood type (the trait) is determined by which two versions of the marker gene are inherited. It is possible to have two copies of the same gene sequence, one on each homologous chromosome (for example, AA, BB, or OO), or two different sequences, such as AB.

Minor variations in traits such as those for blood type, eye color, and height contribute to the natural variation found within a species. The sex chromosomes, X and Y, are the single exception to the rule of homologous chromosomes; other than a small amount of homology that is necessary to reliably produce gametes, the genes found on the X and Y chromosomes are not the same.

Download for free at https://openstax.org/details/books/concepts-biology

6.2 | The Cell Cycle

By the end of this section, you will be able to:

- Describe the three stages of interphase
- Discuss the behavior of chromosomes during mitosis and how the cytoplasmic content divides during cytokinesis
- Define the quiescent G_0 phase
- Explain how the three internal control checkpoints occur at the end of G_1, at the G_2–M transition, and during metaphase

The **cell cycle** is an ordered series of events involving cell growth and cell division that produces two new daughter cells. Cells on the path to cell division proceed through a series of precisely timed and carefully regulated stages of growth, DNA replication, and division that produce two genetically identical cells. The cell cycle has two major phases: interphase and the mitotic phase (Figure 6.3). During **interphase**, the cell grows and DNA is replicated. During the **mitotic phase**, the replicated DNA and cytoplasmic contents are separated and the cell divides. Watch this video about the cell cycle: https://www.youtube.com/watch?v=Wy3N5NCZBHQ (https://www.youtube.com/watch?v=Wy3N5NCZBHQ)

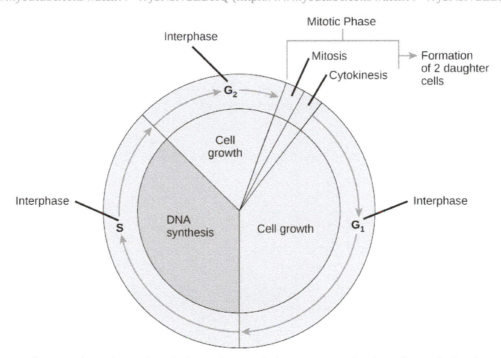

Figure 6.3 A cell moves through a series of phases in an orderly manner. During interphase, G_1 involves cell growth and protein synthesis, the S phase involves DNA replication and the replication of the centrosome, and G_2 involves further growth and protein synthesis. The mitotic phase follows interphase. Mitosis is nuclear division during which duplicated chromosomes are segregated and distributed into daughter nuclei. Usually the cell will divide after mitosis in a process called cytokinesis in which the cytoplasm is divided and two daughter cells are formed.

Interphase

During interphase, the cell undergoes normal processes while also preparing for cell division. For a cell to move from interphase to the mitotic phase, many internal and external conditions must be met. The three stages of interphase are called G_1, S, and G_2.

G_1 Phase

The first stage of interphase is called the **G_1 phase**, or first gap, because little change is visible. However, during the G_1 stage, the cell is quite active at the biochemical level. The cell is accumulating the building blocks of chromosomal

DNA and the associated proteins, as well as accumulating enough energy reserves to complete the task of replicating each chromosome in the nucleus.

S Phase

Throughout interphase, nuclear DNA remains in a semi-condensed chromatin configuration. In the **S phase** (synthesis phase), DNA replication results in the formation of two identical copies of each chromosome—sister chromatids—that are firmly attached at the centromere region. At this stage, each chromosome is made of two sister chromatids and is a duplicated chromosome. The centrosome is duplicated during the S phase. The two centrosomes will give rise to the **mitotic spindle**, the apparatus that orchestrates the movement of chromosomes during mitosis. The centrosome consists of a pair of rod-like **centrioles** at right angles to each other. Centrioles help organize cell division. Centrioles are not present in the centrosomes of many eukaryotic species, such as plants and most fungi.

G₂ Phase

In the **G2 phase**, or second gap, the cell replenishes its energy stores and synthesizes the proteins necessary for chromosome manipulation. Some cell organelles are duplicated, and the cytoskeleton is dismantled to provide resources for the mitotic spindle. There may be additional cell growth during G_2. The final preparations for the mitotic phase must be completed before the cell is able to enter the first stage of mitosis.

The Mitotic Phase

To make two daughter cells, the contents of the nucleus and the cytoplasm must be divided. The mitotic phase is a multistep process during which the duplicated chromosomes are aligned, separated, and moved to opposite poles of the cell, and then the cell is divided into two new identical daughter cells. The first portion of the mitotic phase, **mitosis**, is composed of five stages, which accomplish nuclear division. The second portion of the mitotic phase, called cytokinesis, is the physical separation of the cytoplasmic components into two daughter cells.

Mitosis

Mitosis is divided into a series of phases—prophase, prometaphase, metaphase, anaphase, and telophase—that result in the division of the cell nucleus (Figure 6.4).

art CONNECTION

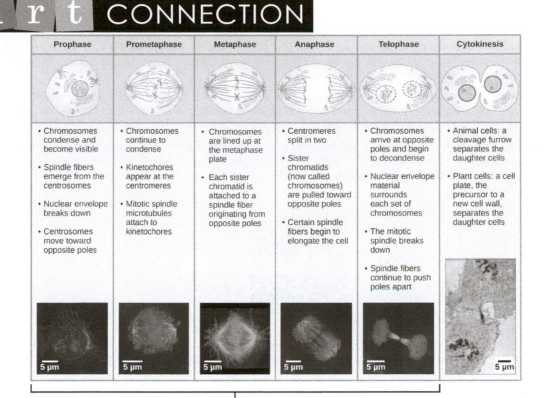

Prophase	Prometaphase	Metaphase	Anaphase	Telophase	Cytokinesis
• Chromosomes condense and become visible • Spindle fibers emerge from the centrosomes • Nuclear envelope breaks down • Centrosomes move toward opposite poles	• Chromosomes continue to condense • Kinetochores appear at the centromeres • Mitotic spindle microtubules attach to kinetochores	• Chromosomes are lined up at the metaphase plate • Each sister chromatid is attached to a spindle fiber originating from opposite poles	• Centromeres split in two • Sister chromatids (now called chromosomes) are pulled toward opposite poles • Certain spindle fibers begin to elongate the cell	• Chromosomes arrive at opposite poles and begin to decondense • Nuclear envelope material surrounds each set of chromosomes • The mitotic spindle breaks down • Spindle fibers continue to push poles apart	• Animal cells: a cleavage furrow separates the daughter cells • Plant cells: a cell plate, the precursor to a new cell wall, separates the daughter cells

MITOSIS

Figure 6.4 Animal cell mitosis is divided into five stages—prophase, prometaphase, metaphase, anaphase, and telophase—visualized here by light microscopy with fluorescence. Mitosis is usually accompanied by cytokinesis, shown here by a transmission electron microscope. (credit "diagrams": modification of work by Mariana Ruiz Villareal; credit "mitosis micrographs": modification of work by Roy van Heesbeen; credit "cytokinesis micrograph": modification of work by the Wadsworth Center, NY State Department of Health; donated to the Wikimedia foundation; scale-bar data from Matt Russell)

Which of the following is the correct order of events in mitosis?

a. Sister chromatids line up at the metaphase plate. The kinetochore becomes attached to the mitotic spindle. The nucleus re-forms and the cell divides. The sister chromatids separate.

b. The kinetochore becomes attached to the mitotic spindle. The sister chromatids separate. Sister chromatids line up at the metaphase plate. The nucleus re-forms and the cell divides.

c. The kinetochore becomes attached to metaphase plate. Sister chromatids line up at the metaphase plate. The kinetochore breaks down and the sister chromatids separate. The nucleus re-forms and the cell divides.

d. The kinetochore becomes attached to the mitotic spindle. Sister chromatids line up at the metaphase plate. The kinetochore breaks apart and the sister chromatids separate. The nucleus re-forms and the cell divides.

During **prophase**, the "first phase," several events must occur to provide access to the chromosomes in the nucleus. The nuclear envelope starts to break into small vesicles, and the Golgi apparatus and endoplasmic reticulum fragment and disperse to the periphery of the cell. The nucleolus disappears. The centrosomes begin to move to opposite poles of the cell. The microtubules that form the basis of the mitotic spindle extend between the centrosomes, pushing them farther apart as the microtubule fibers lengthen. The sister chromatids begin to coil more tightly and become visible under a light microscope.

During **prometaphase**, many processes that were begun in prophase continue to advance and culminate in the formation of a connection between the chromosomes and cytoskeleton. The remnants of the nuclear envelope disappear. The mitotic spindle continues to develop as more microtubules assemble and stretch across the length of the former nuclear area. Chromosomes become more condensed and visually discrete. Each sister chromatid attaches to spindle microtubules at the centromere via a protein complex called the **kinetochore**.

During **metaphase**, all of the chromosomes are aligned in a plane called the **metaphase plate**, or the equatorial plane, midway between the two poles of the cell. The sister chromatids are still tightly attached to each other. At this time, the chromosomes are maximally condensed.

During **anaphase**, the sister chromatids at the equatorial plane are split apart at the centromere. Each chromatid, now called a chromosome, is pulled rapidly toward the centrosome to which its microtubule was attached. The cell becomes visibly elongated as the non-kinetochore microtubules slide against each other at the metaphase plate where they overlap.

During **telophase**, all of the events that set up the duplicated chromosomes for mitosis during the first three phases are reversed. The chromosomes reach the opposite poles and begin to decondense (unravel). The mitotic spindles are broken down into monomers that will be used to assemble cytoskeleton components for each daughter cell. Nuclear envelopes form around chromosomes.

This page of movies (http://openstaxcollege.org/l/divisn_newtcell) illustrates different aspects of mitosis. Watch the movie entitled "DIC microscopy of cell division in a newt lung cell" and identify the phases of mitosis.

Cytokinesis

Cytokinesis is the second part of the mitotic phase during which cell division is completed by the physical separation of the cytoplasmic components into two daughter cells. Although the stages of mitosis are similar for most eukaryotes, the process of cytokinesis is quite different for eukaryotes that have cell walls, such as plant cells.

In cells such as animal cells that lack cell walls, cytokinesis begins following the onset of anaphase. A contractile ring composed of actin filaments forms just inside the plasma membrane at the former metaphase plate. The actin filaments pull the equator of the cell inward, forming a fissure. This fissure, or "crack," is called the **cleavage furrow**. The furrow deepens as the actin ring contracts, and eventually the membrane and cell are cleaved in two (Figure 6.5).

In plant cells, a cleavage furrow is not possible because of the rigid cell walls surrounding the plasma membrane. A new cell wall must form between the daughter cells. During interphase, the Golgi apparatus accumulates enzymes, structural proteins, and glucose molecules prior to breaking up into vesicles and dispersing throughout the dividing cell. During telophase, these Golgi vesicles move on microtubules to collect at the metaphase plate. There, the vesicles fuse from the center toward the cell walls; this structure is called a **cell plate**. As more vesicles fuse, the cell plate enlarges until it merges with the cell wall at the periphery of the cell. Enzymes use the glucose that has accumulated between the membrane layers to build a new cell wall of cellulose. The Golgi membranes become the plasma membrane on either side of the new cell wall (Figure 6.5).

Download for free at https://openstax.org/details/books/concepts-biology

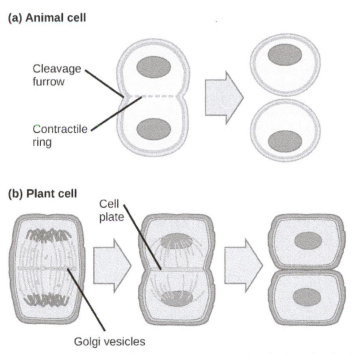

Figure 6.5 In part (a), a cleavage furrow forms at the former metaphase plate in the animal cell. The plasma membrane is drawn in by a ring of actin fibers contracting just inside the membrane. The cleavage furrow deepens until the cells are pinched in two. In part (b), Golgi vesicles coalesce at the former metaphase plate in a plant cell. The vesicles fuse and form the cell plate. The cell plate grows from the center toward the cell walls. New cell walls are made from the vesicle contents.

G_0 Phase

Not all cells adhere to the classic cell-cycle pattern in which a newly formed daughter cell immediately enters interphase, closely followed by the mitotic phase. Cells in the **G_0 phase** are not actively preparing to divide. The cell is in a quiescent (inactive) stage, having exited the cell cycle. Some cells enter G_0 temporarily until an external signal triggers the onset of G_1. Other cells that never or rarely divide, such as mature cardiac muscle and nerve cells, remain in G_0 permanently (Figure 6.6).

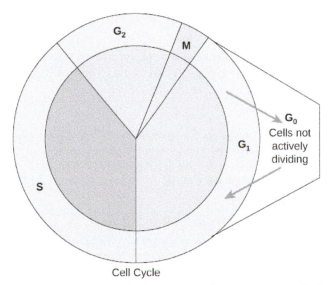

Figure 6.6 Cells that are not actively preparing to divide enter an alternate phase called G_0. In some cases, this is a temporary condition until triggered to enter G_1. In other cases, the cell will remain in G_0 permanently.

Control of the Cell Cycle

The length of the cell cycle is highly variable even within the cells of an individual organism. In humans, the frequency of cell turnover ranges from a few hours in early embryonic development to an average of two to five days for epithelial cells, or to an entire human lifetime spent in G_0 by specialized cells such as cortical neurons or cardiac muscle cells. There is also variation in the time that a cell spends in each phase of the cell cycle. When fast-dividing mammalian cells are grown in culture (outside the body under optimal growing conditions), the length of the cycle is approximately 24 hours. In rapidly dividing human cells with a 24-hour cell cycle, the G_1 phase lasts approximately 11 hours. The timing of events in the cell cycle is controlled by mechanisms that are both internal and external to the cell.

Regulation at Internal Checkpoints

It is essential that daughter cells be exact duplicates of the parent cell. Mistakes in the duplication or distribution of the chromosomes lead to mutations that may be passed forward to every new cell produced from the abnormal cell. To prevent a compromised cell from continuing to divide, there are internal control mechanisms that operate at three main **cell cycle checkpoints** at which the cell cycle can be stopped until conditions are favorable. These checkpoints occur near the end of G_1, at the G_2–M transition, and during metaphase (Figure 6.7).

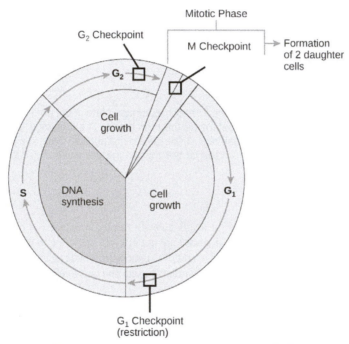

Figure 6.7 The cell cycle is controlled at three checkpoints. Integrity of the DNA is assessed at the G_1 checkpoint. Proper chromosome duplication is assessed at the G_2 checkpoint. Attachment of each kinetochore to a spindle fiber is assessed at the M checkpoint.

The G1 Checkpoint

The G_1 checkpoint determines whether all conditions are favorable for cell division to proceed. The G_1 checkpoint, also called the restriction point, is the point at which the cell irreversibly commits to the cell-division process. In addition to adequate reserves and cell size, there is a check for damage to the genomic DNA at the G_1 checkpoint. A cell that does not meet all the requirements will not be released into the S phase.

The G2 Checkpoint

The G_2 checkpoint bars the entry to the mitotic phase if certain conditions are not met. As in the G_1 checkpoint, cell size and protein reserves are assessed. However, the most important role of the G_2 checkpoint is to ensure that all of the chromosomes have been replicated and that the replicated DNA is not damaged.

The M Checkpoint

The M checkpoint occurs near the end of the metaphase stage of mitosis. The M checkpoint is also known as the spindle checkpoint because it determines if all the sister chromatids are correctly attached to the spindle microtubules. Because the separation of the sister chromatids during anaphase is an irreversible step, the cycle will not proceed until the kinetochores of each pair of sister chromatids are firmly anchored to spindle fibers arising from opposite poles of the cell.

Watch what occurs at the G_1, G_2, and M checkpoints by visiting this animation (http://openstaxcollege.org/l/ cell_checkpnts2) of the cell cycle.

6.3 | Cancer and the Cell Cycle

By the end of this section, you will be able to:

- Explain how cancer is caused by uncontrolled cell division
- Understand how proto-oncogenes are normal cell genes that, when mutated, become oncogenes
- Describe how tumor suppressors function to stop the cell cycle until certain events are completed
- Explain how mutant tumor suppressors cause cancer

Cancer is a collective name for many different diseases caused by a common mechanism: uncontrolled cell division. Despite the redundancy and overlapping levels of cell-cycle control, errors occur. One of the critical processes monitored by the cell-cycle checkpoint surveillance mechanism is the proper replication of DNA during the S phase. Even when all of the cell-cycle controls are fully functional, a small percentage of replication errors (mutations) will be passed on to the daughter cells. If one of these changes to the DNA nucleotide sequence occurs within a gene, a gene mutation results. All cancers begin when a gene mutation gives rise to a faulty protein that participates in the process of cell reproduction. The change in the cell that results from the malformed protein may be minor. Even minor mistakes, however, may allow subsequent mistakes to occur more readily. Over and over, small, uncorrected errors are passed from parent cell to daughter cells and accumulate as each generation of cells produces more non-functional proteins from uncorrected DNA damage. Eventually, the pace of the cell cycle speeds up as the effectiveness of the control and repair mechanisms decreases. Uncontrolled growth of the mutated cells outpaces the growth of normal cells in the area, and a tumor can result.

Proto-oncogenes

The genes that code for the positive cell-cycle regulators are called **proto-oncogenes**. Proto-oncogenes are normal genes that, when mutated, become **oncogenes**—genes that cause a cell to become cancerous. Consider what might happen to the cell cycle in a cell with a recently acquired oncogene. In most instances, the alteration of the DNA sequence will result in a less functional (or non-functional) protein. The result is detrimental to the cell and will likely prevent the cell from completing the cell cycle; however, the organism is not harmed because the mutation will not be carried forward. If a cell cannot reproduce, the mutation is not propagated and the damage is minimal. Occasionally, however, a gene mutation causes a change that increases the activity of a positive regulator. For example, a mutation that allows Cdk, a protein involved in cell-cycle regulation, to be activated before it should be could push the cell cycle past a checkpoint before all of the required conditions are met. If the resulting daughter cells are too damaged to undertake further cell divisions, the mutation would not be propagated and no harm comes to the organism. However, if the atypical daughter cells are able to divide further, the subsequent generation of cells will likely accumulate even more mutations, some possibly in additional genes that regulate the cell cycle.

The Cdk example is only one of many genes that are considered proto-oncogenes. In addition to the cell-cycle regulatory proteins, any protein that influences the cycle can be altered in such a way as to override cell-cycle checkpoints. Once a proto-oncogene has been altered such that there is an increase in the rate of the cell cycle, it is then called an oncogene.

Tumor Suppressor Genes

Like proto-oncogenes, many of the negative cell-cycle regulatory proteins were discovered in cells that had become cancerous. **Tumor suppressor genes** are genes that code for the negative regulator proteins, the type of regulator that—when activated—can prevent the cell from undergoing uncontrolled division. The collective function of the best-understood tumor suppressor gene proteins, retinoblastoma protein (RB1), p53, and p21, is to put up a roadblock to cell-cycle progress until certain events are completed. A cell that carries a mutated form of a negative regulator might not be able to halt the cell cycle if there is a problem.

Mutated p53 genes have been identified in more than half of all human tumor cells. This discovery is not surprising in light of the multiple roles that the p53 protein plays at the G_1 checkpoint. The p53 protein activates other genes whose products halt the cell cycle (allowing time for DNA repair), activates genes whose products participate in DNA repair, or activates genes that initiate cell death when DNA damage cannot be repaired. A damaged p53 gene can result in the cell behaving as if there are no mutations (Figure 6.8). This allows cells to divide, propagating the mutation in daughter cells and allowing the accumulation of new mutations. In addition, the damaged version of p53 found in cancer cells cannot trigger cell death.

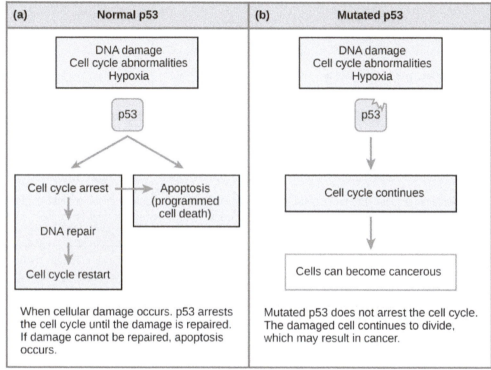

Figure 6.8 (a) The role of p53 is to monitor DNA. If damage is detected, p53 triggers repair mechanisms. If repairs are unsuccessful, p53 signals apoptosis. (b) A cell with an abnormal p53 protein cannot repair damaged DNA and cannot signal apoptosis. Cells with abnormal p53 can become cancerous. (credit: modification of work by Thierry Soussi)

Go to this website (http://openstaxcollege.org/l/cancer2) to watch an animation of how cancer results from errors in the cell cycle.

6.4 | Prokaryotic Cell Division

By the end of this section, you will be able to:

- Describe the process of binary fission in prokaryotes
- Explain how FtsZ and tubulin proteins are examples of homology

Prokaryotes such as bacteria propagate by binary fission. For unicellular organisms, cell division is the only method to produce new individuals. In both prokaryotic and eukaryotic cells, the outcome of cell reproduction is a pair of daughter cells that are genetically identical to the parent cell. In unicellular organisms, daughter cells are individuals.

To achieve the outcome of identical daughter cells, some steps are essential. The genomic DNA must be replicated and then allocated into the daughter cells; the cytoplasmic contents must also be divided to give both new cells the machinery to sustain life. In bacterial cells, the genome consists of a single, circular DNA chromosome; therefore, the process of cell division is simplified. Mitosis is unnecessary because there is no nucleus or multiple chromosomes. This type of cell division is called binary fission.

Binary Fission

The cell division process of prokaryotes, called **binary fission**, is a less complicated and much quicker process than cell division in eukaryotes. Because of the speed of bacterial cell division, populations of bacteria can grow very rapidly. The single, circular DNA chromosome of bacteria is not enclosed in a nucleus, but instead occupies a specific location, the nucleoid, within the cell. As in eukaryotes, the DNA of the nucleoid is associated with proteins that aid in packaging the molecule into a compact size. The packing proteins of bacteria are, however, related to some of the proteins involved in the chromosome compaction of eukaryotes.

The starting point of replication, the **origin**, is close to the binding site of the chromosome to the plasma membrane (Figure 6.9). Replication of the DNA is bidirectional—moving away from the origin on both strands of the DNA loop simultaneously. As the new double strands are formed, each origin point moves away from the cell-wall attachment toward opposite ends of the cell. As the cell elongates, the growing membrane aids in the transport of the chromosomes. After the chromosomes have cleared the midpoint of the elongated cell, cytoplasmic separation begins. A **septum** is formed between the nucleoids from the periphery toward the center of the cell. When the new cell walls are in place, the daughter cells separate.

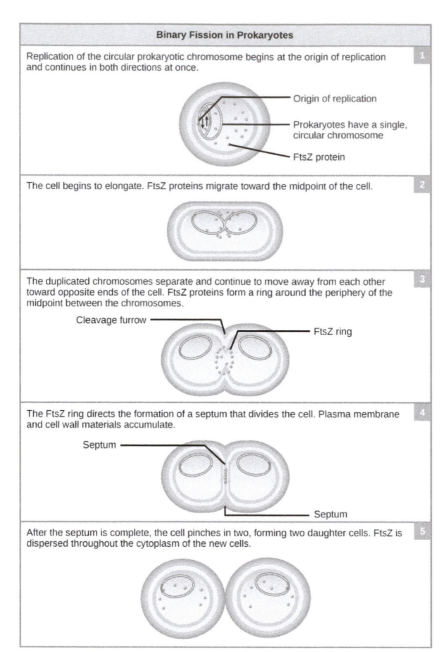

Binary Fission in Prokaryotes

Replication of the circular prokaryotic chromosome begins at the origin of replication and continues in both directions at once. **1**

Origin of replication

Prokaryotes have a single, circular chromosome

FtsZ protein

The cell begins to elongate. FtsZ proteins migrate toward the midpoint of the cell. **2**

The duplicated chromosomes separate and continue to move away from each other toward opposite ends of the cell. FtsZ proteins form a ring around the periphery of the midpoint between the chromosomes. **3**

Cleavage furrow

FtsZ ring

The FtsZ ring directs the formation of a septum that divides the cell. Plasma membrane and cell wall materials accumulate. **4**

Septum

Septum

After the septum is complete, the cell pinches in two, forming two daughter cells. FtsZ is dispersed throughout the cytoplasm of the new cells. **5**

Figure 6.9 The binary fission of a bacterium is outlined in five steps. (credit: modification of work by "Mcstrother"/Wikimedia Commons)

e/olution IN ACTION

Mitotic Spindle Apparatus

The precise timing and formation of the mitotic spindle is critical to the success of eukaryotic cell division. Prokaryotic cells, on the other hand, do not undergo mitosis and therefore have no need for a mitotic spindle. However, the FtsZ protein that plays such a vital role in prokaryotic cytokinesis is structurally and functionally very similar to tubulin, the building block of the microtubules that make up the mitotic spindle fibers that are necessary for eukaryotes. The formation of a ring composed of repeating units of a protein called **FtsZ** directs the partition between the nucleoids in prokaryotes. Formation of the FtsZ ring triggers the accumulation of other proteins that work together to recruit new membrane and cell-wall materials to the site. FtsZ proteins can form filaments, rings, and other three-dimensional structures resembling the way tubulin forms microtubules, centrioles, and various cytoskeleton components. In addition, both FtsZ and tubulin employ the same energy source, GTP (guanosine triphosphate), to rapidly assemble and disassemble complex structures.

FtsZ and tubulin are an example of homology, structures derived from the same evolutionary origins. In this example, FtsZ is presumed to be similar to the ancestor protein to both the modern FtsZ and tubulin. While both proteins are found in extant organisms, tubulin function has evolved and diversified tremendously since the evolution from its FtsZ-like prokaryotic origin. A survey of cell-division machinery in present-day unicellular eukaryotes reveals crucial intermediary steps to the complex mitotic machinery of multicellular eukaryotes (Table 6.1).

Mitotic Spindle Evolution

	Structure of genetic material	Division of nuclear material	Separation of daughter cells
Prokaryotes	There is no nucleus. The single, circular chromosome exists in a region of cytoplasm called the nucleoid.	Occurs through binary fission. As the chromosome is replicated, the two copies move to opposite ends of the cell by an unknown mechanism.	FtsZ proteins assemble into a ring that pinches the cell in two.
Some protists	Linear chromosomes exist in the nucleus.	Chromosomes attach to the nuclear envelope, which remains intact. The mitotic spindle passes through the envelope and elongates the cell. No centrioles exist.	Microfilaments form a cleavage furrow that pinches the cell in two.
Other protists	Linear chromosomes exist in the nucleus.	A mitotic spindle forms from the centrioles and passes through the nuclear membrane, which remains intact. Chromosomes attach to the mitotic spindle. The mitotic spindle separates the chromosomes and elongates the cell.	Microfilaments form a cleavage furrow that pinches the cell in two.

Table 6.1 The mitotic spindle fibers of eukaryotes are composed of microtubules. Microtubules are polymers of the protein tubulin. The FtsZ protein active in prokaryote cell division is very similar to tubulin in the structures it can form and its energy source. Single-celled eukaryotes (such as yeast) display possible intermediary steps between FtsZ activity during binary fission in prokaryotes and the mitotic spindle in multicellular eukaryotes, during which the nucleus breaks down and is reformed.

Mitotic Spindle Evolution

	Structure of genetic material	Division of nuclear material	Separation of daughter cells
Animal cells	Linear chromosomes exist in the nucleus.	A mitotic spindle forms from the centrioles. The nuclear envelope dissolves. Chromosomes attach to the mitotic spindle, which separates them and elongates the cell.	Microfilaments form a cleavage furrow that pinches the cell in two.

Table 6.1 The mitotic spindle fibers of eukaryotes are composed of microtubules. Microtubules are polymers of the protein tubulin. The FtsZ protein active in prokaryote cell division is very similar to tubulin in the structures it can form and its energy source. Single-celled eukaryotes (such as yeast) display possible intermediary steps between FtsZ activity during binary fission in prokaryotes and the mitotic spindle in multicellular eukaryotes, during which the nucleus breaks down and is reformed.

KEY TERMS

anaphase the stage of mitosis during which sister chromatids are separated from each other

binary fission the process of prokaryotic cell division

cell cycle the ordered sequence of events that a cell passes through between one cell division and the next

cell cycle checkpoints mechanisms that monitor the preparedness of a eukaryotic cell to advance through the various cell cycle stages

cell plate a structure formed during plant-cell cytokinesis by Golgi vesicles fusing at the metaphase plate; will ultimately lead to formation of a cell wall to separate the two daughter cells

centriole a paired rod-like structure constructed of microtubules at the center of each animal cell centrosome

cleavage furrow a constriction formed by the actin ring during animal-cell cytokinesis that leads to cytoplasmic division

cytokinesis the division of the cytoplasm following mitosis to form two daughter cells

diploid describes a cell, nucleus, or organism containing two sets of chromosomes ($2n$)

FtsZ a tubulin-like protein component of the prokaryotic cytoskeleton that is important in prokaryotic cytokinesis (name origin: **F**ilamenting **t**emperature-**s**ensitive mutant **Z**)

G$_0$ phase a cell-cycle phase distinct from the G$_1$ phase of interphase; a cell in G$_0$ is not preparing to divide

G$_1$ phase (also, first gap) a cell-cycle phase; first phase of interphase centered on cell growth during mitosis

G$_2$ phase (also, second gap) a cell-cycle phase; third phase of interphase where the cell undergoes the final preparations for mitosis

gamete a haploid reproductive cell or sex cell (sperm or egg)

gene the physical and functional unit of heredity; a sequence of DNA that codes for a specific peptide or RNA molecule

genome the entire genetic complement (DNA) of an organism

haploid describes a cell, nucleus, or organism containing one set of chromosomes (n)

homologous chromosomes chromosomes of the same length with genes in the same location; diploid organisms have pairs of homologous chromosomes, and the members of each pair come from different parents

interphase the period of the cell cycle leading up to mitosis; includes G$_1$, S, and G$_2$ phases; the interim between two consecutive cell divisions

kinetochore a protein structure in the centromere of each sister chromatid that attracts and binds spindle microtubules during prometaphase

locus the position of a gene on a chromosome

metaphase the stage of mitosis during which chromosomes are lined up at the metaphase plate

metaphase plate the equatorial plane midway between two poles of a cell where the chromosomes align during metaphase

mitosis the period of the cell cycle at which the duplicated chromosomes are separated into identical nuclei; includes prophase, prometaphase, metaphase, anaphase, and telophase

mitotic phase the period of the cell cycle when duplicated chromosomes are distributed into two nuclei and the cytoplasmic contents are divided; includes mitosis and cytokinesis

mitotic spindle the microtubule apparatus that orchestrates the movement of chromosomes during mitosis

oncogene a mutated version of a proto-oncogene, which allows for uncontrolled progression of the cell cycle, or uncontrolled cell reproduction

origin the region of the prokaryotic chromosome at which replication begins

prometaphase the stage of mitosis during which mitotic spindle fibers attach to kinetochores

prophase the stage of mitosis during which chromosomes condense and the mitotic spindle begins to form

proto-oncogene a normal gene that controls cell division by regulating the cell cycle that becomes an oncogene if it is mutated

quiescent describes a cell that is performing normal cell functions and has not initiated preparations for cell division

S phase the second, or synthesis phase, of interphase during which DNA replication occurs

septum a wall formed between bacterial daughter cells as a precursor to cell separation

telophase the stage of mitosis during which chromosomes arrive at opposite poles, decondense, and are surrounded by new nuclear envelopes

tumor suppressor gene a gene that codes for regulator proteins that prevent the cell from undergoing uncontrolled division

CHAPTER SUMMARY

6.1 The Genome

Prokaryotes have a single loop chromosome, whereas eukaryotes have multiple, linear chromosomes surrounded by a nuclear membrane. Human somatic cells have 46 chromosomes consisting of two sets of 22 homologous chromosomes and a pair of nonhomologous sex chromosomes. This is the $2n$, or diploid, state. Human gametes have 23 chromosomes or one complete set of chromosomes. This is the n, or haploid, state. Genes are segments of DNA that code for a specific protein or RNA molecule. An organism's traits are determined in large part by the genes inherited from each parent, but also by the environment that they experience. Genes are expressed as characteristics of the organism and each characteristic may have different variants called traits that are caused by differences in the DNA sequence for a gene.

6.2 The Cell Cycle

The cell cycle is an orderly sequence of events. Cells on the path to cell division proceed through a series of precisely timed and carefully regulated stages. In eukaryotes, the cell cycle consists of a long preparatory period, called interphase. Interphase is divided into G_1, S, and G_2 phases. Mitosis consists of five stages: prophase, prometaphase, metaphase, anaphase, and telophase. Mitosis is usually accompanied by cytokinesis, during which the cytoplasmic components of the daughter cells are separated either by an actin ring (animal cells) or by cell plate formation (plant cells).

Each step of the cell cycle is monitored by internal controls called checkpoints. There are three major checkpoints in the cell cycle: one near the end of G_1, a second at the G_2–M transition, and the third during metaphase.

6.3 Cancer and the Cell Cycle

Cancer is the result of unchecked cell division caused by a breakdown of the mechanisms regulating the cell cycle. The loss of control begins with a change in the DNA sequence of a gene that codes for one of the regulatory molecules. Faulty instructions lead to a protein that does not function as it should. Any disruption of the monitoring system can allow other mistakes to be passed on to the daughter cells. Each successive cell division will give rise to daughter cells with even more accumulated damage. Eventually, all checkpoints become nonfunctional, and rapidly reproducing cells crowd out normal cells, resulting in tumorous growth.

6.4 Prokaryotic Cell Division

In both prokaryotic and eukaryotic cell division, the genomic DNA is replicated and each copy is allocated into a daughter cell. The cytoplasmic contents are also divided evenly to the new cells. However, there are many differences between prokaryotic and eukaryotic cell division. Bacteria have a single, circular DNA chromosome and no nucleus. Therefore, mitosis is not necessary in bacterial cell division. Bacterial cytokinesis is directed by a ring composed of a protein called

FtsZ. Ingrowth of membrane and cell-wall material from the periphery of the cells results in a septum that eventually forms the separate cell walls of the daughter cells.

ART CONNECTION QUESTIONS

1. Figure 6.4 Which of the following is the correct order of events in mitosis?
  a. Sister chromatids line up at the metaphase plate. The kinetochore becomes attached to the mitotic spindle. The nucleus re-forms and the cell divides. The sister chromatids separate.
  b. The kinetochore becomes attached to the mitotic spindle. The sister chromatids separate. Sister chromatids line up at the metaphase plate. The nucleus re-forms and the cell divides.

  c. The kinetochore becomes attached to metaphase plate. Sister chromatids line up at the metaphase plate. The kinetochore breaks down and the sister chromatids separate. The nucleus re-forms and the cell divides.
  d. The kinetochore becomes attached to the mitotic spindle. Sister chromatids line up at the metaphase plate. The kinetochore breaks apart and the sister chromatids separate. The nucleus re-forms and the cell divides.

REVIEW QUESTIONS

2. A diploid cell has _____ the number of chromosomes as a haploid cell.
  a. one-fourth
  b. one-half
  c. twice
  d. four times

3. An organism's traits are determined by the specific combination of inherited _____.
  a. cells
  b. genes
  c. proteins
  d. chromatids

4. Chromosomes are duplicated during what portion of the cell cycle?
  a. G_1 phase
  b. S phase
  c. prophase
  d. prometaphase

5. Separation of the sister chromatids is a characteristic of which stage of mitosis?
  a. prometaphase
  b. metaphase
  c. anaphase
  d. telophase

6. The individual chromosomes become visible with a light microscope during which stage of mitosis?
  a. prophase
  b. prometaphase
  c. metaphase
  d. anaphase

7. What is necessary for a cell to pass the G_2 checkpoint?

  a. cell has reached a sufficient size
  b. an adequate stockpile of nucleotides
  c. accurate and complete DNA replication
  d. proper attachment of mitotic spindle fibers to kinetochores

8. _____ are changes to the nucleotides in a segment of DNA that codes for a protein.
  a. Proto-oncogenes
  b. Tumor suppressor genes
  c. Gene mutations
  d. Negative regulators

9. A gene that codes for a positive cell cycle regulator is called a(n) _____.
  a. kinase inhibitor
  b. tumor suppressor gene
  c. proto-oncogene
  d. oncogene

10. Which eukaryotic cell-cycle event is missing in binary fission?
  a. cell growth
  b. DNA duplication
  c. mitosis
  d. cytokinesis

11. FtsZ proteins direct the formation of a _____ that will eventually form the new cell walls of the daughter cells.
  a. contractile ring
  b. cell plate
  c. cytoskeleton
  d. septum

CRITICAL THINKING QUESTIONS

12. Compare and contrast a human somatic cell to a human gamete.

13. Describe the similarities and differences between the cytokinesis mechanisms found in animal cells versus those in plant cells.

14. Outline the steps that lead to a cell becoming cancerous.

15. Explain the difference between a proto-oncogene and a tumor suppressor gene.

16. Name the common components of eukaryotic cell division and binary fission.

7 | THE CELLULAR BASIS OF INHERITANCE

(a)   (b)   (c)

Figure 7.1 Each of us, like these other large multicellular organisms, begins life as a fertilized egg. After trillions of cell divisions, each of us develops into a complex, multicellular organism. (credit a: modification of work by Frank Wouters; credit b: modification of work by Ken Cole, USGS; credit c: modification of work by Martin Pettitt)

Chapter Outline
7.1: Sexual Reproduction
7.2: Meiosis
7.3: Errors in Meiosis

Introduction

The ability to reproduce *in kind* is a basic characteristic of all living things. *In kind* means that the offspring of any organism closely resembles its parent or parents. Hippopotamuses give birth to hippopotamus calves; Monterey pine trees produce seeds from which Monterey pine seedlings emerge; and adult flamingos lay eggs that hatch into flamingo chicks. *In kind* does not generally mean *exactly the same*. While many single-celled organisms and a few multicellular organisms can produce genetically identical clones of themselves through mitotic cell division, many single-celled organisms and most multicellular organisms reproduce regularly using another method.

Sexual reproduction is the production by parents of haploid cells and the fusion of a haploid cell from each parent to form a single, unique diploid cell. In multicellular organisms, the new diploid cell will then undergo mitotic cell divisions to develop into an adult organism. A type of cell division called meiosis leads to the haploid cells that are part of the sexual reproductive cycle. Sexual reproduction, specifically meiosis and fertilization, introduces variation into offspring that may account for the evolutionary success of sexual reproduction. The vast majority of eukaryotic organisms can or must employ some form of meiosis and fertilization to reproduce.

7.1 | Sexual Reproduction

By the end of this section, you will be able to:

- Explain that variation among offspring is a potential evolutionary advantage resulting from sexual reproduction
- Describe the three different life-cycle strategies among sexual multicellular organisms and their commonalities

Sexual reproduction was an early evolutionary innovation after the appearance of eukaryotic cells. The fact that most eukaryotes reproduce sexually is evidence of its evolutionary success. In many animals, it is the only mode of reproduction.

And yet, scientists recognize some real disadvantages to sexual reproduction. On the surface, offspring that are genetically identical to the parent may appear to be more advantageous. If the parent organism is successfully occupying a habitat, offspring with the same traits would be similarly successful. There is also the obvious benefit to an organism that can produce offspring by asexual budding, fragmentation, or asexual eggs. These methods of reproduction do not require another organism of the opposite sex. There is no need to expend energy finding or attracting a mate. That energy can be spent on producing more offspring. Indeed, some organisms that lead a solitary lifestyle have retained the ability to reproduce asexually. In addition, asexual populations only have female individuals, so every individual is capable of reproduction. In contrast, the males in sexual populations (half the population) are not producing offspring themselves. Because of this, an asexual population can grow twice as fast as a sexual population in theory. This means that in competition, the asexual population would have the advantage. All of these advantages to asexual reproduction, which are also disadvantages to sexual reproduction, should mean that the number of species with asexual reproduction should be more common.

However, multicellular organisms that exclusively depend on asexual reproduction are exceedingly rare. Why is sexual reproduction so common? This is one of the important questions in biology and has been the focus of much research from the latter half of the twentieth century until now. A likely explanation is that the variation that sexual reproduction creates among offspring is very important to the survival and reproduction of those offspring. The only source of variation in asexual organisms is mutation. This is the ultimate source of variation in sexual organisms. In addition, those different mutations are continually reshuffled from one generation to the next when different parents combine their unique genomes, and the genes are mixed into different combinations by the process of **meiosis**. Meiosis is the division of the contents of the nucleus that divides the chromosomes among gametes. Variation is introduced during meiosis, as well as when the gametes combine in fertilization.

eVolution IN ACTION

The Red Queen Hypothesis

There is no question that sexual reproduction provides evolutionary advantages to organisms that employ this mechanism to produce offspring. The problematic question is why, even in the face of fairly stable conditions, sexual reproduction persists when it is more difficult and produces fewer offspring for individual organisms? Variation is the outcome of sexual reproduction, but why are ongoing variations necessary?

Enter the Red Queen hypothesis, first proposed by Leigh Van Valen in 1973.[1] The concept was named in reference to the Red Queen's race in Lewis Carroll's book, *Through the Looking-Glass,* in which the Red Queen says one must run at full speed just to stay where one is.

All species coevolve with other organisms. For example, predators coevolve with their prey, and parasites coevolve with their hosts. A remarkable example of coevolution between predators and their prey is the unique coadaptation of night flying bats and their moth prey. Bats find their prey by emitting high-pitched clicks, but moths have evolved simple ears to hear these clicks so they can avoid the bats. The moths have also adapted behaviors, such as flying away from the bat when they first hear it, or dropping suddenly to the ground when the bat is upon them. Bats have evolved "quiet" clicks in an attempt to evade the moth's hearing. Some moths have evolved the ability to respond to the bats' clicks with their own clicks as a strategy to confuse the bats echolocation abilities.

Each tiny advantage gained by favorable variation gives a species an edge over close competitors, predators, parasites, or even prey. The only method that will allow a coevolving species to keep its own share of the resources is also to continually improve its ability to survive and produce offspring. As one species gains an advantage, other species must also develop an advantage or they will be outcompeted. No single species progresses too far ahead because genetic variation among progeny of sexual reproduction provides all species with a mechanism to produce adapted individuals. Species whose individuals cannot keep up become extinct. The Red Queen's catchphrase was, "It takes all the running you can do to stay in the same place." This is an apt description of coevolution between competing species.

Life Cycles of Sexually Reproducing Organisms

Fertilization and meiosis alternate in sexual **life cycles**. What happens between these two events depends on the organism. The process of meiosis reduces the resulting gamete's chromosome number by half. Fertilization, the joining of two haploid

---

1. Leigh Van Valen, "A new evolutionary law," Evolutionary Theory 1 (1973): 1–30.

gametes, restores the diploid condition. There are three main categories of life cycles in multicellular organisms: **diploid-dominant**, in which the multicellular diploid stage is the most obvious life stage (and there is no multicellular haploid stage), as with most animals including humans; **haploid-dominant**, in which the multicellular haploid stage is the most obvious life stage (and there is no multicellular diploid stage), as with all fungi and some algae; and **alternation of generations**, in which the two stages, haploid and diploid, are apparent to one degree or another depending on the group, as with plants and some algae.

Nearly all animals employ a diploid-dominant life-cycle strategy in which the only haploid cells produced by the organism are the gametes. The gametes are produced from diploid **germ cells**, a special cell line that only produces gametes. Once the haploid gametes are formed, they lose the ability to divide again. There is no multicellular haploid life stage. Fertilization occurs with the fusion of two gametes, usually from different individuals, restoring the diploid state (Figure 7.2**a**).

a r t CONNECTION

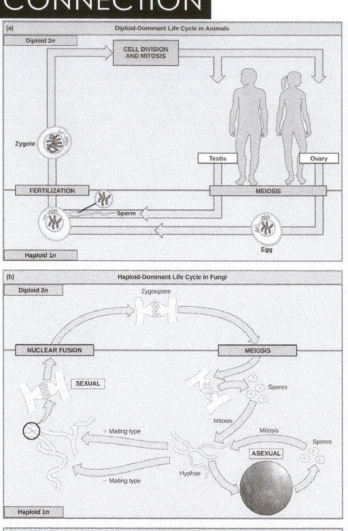

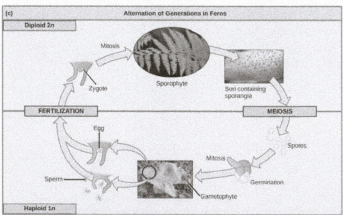

Figure 7.2 (a) In animals, sexually reproducing adults form haploid gametes from diploid germ cells. (b) Fungi, such as black bread mold (*Rhizopus nigricans*), have haploid-dominant life cycles. (c) Plants have a life cycle that alternates between a multicellular haploid organism and a multicellular diploid organism. (credit c "fern": modification of work by Cory Zanker; credit c "gametophyte": modification of work by "Vlmastra"/Wikimedia Commons)

If a mutation occurs so that a fungus is no longer able to produce a minus mating type, will it still be able to reproduce?

Most fungi and algae employ a life-cycle strategy in which the multicellular "body" of the organism is haploid. During sexual reproduction, specialized haploid cells from two individuals join to form a diploid zygote. The zygote immediately undergoes meiosis to form four haploid cells called spores (Figure 7.2**b**).

The third life-cycle type, employed by some algae and all plants, is called alternation of generations. These species have both haploid and diploid multicellular organisms as part of their life cycle. The haploid multicellular plants are called **gametophytes** because they produce gametes. Meiosis is not involved in the production of gametes in this case, as the organism that produces gametes is already haploid. Fertilization between the gametes forms a diploid zygote. The zygote will undergo many rounds of mitosis and give rise to a diploid multicellular plant called a **sporophyte**. Specialized cells of the sporophyte will undergo meiosis and produce haploid spores. The spores will develop into the gametophytes (Figure 7.2**c**).

7.2 | Meiosis

By the end of this section, you will be able to:

- Describe the behavior of chromosomes during meiosis
- Describe cellular events during meiosis
- Explain the differences between meiosis and mitosis
- Explain the mechanisms within meiosis that generate genetic variation among the products of meiosis

Sexual reproduction requires **fertilization**, a union of two cells from two individual organisms. If those two cells each contain one set of chromosomes, then the resulting cell contains two sets of chromosomes. The number of sets of chromosomes in a cell is called its ploidy level. Haploid cells contain one set of chromosomes. Cells containing two sets of chromosomes are called diploid. If the reproductive cycle is to continue, the diploid cell must somehow reduce its number of chromosome sets before fertilization can occur again, or there will be a continual doubling in the number of chromosome sets in every generation. So, in addition to fertilization, sexual reproduction includes a nuclear division, known as meiosis, that reduces the number of chromosome sets.

Most animals and plants are diploid, containing two sets of chromosomes; in each **somatic cell** (the nonreproductive cells of a multicellular organism), the nucleus contains two copies of each chromosome that are referred to as homologous chromosomes. Somatic cells are sometimes referred to as "body" cells. Homologous chromosomes are matched pairs containing genes for the same traits in identical locations along their length. Diploid organisms inherit one copy of each homologous chromosome from each parent; all together, they are considered a full set of chromosomes. In animals, haploid cells containing a single copy of each homologous chromosome are found only within gametes. Gametes fuse with another haploid gamete to produce a diploid cell.

The nuclear division that forms haploid cells, which is called meiosis, is related to mitosis. As you have learned, mitosis is part of a cell reproduction cycle that results in identical daughter nuclei that are also genetically identical to the original parent nucleus. In mitosis, both the parent and the daughter nuclei contain the same number of chromosome sets—diploid for most plants and animals. Meiosis employs many of the same mechanisms as mitosis. However, the starting nucleus is always diploid and the nuclei that result at the end of a meiotic cell division are haploid. To achieve the reduction in chromosome number, meiosis consists of one round of chromosome duplication and two rounds of nuclear division. Because the events that occur during each of the division stages are analogous to the events of mitosis, the same stage names are assigned. However, because there are two rounds of division, the stages are designated with a "I" or "II." Thus, **meiosis I** is the first round of meiotic division and consists of prophase I, prometaphase I, and so on. Meiosis I reduces the number of chromosome sets from two to one. The genetic information is also mixed during this division to create unique recombinant chromosomes. **Meiosis II**, in which the second round of meiotic division takes place in a way that is similar to mitosis, includes prophase II, prometaphase II, and so on.

Interphase

Meiosis is preceded by an interphase consisting of the G_1, S, and G_2 phases, which are nearly identical to the phases preceding mitosis. The G_1 phase is the first phase of interphase and is focused on cell growth. In the S phase, the DNA of the chromosomes is replicated. Finally, in the G_2 phase, the cell undergoes the final preparations for meiosis.

During DNA duplication of the S phase, each chromosome becomes composed of two identical copies (called sister chromatids) that are held together at the centromere until they are pulled apart during meiosis II. In an animal cell, the centrosomes that organize the microtubules of the meiotic spindle also replicate. This prepares the cell for the first meiotic phase.

Meiosis I

Early in prophase I, the chromosomes can be seen clearly microscopically. As the nuclear envelope begins to break down, the proteins associated with homologous chromosomes bring the pair close to each other. The tight pairing of the homologous chromosomes is called **synapsis**. In synapsis, the genes on the chromatids of the homologous chromosomes are precisely aligned with each other. An exchange of chromosome segments between non-sister homologous chromatids occurs and is called **crossing over**. This process is revealed visually after the exchange as **chiasmata** (singular = *chiasma*) (Figure 7.3).

As prophase I progresses, the close association between homologous chromosomes begins to break down, and the chromosomes continue to condense, although the homologous chromosomes remain attached to each other at chiasmata. The number of chiasmata varies with the species and the length of the chromosome. At the end of prophase I, the pairs are held together only at chiasmata (Figure 7.3) and are called **tetrads** because the four sister chromatids of each pair of homologous chromosomes are now visible.

The crossover events are the first source of genetic variation produced by meiosis. A single crossover event between homologous non-sister chromatids leads to a reciprocal exchange of equivalent DNA between a maternal chromosome and a paternal chromosome. Now, when that sister chromatid is moved into a gamete, it will carry some DNA from one parent of the individual and some DNA from the other parent. The **recombinant** sister chromatid has a combination of maternal and paternal genes that did not exist before the crossover.

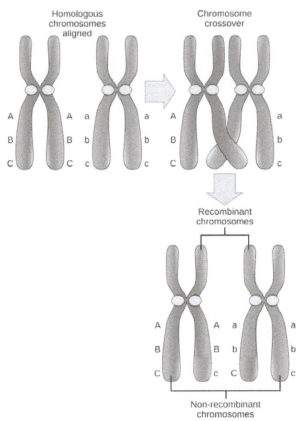

Figure 7.3 In this illustration of the effects of crossing over, the blue chromosome came from the individual's father and the red chromosome came from the individual's mother. Crossover occurs between non-sister chromatids of homologous chromosomes. The result is an exchange of genetic material between homologous chromosomes. The chromosomes that have a mixture of maternal and paternal sequence are called recombinant and the chromosomes that are completely paternal or maternal are called non-recombinant.

The key event in prometaphase I is the attachment of the spindle fiber microtubules to the kinetochore proteins at the centromeres. The microtubules assembled from centrosomes at opposite poles of the cell grow toward the middle of the cell. At the end of prometaphase I, each tetrad is attached to microtubules from both poles, with one homologous chromosome attached at one pole and the other homologous chromosome attached to the other pole. The homologous chromosomes are still held together at chiasmata. In addition, the nuclear membrane has broken down entirely.

During metaphase I, the homologous chromosomes are arranged in the center of the cell with the kinetochores facing opposite poles. The orientation of each pair of homologous chromosomes at the center of the cell is random.

This randomness, called independent assortment, is the physical basis for the generation of the second form of genetic variation in offspring. Consider that the homologous chromosomes of a sexually reproducing organism are originally inherited as two separate sets, one from each parent. Using humans as an example, one set of 23 chromosomes is present in the egg donated by the mother. The father provides the other set of 23 chromosomes in the sperm that fertilizes the egg. In metaphase I, these pairs line up at the midway point between the two poles of the cell. Because there is an equal chance that a microtubule fiber will encounter a maternally or paternally inherited chromosome, the arrangement of the tetrads at the metaphase plate is random. Any maternally inherited chromosome may face either pole. Any paternally inherited chromosome may also face either pole. The orientation of each tetrad is independent of the orientation of the other 22 tetrads.

In each cell that undergoes meiosis, the arrangement of the tetrads is different. The number of variations depends on the number of chromosomes making up a set. There are two possibilities for orientation (for each tetrad); thus, the possible number of alignments equals 2^n where n is the number of chromosomes per set. Humans have 23 chromosome pairs, which results in over eight million (2^{23}) possibilities. This number does not include the variability previously created in the sister chromatids by crossover. Given these two mechanisms, it is highly unlikely that any two haploid cells resulting from meiosis will have the same genetic composition (Figure 7.4).

To summarize the genetic consequences of meiosis I: the maternal and paternal genes are recombined by crossover events occurring on each homologous pair during prophase I; in addition, the random assortment of tetrads at metaphase produces a unique combination of maternal and paternal chromosomes that will make their way into the gametes.

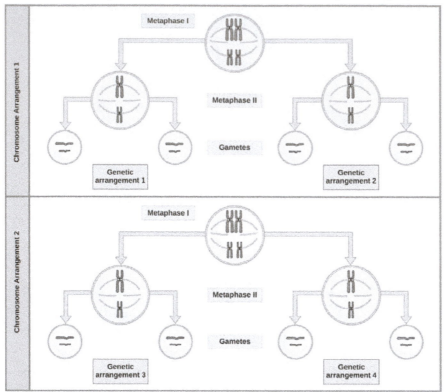

Figure 7.4 To demonstrate random, independent assortment at metaphase I, consider a cell with *n* = 2. In this case, there are two possible arrangements at the equatorial plane in metaphase I, as shown in the upper cell of each panel. These two possible orientations lead to the production of genetically different gametes. With more chromosomes, the number of possible arrangements increases dramatically.

In anaphase I, the spindle fibers pull the linked chromosomes apart. The sister chromatids remain tightly bound together at the centromere. It is the chiasma connections that are broken in anaphase I as the fibers attached to the fused kinetochores pull the homologous chromosomes apart (Figure 7.5).

In telophase I, the separated chromosomes arrive at opposite poles. The remainder of the typical telophase events may or may not occur depending on the species. In some organisms, the chromosomes decondense and nuclear envelopes form around the chromatids in telophase I.

Cytokinesis, the physical separation of the cytoplasmic components into two daughter cells, occurs without reformation of the nuclei in other organisms. In nearly all species, cytokinesis separates the cell contents by either a cleavage furrow (in animals and some fungi), or a cell plate that will ultimately lead to formation of cell walls that separate the two daughter cells (in plants). At each pole, there is just one member of each pair of the homologous chromosomes, so only one full set of the chromosomes is present. This is why the cells are considered haploid—there is only one chromosome set, even though there are duplicate copies of the set because each homolog still consists of two sister chromatids that are still attached to each other. However, although the sister chromatids were once duplicates of the same chromosome, they are no longer identical at this stage because of crossovers.

Review the process of meiosis, observing how chromosomes align and migrate, at this site (http://openstaxcollege.org/l/animal_meiosis2) .

Meiosis II

In meiosis II, the connected sister chromatids remaining in the haploid cells from meiosis I will be split to form four haploid cells. In some species, cells enter a brief interphase, or **interkinesis**, that lacks an S phase, before entering meiosis II. Chromosomes are not duplicated during interkinesis. The two cells produced in meiosis I go through the events of meiosis II in synchrony. Overall, meiosis II resembles the mitotic division of a haploid cell.

In prophase II, if the chromosomes decondensed in telophase I, they condense again. If nuclear envelopes were formed, they fragment into vesicles. The centrosomes duplicated during interkinesis move away from each other toward opposite poles, and new spindles are formed. In prometaphase II, the nuclear envelopes are completely broken down, and the spindle is fully formed. Each sister chromatid forms an individual kinetochore that attaches to microtubules from opposite poles. In metaphase II, the sister chromatids are maximally condensed and aligned at the center of the cell. In anaphase II, the sister chromatids are pulled apart by the spindle fibers and move toward opposite poles.

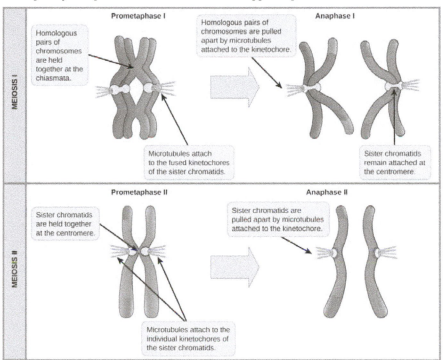

Figure 7.5 In prometaphase I, microtubules attach to the fused kinetochores of homologous chromosomes. In anaphase I, the homologous chromosomes are separated. In prometaphase II, microtubules attach to individual kinetochores of sister chromatids. In anaphase II, the sister chromatids are separated.

In telophase II, the chromosomes arrive at opposite poles and begin to decondense. Nuclear envelopes form around the chromosomes. Cytokinesis separates the two cells into four genetically unique haploid cells. At this point, the nuclei in the newly produced cells are both haploid and have only one copy of the single set of chromosomes. The cells produced are genetically unique because of the random assortment of paternal and maternal homologs and because of the recombination of maternal and paternal segments of chromosomes—with their sets of genes—that occurs during crossover.

Comparing Meiosis and Mitosis

Mitosis and meiosis, which are both forms of division of the nucleus in eukaryotic cells, share some similarities, but also exhibit distinct differences that lead to their very different outcomes. Mitosis is a single nuclear division that results in two nuclei, usually partitioned into two new cells. The nuclei resulting from a mitotic division are genetically identical to the original. They have the same number of sets of chromosomes: one in the case of haploid cells, and two in the case of diploid cells. On the other hand, meiosis is two nuclear divisions that result in four nuclei, usually partitioned into four new cells. The nuclei resulting from meiosis are never genetically identical, and they contain one chromosome set only—this is half the number of the original cell, which was diploid (Figure 7.6).

The differences in the outcomes of meiosis and mitosis occur because of differences in the behavior of the chromosomes during each process. Most of these differences in the processes occur in meiosis I, which is a very different nuclear division than mitosis. In meiosis I, the homologous chromosome pairs become associated with each other, are bound together, experience chiasmata and crossover between sister chromatids, and line up along the metaphase plate in tetrads with spindle fibers from opposite spindle poles attached to each kinetochore of a homolog in a tetrad. All of these events occur only in meiosis I, never in mitosis.

Homologous chromosomes move to opposite poles during meiosis I so the number of sets of chromosomes in each nucleus-to-be is reduced from two to one. For this reason, meiosis I is referred to as a **reduction division**. There is no such reduction in ploidy level in mitosis.

Meiosis II is much more analogous to a mitotic division. In this case, duplicated chromosomes (only one set of them) line up at the center of the cell with divided kinetochores attached to spindle fibers from opposite poles. During anaphase II, as in mitotic anaphase, the kinetochores divide and one sister chromatid is pulled to one pole and the other sister chromatid is pulled to the other pole. If it were not for the fact that there had been crossovers, the two products of each meiosis II division would be identical as in mitosis; instead, they are different because there has always been at least one crossover per chromosome. Meiosis II is not a reduction division because, although there are fewer copies of the genome in the resulting cells, there is still one set of chromosomes, as there was at the end of meiosis I.

Cells produced by mitosis will function in different parts of the body as a part of growth or replacing dead or damaged cells. They may even be involved in asexual reproduction in some organisms. Cells produced by meiosis in a diploid-dominant organism such as an animal will only participate in sexual reproduction.

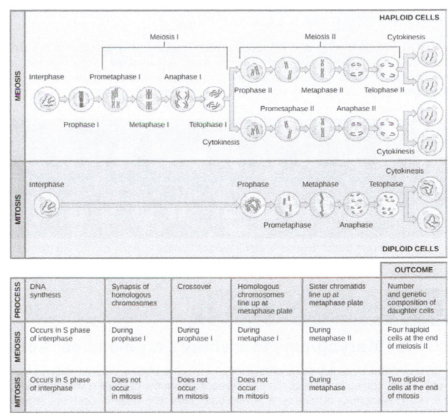

	PROCESS					OUTCOME
PROCESS	DNA synthesis	Synapsis of homologous chromosomes	Crossover	Homologous chromosomes line up at metaphase plate	Sister chromatids line up at metaphase plate	Number and genetic composition of daughter cells
MEIOSIS	Occurs in S phase of interphase	During prophase I	During prophase I	During metaphase I	During metaphase II	Four haploid cells at the end of meiosis II
MITOSIS	Occurs in S phase of interphase	Does not occur in mitosis	Does not occur in mitosis	Does not occur in mitosis	During metaphase	Two diploid cells at the end of mitosis

Figure 7.6 Meiosis and mitosis are both preceded by one round of DNA replication; however, meiosis includes two nuclear divisions. The four daughter cells resulting from meiosis are haploid and genetically distinct. The daughter cells resulting from mitosis are diploid and identical to the parent cell.

For an animation comparing mitosis and meiosis, go to this website (http://openstaxcollege.org/l/how_cells_dvid2) .

7.3 | Errors in Meiosis

By the end of this section, you will be able to:

- Explain how nondisjunction leads to disorders in chromosome number
- Describe how errors in chromosome structure occur through inversions and translocations

Inherited disorders can arise when chromosomes behave abnormally during meiosis. Chromosome disorders can be divided into two categories: abnormalities in chromosome number and chromosome structural rearrangements. Because even small segments of chromosomes can span many genes, chromosomal disorders are characteristically dramatic and often fatal.

Disorders in Chromosome Number

The isolation and microscopic observation of chromosomes forms the basis of cytogenetics and is the primary method by which clinicians detect chromosomal abnormalities in humans. A **karyotype** is the number and appearance of chromosomes, including their length, banding pattern, and centromere position. To obtain a view of an individual's karyotype, cytologists photograph the chromosomes and then cut and paste each chromosome into a chart, or **karyogram** (Figure 7.7).

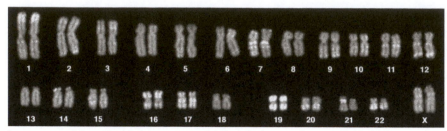

Figure 7.7 This karyogram shows the chromosomes of a female human immune cell during mitosis. (credit: Andreas Bolzer, et al)

careers IN ACTION

Geneticists Use Karyograms to Identify Chromosomal Aberrations

The karyotype is a method by which traits characterized by chromosomal abnormalities can be identified from a single cell. To observe an individual's karyotype, a person's cells (like white blood cells) are first collected from a blood sample or other tissue. In the laboratory, the isolated cells are stimulated to begin actively dividing. A chemical is then applied to the cells to arrest mitosis during metaphase. The cells are then fixed to a slide.

The geneticist then stains chromosomes with one of several dyes to better visualize the distinct and reproducible banding patterns of each chromosome pair. Following staining, chromosomes are viewed using bright-field microscopy. An experienced cytogeneticist can identify each band. In addition to the banding patterns, chromosomes are further identified on the basis of size and centromere location. To obtain the classic depiction of the karyotype in which homologous pairs of chromosomes are aligned in numerical order from longest to shortest, the geneticist obtains a digital image, identifies each chromosome, and manually arranges the chromosomes into this pattern (Figure 7.7).

At its most basic, the karyogram may reveal genetic abnormalities in which an individual has too many or too few chromosomes per cell. Examples of this are Down syndrome, which is identified by a third copy of chromosome 21, and Turner syndrome, which is characterized by the presence of only one X chromosome in women instead of two. Geneticists can also identify large deletions or insertions of DNA. For instance, Jacobsen syndrome, which involves distinctive facial features as well as heart and bleeding defects, is identified by a deletion on chromosome 11. Finally, the karyotype can pinpoint **translocations**, which occur when a segment of genetic material breaks from one chromosome and reattaches to another chromosome or to a different part of the same chromosome. Translocations are implicated in certain cancers, including chronic myelogenous leukemia.

By observing a karyogram, geneticists can actually visualize the chromosomal composition of an individual to confirm or predict genetic abnormalities in offspring even before birth.

Nondisjunctions, Duplications, and Deletions

Of all the chromosomal disorders, abnormalities in chromosome number are the most easily identifiable from a karyogram. Disorders of chromosome number include the duplication or loss of entire chromosomes, as well as changes in the number of complete sets of chromosomes. They are caused by **nondisjunction**, which occurs when pairs of homologous chromosomes or sister chromatids fail to separate during meiosis. The risk of nondisjunction increases with the age of the parents.

Nondisjunction can occur during either meiosis I or II, with different results (Figure 7.8). If homologous chromosomes fail to separate during meiosis I, the result is two gametes that lack that chromosome and two gametes with two copies of the chromosome. If sister chromatids fail to separate during meiosis II, the result is one gamete that lacks that chromosome, two normal gametes with one copy of the chromosome, and one gamete with two copies of the chromosome.

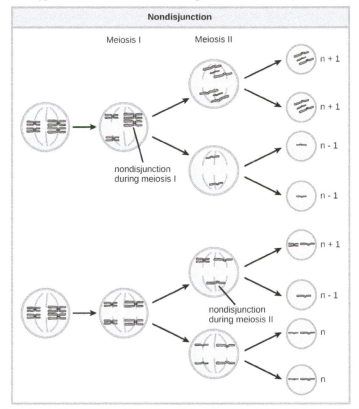

Figure 7.8 Following meiosis, each gamete has one copy of each chromosome. Nondisjunction occurs when homologous chromosomes (meiosis I) or sister chromatids (meiosis II) fail to separate during meiosis.

An individual with the appropriate number of chromosomes for their species is called **euploid**; in humans, euploidy corresponds to 22 pairs of **autosomes** and one pair of sex chromosomes. An individual with an error in chromosome number is described as **aneuploid**, a term that includes **monosomy** (loss of one chromosome) or **trisomy** (gain of an extraneous chromosome). Monosomic human zygotes missing any one copy of an autosome invariably fail to develop to birth because they have only one copy of essential genes. Most autosomal trisomies also fail to develop to birth; however, duplications of some of the smaller chromosomes (13, 15, 18, 21, or 22) can result in offspring that survive for several weeks to many years. Trisomic individuals suffer from a different type of genetic imbalance: an excess in gene dose. Cell functions are calibrated to the amount of gene product produced by two copies (doses) of each gene; adding a third copy (dose) disrupts this balance. The most common trisomy is that of chromosome 21, which leads to Down syndrome. Individuals with this inherited disorder have characteristic physical features and developmental delays in growth and cognition. The incidence of Down syndrome is correlated with maternal age, such that older women are more likely to give birth to children with Down syndrome (Figure 7.9).

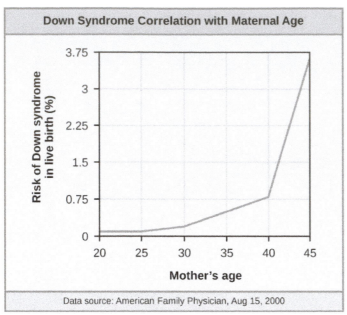

Figure 7.9 The incidence of having a fetus with trisomy 21 increases dramatically with maternal age.

Visualize the addition of a chromosome that leads to Down syndrome in this video simulation (http://openstaxcollege.org/l/down_syndrome2) .

Humans display dramatic deleterious effects with autosomal trisomies and monosomies. Therefore, it may seem counterintuitive that human females and males can function normally, despite carrying different numbers of the X chromosome. In part, this occurs because of a process called **X inactivation**. Early in development, when female mammalian embryos consist of just a few thousand cells, one X chromosome in each cell inactivates by condensing into a structure called a Barr body. The genes on the inactive X chromosome are not expressed. The particular X chromosome (maternally or paternally derived) that is inactivated in each cell is random, but once the inactivation occurs, all cells descended from that cell will have the same inactive X chromosome. By this process, females compensate for their double genetic dose of X chromosome.

In so-called "tortoiseshell" cats, X inactivation is observed as coat-color variegation (Figure 7.10). Females heterozygous for an X-linked coat color gene will express one of two different coat colors over different regions of their body, corresponding to whichever X chromosome is inactivated in the embryonic cell progenitor of that region. When you see a tortoiseshell cat, you will know that it has to be a female.

Figure 7.10 Embryonic inactivation of one of two different X chromosomes encoding different coat colors gives rise to the tortoiseshell phenotype in cats. (credit: Michael Bodega)

In an individual carrying an abnormal number of X chromosomes, cellular mechanisms will inactivate all but one X in each of her cells. As a result, X-chromosomal abnormalities are typically associated with mild mental and physical defects, as well as sterility. If the X chromosome is absent altogether, the individual will not develop.

Several errors in sex chromosome number have been characterized. Individuals with three X chromosomes, called triplo-X, appear female but express developmental delays and reduced fertility. The XXY chromosome complement, corresponding to one type of Klinefelter syndrome, corresponds to male individuals with small testes, enlarged breasts, and reduced body hair. The extra X chromosome undergoes inactivation to compensate for the excess genetic dosage. Turner syndrome, characterized as an X0 chromosome complement (i.e., only a single sex chromosome), corresponds to a female individual with short stature, webbed skin in the neck region, hearing and cardiac impairments, and sterility.

An individual with more than the correct number of chromosome sets (two for diploid species) is called **polyploid**. For instance, fertilization of an abnormal diploid egg with a normal haploid sperm would yield a triploid zygote. Polyploid animals are extremely rare, with only a few examples among the flatworms, crustaceans, amphibians, fish, and lizards. Triploid animals are sterile because meiosis cannot proceed normally with an odd number of chromosome sets. In contrast, polyploidy is very common in the plant kingdom, and polyploid plants tend to be larger and more robust than euploids of their species.

Chromosome Structural Rearrangements

Cytologists have characterized numerous structural rearrangements in chromosomes, including partial duplications, deletions, inversions, and translocations. Duplications and deletions often produce offspring that survive but exhibit physical and mental abnormalities. Cri-du-chat (from the French for "cry of the cat") is a syndrome associated with nervous system abnormalities and identifiable physical features that results from a deletion of most of the small arm of chromosome 5 (Figure 7.11). Infants with this genotype emit a characteristic high-pitched cry upon which the disorder's name is based.

Figure 7.11 This individual with cri-du-chat syndrome is shown at various ages: (A) age two, (B) age four, (C) age nine, and (D) age 12. (credit: Paola Cerruti Mainardi)

Chromosome inversions and translocations can be identified by observing cells during meiosis because homologous chromosomes with a rearrangement in one of the pair must contort to maintain appropriate gene alignment and pair effectively during prophase I.

A **chromosome inversion** is the detachment, 180° rotation, and reinsertion of part of a chromosome (Figure 7.12). Unless they disrupt a gene sequence, inversions only change the orientation of genes and are likely to have more mild effects than aneuploid errors.

e**I**olution IN ACTION

The Chromosome 18 Inversion

Not all structural rearrangements of chromosomes produce nonviable, impaired, or infertile individuals. In rare instances, such a change can result in the evolution of a new species. In fact, an inversion in chromosome 18 appears to have contributed to the evolution of humans. This inversion is not present in our closest genetic relatives, the chimpanzees.

The chromosome 18 inversion is believed to have occurred in early humans following their divergence from a common ancestor with chimpanzees approximately five million years ago. Researchers have suggested that a long stretch of DNA was duplicated on chromosome 18 of an ancestor to humans, but that during the duplication it was inverted (inserted into the chromosome in reverse orientation.

A comparison of human and chimpanzee genes in the region of this inversion indicates that two genes—*ROCK1* and *USP14*—are farther apart on human chromosome 18 than they are on the corresponding chimpanzee chromosome. This suggests that one of the inversion breakpoints occurred between these two genes. Interestingly, humans and chimpanzees express *USP14* at distinct levels in specific cell types, including cortical cells and fibroblasts. Perhaps the chromosome 18 inversion in an ancestral human repositioned specific genes and reset their expression levels in a useful way. Because both *ROCK1* and *USP14* code for enzymes, a change in their expression could alter cellular function. It is not known how this inversion contributed to hominid evolution, but it appears to be a significant factor in the divergence of humans from other primates.[2]

A translocation occurs when a segment of a chromosome dissociates and reattaches to a different, nonhomologous chromosome. Translocations can be benign or have devastating effects, depending on how the positions of genes are altered with respect to regulatory sequences. Notably, specific translocations have been associated with several cancers and with schizophrenia. Reciprocal translocations result from the exchange of chromosome segments between two nonhomologous chromosomes such that there is no gain or loss of genetic information (Figure 7.12).

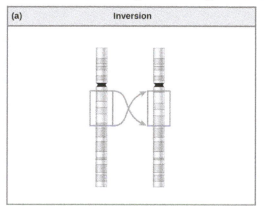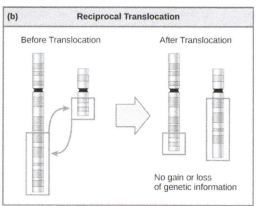

Figure 7.12 An (a) inversion occurs when a chromosome segment breaks from the chromosome, reverses its orientation, and then reattaches in the original position. A (b) reciprocal translocation occurs between two nonhomologous chromosomes and does not cause any genetic information to be lost or duplicated. (credit: modification of work by National Human Genome Research Institute (USA)

---

2. V Goidts, et al., "Segmental duplication associated with the human-specific inversion of chromosome 18: a further example of the impact of segmental duplications on karyotype and genome evolution in primates," *Human Genetics*, 115 (2004):116–22.

KEY TERMS

alternation of generations a life-cycle type in which the diploid and haploid stages alternate

aneuploid an individual with an error in chromosome number; includes deletions and duplications of chromosome segments

autosome any of the non-sex chromosomes

chiasmata (singular = *chiasma*) the structure that forms at the crossover points after genetic material is exchanged

chromosome inversion the detachment, 180° rotation, and reinsertion of a chromosome arm

crossing over (also, recombination) the exchange of genetic material between homologous chromosomes resulting in chromosomes that incorporate genes from both parents of the organism forming reproductive cells

diploid-dominant a life-cycle type in which the multicellular diploid stage is prevalent

euploid an individual with the appropriate number of chromosomes for their species

fertilization the union of two haploid cells typically from two individual organisms

gametophyte a multicellular haploid life-cycle stage that produces gametes

germ cell a specialized cell that produces gametes, such as eggs or sperm

haploid-dominant a life-cycle type in which the multicellular haploid stage is prevalent

interkinesis a period of rest that may occur between meiosis I and meiosis II; there is no replication of DNA during interkinesis

karyogram the photographic image of a karyotype

karyotype the number and appearance of an individuals chromosomes, including the size, banding patterns, and centromere position

life cycle the sequence of events in the development of an organism and the production of cells that produce offspring

meiosis a nuclear division process that results in four haploid cells

meiosis I the first round of meiotic cell division; referred to as reduction division because the resulting cells are haploid

meiosis II the second round of meiotic cell division following meiosis I; sister chromatids are separated from each other, and the result is four unique haploid cells

monosomy an otherwise diploid genotype in which one chromosome is missing

nondisjunction the failure of synapsed homologs to completely separate and migrate to separate poles during the first cell division of meiosis

polyploid an individual with an incorrect number of chromosome sets

recombinant describing something composed of genetic material from two sources, such as a chromosome with both maternal and paternal segments of DNA

reduction division a nuclear division that produces daughter nuclei each having one-half as many chromosome sets as the parental nucleus; meiosis I is a reduction division

somatic cell all the cells of a multicellular organism except the gamete-forming cells

sporophyte a multicellular diploid life-cycle stage that produces spores

synapsis the formation of a close association between homologous chromosomes during prophase I

Download for free at https://openstax.org/details/books/concepts-biology

tetrad two duplicated homologous chromosomes (four chromatids) bound together by chiasmata during prophase I

translocation the process by which one segment of a chromosome dissociates and reattaches to a different, nonhomologous chromosome

trisomy an otherwise diploid genotype in which one entire chromosome is duplicated

X inactivation the condensation of X chromosomes into Barr bodies during embryonic development in females to compensate for the double genetic dose

CHAPTER SUMMARY

7.1 Sexual Reproduction

Nearly all eukaryotes undergo sexual reproduction. The variation introduced into the reproductive cells by meiosis appears to be one of the advantages of sexual reproduction that has made it so successful. Meiosis and fertilization alternate in sexual life cycles. The process of meiosis produces genetically unique reproductive cells called gametes, which have half the number of chromosomes as the parent cell. Fertilization, the fusion of haploid gametes from two individuals, restores the diploid condition. Thus, sexually reproducing organisms alternate between haploid and diploid stages. However, the ways in which reproductive cells are produced and the timing between meiosis and fertilization vary greatly. There are three main categories of life cycles: diploid-dominant, demonstrated by most animals; haploid-dominant, demonstrated by all fungi and some algae; and alternation of generations, demonstrated by plants and some algae.

7.2 Meiosis

Sexual reproduction requires that diploid organisms produce haploid cells that can fuse during fertilization to form diploid offspring. The process that results in haploid cells is called meiosis. Meiosis is a series of events that arrange and separate chromosomes into daughter cells. During the interphase of meiosis, each chromosome is duplicated. In meiosis, there are two rounds of nuclear division resulting in four nuclei and usually four haploid daughter cells, each with half the number of chromosomes as the parent cell. During meiosis, variation in the daughter nuclei is introduced because of crossover in prophase I and random alignment at metaphase I. The cells that are produced by meiosis are genetically unique.

Meiosis and mitosis share similarities, but have distinct outcomes. Mitotic divisions are single nuclear divisions that produce daughter nuclei that are genetically identical and have the same number of chromosome sets as the original cell. Meiotic divisions are two nuclear divisions that produce four daughter nuclei that are genetically different and have one chromosome set rather than the two sets the parent cell had. The main differences between the processes occur in the first division of meiosis. The homologous chromosomes separate into different nuclei during meiosis I causing a reduction of ploidy level. The second division of meiosis is much more similar to a mitotic division.

7.3 Errors in Meiosis

The number, size, shape, and banding pattern of chromosomes make them easily identifiable in a karyogram and allow for the assessment of many chromosomal abnormalities. Disorders in chromosome number, or aneuploidies, are typically lethal to the embryo, although a few trisomic genotypes are viable. Because of X inactivation, aberrations in sex chromosomes typically have milder effects on an individual. Aneuploidies also include instances in which segments of a chromosome are duplicated or deleted. Chromosome structures also may be rearranged, for example by inversion or translocation. Both of these aberrations can result in negative effects on development, or death. Because they force chromosomes to assume contorted pairings during meiosis I, inversions and translocations are often associated with reduced fertility because of the likelihood of nondisjunction.

ART CONNECTION QUESTIONS

1. Figure 7.2 If a mutation occurs so that a fungus is no longer able to produce a minus mating type, will it still be able to reproduce?

REVIEW QUESTIONS

2. What is a likely evolutionary advantage of sexual reproduction over asexual reproduction?

a. sexual reproduction involves fewer steps

   b. less chance of using up the resources in a given environment
   c. sexual reproduction results in greater variation in the offspring
   d. sexual reproduction is more cost-effective

3. Which type of life cycle has both a haploid and diploid multicellular stage?
   a. an asexual life cycle
   b. diploid-dominant
   c. haploid-dominant
   d. alternation of generations

4. Which event leads to a diploid cell in a life cycle?

   a. meiosis
   b. fertilization
   c. alternation of generations
   d. mutation

5. Meiosis produces _____ daughter cells.
   a. two haploid
   b. two diploid
   c. four haploid
   d. four diploid

6. At which stage of meiosis are sister chromatids separated from each other?
   a. prophase I
   b. prophase II
   c. anaphase I
   d. anaphase II

7. The part of meiosis that is similar to mitosis is _____.

   a. meiosis I
   b. anaphase I
   c. meiosis II
   d. interkinesis

8. If a muscle cell of a typical organism has 32 chromosomes, how many chromosomes will be in a gamete of that same organism?
   a. 8
   b. 16
   c. 32
   d. 64

9. The genotype XXY corresponds to:
   a. Klinefelter syndrome
   b. Turner syndrome
   c. Triplo-X
   d. Jacob syndrome

10. Abnormalities in the number of X chromosomes tend to be milder than the same abnormalities in autosomes because of _____.
   a. deletions
   b. nonhomologous recombination
   c. synapsis
   d. X inactivation

11. Aneuploidies are deleterious for the individual because of what phenomenon?
   a. nondisjunction
   b. gene dosage
   c. meiotic errors
   d. X inactivation

CRITICAL THINKING QUESTIONS

12. Explain the advantage that populations of sexually reproducing organisms have over asexually reproducing organisms?

13. Describe the two events that are common to all sexually reproducing organisms and how they fit into the different life cycles of those organisms.

14. Explain how the random alignment of homologous chromosomes during metaphase I contributes to variation in gametes produced by meiosis.

15. In what ways is meiosis II similar to and different from mitosis of a diploid cell?

16. Individuals with trisomy 21 are more likely to survive to adulthood than individuals with trisomy 18. Based on what you know about aneuploidies from this module, what can you hypothesize about chromosomes 21 and 18?

8 | PATTERNS OF INHERITANCE

Figure 8.1 Experimenting with thousands of garden peas, Mendel uncovered the fundamentals of genetics. (credit: modification of work by Jerry Kirkhart)

Chapter Outline

8.1: Mendel's Experiments

8.2: Laws of Inheritance

8.3: Extensions of the Laws of Inheritance

Introduction

Genetics is the study of heredity. Johann Gregor Mendel set the framework for genetics long before chromosomes or genes had been identified, at a time when meiosis was not well understood. Mendel selected a simple biological system and conducted methodical, quantitative analyses using large sample sizes. Because of Mendel's work, the fundamental principles of heredity were revealed. We now know that genes, carried on chromosomes, are the basic functional units of heredity with the ability to be replicated, expressed, or mutated. Today, the postulates put forth by Mendel form the basis of classical, or Mendelian, genetics. Not all genes are transmitted from parents to offspring according to Mendelian genetics, but Mendel's experiments serve as an excellent starting point for thinking about inheritance.

8.1 | Mendel's Experiments

By the end of this section, you will be able to:

- Explain the scientific reasons for the success of Mendel's experimental work
- Describe the expected outcomes of monohybrid crosses involving dominant and recessive alleles

Figure 8.2 Johann Gregor Mendel set the framework for the study of genetics.

Johann Gregor Mendel (1822–1884) (Figure 8.2) was a lifelong learner, teacher, scientist, and man of faith. As a young adult, he joined the Augustinian Abbey of St. Thomas in Brno in what is now the Czech Republic. Supported by the monastery, he taught physics, botany, and natural science courses at the secondary and university levels. In 1856, he began a decade-long research pursuit involving inheritance patterns in honeybees and plants, ultimately settling on pea plants as his primary **model system** (a system with convenient characteristics that is used to study a specific biological phenomenon to gain understanding to be applied to other systems). In 1865, Mendel presented the results of his experiments with nearly 30,000 pea plants to the local natural history society. He demonstrated that traits are transmitted faithfully from parents to offspring in specific patterns. In 1866, he published his work, *Experiments in Plant Hybridization*,[1] in the proceedings of the Natural History Society of Brünn.

Mendel's work went virtually unnoticed by the scientific community, which incorrectly believed that the process of inheritance involved a blending of parental traits that produced an intermediate physical appearance in offspring. This hypothetical process appeared to be correct because of what we know now as continuous variation. **Continuous variation** is the range of small differences we see among individuals in a characteristic like human height. It does appear that offspring are a "blend" of their parents' traits when we look at characteristics that exhibit continuous variation. Mendel worked instead with traits that show **discontinuous variation**. Discontinuous variation is the variation seen among individuals when each individual shows one of two—or a very few—easily distinguishable traits, such as violet or white flowers. Mendel's choice of these kinds of traits allowed him to see experimentally that the traits were not blended in the offspring as would have been expected at the time, but that they were inherited as distinct traits. In 1868, Mendel became abbot of the monastery and exchanged his scientific pursuits for his pastoral duties. He was not recognized for his extraordinary scientific contributions during his lifetime; in fact, it was not until 1900 that his work was rediscovered, reproduced, and revitalized by scientists on the brink of discovering the chromosomal basis of heredity.

Mendel's Crosses

Mendel's seminal work was accomplished using the garden pea, *Pisum sativum*, to study inheritance. This species naturally self-fertilizes, meaning that pollen encounters ova within the same flower. The flower petals remain sealed tightly until pollination is completed to prevent the pollination of other plants. The result is highly inbred, or "true-breeding," pea plants.

1. Johann Gregor Mendel, "Versuche über Pflanzenhybriden." *Verhandlungen des naturforschenden Vereines in Brünn*, Bd. IV für das Jahr, 1865 Abhandlungen (1866):3–47. [for English translation, see http://www.mendelweb.org/Mendel.plain.html]

These are plants that always produce offspring that look like the parent. By experimenting with true-breeding pea plants, Mendel avoided the appearance of unexpected traits in offspring that might occur if the plants were not true breeding. The garden pea also grows to maturity within one season, meaning that several generations could be evaluated over a relatively short time. Finally, large quantities of garden peas could be cultivated simultaneously, allowing Mendel to conclude that his results did not come about simply by chance.

Mendel performed **hybridizations**, which involve mating two true-breeding individuals that have different traits. In the pea, which is naturally self-pollinating, this is done by manually transferring pollen from the anther of a mature pea plant of one variety to the stigma of a separate mature pea plant of the second variety.

Plants used in first-generation crosses were called **P**, or parental generation, plants (Figure 8.3). Mendel collected the seeds produced by the P plants that resulted from each cross and grew them the following season. These offspring were called the **F_1**, or the first filial (filial = daughter or son), generation. Once Mendel examined the characteristics in the F_1 generation of plants, he allowed them to self-fertilize naturally. He then collected and grew the seeds from the F_1 plants to produce the **F_2**, or second filial, generation. Mendel's experiments extended beyond the F_2 generation to the F_3 generation, F_4 generation, and so on, but it was the ratio of characteristics in the P, F_1, and F_2 generations that were the most intriguing and became the basis of Mendel's postulates.

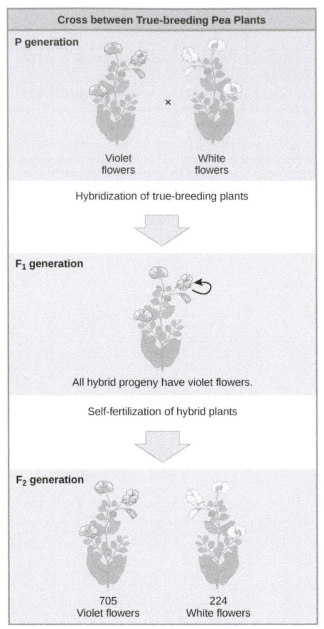

Figure 8.3 Mendel's process for performing crosses included examining flower color.

Garden Pea Characteristics Revealed the Basics of Heredity

In his 1865 publication, Mendel reported the results of his crosses involving seven different characteristics, each with two contrasting traits. A **trait** is defined as a variation in the physical appearance of a heritable characteristic. The characteristics included plant height, seed texture, seed color, flower color, pea-pod size, pea-pod color, and flower position. For the characteristic of flower color, for example, the two contrasting traits were white versus violet. To fully examine each characteristic, Mendel generated large numbers of F_1 and F_2 plants and reported results from thousands of F_2 plants.

What results did Mendel find in his crosses for flower color? First, Mendel confirmed that he was using plants that bred true for white or violet flower color. Irrespective of the number of generations that Mendel examined, all self-crossed offspring of parents with white flowers had white flowers, and all self-crossed offspring of parents with violet flowers had violet flowers. In addition, Mendel confirmed that, other than flower color, the pea plants were physically identical. This was an important check to make sure that the two varieties of pea plants only differed with respect to one trait, flower color.

Once these validations were complete, Mendel applied the pollen from a plant with violet flowers to the stigma of a plant with white flowers. After gathering and sowing the seeds that resulted from this cross, Mendel found that 100 percent of the F$_1$ hybrid generation had violet flowers. Conventional wisdom at that time would have predicted the hybrid flowers to be pale violet or for hybrid plants to have equal numbers of white and violet flowers. In other words, the contrasting parental traits were expected to blend in the offspring. Instead, Mendel's results demonstrated that the white flower trait had completely disappeared in the F$_1$ generation.

Importantly, Mendel did not stop his experimentation there. He allowed the F$_1$ plants to self-fertilize and found that 705 plants in the F$_2$ generation had violet flowers and 224 had white flowers. This was a ratio of 3.15 violet flowers to one white flower, or approximately 3:1. When Mendel transferred pollen from a plant with violet flowers to the stigma of a plant with white flowers and vice versa, he obtained approximately the same ratio irrespective of which parent—male or female—contributed which trait. This is called a **reciprocal cross**—a paired cross in which the respective traits of the male and female in one cross become the respective traits of the female and male in the other cross. For the other six characteristics that Mendel examined, the F$_1$ and F$_2$ generations behaved in the same way that they behaved for flower color. One of the two traits would disappear completely from the F$_1$ generation, only to reappear in the F$_2$ generation at a ratio of roughly 3:1 (Figure 8.4).

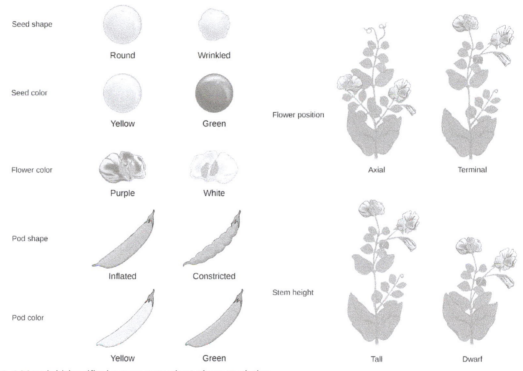

Figure 8.4 Mendel identified seven pea plant characteristics.

Upon compiling his results for many thousands of plants, Mendel concluded that the characteristics could be divided into expressed and latent traits. He called these dominant and recessive traits, respectively. **Dominant** traits are those that are inherited unchanged in a hybridization. **Recessive** traits become latent, or disappear in the offspring of a hybridization. The recessive trait does, however, reappear in the progeny of the hybrid offspring. An example of a dominant trait is the violet-colored flower trait. For this same characteristic (flower color), white-colored flowers are a recessive trait. The fact that the recessive trait reappeared in the F$_2$ generation meant that the traits remained separate (and were not blended) in the plants of the F$_1$ generation. Mendel proposed that this was because the plants possessed two copies of the trait for the flower-color characteristic, and that each parent transmitted one of their two copies to their offspring, where they came together. Moreover, the physical observation of a dominant trait could mean that the genetic composition of the organism included two dominant versions of the characteristic, or that it included one dominant and one recessive version. Conversely, the observation of a recessive trait meant that the organism lacked any dominant versions of this characteristic.

For an excellent review of Mendel's experiments and to perform your own crosses and identify patterns of inheritance, visit the Mendel's Peas (http://openstaxcollege.org/l/mendels_peas) web lab.

8.2 | Laws of Inheritance

By the end of this section, you will be able to:

- Explain the relationship between genotypes and phenotypes in dominant and recessive gene systems
- Use a Punnett square to calculate the expected proportions of genotypes and phenotypes in a monohybrid cross
- Explain Mendel's law of segregation and independent assortment in terms of genetics and the events of meiosis
- Explain the purpose and methods of a test cross

The seven characteristics that Mendel evaluated in his pea plants were each expressed as one of two versions, or traits. Mendel deduced from his results that each individual had two discrete copies of the characteristic that are passed individually to offspring. We now call those two copies genes, which are carried on chromosomes. The reason we have two copies of each gene is that we inherit one from each parent. In fact, it is the chromosomes we inherit and the two copies of each gene are located on paired chromosomes. Recall that in meiosis these chromosomes are separated out into haploid gametes. This separation, or segregation, of the homologous chromosomes means also that only one of the copies of the gene gets moved into a gamete. The offspring are formed when that gamete unites with one from another parent and the two copies of each gene (and chromosome) are restored.

For cases in which a single gene controls a single characteristic, a diploid organism has two genetic copies that may or may not encode the same version of that characteristic. For example, one individual may carry a gene that determines white flower color and a gene that determines violet flower color. Gene variants that arise by mutation and exist at the same relative locations on homologous chromosomes are called **alleles**. Mendel examined the inheritance of genes with just two allele forms, but it is common to encounter more than two alleles for any given gene in a natural population.

Phenotypes and Genotypes

Two alleles for a given gene in a diploid organism are expressed and interact to produce physical characteristics. The observable traits expressed by an organism are referred to as its **phenotype**. An organism's underlying genetic makeup, consisting of both the physically visible and the non-expressed alleles, is called its **genotype**. Mendel's hybridization experiments demonstrate the difference between phenotype and genotype. For example, the phenotypes that Mendel observed in his crosses between pea plants with differing traits are connected to the diploid genotypes of the plants in the P, F_1, and F_2 generations. We will use a second trait that Mendel investigated, seed color, as an example. Seed color is governed by a single gene with two alleles. The yellow-seed allele is dominant and the green-seed allele is recessive. When true-breeding plants were cross-fertilized, in which one parent had yellow seeds and one had green seeds, all of the F_1 hybrid offspring had yellow seeds. That is, the hybrid offspring were phenotypically identical to the true-breeding parent with yellow seeds. However, we know that the allele donated by the parent with green seeds was not simply lost because it reappeared in some of the F_2 offspring (Figure 8.5). Therefore, the F_1 plants must have been genotypically different from the parent with yellow seeds.

The P plants that Mendel used in his experiments were each homozygous for the trait he was studying. Diploid organisms that are **homozygous** for a gene have two identical alleles, one on each of their homologous chromosomes. The genotype is often written as YY or yy, for which each letter represents one of the two alleles in the genotype. The dominant allele is capitalized and the recessive allele is lower case. The letter used for the gene (seed color in this case) is usually related to the dominant trait (yellow allele, in this case, or "Y"). Mendel's parental pea plants always bred true because both produced gametes carried the same allele. When P plants with contrasting traits were cross-fertilized, all of the offspring were **heterozygous** for the contrasting trait, meaning their genotype had different alleles for the gene being examined. For example, the F_1 yellow plants that received a Y allele from their yellow parent and a y allele from their green parent had the genotype Yy.

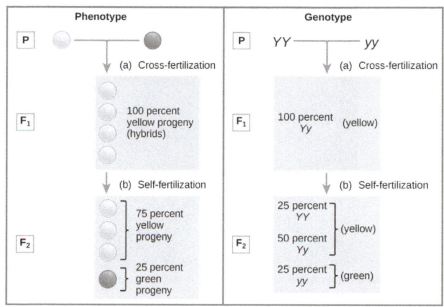

Figure 8.5 Phenotypes are physical expressions of traits that are transmitted by alleles. Capital letters represent dominant alleles and lowercase letters represent recessive alleles. The phenotypic ratios are the ratios of visible characteristics. The genotypic ratios are the ratios of gene combinations in the offspring, and these are not always distinguishable in the phenotypes.

Law of Dominance

Our discussion of homozygous and heterozygous organisms brings us to why the F_1 heterozygous offspring were identical to one of the parents, rather than expressing both alleles. In all seven pea-plant characteristics, one of the two contrasting alleles was dominant, and the other was recessive. Mendel called the dominant allele the expressed unit factor; the recessive allele was referred to as the latent unit factor. We now know that these so-called unit factors are actually genes on homologous chromosomes. For a gene that is expressed in a dominant and recessive pattern, homozygous dominant and heterozygous organisms will look identical (that is, they will have different genotypes but the same phenotype), and the recessive allele will only be observed in homozygous recessive individuals (Table 8.1).

Correspondence between Genotype and Phenotype for a Dominant-Recessive Characteristic.

	Homozygous	Heterozygous	Homozygous
Genotype	YY	Yy	yy
Phenotype	yellow	yellow	green

Table 8.1

Mendel's **law of dominance** states that in a heterozygote, one trait will conceal the presence of another trait for the same characteristic. For example, when crossing true-breeding violet-flowered plants with true-breeding white-flowered plants, all of the offspring were violet-flowered, even though they all had one allele for violet and one allele for white. Rather than both alleles contributing to a phenotype, the dominant allele will be expressed exclusively. The recessive allele will remain latent, but will be transmitted to offspring in the same manner as that by which the dominant allele is transmitted. The recessive trait will only be expressed by offspring that have two copies of this allele (Figure 8.6), and these offspring will breed true when self-crossed.

Figure 8.6 The allele for albinism, expressed here in humans, is recessive. Both of this child's parents carried the recessive allele.

Monohybrid Cross and the Punnett Square

When fertilization occurs between two true-breeding parents that differ by only the characteristic being studied, the process is called a **monohybrid** cross, and the resulting offspring are called monohybrids. Mendel performed seven types of monohybrid crosses, each involving contrasting traits for different characteristics. Out of these crosses, all of the F_1 offspring had the phenotype of one parent, and the F_2 offspring had a 3:1 phenotypic ratio. On the basis of these results, Mendel postulated that each parent in the monohybrid cross contributed one of two paired unit factors to each offspring, and every possible combination of unit factors was equally likely.

The results of Mendel's research can be explained in terms of probabilities, which are mathematical measures of likelihood. The probability of an event is calculated by the number of times the event occurs divided by the total number of opportunities for the event to occur. A probability of one (100 percent) for some event indicates that it is guaranteed to occur, whereas a probability of zero (0 percent) indicates that it is guaranteed to not occur, and a probability of 0.5 (50 percent) means it has an equal chance of occurring or not occurring.

To demonstrate this with a monohybrid cross, consider the case of true-breeding pea plants with yellow versus green seeds. The dominant seed color is yellow; therefore, the parental genotypes were YY for the plants with yellow seeds and yy for the plants with green seeds. A **Punnett square**, devised by the British geneticist Reginald Punnett, is useful for determining probabilities because it is drawn to predict all possible outcomes of all possible random fertilization events and their expected frequencies. Figure 8.9 shows a Punnett square for a cross between a plant with yellow peas and one with green peas. To prepare a Punnett square, all possible combinations of the parental alleles (the genotypes of the gametes) are listed along the top (for one parent) and side (for the other parent) of a grid. The combinations of egg and sperm gametes are then made in the boxes in the table on the basis of which alleles are combining. Each box then represents the diploid genotype of a zygote, or fertilized egg. Because each possibility is equally likely, genotypic ratios can be determined from a Punnett square. If the pattern of inheritance (dominant and recessive) is known, the phenotypic ratios can be inferred as well. For a monohybrid cross of two true-breeding parents, each parent contributes one type of allele. In this case, only one genotype is possible in the F_1 offspring. All offspring are Yy and have yellow seeds.

When the F_1 offspring are crossed with each other, each has an equal probability of contributing either a Y or a y to the F_2 offspring. The result is a 1 in 4 (25 percent) probability of both parents contributing a Y, resulting in an offspring with a yellow phenotype; a 25 percent probability of parent A contributing a Y and parent B a y, resulting in offspring with a yellow phenotype; a 25 percent probability of parent A contributing a y and parent B a Y, also resulting in a yellow phenotype; and a (25 percent) probability of both parents contributing a y, resulting in a green phenotype. When counting all four possible outcomes, there is a 3 in 4 probability of offspring having the yellow phenotype and a 1 in 4 probability of offspring having the green phenotype. This explains why the results of Mendel's F_2 generation occurred in a 3:1 phenotypic ratio. Using large numbers of crosses, Mendel was able to calculate probabilities, found that they fit the model of inheritance, and use these to predict the outcomes of other crosses.

Law of Segregation

Observing that true-breeding pea plants with contrasting traits gave rise to F_1 generations that all expressed the dominant trait and F_2 generations that expressed the dominant and recessive traits in a 3:1 ratio, Mendel proposed the **law of segregation**. This law states that paired unit factors (genes) must segregate equally into gametes such that offspring have an equal likelihood of inheriting either factor. For the F_2 generation of a monohybrid cross, the following three possible combinations of genotypes result: homozygous dominant, heterozygous, or homozygous recessive. Because heterozygotes could arise from two different pathways (receiving one dominant and one recessive allele from either parent), and because heterozygotes and homozygous dominant individuals are phenotypically identical, the law supports Mendel's observed 3:1 phenotypic ratio. The equal segregation of alleles is the reason we can apply the Punnett square to accurately predict the offspring of parents with known genotypes. The physical basis of Mendel's law of segregation is the first division of meiosis in which the homologous chromosomes with their different versions of each gene are segregated into daughter nuclei. This process was not understood by the scientific community during Mendel's lifetime (Figure 8.7).

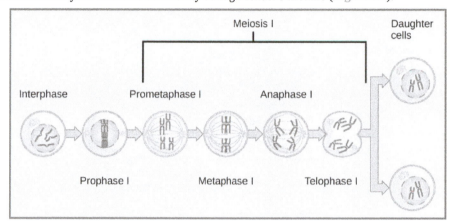

Figure 8.7 The first division in meiosis is shown.

Test Cross

Beyond predicting the offspring of a cross between known homozygous or heterozygous parents, Mendel also developed a way to determine whether an organism that expressed a dominant trait was a heterozygote or a homozygote. Called the **test cross**, this technique is still used by plant and animal breeders. In a test cross, the dominant-expressing organism is crossed with an organism that is homozygous recessive for the same characteristic. If the dominant-expressing organism is a homozygote, then all F_1 offspring will be heterozygotes expressing the dominant trait (Figure 8.8). Alternatively, if the dominant-expressing organism is a heterozygote, the F_1 offspring will exhibit a 1:1 ratio of heterozygotes and recessive homozygotes (Figure 8.8). The test cross further validates Mendel's postulate that pairs of unit factors segregate equally.

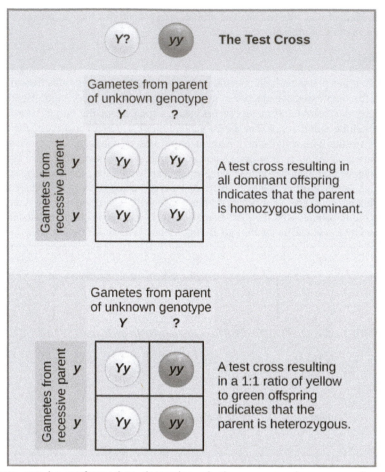

Figure 8.8 A test cross can be performed to determine whether an organism expressing a dominant trait is a homozygote or a heterozygote.

a r t CONNECTION

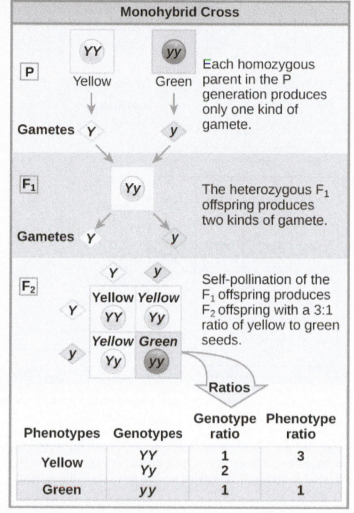

Monohybrid Cross

P — YY (Yellow)   yy (Green)   Each homozygous parent in the P generation produces only one kind of gamete.

Gametes   Y   y

F₁ — Yy   The heterozygous F₁ offspring produces two kinds of gamete.

Gametes   Y   y

F₂ —

	Y	y
Y	Yellow YY	Yellow Yy
y	Yellow Yy	Green yy

Self-pollination of the F₁ offspring produces F₂ offspring with a 3:1 ratio of yellow to green seeds.

Ratios

Phenotypes	Genotypes	Genotype ratio	Phenotype ratio
Yellow	YY Yy	1 2	3
Green	yy	1	1

Figure 8.9 This Punnett square shows the cross between plants with yellow seeds and green seeds. The cross between the true-breeding P plants produces F₁ heterozygotes that can be self-fertilized. The self-cross of the F₁ generation can be analyzed with a Punnett square to predict the genotypes of the F₂ generation. Given an inheritance pattern of dominant–recessive, the genotypic and phenotypic ratios can then be determined.

In pea plants, round peas (*R*) are dominant to wrinkled peas (*r*). You do a test cross between a pea plant with wrinkled peas (genotype *rr*) and a plant of unknown genotype that has round peas. You end up with three plants, all which have round peas. From this data, can you tell if the parent plant is homozygous dominant or heterozygous?

Law of Independent Assortment

Mendel's **law of independent assortment** states that genes do not influence each other with regard to the sorting of alleles into gametes, and every possible combination of alleles for every gene is equally likely to occur. Independent assortment of genes can be illustrated by the **dihybrid** cross, a cross between two true-breeding parents that express different traits for two characteristics. Consider the characteristics of seed color and seed texture for two pea plants, one that has wrinkled, green seeds (*rryy*) and another that has round, yellow seeds (*RRYY*). Because each parent is homozygous, the law of segregation indicates that the gametes for the wrinkled–green plant all are *ry*, and the gametes for the round–yellow plant are all *RY*. Therefore, the F₁ generation of offspring all are *RrYy* (Figure 8.10).

art CONNECTION

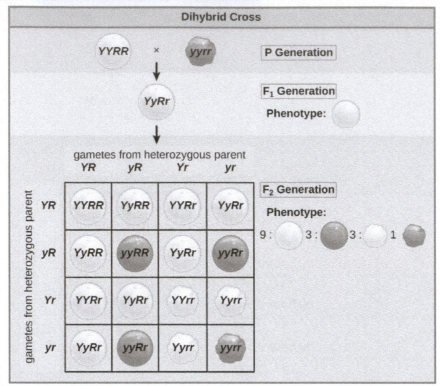

Figure 8.10 A dihybrid cross in pea plants involves the genes for seed color and texture. The P cross produces F_1 offspring that are all heterozygous for both characteristics. The resulting 9:3:3:1 F_2 phenotypic ratio is obtained using a Punnett square.

In pea plants, purple flowers (*P*) are dominant to white (*p*), and yellow peas (*Y*) are dominant to green (*y*). What are the possible genotypes and phenotypes for a cross between *PpYY* and *ppYy* pea plants? How many squares would you need to complete a Punnett square analysis of this cross?

The gametes produced by the F_1 individuals must have one allele from each of the two genes. For example, a gamete could get an *R* allele for the seed shape gene and either a *Y* or a *y* allele for the seed color gene. It cannot get both an *R* and an *r* allele; each gamete can have only one allele per gene. The law of independent assortment states that a gamete into which an *r* allele is sorted would be equally likely to contain either a *Y* or a *y* allele. Thus, there are four equally likely gametes that can be formed when the *RrYy* heterozygote is self-crossed, as follows: *RY*, *rY*, *Ry*, and *ry*. Arranging these gametes along the top and left of a 4 × 4 Punnett square (Figure 8.10) gives us 16 equally likely genotypic combinations. From these genotypes, we find a phenotypic ratio of 9 round–yellow:3 round–green:3 wrinkled–yellow:1 wrinkled–green (Figure 8.10). These are the offspring ratios we would expect, assuming we performed the crosses with a large enough sample size.

The physical basis for the law of independent assortment also lies in meiosis I, in which the different homologous pairs line up in random orientations. Each gamete can contain any combination of paternal and maternal chromosomes (and therefore the genes on them) because the orientation of tetrads on the metaphase plane is random (Figure 8.11).

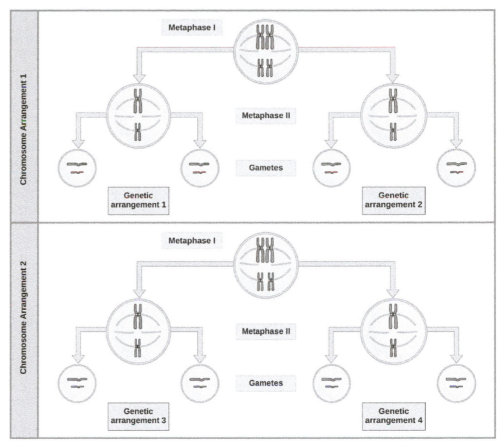

Figure 8.11 The random segregation into daughter nuclei that happens during the first division in meiosis can lead to a variety of possible genetic arrangements.

8.3 | Extensions of the Laws of Inheritance

By the end of this section, you will be able to:

- Identify non-Mendelian inheritance patterns such as incomplete dominance, codominance, multiple alleles, and sex linkage from the results of crosses
- Explain the effect of linkage and recombination on gamete genotypes
- Explain the phenotypic outcomes of epistatic effects among genes

Mendel studied traits with only one mode of inheritance in pea plants. The inheritance of the traits he studied all followed the relatively simple pattern of dominant and recessive alleles for a single characteristic. There are several important modes of inheritance, discovered after Mendel's work, that do not follow the dominant and recessive, single-gene model.

Alternatives to Dominance and Recessiveness

Mendel's experiments with pea plants suggested that: 1) two types of "units" or alleles exist for every gene; 2) alleles maintain their integrity in each generation (no blending); and 3) in the presence of the dominant allele, the recessive allele is hidden, with no contribution to the phenotype. Therefore, recessive alleles can be "carried" and not expressed by individuals. Such heterozygous individuals are sometimes referred to as "carriers." Since then, genetic studies in other organisms have shown that much more complexity exists, but that the fundamental principles of Mendelian genetics still hold true. In the sections to follow, we consider some of the extensions of Mendelism.

Incomplete Dominance

Mendel's results, demonstrating that traits are inherited as dominant and recessive pairs, contradicted the view at that time that offspring exhibited a blend of their parents' traits. However, the heterozygote phenotype occasionally does appear to be intermediate between the two parents. For example, in the snapdragon, *Antirrhinum majus* (Figure 8.12), a cross between a homozygous parent with white flowers ($C^W C^W$) and a homozygous parent with red flowers ($C^R C^R$) will produce offspring with pink flowers ($C^R C^W$). (Note that different genotypic abbreviations are used for Mendelian extensions to distinguish these patterns from simple dominance and recessiveness.) This pattern of inheritance is described as **incomplete dominance**, meaning that one of the alleles appears in the phenotype in the heterozygote, but not to the exclusion of the other, which can also be seen. The allele for red flowers is incompletely dominant over the allele for white flowers. However, the results of a heterozygote self-cross can still be predicted, just as with Mendelian dominant and recessive crosses. In this case, the genotypic ratio would be 1 $C^R C^R$:2 $C^R C^W$:1 $C^W C^W$, and the phenotypic ratio would be 1:2:1 for red:pink:white. The basis for the intermediate color in the heterozygote is simply that the pigment produced by the red allele (anthocyanin) is diluted in the heterozygote and therefore appears pink because of the white background of the flower petals.

Figure 8.12 These pink flowers of a heterozygote snapdragon result from incomplete dominance. (credit: "storebukkebruse"/Flickr)

Codominance

A variation on incomplete dominance is **codominance**, in which both alleles for the same characteristic are simultaneously expressed in the heterozygote. An example of codominance occurs in the ABO blood groups of humans. The A and B alleles are expressed in the form of A or B molecules present on the surface of red blood cells. Homozygotes ($I^A I^A$ and $I^B I^B$) express either the A or the B phenotype, and heterozygotes ($I^A I^B$) express both phenotypes equally. The $I^A I^B$ individual has blood type AB. In a self-cross between heterozygotes expressing a codominant trait, the three possible offspring genotypes are phenotypically distinct. However, the 1:2:1 genotypic ratio characteristic of a Mendelian monohybrid cross still applies (Figure 8.13).

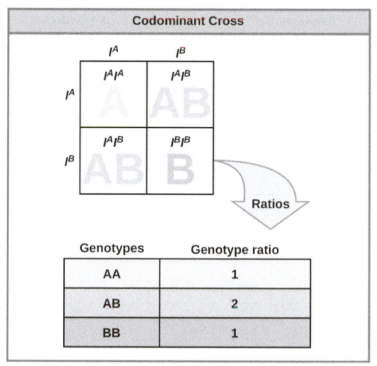

Figure 8.13 This Punnet square shows an AB/AB blood type cross

Multiple Alleles

Mendel implied that only two alleles, one dominant and one recessive, could exist for a given gene. We now know that this is an oversimplification. Although individual humans (and all diploid organisms) can only have two alleles for a given gene, multiple alleles may exist at the population level, such that many combinations of two alleles are observed. Note that when many alleles exist for the same gene, the convention is to denote the most common phenotype or genotype in the natural population as the **wild type** (often abbreviated "+"). All other phenotypes or genotypes are considered variants (mutants) of this typical form, meaning they deviate from the wild type. The variant may be recessive or dominant to the wild-type allele.

An example of multiple alleles is the ABO blood-type system in humans. In this case, there are three alleles circulating in the population. The I^A allele codes for A molecules on the red blood cells, the I^B allele codes for B molecules on the surface of red blood cells, and the i allele codes for no molecules on the red blood cells. In this case, the I^A and I^B alleles are codominant with each other and are both dominant over the i allele. Although there are three alleles present in a population, each individual only gets two of the alleles from their parents. This produces the genotypes and phenotypes shown in Figure 8.14. Notice that instead of three genotypes, there are six different genotypes when there are three alleles. The number of possible phenotypes depends on the dominance relationships between the three alleles.

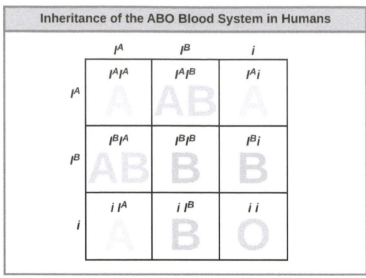

Figure 8.14 Inheritance of the ABO blood system in humans is shown.

Multiple Alleles Confer Drug Resistance in the Malaria Parasite

Malaria is a parasitic disease in humans that is transmitted by infected female mosquitoes, including *Anopheles gambiae*, and is characterized by cyclic high fevers, chills, flu-like symptoms, and severe anemia. *Plasmodium falciparum* and *P. vivax* are the most common causative agents of malaria, and *P. falciparum* is the most deadly. When promptly and correctly treated, *P. falciparum* malaria has a mortality rate of 0.1 percent. However, in some parts of the world, the parasite has evolved resistance to commonly used malaria treatments, so the most effective malarial treatments can vary by geographic region.

In Southeast Asia, Africa, and South America, *P. falciparum* has developed resistance to the anti-malarial drugs chloroquine, mefloquine, and sulfadoxine-pyrimethamine. *P. falciparum*, which is haploid during the life stage in which it is infective to humans, has evolved multiple drug-resistant mutant alleles of the *dhps* gene. Varying degrees of sulfadoxine resistance are associated with each of these alleles. Being haploid, *P. falciparum* needs only one drug-resistant allele to express this trait.

In Southeast Asia, different sulfadoxine-resistant alleles of the *dhps* gene are localized to different geographic regions. This is a common evolutionary phenomenon that comes about because drug-resistant mutants arise in a population and interbreed with other *P. falciparum* isolates in close proximity. Sulfadoxine-resistant parasites cause considerable human hardship in regions in which this drug is widely used as an over-the-counter malaria remedy. As is common with pathogens that multiply to large numbers within an infection cycle, *P. falciparum* evolves relatively rapidly (over a decade or so) in response to the selective pressure of commonly used anti-malarial drugs. For this reason, scientists must constantly work to develop new drugs or drug combinations to combat the worldwide malaria burden.[2]

Sex-Linked Traits

In humans, as well as in many other animals and some plants, the sex of the individual is determined by sex chromosomes—one pair of non-homologous chromosomes. Until now, we have only considered inheritance patterns among non-sex chromosomes, or autosomes. In addition to 22 homologous pairs of autosomes, human females have a homologous pair of X chromosomes, whereas human males have an XY chromosome pair. Although the Y chromosome contains a small region of similarity to the X chromosome so that they can pair during meiosis, the Y chromosome is much shorter and contains fewer genes. When a gene being examined is present on the X, but not the Y, chromosome, it is **X-linked**.

Eye color in *Drosophila*, the common fruit fly, was the first X-linked trait to be identified. Thomas Hunt Morgan mapped this trait to the X chromosome in 1910. Like humans, *Drosophila* males have an XY chromosome pair, and females are XX.

---

2. Sumiti Vinayak et al., "Origin and Evolution of Sulfadoxine Resistant *Plasmodium falciparum*," *PLoS Pathogens* 6 (2010): e1000830.

In flies the wild-type eye color is red (X^W) and is dominant to white eye color (X^w) (Figure 8.15). Because of the location of the eye-color gene, reciprocal crosses do not produce the same offspring ratios. Males are said to be **hemizygous**, in that they have only one allele for any X-linked characteristic. Hemizygosity makes descriptions of dominance and recessiveness irrelevant for XY males. *Drosophila* males lack the white gene on the Y chromosome; that is, their genotype can only be $X^W Y$ or $X^w Y$. In contrast, females have two allele copies of this gene and can be $X^W X^W$, $X^W X^w$, or $X^w X^w$.

Figure 8.15 In *Drosophila*, the gene for eye color is located on the X chromosome. Red eye color is wild-type and is dominant to white eye color.

In an X-linked cross, the genotypes of F_1 and F_2 offspring depend on whether the recessive trait was expressed by the male or the female in the P generation. With respect to *Drosophila* eye color, when the P male expresses the white-eye phenotype and the female is homozygously red-eyed, all members of the F_1 generation exhibit red eyes (Figure 8.16). The F_1 females are heterozygous ($X^W X^w$), and the males are all $X^W Y$, having received their X chromosome from the homozygous dominant P female and their Y chromosome from the P male. A subsequent cross between the $X^W X^w$ female and the $X^W Y$ male would produce only red-eyed females (with $X^W X^W$ or $X^W X^w$ genotypes) and both red- and white-eyed males (with $X^W Y$ or $X^w Y$ genotypes). Now, consider a cross between a homozygous white-eyed female and a male with red eyes. The F_1 generation would exhibit only heterozygous red-eyed females ($X^W X^w$) and only white-eyed males ($X^w Y$). Half of the F_2 females would be red-eyed ($X^W X^w$) and half would be white-eyed ($X^w X^w$). Similarly, half of the F_2 males would be red-eyed ($X^W Y$) and half would be white-eyed ($X^w Y$).

art CONNECTION

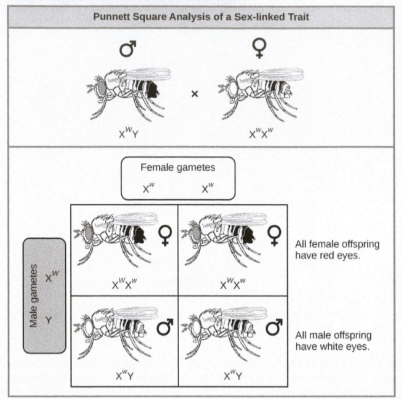

Figure 8.16 Crosses involving sex-linked traits often give rise to different phenotypes for the different sexes of offspring, as is the case for this cross involving red and white eye color in *Drosophila*. In the diagram, *w* is the white-eye mutant allele and *W* is the wild-type, red-eye allele.

What ratio of offspring would result from a cross between a white-eyed male and a female that is heterozygous for red eye color?

Discoveries in fruit fly genetics can be applied to human genetics. When a female parent is homozygous for a recessive X-linked trait, she will pass the trait on to 100 percent of her male offspring, because the males will receive the Y chromosome from the male parent. In humans, the alleles for certain conditions (some color-blindness, hemophilia, and muscular dystrophy) are X-linked. Females who are heterozygous for these diseases are said to be carriers and may not exhibit any phenotypic effects. These females will pass the disease to half of their sons and will pass carrier status to half of their daughters; therefore, X-linked traits appear more frequently in males than females.

In some groups of organisms with sex chromosomes, the sex with the non-homologous sex chromosomes is the female rather than the male. This is the case for all birds. In this case, sex-linked traits will be more likely to appear in the female, in whom they are hemizygous.

Watch this video (http://openstaxcollege.org/l/sex-linked_trts) to learn more about sex-linked traits.

Linked Genes Violate the Law of Independent Assortment

Although all of Mendel's pea plant characteristics behaved according to the law of independent assortment, we now know that some allele combinations are not inherited independently of each other. Genes that are located on separate, non-homologous chromosomes will always sort independently. However, each chromosome contains hundreds or thousands of genes, organized linearly on chromosomes like beads on a string. The segregation of alleles into gametes can be influenced by **linkage**, in which genes that are located physically close to each other on the same chromosome are more likely to be inherited as a pair. However, because of the process of recombination, or "crossover," it is possible for two genes on the same chromosome to behave independently, or as if they are not linked. To understand this, let us consider the biological basis of gene linkage and recombination.

Homologous chromosomes possess the same genes in the same order, though the specific alleles of the gene can be different on each of the two chromosomes. Recall that during interphase and prophase I of meiosis, homologous chromosomes first replicate and then synapse, with like genes on the homologs aligning with each other. At this stage, segments of homologous chromosomes exchange linear segments of genetic material (Figure 8.17). This process is called **recombination**, or crossover, and it is a common genetic process. Because the genes are aligned during recombination, the gene order is not altered. Instead, the result of recombination is that maternal and paternal alleles are combined onto the same chromosome. Across a given chromosome, several recombination events may occur, causing extensive shuffling of alleles.

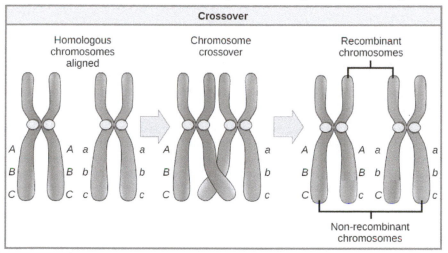

Figure 8.17 The process of crossover, or recombination, occurs when two homologous chromosomes align and exchange a segment of genetic material.

When two genes are located on the same chromosome, they are considered linked, and their alleles tend to be transmitted through meiosis together. To exemplify this, imagine a dihybrid cross involving flower color and plant height in which the genes are next to each other on the chromosome. If one homologous chromosome has alleles for tall plants and red flowers, and the other chromosome has genes for short plants and yellow flowers, then when the gametes are formed, the tall and red alleles will tend to go together into a gamete and the short and yellow alleles will go into other gametes. These are called the parental genotypes because they have been inherited intact from the parents of the individual producing gametes. But unlike if the genes were on different chromosomes, there will be no gametes with tall and yellow alleles and no gametes with short and red alleles. If you create a Punnett square with these gametes, you will see that the classical Mendelian prediction of a 9:3:3:1 outcome of a dihybrid cross would not apply. As the distance between two genes increases, the probability of one or more crossovers between them increases and the genes behave more like they are on separate chromosomes. Geneticists have used the proportion of recombinant gametes (the ones not like the parents) as a measure of how far apart genes are on a chromosome. Using this information, they have constructed linkage maps of genes on chromosomes for well-studied organisms, including humans.

Mendel's seminal publication makes no mention of linkage, and many researchers have questioned whether he encountered linkage but chose not to publish those crosses out of concern that they would invalidate his independent assortment postulate. The garden pea has seven chromosomes, and some have suggested that his choice of seven characteristics was not a coincidence. However, even if the genes he examined were not located on separate chromosomes, it is possible that he simply did not observe linkage because of the extensive shuffling effects of recombination.

Epistasis

Mendel's studies in pea plants implied that the sum of an individual's phenotype was controlled by genes (or as he called them, unit factors), such that every characteristic was distinctly and completely controlled by a single gene. In fact, single observable characteristics are almost always under the influence of multiple genes (each with two or more alleles) acting in unison. For example, at least eight genes contribute to eye color in humans.

Eye color in humans is determined by multiple alleles. Use the Eye Color Calculator (http://openstaxcollege.org/l/ eye_color_calc) to predict the eye color of children from parental eye color.

In some cases, several genes can contribute to aspects of a common phenotype without their gene products ever directly interacting. In the case of organ development, for instance, genes may be expressed sequentially, with each gene adding to the complexity and specificity of the organ. Genes may function in complementary or synergistic fashions, such that two or more genes expressed simultaneously affect a phenotype. An apparent example of this occurs with human skin color, which appears to involve the action of at least three (and probably more) genes. Cases in which inheritance for a characteristic like skin color or human height depend on the combined effects of numerous genes are called polygenic inheritance.

Genes may also oppose each other, with one gene suppressing the expression of another. In **epistasis**, the interaction between genes is antagonistic, such that one gene masks or interferes with the expression of another. "Epistasis" is a word composed of Greek roots meaning "standing upon." The alleles that are being masked or silenced are said to be hypostatic to the epistatic alleles that are doing the masking. Often the biochemical basis of epistasis is a gene pathway in which expression of one gene is dependent on the function of a gene that precedes or follows it in the pathway.

An example of epistasis is pigmentation in mice. The wild-type coat color, agouti (AA) is dominant to solid-colored fur (aa). However, a separate gene C, when present as the recessive homozygote (cc), negates any expression of pigment from the A gene and results in an albino mouse (Figure 8.18). Therefore, the genotypes *AAcc*, *Aacc*, and *aacc* all produce the same albino phenotype. A cross between heterozygotes for both genes (*AaCc* x *AaCc*) would generate offspring with a phenotypic ratio of 9 agouti:3 black:4 albino (Figure 8.18). In this case, the *C* gene is epistatic to the *A* gene.

Download for free at https://openstax.org/details/books/concepts-biology

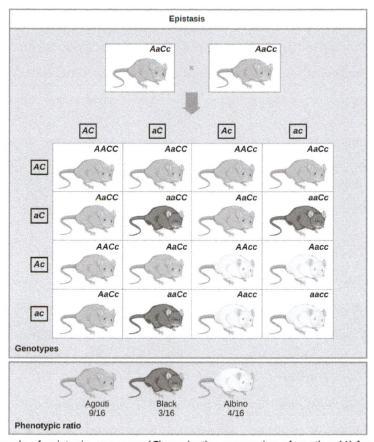

Figure 8.18 In this example of epistasis, one gene (*C*) masks the expression of another (*A*) for coat color. When the *C* allele is present, coat color is expressed; when it is absent (*cc*), no coat color is expressed. Coat color depends on the *A* gene, which shows dominance, with the recessive homozygote showing a different phenotype than the heterozygote or dominant homozygote.

KEY TERMS

allele one of two or more variants of a gene that determines a particular trait for a characteristic

codominance in a heterozygote, complete and simultaneous expression of both alleles for the same characteristic

continuous variation a variation in a characteristic in which individuals show a range of traits with small differences between them

dihybrid the result of a cross between two true-breeding parents that express different traits for two characteristics

discontinuous variation a variation in a characteristic in which individuals show two, or a few, traits with large differences between them

dominant describes a trait that masks the expression of another trait when both versions of the gene are present in an individual

epistasis an interaction between genes such that one gene masks or interferes with the expression of another

F$_1$ the first filial generation in a cross; the offspring of the parental generation

F$_2$ the second filial generation produced when F$_1$ individuals are self-crossed or fertilized with each other

genotype the underlying genetic makeup, consisting of both physically visible and non-expressed alleles, of an organism

hemizygous the presence of only one allele for a characteristic, as in X-linkage; hemizygosity makes descriptions of dominance and recessiveness irrelevant

heterozygous having two different alleles for a given gene on the homologous chromosomes

homozygous having two identical alleles for a given gene on the homologous chromosomes

hybridization the process of mating two individuals that differ, with the goal of achieving a certain characteristic in their offspring

incomplete dominance in a heterozygote, expression of two contrasting alleles such that the individual displays an intermediate phenotype

law of dominance in a heterozygote, one trait will conceal the presence of another trait for the same characteristic

law of independent assortment genes do not influence each other with regard to sorting of alleles into gametes; every possible combination of alleles is equally likely to occur

law of segregation paired unit factors (i.e., genes) segregate equally into gametes such that offspring have an equal likelihood of inheriting any combination of factors

linkage a phenomenon in which alleles that are located in close proximity to each other on the same chromosome are more likely to be inherited together

model system a species or biological system used to study a specific biological phenomenon to gain understanding that will be applied to other species

monohybrid the result of a cross between two true-breeding parents that express different traits for only one characteristic

P the parental generation in a cross

phenotype the observable traits expressed by an organism

Punnett square a visual representation of a cross between two individuals in which the gametes of each individual are denoted along the top and side of a grid, respectively, and the possible zygotic genotypes are recombined at each box in the grid

recessive describes a trait whose expression is masked by another trait when the alleles for both traits are present in an individual

reciprocal cross a paired cross in which the respective traits of the male and female in one cross become the respective traits of the female and male in the other cross

recombination the process during meiosis in which homologous chromosomes exchange linear segments of genetic material, thereby dramatically increasing genetic variation in the offspring and separating linked genes

test cross a cross between a dominant expressing individual with an unknown genotype and a homozygous recessive individual; the offspring phenotypes indicate whether the unknown parent is heterozygous or homozygous for the dominant trait

trait a variation in an inherited characteristic

wild type the most commonly occurring genotype or phenotype for a given characteristic found in a population

X-linked a gene present on the X chromosome, but not the Y chromosome

CHAPTER SUMMARY

8.1 Mendel's Experiments

Working with garden pea plants, Mendel found that crosses between parents that differed for one trait produced F_1 offspring that all expressed one parent's traits. The traits that were visible in the F_1 generation are referred to as dominant, and traits that disappear in the F_1 generation are described as recessive. When the F_1 plants in Mendel's experiment were self-crossed, the F_2 offspring exhibited the dominant trait or the recessive trait in a 3:1 ratio, confirming that the recessive trait had been transmitted faithfully from the original P parent. Reciprocal crosses generated identical F_1 and F_2 offspring ratios. By examining sample sizes, Mendel showed that traits were inherited as independent events.

8.2 Laws of Inheritance

When true-breeding, or homozygous, individuals that differ for a certain trait are crossed, all of the offspring will be heterozygous for that trait. If the traits are inherited as dominant and recessive, the F_1 offspring will all exhibit the same phenotype as the parent homozygous for the dominant trait. If these heterozygous offspring are self-crossed, the resulting F_2 offspring will be equally likely to inherit gametes carrying the dominant or recessive trait, giving rise to offspring of which one quarter are homozygous dominant, half are heterozygous, and one quarter are homozygous recessive. Because homozygous dominant and heterozygous individuals are phenotypically identical, the observed traits in the F_2 offspring will exhibit a ratio of three dominant to one recessive.

Mendel postulated that genes (characteristics) are inherited as pairs of alleles (traits) that behave in a dominant and recessive pattern. Alleles segregate into gametes such that each gamete is equally likely to receive either one of the two alleles present in a diploid individual. In addition, genes are assorted into gametes independently of one another. That is, in general, alleles are not more likely to segregate into a gamete with a particular allele of another gene.

8.3 Extensions of the Laws of Inheritance

Alleles do not always behave in dominant and recessive patterns. Incomplete dominance describes situations in which the heterozygote exhibits a phenotype that is intermediate between the homozygous phenotypes. Codominance describes the simultaneous expression of both of the alleles in the heterozygote. Although diploid organisms can only have two alleles for any given gene, it is common for more than two alleles for a gene to exist in a population. In humans, as in many animals and some plants, females have two X chromosomes and males have one X and one Y chromosome. Genes that are present on the X but not the Y chromosome are said to be X-linked, such that males only inherit one allele for the gene, and females inherit two.

According to Mendel's law of independent assortment, genes sort independently of each other into gametes during meiosis. This occurs because chromosomes, on which the genes reside, assort independently during meiosis and crossovers cause most genes on the same chromosomes to also behave independently. When genes are located in close proximity on the same chromosome, their alleles tend to be inherited together. This results in offspring ratios that violate Mendel's law of independent assortment. However, recombination serves to exchange genetic material on homologous chromosomes such that maternal and paternal alleles may be recombined on the same chromosome. This is why alleles on a given chromosome are not always inherited together. Recombination is a random event occurring anywhere on a chromosome. Therefore, genes that are far apart on the same chromosome are likely to still assort independently because of recombination events that occurred in the intervening chromosomal space.

Whether or not they are sorting independently, genes may interact at the level of gene products, such that the expression of an allele for one gene masks or modifies the expression of an allele for a different gene. This is called epistasis.

ART CONNECTION QUESTIONS

1. Figure 8.9 In pea plants, round peas (*R*) are dominant to wrinkled peas (*r*). You do a test cross between a pea plant with wrinkled peas (genotype *rr*) and a plant of unknown genotype that has round peas. You end up with three plants, all which have round peas. From this data, can you tell if the parent plant is homozygous dominant or heterozygous?

2. Figure 8.10 In pea plants, purple flowers (*P*) are dominant to white (*p*), and yellow peas (*Y*) are dominant to

green (*y*). What are the possible genotypes and phenotypes for a cross between *PpYY* and *ppYy* pea plants? How many squares would you need to complete a Punnett square analysis of this cross?

3. Figure 8.16 What ratio of offspring would result from a cross between a white-eyed male and a female that is heterozygous for red eye color?

REVIEW QUESTIONS

4. Imagine that you are performing a cross involving seed color in garden pea plants. What traits would you expect to observe in the F₁ offspring if you cross true-breeding parents with green seeds and yellow seeds? Yellow seed color is dominant over green.
    a. only yellow-green seeds
    b. only yellow seeds
    c. 1:1 yellow seeds:green seeds
    d. 1:3 green seeds:yellow seeds

5. Imagine that you are performing a cross involving seed texture in garden pea plants. You cross true-breeding round and wrinkled parents to obtain F1 offspring. Which of the following experimental results in terms of numbers of plants are closest to what you expect in the F2 progeny?

    a. 810 round seeds
    b. 810 wrinkled seeds
    c. 405:395 round seeds:wrinkled seeds
    d. 610:190 round seeds:wrinkled seeds

6. The observable traits expressed by an organism are described as its _____.
    a. phenotype
    b. genotype
    c. alleles
    d. zygote

7. A recessive trait will be observed in individuals that are _____ for that trait.
    a. heterozygous
    b. homozygous or heterozygous
    c. homozygous
    d. diploid

8. What are the types of gametes that can be produced by an individual with the genotype *AaBb*?
    a. *Aa, Bb*
    b. *AA, aa, BB, bb*

    c. *AB, Ab, aB, ab*
    d. *AB, ab*

9. What is the reason for doing a test cross?
    a. to identify heterozygous individuals with the dominant phenotype
    b. to determine which allele is dominant and which is recessive
    c. to identify homozygous recessive individuals in the F2
    d. to determine if two genes assort independently

10. If black and white true-breeding mice are mated and the result is all gray offspring, what inheritance pattern would this be indicative of?
    a. dominance
    b. codominance
    c. multiple alleles
    d. incomplete dominance

11. The ABO blood groups in humans are expressed as the I^A, I^B, and *i* alleles. The I^A allele encodes the A blood group antigen, I^B encodes B, and *i* encodes O. Both A and B are dominant to O. If a heterozygous blood type A parent ($I^A i$) and a heterozygous blood type B parent ($I^B i$) mate, one quarter of their offspring are expected to have the AB blood type ($I^A I^B$) in which both antigens are expressed equally. Therefore, ABO blood groups are an example of:
    a. multiple alleles and incomplete dominance
    b. codominance and incomplete dominance
    c. incomplete dominance only
    d. multiple alleles and codominance

12. In a cross between a homozygous red-eyed female fruit fly and a white-eyed male fruit fly, what is the expected outcome?
    a. all white-eyed male offspring

b. all white-eyed female offspring

c. all red-eyed offspring

d. half white-eyed make offspring

a. 3

b. 6

c. 10

d. 16

13. When a population has a gene with four alleles circulating, how many possible genotypes are there?

CRITICAL THINKING QUESTIONS

14. Describe one of the reasons that made the garden pea an excellent choice of model system for studying inheritance.

15. Use a Punnett square to predict the offspring in a cross between a dwarf pea plant (homozygous recessive) and a tall pea plant (heterozygous). What is the phenotypic ratio of the offspring?

16. Use a Punnett square to predict the offspring in a cross between a tall pea plant (heterozygous) and a tall pea plant (heterozygous). What is the genotypic ratio of the offspring?

17. Can a male be a carrier of red-green color blindness?

18. Could an individual with blood type O (genotype *ii*) be a legitimate child of parents in which one parent had blood type A and the other parent had blood type B?

9 | MOLECULAR BIOLOGY

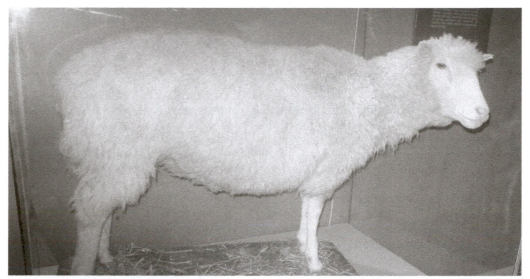

Figure 9.1 Dolly the sheep was the first cloned mammal.

Chapter Outline

9.1: The Structure of DNA

9.2: DNA Replication

9.3: Transcription

9.4: Translation

9.5: How Genes Are Regulated

Introduction

The three letters "DNA" have now become associated with crime solving, paternity testing, human identification, and genetic testing. DNA can be retrieved from hair, blood, or saliva. With the exception of identical twins, each person's DNA is unique and it is possible to detect differences between human beings on the basis of their unique DNA sequence.

DNA analysis has many practical applications beyond forensics and paternity testing. DNA testing is used for tracing genealogy and identifying pathogens. In the medical field, DNA is used in diagnostics, new vaccine development, and cancer therapy. It is now possible to determine predisposition to many diseases by analyzing genes.

DNA is the genetic material passed from parent to offspring for all life on Earth. The technology of molecular genetics developed in the last half century has enabled us to see deep into the history of life to deduce the relationships between living things in ways never thought possible. It also allows us to understand the workings of evolution in populations of organisms. Over a thousand species have had their entire genome sequenced, and there have been thousands of individual human genome sequences completed. These sequences will allow us to understand human disease and the relationship of humans to the rest of the tree of life. Finally, molecular genetics techniques have revolutionized plant and animal breeding for human agricultural needs. All of these advances in biotechnology depended on basic research leading to the discovery of the structure of DNA in 1953, and the research since then that has uncovered the details of DNA replication and the complex process leading to the expression of DNA in the form of proteins in the cell.

9.1 | The Structure of DNA

By the end of this section, you will be able to:

- Describe the structure of DNA
- Describe how eukaryotic and prokaryotic DNA is arranged in the cell

In the 1950s, Francis Crick and James Watson worked together at the University of Cambridge, England, to determine the structure of DNA. Other scientists, such as Linus Pauling and Maurice Wilkins, were also actively exploring this field. Pauling had discovered the secondary structure of proteins using X-ray crystallography. X-ray crystallography is a method for investigating molecular structure by observing the patterns formed by X-rays shot through a crystal of the substance. The patterns give important information about the structure of the molecule of interest. In Wilkins' lab, researcher Rosalind Franklin was using X-ray crystallography to understand the structure of DNA. Watson and Crick were able to piece together the puzzle of the DNA molecule using Franklin's data (Figure 9.2). Watson and Crick also had key pieces of information available from other researchers such as Chargaff's rules. Chargaff had shown that of the four kinds of monomers (nucleotides) present in a DNA molecule, two types were always present in equal amounts and the remaining two types were also always present in equal amounts. This meant they were always paired in some way. In 1962, James Watson, Francis Crick, and Maurice Wilkins were awarded the Nobel Prize in Medicine for their work in determining the structure of DNA.

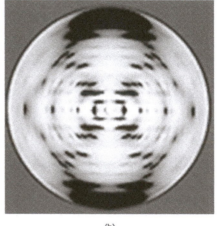

(a)   (b)

Figure 9.2 Pioneering scientists (a) James Watson and Francis Crick are pictured here with American geneticist Maclyn McCarty. Scientist Rosalind Franklin discovered (b) the X-ray diffraction pattern of DNA, which helped to elucidate its double helix structure. (credit a: modification of work by Marjorie McCarty; b: modification of work by NIH)

Now let's consider the structure of the two types of nucleic acids, deoxyribonucleic acid (DNA) and ribonucleic acid (RNA). The building blocks of DNA are nucleotides, which are made up of three parts: a **deoxyribose** (5-carbon sugar), a **phosphate group**, and a **nitrogenous base** (Figure 9.3). There are four types of nitrogenous bases in DNA. Adenine (A) and guanine (G) are double-ringed purines, and cytosine (C) and thymine (T) are smaller, single-ringed pyrimidines. The nucleotide is named according to the nitrogenous base it contains.

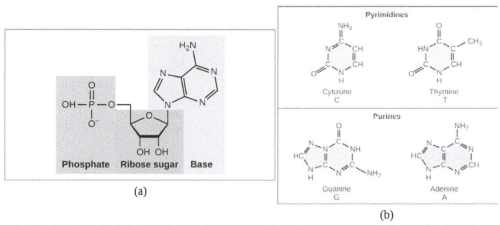

(a)

(b)

Figure 9.3 (a) Each DNA nucleotide is made up of a sugar, a phosphate group, and a base. (b) Cytosine and thymine are pyrimidines. Guanine and adenine are purines.

The phosphate group of one nucleotide bonds covalently with the sugar molecule of the next nucleotide, and so on, forming a long polymer of nucleotide monomers. The sugar–phosphate groups line up in a "backbone" for each single strand of DNA, and the nucleotide bases stick out from this backbone. The carbon atoms of the five-carbon sugar are numbered clockwise from the oxygen as 1', 2', 3', 4', and 5' (1' is read as "one prime"). The phosphate group is attached to the 5' carbon of one nucleotide and the 3' carbon of the next nucleotide. In its natural state, each DNA molecule is actually composed of two single strands held together along their length with hydrogen bonds between the bases.

Watson and Crick proposed that the DNA is made up of two strands that are twisted around each other to form a right-handed helix, called a **double helix**. Base-pairing takes place between a purine and pyrimidine: namely, A pairs with T, and G pairs with C. In other words, adenine and thymine are complementary base pairs, and cytosine and guanine are also complementary base pairs. This is the basis for Chargaff's rule; because of their complementarity, there is as much adenine as thymine in a DNA molecule and as much guanine as cytosine. Adenine and thymine are connected by two hydrogen bonds, and cytosine and guanine are connected by three hydrogen bonds. The two strands are anti-parallel in nature; that is, one strand will have the 3' carbon of the sugar in the "upward" position, whereas the other strand will have the 5' carbon in the upward position. The diameter of the DNA double helix is uniform throughout because a purine (two rings) always pairs with a pyrimidine (one ring) and their combined lengths are always equal. (Figure 9.4).

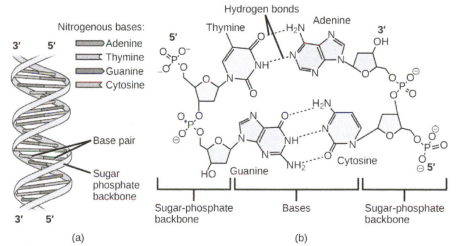

(a)   (b)

Figure 9.4 DNA (a) forms a double stranded helix, and (b) adenine pairs with thymine and cytosine pairs with guanine. (credit a: modification of work by Jerome Walker, Dennis Myts)

The Structure of RNA

There is a second nucleic acid in all cells called ribonucleic acid, or RNA. Like DNA, RNA is a polymer of nucleotides. Each of the nucleotides in RNA is made up of a nitrogenous base, a five-carbon sugar, and a phosphate group. In the case of RNA, the five-carbon sugar is ribose, not deoxyribose. Ribose has a hydroxyl group at the 2' carbon, unlike deoxyribose, which has only a hydrogen atom (Figure 9.5).

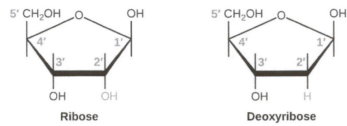

Ribose   **Deoxyribose**

Figure 9.5 The difference between the ribose found in RNA and the deoxyribose found in DNA is that ribose has a hydroxyl group at the 2' carbon.

RNA nucleotides contain the nitrogenous bases adenine, cytosine, and guanine. However, they do not contain thymine, which is instead replaced by uracil, symbolized by a "U." RNA exists as a single-stranded molecule rather than a double-stranded helix. Molecular biologists have named several kinds of RNA on the basis of their function. These include messenger RNA (mRNA), transfer RNA (tRNA), and ribosomal RNA (rRNA)—molecules that are involved in the production of proteins from the DNA code.

How DNA Is Arranged in the Cell

DNA is a working molecule; it must be replicated when a cell is ready to divide, and it must be "read" to produce the molecules, such as proteins, to carry out the functions of the cell. For this reason, the DNA is protected and packaged in very specific ways. In addition, DNA molecules can be very long. Stretched end-to-end, the DNA molecules in a single human cell would come to a length of about 2 meters. Thus, the DNA for a cell must be packaged in a very ordered way to fit and function within a structure (the cell) that is not visible to the naked eye. The chromosomes of prokaryotes are much simpler than those of eukaryotes in many of their features (Figure 9.6). Most prokaryotes contain a single, circular chromosome that is found in an area in the cytoplasm called the nucleoid.

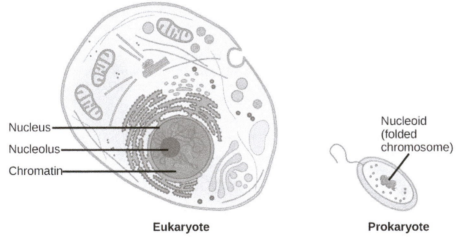

Eukaryote   **Prokaryote**

Figure 9.6 A eukaryote contains a well-defined nucleus, whereas in prokaryotes, the chromosome lies in the cytoplasm in an area called the nucleoid.

The size of the genome in one of the most well-studied prokaryotes, *Escherichia coli*, is 4.6 million base pairs, which would extend a distance of about 1.6 mm if stretched out. So how does this fit inside a small bacterial cell? The DNA is twisted beyond the double helix in what is known as supercoiling. Some proteins are known to be involved in the supercoiling; other proteins and enzymes help in maintaining the supercoiled structure.

Eukaryotes, whose chromosomes each consist of a linear DNA molecule, employ a different type of packing strategy to fit their DNA inside the nucleus (Figure 9.7). At the most basic level, DNA is wrapped around proteins known as histones to form structures called nucleosomes. The DNA is wrapped tightly around the histone core. This nucleosome is linked to the next one by a short strand of DNA that is free of histones. This is also known as the "beads on a string" structure; the nucleosomes are the "beads" and the short lengths of DNA between them are the "string." The nucleosomes, with their DNA coiled around them, stack compactly onto each other to form a 30-nm–wide fiber. This fiber is further coiled into a thicker and more compact structure. At the metaphase stage of mitosis, when the chromosomes are lined up in the center of the cell, the chromosomes are at their most compacted. They are approximately 700 nm in width, and are found in association with scaffold proteins.

In interphase, the phase of the cell cycle between mitoses at which the chromosomes are decondensed, eukaryotic chromosomes have two distinct regions that can be distinguished by staining. There is a tightly packaged region that stains darkly, and a less dense region. The darkly staining regions usually contain genes that are not active, and are found in the regions of the centromere and telomeres. The lightly staining regions usually contain genes that are active, with DNA packaged around nucleosomes but not further compacted.

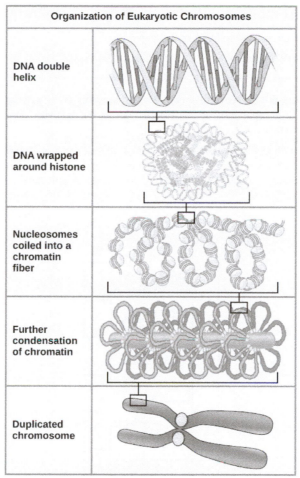

Figure 9.7 These figures illustrate the compaction of the eukaryotic chromosome.

Watch this animation (http://openstaxcollege.org/l/DNA_packaging) of DNA packaging.

9.2 | DNA Replication

By the end of this section, you will be able to:

- Explain the process of DNA replication
- Explain the importance of telomerase to DNA replication
- Describe mechanisms of DNA repair

When a cell divides, it is important that each daughter cell receives an identical copy of the DNA. This is accomplished by the process of DNA replication. The replication of DNA occurs during the synthesis phase, or S phase, of the cell cycle, before the cell enters mitosis or meiosis.

The elucidation of the structure of the double helix provided a hint as to how DNA is copied. Recall that adenine nucleotides pair with thymine nucleotides, and cytosine with guanine. This means that the two strands are complementary to each other. For example, a strand of DNA with a nucleotide sequence of AGTCATGA will have a complementary strand with the sequence TCAGTACT (Figure 9.8).

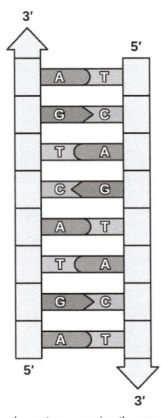

Figure 9.8 The two strands of DNA are complementary, meaning the sequence of bases in one strand can be used to create the correct sequence of bases in the other strand.

Because of the complementarity of the two strands, having one strand means that it is possible to recreate the other strand. This model for replication suggests that the two strands of the double helix separate during replication, and each strand serves as a template from which the new complementary strand is copied (Figure 9.9).

Download for free at https://openstax.org/details/books/concepts-biology

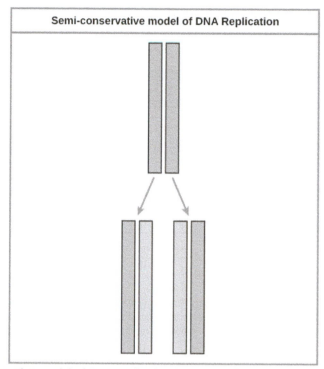

Figure 9.9 The semiconservative model of DNA replication is shown. Gray indicates the original DNA strands, and blue indicates newly synthesized DNA.

During DNA replication, each of the two strands that make up the double helix serves as a template from which new strands are copied. The new strand will be complementary to the parental or "old" strand. Each new double strand consists of one parental strand and one new daughter strand. This is known as **semiconservative replication**. When two DNA copies are formed, they have an identical sequence of nucleotide bases and are divided equally into two daughter cells.

DNA Replication in Eukaryotes

Because eukaryotic genomes are very complex, DNA replication is a very complicated process that involves several enzymes and other proteins. It occurs in three main stages: initiation, elongation, and termination.

Recall that eukaryotic DNA is bound to proteins known as histones to form structures called nucleosomes. During initiation, the DNA is made accessible to the proteins and enzymes involved in the replication process. How does the replication machinery know where on the DNA double helix to begin? It turns out that there are specific nucleotide sequences called origins of replication at which replication begins. Certain proteins bind to the origin of replication while an enzyme called **helicase** unwinds and opens up the DNA helix. As the DNA opens up, Y-shaped structures called **replication forks** are formed (Figure 9.10). Two replication forks are formed at the origin of replication, and these get extended in both directions as replication proceeds. There are multiple origins of replication on the eukaryotic chromosome, such that replication can occur simultaneously from several places in the genome.

During elongation, an enzyme called **DNA polymerase** adds DNA nucleotides to the 3' end of the template. Because DNA polymerase can only add new nucleotides at the end of a backbone, a **primer** sequence, which provides this starting point, is added with complementary RNA nucleotides. This primer is removed later, and the nucleotides are replaced with DNA nucleotides. One strand, which is complementary to the parental DNA strand, is synthesized continuously toward the replication fork so the polymerase can add nucleotides in this direction. This continuously synthesized strand is known as the **leading strand**. Because DNA polymerase can only synthesize DNA in a 5' to 3' direction, the other new strand is put together in short pieces called **Okazaki fragments**. The Okazaki fragments each require a primer made of RNA to start the synthesis. The strand with the Okazaki fragments is known as the **lagging strand**. As synthesis proceeds, an enzyme removes the RNA primer, which is then replaced with DNA nucleotides, and the gaps between fragments are sealed by an enzyme called **DNA ligase**.

The process of DNA replication can be summarized as follows:

1. DNA unwinds at the origin of replication.

2.   New bases are added to the complementary parental strands. One new strand is made continuously, while the other strand is made in pieces.

3.   Primers are removed, new DNA nucleotides are put in place of the primers and the backbone is sealed by DNA ligase.

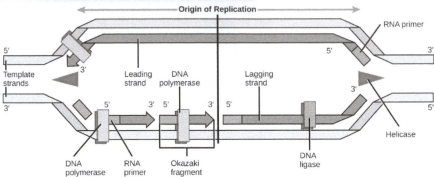

Figure 9.10 A replication fork is formed by the opening of the origin of replication, and helicase separates the DNA strands. An RNA primer is synthesized, and is elongated by the DNA polymerase. On the leading strand, DNA is synthesized continuously, whereas on the lagging strand, DNA is synthesized in short stretches. The DNA fragments are joined by DNA ligase (not shown).

You isolate a cell strain in which the joining together of Okazaki fragments is impaired and suspect that a mutation has occurred in an enzyme found at the replication fork. Which enzyme is most likely to be mutated?

Telomere Replication

Because eukaryotic chromosomes are linear, DNA replication comes to the end of a line in eukaryotic chromosomes. As you have learned, the DNA polymerase enzyme can add nucleotides in only one direction. In the leading strand, synthesis continues until the end of the chromosome is reached; however, on the lagging strand there is no place for a primer to be made for the DNA fragment to be copied at the end of the chromosome. This presents a problem for the cell because the ends remain unpaired, and over time these ends get progressively shorter as cells continue to divide. The ends of the linear chromosomes are known as **telomeres**, which have repetitive sequences that do not code for a particular gene. As a consequence, it is telomeres that are shortened with each round of DNA replication instead of genes. For example, in humans, a six base-pair sequence, TTAGGG, is repeated 100 to 1000 times. The discovery of the enzyme **telomerase** (Figure 9.11) helped in the understanding of how chromosome ends are maintained. The telomerase attaches to the end of the chromosome, and complementary bases to the RNA template are added on the end of the DNA strand. Once the lagging strand template is sufficiently elongated, DNA polymerase can now add nucleotides that are complementary to the ends of the chromosomes. Thus, the ends of the chromosomes are replicated.

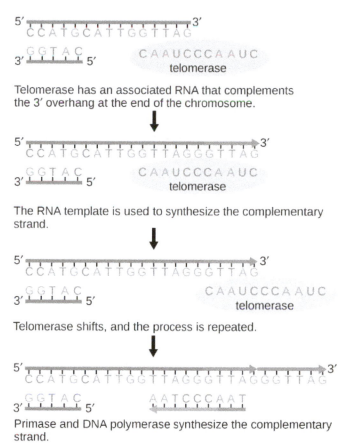

Figure 9.11 The ends of linear chromosomes are maintained by the action of the telomerase enzyme.

Telomerase is typically found to be active in germ cells, adult stem cells, and some cancer cells. For her discovery of telomerase and its action, Elizabeth Blackburn (Figure 9.12) received the Nobel Prize for Medicine and Physiology in 2009.

Figure 9.12 Elizabeth Blackburn, 2009 Nobel Laureate, was the scientist who discovered how telomerase works. (credit: U.S. Embassy, Stockholm, Sweden)

Telomerase is not active in adult somatic cells. Adult somatic cells that undergo cell division continue to have their telomeres shortened. This essentially means that telomere shortening is associated with aging. In 2010, scientists found that telomerase can reverse some age-related conditions in mice, and this may have potential in regenerative medicine.[1]

1. Mariella Jaskelioff, et al., "Telomerase reactivation reverses tissue degeneration in aged telomerase-deficient mice," *Nature*, 469 (2011):102–7.

Telomerase-deficient mice were used in these studies; these mice have tissue atrophy, stem-cell depletion, organ system failure, and impaired tissue injury responses. Telomerase reactivation in these mice caused extension of telomeres, reduced DNA damage, reversed neurodegeneration, and improved functioning of the testes, spleen, and intestines. Thus, telomere reactivation may have potential for treating age-related diseases in humans.

DNA Replication in Prokaryotes

Recall that the prokaryotic chromosome is a circular molecule with a less extensive coiling structure than eukaryotic chromosomes. The eukaryotic chromosome is linear and highly coiled around proteins. While there are many similarities in the DNA replication process, these structural differences necessitate some differences in the DNA replication process in these two life forms.

DNA replication has been extremely well-studied in prokaryotes, primarily because of the small size of the genome and large number of variants available. *Escherichia coli* has 4.6 million base pairs in a single circular chromosome, and all of it gets replicated in approximately 42 minutes, starting from a single origin of replication and proceeding around the chromosome in both directions. This means that approximately 1000 nucleotides are added per second. The process is much more rapid than in eukaryotes. Table 9.1 summarizes the differences between prokaryotic and eukaryotic replications.

Differences between Prokaryotic and Eukaryotic Replications

Property	Prokaryotes	Eukaryotes
Origin of replication	Single	Multiple
Rate of replication	1000 nucleotides/s	50 to 100 nucleotides/s
Chromosome structure	circular	linear
Telomerase	Not present	Present

Table 9.1

Click through a tutorial (http://openstaxcollege.org/l/DNA_replicatio2) on DNA replication.

DNA Repair

DNA polymerase can make mistakes while adding nucleotides. It edits the DNA by proofreading every newly added base. Incorrect bases are removed and replaced by the correct base, and then polymerization continues (Figure 9.13a). Most mistakes are corrected during replication, although when this does not happen, the **mismatch repair** mechanism is employed. Mismatch repair enzymes recognize the wrongly incorporated base and excise it from the DNA, replacing it with the correct base (Figure 9.13b). In yet another type of repair, **nucleotide excision repair**, the DNA double strand is unwound and separated, the incorrect bases are removed along with a few bases on the 5' and 3' end, and these are replaced by copying the template with the help of DNA polymerase (Figure 9.13c). Nucleotide excision repair is particularly important in correcting thymine dimers, which are primarily caused by ultraviolet light. In a thymine dimer, two thymine nucleotides adjacent to each other on one strand are covalently bonded to each other rather than their complementary bases. If the dimer is not removed and repaired it will lead to a mutation. Individuals with flaws in their nucleotide excision repair genes show extreme sensitivity to sunlight and develop skin cancers early in life.

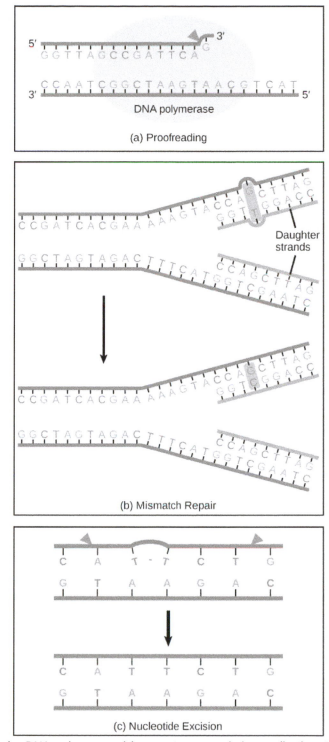

Figure 9.13 Proofreading by DNA polymerase (a) corrects errors during replication. In mismatch repair (b), the incorrectly added base is detected after replication. The mismatch repair proteins detect this base and remove it from the newly synthesized strand by nuclease action. The gap is now filled with the correctly paired base. Nucleotide excision (c) repairs thymine dimers. When exposed to UV, thymines lying adjacent to each other can form thymine dimers. In normal cells, they are excised and replaced.

Most mistakes are corrected; if they are not, they may result in a **mutation**—defined as a permanent change in the DNA sequence. Mutations in repair genes may lead to serious consequences like cancer.

9.3 | Transcription

By the end of this section, you will be able to:

- Explain the central dogma
- Explain the main steps of transcription
- Describe how eukaryotic mRNA is processed

In both prokaryotes and eukaryotes, the second function of DNA (the first was replication) is to provide the information needed to construct the proteins necessary so that the cell can perform all of its functions. To do this, the DNA is "read" or transcribed into an **mRNA** molecule. The mRNA then provides the code to form a protein by a process called translation. Through the processes of transcription and translation, a protein is built with a specific sequence of amino acids that was originally encoded in the DNA. This module discusses the details of transcription.

The Central Dogma: DNA Encodes RNA; RNA Encodes Protein

The flow of genetic information in cells from DNA to mRNA to protein is described by the central dogma (Figure 9.14), which states that genes specify the sequences of mRNAs, which in turn specify the sequences of proteins.

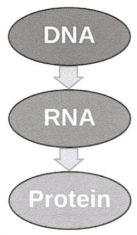

Figure 9.14 The central dogma states that DNA encodes RNA, which in turn encodes protein.

The copying of DNA to mRNA is relatively straightforward, with one nucleotide being added to the mRNA strand for every complementary nucleotide read in the DNA strand. The translation to protein is more complex because groups of three mRNA nucleotides correspond to one amino acid of the protein sequence. However, as we shall see in the next module, the translation to protein is still systematic, such that nucleotides 1 to 3 correspond to amino acid 1, nucleotides 4 to 6 correspond to amino acid 2, and so on.

Transcription: from DNA to mRNA

Both prokaryotes and eukaryotes perform fundamentally the same process of transcription, with the important difference of the membrane-bound nucleus in eukaryotes. With the genes bound in the nucleus, transcription occurs in the nucleus of the cell and the mRNA transcript must be transported to the cytoplasm. The prokaryotes, which include bacteria and archaea, lack membrane-bound nuclei and other organelles, and transcription occurs in the cytoplasm of the cell. In both prokaryotes and eukaryotes, transcription occurs in three main stages: initiation, elongation, and termination.

Initiation

Transcription requires the DNA double helix to partially unwind in the region of mRNA synthesis. The region of unwinding is called a **transcription bubble**. The DNA sequence onto which the proteins and enzymes involved in transcription bind to initiate the process is called a **promoter**. In most cases, promoters exist upstream of the genes they regulate. The specific sequence of a promoter is very important because it determines whether the corresponding gene is transcribed all of the time, some of the time, or hardly at all (Figure 9.15).

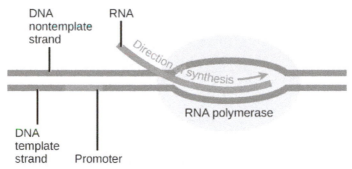

Figure 9.15 The initiation of transcription begins when DNA is unwound, forming a transcription bubble. Enzymes and other proteins involved in transcription bind at the promoter.

Elongation

Transcription always proceeds from one of the two DNA strands, which is called the **template strand**. The mRNA product is complementary to the template strand and is almost identical to the other DNA strand, called the **nontemplate strand**, with the exception that RNA contains a uracil (U) in place of the thymine (T) found in DNA. During elongation, an enzyme called **RNA polymerase** proceeds along the DNA template adding nucleotides by base pairing with the DNA template in a manner similar to DNA replication, with the difference that an RNA strand is being synthesized that does not remain bound to the DNA template. As elongation proceeds, the DNA is continuously unwound ahead of the core enzyme and rewound behind it (Figure 9.16).

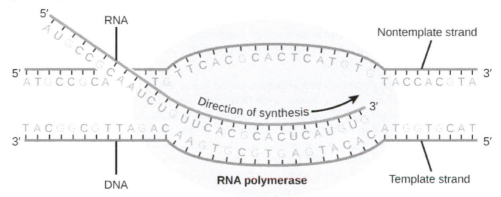

Figure 9.16 During elongation, RNA polymerase tracks along the DNA template, synthesizes mRNA in the 5' to 3' direction, and unwinds then rewinds the DNA as it is read.

Termination

Once a gene is transcribed, the prokaryotic polymerase needs to be instructed to dissociate from the DNA template and liberate the newly made mRNA. Depending on the gene being transcribed, there are two kinds of termination signals, but both involve repeated nucleotide sequences in the DNA template that result in RNA polymerase stalling, leaving the DNA template, and freeing the mRNA transcript.

On termination, the process of transcription is complete. In a prokaryotic cell, by the time termination occurs, the transcript would already have been used to partially synthesize numerous copies of the encoded protein because these processes can occur concurrently using multiple ribosomes (polyribosomes) (Figure 9.17). In contrast, the presence of a nucleus in eukaryotic cells precludes simultaneous transcription and translation.

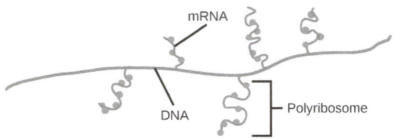

Figure 9.17 Multiple polymerases can transcribe a single bacterial gene while numerous ribosomes concurrently translate the mRNA transcripts into polypeptides. In this way, a specific protein can rapidly reach a high concentration in the bacterial cell.

Eukaryotic RNA Processing

The newly transcribed eukaryotic mRNAs must undergo several processing steps before they can be transferred from the nucleus to the cytoplasm and translated into a protein. The additional steps involved in eukaryotic mRNA maturation create a molecule that is much more stable than a prokaryotic mRNA. For example, eukaryotic mRNAs last for several hours, whereas the typical prokaryotic mRNA lasts no more than five seconds.

The mRNA transcript is first coated in RNA-stabilizing proteins to prevent it from degrading while it is processed and exported out of the nucleus. This occurs while the pre-mRNA still is being synthesized by adding a special nucleotide "cap" to the 5' end of the growing transcript. In addition to preventing degradation, factors involved in protein synthesis recognize the cap to help initiate translation by ribosomes.

Once elongation is complete, an enzyme then adds a string of approximately 200 adenine residues to the 3' end, called the poly-A tail. This modification further protects the pre-mRNA from degradation and signals to cellular factors that the transcript needs to be exported to the cytoplasm.

Eukaryotic genes are composed of protein-coding sequences called **exons** (*ex*-on signifies that they are *ex*pressed) and *int*ervening sequences called **introns** (*int*-ron denotes their *int*ervening role). Introns are removed from the pre-mRNA during processing. Intron sequences in mRNA do not encode functional proteins. It is essential that all of a pre-mRNA's introns be completely and precisely removed before protein synthesis so that the exons join together to code for the correct amino acids. If the process errs by even a single nucleotide, the sequence of the rejoined exons would be shifted, and the resulting protein would be nonfunctional. The process of removing introns and reconnecting exons is called **splicing** (Figure 9.18). Introns are removed and degraded while the pre-mRNA is still in the nucleus.

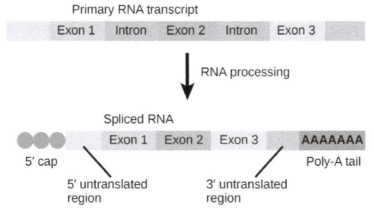

Figure 9.18 Eukaryotic mRNA contains introns that must be spliced out. A 5' cap and 3' tail are also added.

9.4 | Translation

By the end of this section, you will be able to:

- Describe the different steps in protein synthesis
- Discuss the role of ribosomes in protein synthesis
- Describe the genetic code and how the nucleotide sequence determines the amino acid and the protein sequence

The synthesis of proteins is one of a cell's most energy-consuming metabolic processes. In turn, proteins account for more mass than any other component of living organisms (with the exception of water), and proteins perform a wide variety of the functions of a cell. The process of translation, or protein synthesis, involves decoding an mRNA message into a polypeptide product. Amino acids are covalently strung together in lengths ranging from approximately 50 amino acids to more than 1,000.

The Protein Synthesis Machinery

In addition to the mRNA template, many other molecules contribute to the process of translation. The composition of each component may vary across species; for instance, ribosomes may consist of different numbers of ribosomal RNAs (**rRNA**) and polypeptides depending on the organism. However, the general structures and functions of the protein synthesis machinery are comparable from bacteria to human cells. Translation requires the input of an mRNA template, ribosomes, tRNAs, and various enzymatic factors (Figure 9.19).

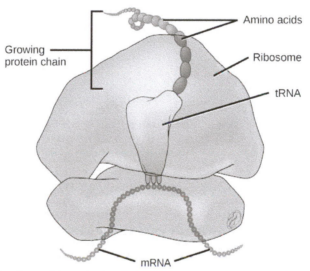

Figure 9.19 The protein synthesis machinery includes the large and small subunits of the ribosome, mRNA, and tRNA. (credit: modification of work by NIGMS, NIH)

In *E. coli*, there are 200,000 ribosomes present in every cell at any given time. A ribosome is a complex macromolecule composed of structural and catalytic rRNAs, and many distinct polypeptides. In eukaryotes, the nucleolus is completely specialized for the synthesis and assembly of rRNAs.

Ribosomes are located in the cytoplasm in prokaryotes and in the cytoplasm and endoplasmic reticulum of eukaryotes. Ribosomes are made up of a large and a small subunit that come together for translation. The small subunit is responsible for binding the mRNA template, whereas the large subunit sequentially binds **tRNAs**, a type of RNA molecule that brings amino acids to the growing chain of the polypeptide. Each mRNA molecule is simultaneously translated by many ribosomes, all synthesizing protein in the same direction.

Depending on the species, 40 to 60 types of tRNA exist in the cytoplasm. Serving as adaptors, specific tRNAs bind to sequences on the mRNA template and add the corresponding amino acid to the polypeptide chain. Therefore, tRNAs are the molecules that actually "translate" the language of RNA into the language of proteins. For each tRNA to function, it must have its specific amino acid bonded to it. In the process of tRNA "charging," each tRNA molecule is bonded to its correct amino acid.

The Genetic Code

To summarize what we know to this point, the cellular process of transcription generates messenger RNA (mRNA), a mobile molecular copy of one or more genes with an alphabet of A, C, G, and uracil (U). Translation of the mRNA template converts nucleotide-based genetic information into a protein product. Protein sequences consist of 20 commonly occurring amino acids; therefore, it can be said that the protein alphabet consists of 20 letters. Each amino acid is defined by a three-nucleotide sequence called the triplet **codon**. The relationship between a nucleotide codon and its corresponding amino acid is called the **genetic code**.

Given the different numbers of "letters" in the mRNA and protein "alphabets," combinations of nucleotides corresponded to single amino acids. Using a three-nucleotide code means that there are a total of 64 (4 × 4 × 4) possible combinations; therefore, a given amino acid is encoded by more than one nucleotide triplet (Figure 9.20).

First letter	Second letter: U	Second letter: C	Second letter: A	Second letter: G	Third letter
U	UUU, UUC: Phe UUA, UUG: Leu	UCU, UCC, UCA, UCG: Ser	UAU, UAC: Tyr UAA Stop UAG Stop	UGU, UGC: Cys UGA Stop UGG Trp	U C A G
C	CUU, CUC, CUA, CUG: Leu	CCU, CCC, CCA, CCG: Pro	CAU, CAC: His CAA, CAG: Gln	CGU, CGC, CGA, CGG: Arg	U C A G
A	AUU, AUC, AUA: Ile AUG Met	ACU, ACC, ACA, ACG: Thr	AAU, AAC: Asn AAA, AAG: Lys	AGU, AGC: Ser AGA, AGG: Arg	U C A G
G	GUU, GUC, GUA, GUG: Val	GCU, GCC, GCA, GCG: Ala	GAU, GAC: Asp GAA, GAG: Glu	GGU, GGC, GGA, GGG: Gly	U C A G

Figure 9.20 This figure shows the genetic code for translating each nucleotide triplet, or codon, in mRNA into an amino acid or a termination signal in a nascent protein. (credit: modification of work by NIH)

Three of the 64 codons terminate protein synthesis and release the polypeptide from the translation machinery. These triplets are called **stop codons**. Another codon, AUG, also has a special function. In addition to specifying the amino acid methionine, it also serves as the **start codon** to initiate translation. The reading frame for translation is set by the AUG start codon near the 5' end of the mRNA. The genetic code is universal. With a few exceptions, virtually all species use the same genetic code for protein synthesis, which is powerful evidence that all life on Earth shares a common origin.

The Mechanism of Protein Synthesis

Just as with mRNA synthesis, protein synthesis can be divided into three phases: initiation, elongation, and termination. The process of translation is similar in prokaryotes and eukaryotes. Here we will explore how translation occurs in *E. coli*, a representative prokaryote, and specify any differences between prokaryotic and eukaryotic translation.

Protein synthesis begins with the formation of an initiation complex. In *E. coli*, this complex involves the small ribosome subunit, the mRNA template, three initiation factors, and a special initiator tRNA. The initiator tRNA interacts with the AUG start codon, and links to a special form of the amino acid methionine that is typically removed from the polypeptide after translation is complete.

In prokaryotes and eukaryotes, the basics of polypeptide elongation are the same, so we will review elongation from the perspective of *E. coli*. The large ribosomal subunit of *E. coli* consists of three compartments: the A site binds incoming charged tRNAs (tRNAs with their attached specific amino acids). The P site binds charged tRNAs carrying amino acids that have formed bonds with the growing polypeptide chain but have not yet dissociated from their corresponding tRNA. The E site releases dissociated tRNAs so they can be recharged with free amino acids. The ribosome shifts one codon at a time, catalyzing each process that occurs in the three sites. With each step, a charged tRNA enters the complex, the polypeptide becomes one amino acid longer, and an uncharged tRNA departs. The energy for each bond between amino acids is derived from GTP, a molecule similar to ATP (Figure 9.21). Amazingly, the *E. coli* translation apparatus takes only 0.05 seconds to add each amino acid, meaning that a 200-amino acid polypeptide could be translated in just 10 seconds.

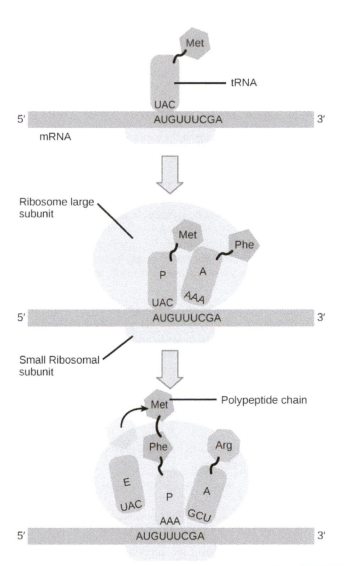

Figure 9.21 Translation begins when a tRNA anticodon recognizes a codon on the mRNA. The large ribosomal subunit joins the small subunit, and a second tRNA is recruited. As the mRNA moves relative to the ribosome, the polypeptide chain is formed. Entry of a release factor into the A site terminates translation and the components dissociate.

Termination of translation occurs when a stop codon (UAA, UAG, or UGA) is encountered. When the ribosome encounters the stop codon, the growing polypeptide is released and the ribosome subunits dissociate and leave the mRNA. After many ribosomes have completed translation, the mRNA is degraded so the nucleotides can be reused in another transcription reaction.

Transcribe a gene and translate it to protein using complementary pairing and the genetic code at this site (http://openstaxcollege.org/l/create_protein2) .

9.5 | How Genes Are Regulated

By the end of this section, you will be able to:

- Discuss why every cell does not express all of its genes
- Describe how prokaryotic gene expression occurs at the transcriptional level
- Understand that eukaryotic gene expression occurs at the epigenetic, transcriptional, post-transcriptional, translational, and post-translational levels

For a cell to function properly, necessary proteins must be synthesized at the proper time. All organisms and cells control or regulate the transcription and translation of their DNA into protein. The process of turning on a gene to produce RNA and protein is called **gene expression**. Whether in a simple unicellular organism or in a complex multicellular organism, each cell controls when and how its genes are expressed. For this to occur, there must be a mechanism to control when a gene is expressed to make RNA and protein, how much of the protein is made, and when it is time to stop making that protein because it is no longer needed.

Cells in multicellular organisms are specialized; cells in different tissues look very different and perform different functions. For example, a muscle cell is very different from a liver cell, which is very different from a skin cell. These differences are a consequence of the expression of different sets of genes in each of these cells. All cells have certain basic functions they must perform for themselves, such as converting the energy in sugar molecules into energy in ATP. Each cell also has many genes that are not expressed, and expresses many that are not expressed by other cells, such that it can carry out its specialized functions. In addition, cells will turn on or off certain genes at different times in response to changes in the environment or at different times during the development of the organism. Unicellular organisms, both eukaryotic and prokaryotic, also turn on and off genes in response to the demands of their environment so that they can respond to special conditions.

The control of gene expression is extremely complex. Malfunctions in this process are detrimental to the cell and can lead to the development of many diseases, including cancer.

Prokaryotic versus Eukaryotic Gene Expression

To understand how gene expression is regulated, we must first understand how a gene becomes a functional protein in a cell. The process occurs in both prokaryotic and eukaryotic cells, just in slightly different fashions.

Because prokaryotic organisms lack a cell nucleus, the processes of transcription and translation occur almost simultaneously. When the protein is no longer needed, transcription stops. As a result, the primary method to control what type and how much protein is expressed in a prokaryotic cell is through the regulation of DNA transcription into RNA. All the subsequent steps happen automatically. When more protein is required, more transcription occurs. Therefore, in prokaryotic cells, the control of gene expression is almost entirely at the transcriptional level.

The first example of such control was discovered using *E. coli* in the 1950s and 1960s by French researchers and is called the *lac* operon. The *lac* operon is a stretch of DNA with three adjacent genes that code for proteins that participate in the absorption and metabolism of lactose, a food source for *E. coli*. When lactose is not present in the bacterium's environment, the *lac* genes are transcribed in small amounts. When lactose is present, the genes are transcribed and the bacterium is able to use the lactose as a food source. The operon also contains a promoter sequence to which the RNA polymerase binds to begin transcription; between the promoter and the three genes is a region called the operator. When there is no lactose present, a protein known as a repressor binds to the operator and prevents RNA polymerase from binding to the promoter, except in rare cases. Thus very little of the protein products of the three genes is made. When lactose is present, an end product of lactose metabolism binds to the repressor protein and prevents it from binding to the operator. This allows RNA polymerase to bind to the promoter and freely transcribe the three genes, allowing the organism to metabolize the lactose.

Eukaryotic cells, in contrast, have intracellular organelles and are much more complex. Recall that in eukaryotic cells, the DNA is contained inside the cell's nucleus and it is transcribed into mRNA there. The newly synthesized mRNA is then transported out of the nucleus into the cytoplasm, where ribosomes translate the mRNA into protein. The processes of transcription and translation are physically separated by the nuclear membrane; transcription occurs only within the nucleus, and translation only occurs outside the nucleus in the cytoplasm. The regulation of gene expression can occur at all stages of the process (Figure 9.22). Regulation may occur when the DNA is uncoiled and loosened from nucleosomes to bind transcription factors (**epigenetic** level), when the RNA is transcribed (transcriptional level), when RNA is processed and exported to the cytoplasm after it is transcribed (**post-transcriptional** level), when the RNA is translated into protein (translational level), or after the protein has been made (**post-translational** level).

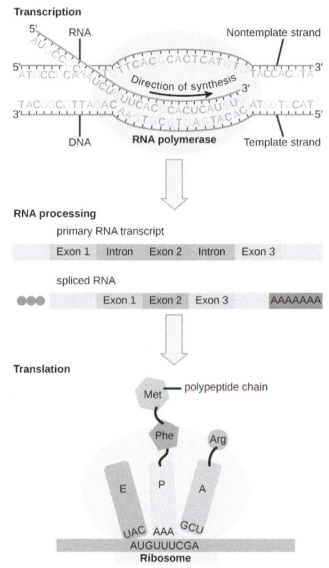

Figure 9.22 Eukaryotic gene expression is regulated during transcription and RNA processing, which take place in the nucleus, as well as during protein translation, which takes place in the cytoplasm. Further regulation may occur through post-translational modifications of proteins.

The differences in the regulation of gene expression between prokaryotes and eukaryotes are summarized in Table 9.2.

Differences in the Regulation of Gene Expression of Prokaryotic and Eukaryotic Organisms

Prokaryotic organisms		Eukaryotic organisms	
Lack nucleus		Contain nucleus	

Table 9.2

Differences in the Regulation of Gene Expression of Prokaryotic and Eukaryotic Organisms

Prokaryotic organisms	Eukaryotic organisms
RNA transcription and protein translation occur almost simultaneously	RNA transcription occurs prior to protein translation, and it takes place in the nucleus. RNA translation to protein occurs in the cytoplasm. RNA post-processing includes addition of a 5' cap, poly-A tail, and excision of introns and splicing of exons.
Gene expression is regulated primarily at the transcriptional level	Gene expression is regulated at many levels (epigenetic, transcriptional, post-transcriptional, translational, and post-translational)

Table 9.2

Alternative RNA Splicing

In the 1970s, genes were first observed that exhibited **alternative RNA splicing**. Alternative RNA splicing is a mechanism that allows different protein products to be produced from one gene when different combinations of introns (and sometimes exons) are removed from the transcript (Figure 9.23). This alternative splicing can be haphazard, but more often it is controlled and acts as a mechanism of gene regulation, with the frequency of different splicing alternatives controlled by the cell as a way to control the production of different protein products in different cells, or at different stages of development. Alternative splicing is now understood to be a common mechanism of gene regulation in eukaryotes; according to one estimate, 70% of genes in humans are expressed as multiple proteins through alternative splicing.

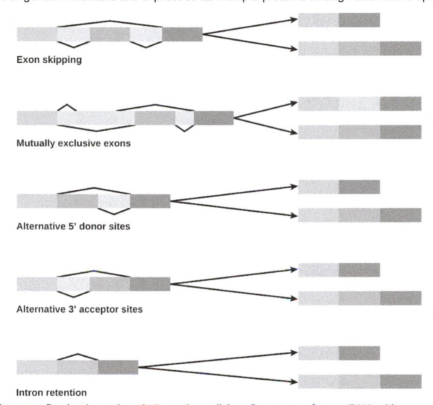

Figure 9.23 There are five basic modes of alternative splicing. Segments of pre-mRNA with exons shown in blue, red, orange, and pink can be spliced to produce a variety of new mature mRNA segments.

How could alternative splicing evolve? Introns have a beginning and ending recognition sequence, and it is easy to imagine the failure of the splicing mechanism to identify the end of an intron and find the end of the next intron, thus removing two introns and the intervening exon. In fact, there are mechanisms in place to prevent such exon skipping, but mutations are likely to lead to their failure. Such "mistakes" would more than likely produce a nonfunctional protein. Indeed, the cause of many genetic diseases is alternative splicing rather than mutations in a sequence. However, alternative splicing would create a protein variant without the loss of the original protein, opening up possibilities for adaptation of the new variant to new functions. Gene duplication has played an important role in the evolution of new functions in a similar way—by providing genes that may evolve without eliminating the original functional protein.

KEY TERMS

alternative RNA splicing a post-transcriptional gene regulation mechanism in eukaryotes in which multiple protein products are produced by a single gene through alternative splicing combinations of the RNA transcript

codon three consecutive nucleotides in mRNA that specify the addition of a specific amino acid or the release of a polypeptide chain during translation

deoxyribose a five-carbon sugar molecule with a hydrogen atom rather than a hydroxyl group in the 2' position; the sugar component of DNA nucleotides

DNA ligase the enzyme that catalyzes the joining of DNA fragments together

DNA polymerase an enzyme that synthesizes a new strand of DNA complementary to a template strand

double helix the molecular shape of DNA in which two strands of nucleotides wind around each other in a spiral shape

epigenetic describing non-genetic regulatory factors, such as changes in modifications to histone proteins and DNA that control accessibility to genes in chromosomes

exon a sequence present in protein-coding mRNA after completion of pre-mRNA splicing

gene expression processes that control whether a gene is expressed

genetic code the amino acids that correspond to three-nucleotide codons of mRNA

helicase an enzyme that helps to open up the DNA helix during DNA replication by breaking the hydrogen bonds

intron non–protein-coding intervening sequences that are spliced from mRNA during processing

lagging strand during replication of the 3' to 5' strand, the strand that is replicated in short fragments and away from the replication fork

leading strand the strand that is synthesized continuously in the 5' to 3' direction that is synthesized in the direction of the replication fork

mismatch repair a form of DNA repair in which non-complementary nucleotides are recognized, excised, and replaced with correct nucleotides

mRNA messenger RNA; a form of RNA that carries the nucleotide sequence code for a protein sequence that is translated into a polypeptide sequence

mutation a permanent variation in the nucleotide sequence of a genome

nitrogenous base a nitrogen-containing molecule that acts as a base; often referring to one of the purine or pyrimidine components of nucleic acids

nontemplate strand the strand of DNA that is not used to transcribe mRNA; this strand is identical to the mRNA except that T nucleotides in the DNA are replaced by U nucleotides in the mRNA

nucleotide excision repair a form of DNA repair in which the DNA molecule is unwound and separated in the region of the nucleotide damage, the damaged nucleotides are removed and replaced with new nucleotides using the complementary strand, and the DNA strand is resealed and allowed to rejoin its complement

Okazaki fragments the DNA fragments that are synthesized in short stretches on the lagging strand

phosphate group a molecular group consisting of a central phosphorus atom bound to four oxygen atoms

post-transcriptional control of gene expression after the RNA molecule has been created but before it is translated into protein

post-translational control of gene expression after a protein has been created

primer a short stretch of RNA nucleotides that is required to initiate replication and allow DNA polymerase to bind and begin replication

promoter a sequence on DNA to which RNA polymerase and associated factors bind and initiate transcription

replication fork the Y-shaped structure formed during the initiation of replication

RNA polymerase an enzyme that synthesizes an RNA strand from a DNA template strand

rRNA ribosomal RNA; molecules of RNA that combine to form part of the ribosome

semiconservative replication the method used to replicate DNA in which the double-stranded molecule is separated and each strand acts as a template for a new strand to be synthesized, so the resulting DNA molecules are composed of one new strand of nucleotides and one old strand of nucleotides

splicing the process of removing introns and reconnecting exons in a pre-mRNA

start codon the AUG (or, rarely GUG) on an mRNA from which translation begins; always specifies methionine

stop codon one of the three mRNA codons that specifies termination of translation

telomerase an enzyme that contains a catalytic part and an inbuilt RNA template; it functions to maintain telomeres at chromosome ends

telomere the DNA at the end of linear chromosomes

template strand the strand of DNA that specifies the complementary mRNA molecule

transcription bubble the region of locally unwound DNA that allows for transcription of mRNA

tRNA transfer RNA; an RNA molecule that contains a specific three-nucleotide anticodon sequence to pair with the mRNA codon and also binds to a specific amino acid

CHAPTER SUMMARY

9.1 The Structure of DNA

The model of the double-helix structure of DNA was proposed by Watson and Crick. The DNA molecule is a polymer of nucleotides. Each nucleotide is composed of a nitrogenous base, a five-carbon sugar (deoxyribose), and a phosphate group. There are four nitrogenous bases in DNA, two purines (adenine and guanine) and two pyrimidines (cytosine and thymine). A DNA molecule is composed of two strands. Each strand is composed of nucleotides bonded together covalently between the phosphate group of one and the deoxyribose sugar of the next. From this backbone extend the bases. The bases of one strand bond to the bases of the second strand with hydrogen bonds. Adenine always bonds with thymine, and cytosine always bonds with guanine. The bonding causes the two strands to spiral around each other in a shape called a double helix. Ribonucleic acid (RNA) is a second nucleic acid found in cells. RNA is a single-stranded polymer of nucleotides. It also differs from DNA in that it contains the sugar ribose, rather than deoxyribose, and the nucleotide uracil rather than thymine. Various RNA molecules function in the process of forming proteins from the genetic code in DNA.

Prokaryotes contain a single, double-stranded circular chromosome. Eukaryotes contain double-stranded linear DNA molecules packaged into chromosomes. The DNA helix is wrapped around proteins to form nucleosomes. The protein coils are further coiled, and during mitosis and meiosis, the chromosomes become even more greatly coiled to facilitate their movement. Chromosomes have two distinct regions which can be distinguished by staining, reflecting different degrees of packaging and determined by whether the DNA in a region is being expressed (euchromatin) or not (heterochromatin).

9.2 DNA Replication

DNA replicates by a semi-conservative method in which each of the two parental DNA strands act as a template for new DNA to be synthesized. After replication, each DNA has one parental or "old" strand, and one daughter or "new" strand.

Replication in eukaryotes starts at multiple origins of replication, while replication in prokaryotes starts from a single origin of replication. The DNA is opened with enzymes, resulting in the formation of the replication fork. Primase synthesizes an RNA primer to initiate synthesis by DNA polymerase, which can add nucleotides in only one direction.

One strand is synthesized continuously in the direction of the replication fork; this is called the leading strand. The other strand is synthesized in a direction away from the replication fork, in short stretches of DNA known as Okazaki fragments. This strand is known as the lagging strand. Once replication is completed, the RNA primers are replaced by DNA nucleotides and the DNA is sealed with DNA ligase.

The ends of eukaryotic chromosomes pose a problem, as polymerase is unable to extend them without a primer. Telomerase, an enzyme with an inbuilt RNA template, extends the ends by copying the RNA template and extending one end of the chromosome. DNA polymerase can then extend the DNA using the primer. In this way, the ends of the chromosomes are protected. Cells have mechanisms for repairing DNA when it becomes damaged or errors are made in replication. These mechanisms include mismatch repair to replace nucleotides that are paired with a non-complementary base and nucleotide excision repair, which removes bases that are damaged such as thymine dimers.

9.3 Transcription

In prokaryotes, mRNA synthesis is initiated at a promoter sequence on the DNA template. Elongation synthesizes new mRNA. Termination liberates the mRNA and occurs by mechanisms that stall the RNA polymerase and cause it to fall off the DNA template. Newly transcribed eukaryotic mRNAs are modified with a cap and a poly-A tail. These structures protect the mature mRNA from degradation and help export it from the nucleus. Eukaryotic mRNAs also undergo splicing, in which introns are removed and exons are reconnected with single-nucleotide accuracy. Only finished mRNAs are exported from the nucleus to the cytoplasm.

9.4 Translation

The central dogma describes the flow of genetic information in the cell from genes to mRNA to proteins. Genes are used to make mRNA by the process of transcription; mRNA is used to synthesize proteins by the process of translation. The genetic code is the correspondence between the three-nucleotide mRNA codon and an amino acid. The genetic code is "translated" by the tRNA molecules, which associate a specific codon with a specific amino acid. The genetic code is degenerate because 64 triplet codons in mRNA specify only 20 amino acids and three stop codons. This means that more than one codon corresponds to an amino acid. Almost every species on the planet uses the same genetic code.

The players in translation include the mRNA template, ribosomes, tRNAs, and various enzymatic factors. The small ribosomal subunit binds to the mRNA template. Translation begins at the initiating AUG on the mRNA. The formation of bonds occurs between sequential amino acids specified by the mRNA template according to the genetic code. The ribosome accepts charged tRNAs, and as it steps along the mRNA, it catalyzes bonding between the new amino acid and the end of the growing polypeptide. The entire mRNA is translated in three-nucleotide "steps" of the ribosome. When a stop codon is encountered, a release factor binds and dissociates the components and frees the new protein.

9.5 How Genes Are Regulated

While all somatic cells within an organism contain the same DNA, not all cells within that organism express the same proteins. Prokaryotic organisms express the entire DNA they encode in every cell, but not necessarily all at the same time. Proteins are expressed only when they are needed. Eukaryotic organisms express a subset of the DNA that is encoded in any given cell. In each cell type, the type and amount of protein is regulated by controlling gene expression. To express a protein, the DNA is first transcribed into RNA, which is then translated into proteins. In prokaryotic cells, these processes occur almost simultaneously. In eukaryotic cells, transcription occurs in the nucleus and is separate from the translation that occurs in the cytoplasm. Gene expression in prokaryotes is regulated only at the transcriptional level, whereas in eukaryotic cells, gene expression is regulated at the epigenetic, transcriptional, post-transcriptional, translational, and post-translational levels.

ART CONNECTION QUESTIONS

1. Figure 9.10 You isolate a cell strain in which the joining together of Okazaki fragments is impaired and suspect that a mutation has occurred in an enzyme found at the replication fork. Which enzyme is most likely to be mutated?

REVIEW QUESTIONS

2. Which of the following does cytosine pair with?

   a. guanine

b. thymine
c. adenine
d. a pyrimidine

3. Prokaryotes contain a _____ chromosome, and eukaryotes contain _____ chromosomes.
   a. single-stranded circular; single-stranded linear
   b. single-stranded linear; single-stranded circular
   c. double-stranded circular; double-stranded linear
   d. double-stranded linear; double-stranded circular

4. DNA replicates by which of the following models?

   a. conservative
   b. semiconservative
   c. dispersive
   d. none of the above

5. The initial mechanism for repairing nucleotide errors in DNA is _____.
   a. mismatch repair
   b. DNA polymerase proofreading
   c. nucleotide excision repair
   d. thymine dimers

6. A promoter is _____.
   a. a specific sequence of DNA nucleotides
   b. a specific sequence of RNA nucleotides
   c. a protein that binds to DNA
   d. an enzyme that synthesizes RNA

7. Portions of eukaryotic mRNA sequence that are removed during RNA processing are _____.
   a. exons
   b. caps
   c. poly-A tails
   d. introns

8. The RNA components of ribosomes are synthesized in the _____.
   a. cytoplasm
   b. nucleus
   c. nucleolus
   d. endoplasmic reticulum

9. How long would the peptide be that is translated from this MRNA sequence: 5'-AUGGGCUACCGA-3'?

   a. 0
   b. 2
   c. 3
   d. 4

10. Control of gene expression in eukaryotic cells occurs at which level(s)?
   a. only the transcriptional level
   b. epigenetic and transcriptional levels
   c. epigenetic, transcriptional, and translational levels
   d. epigenetic, transcriptional, post-transcriptional, translational, and post-translational levels

11. Post-translational control refers to:
   a. regulation of gene expression after transcription
   b. regulation of gene expression after translation
   c. control of epigenetic activation
   d. period between transcription and translation

CRITICAL THINKING QUESTIONS

12. Describe the organization of the eukaryotic chromosome.

13. Describe the structure and complementary base pairing of DNA.

14. How do the linear chromosomes in eukaryotes ensure that its ends are replicated completely?

15. Transcribe and translate the following DNA sequence (nontemplate strand): 5'-ATGGCCGGTTATTAAGCA-3'

16. Describe how controlling gene expression will alter the overall protein levels in the cell.

10 | BIOTECHNOLOGY

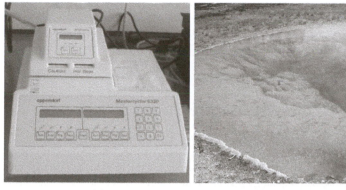

(a)   (b)

Figure 10.1 (a) A thermal cycler, such as the one shown here, is a basic tool used to study DNA in a process called the polymerase chain reaction (PCR). The polymerase enzyme most often used with PCR comes from a strain of bacteria that lives in (b) the hot springs of Yellowstone National Park. (credit a: modification of work by Magnus Manske; credit b: modification of work by Jon Sullivan)

Chapter Outline
10.1: Cloning and Genetic Engineering
10.2: Biotechnology in Medicine and Agriculture
10.3: Genomics and Proteomics

Introduction

The latter half of the twentieth century began with the discovery of the structure of DNA, then progressed to the development of the basic tools used to study and manipulate DNA. These advances, as well as advances in our understanding of and ability to manipulate cells, have led some to refer to the twenty-first century as the biotechnology century. The rate of discovery and of the development of new applications in medicine, agriculture, and energy is expected to accelerate, bringing huge benefits to humankind and perhaps also significant risks. Many of these developments are expected to raise significant ethical and social questions that human societies have not yet had to consider.

10.1 | Cloning and Genetic Engineering

By the end of this section, you will be able to:

- Explain the basic techniques used to manipulate genetic material
- Explain molecular and reproductive cloning

Biotechnology is the use of artificial methods to modify the genetic material of living organisms or cells to produce novel compounds or to perform new functions. Biotechnology has been used for improving livestock and crops since the beginning of agriculture through selective breeding. Since the discovery of the structure of DNA in 1953, and particularly since the development of tools and methods to manipulate DNA in the 1970s, biotechnology has become synonymous with the manipulation of organisms' DNA at the molecular level. The primary applications of this technology are in medicine (for the production of vaccines and antibiotics) and in agriculture (for the genetic modification of crops). Biotechnology also has many industrial applications, such as fermentation, the treatment of oil spills, and the production of biofuels, as well as many household applications such as the use of enzymes in laundry detergent.

Manipulating Genetic Material

To accomplish the applications described above, biotechnologists must be able to extract, manipulate, and analyze nucleic acids.

Review of Nucleic Acid Structure

To understand the basic techniques used to work with nucleic acids, remember that nucleic acids are macromolecules made of nucleotides (a sugar, a phosphate, and a nitrogenous base). The phosphate groups on these molecules each have a net negative charge. An entire set of DNA molecules in the nucleus of eukaryotic organisms is called the genome. DNA has two complementary strands linked by hydrogen bonds between the paired bases.

Unlike DNA in eukaryotic cells, RNA molecules leave the nucleus. Messenger RNA (mRNA) is analyzed most frequently because it represents the protein-coding genes that are being expressed in the cell.

Isolation of Nucleic Acids

To study or manipulate nucleic acids, the DNA must first be extracted from cells. Various techniques are used to extract different types of DNA (Figure 10.2). Most nucleic acid extraction techniques involve steps to break open the cell, and then the use of enzymatic reactions to destroy all undesired macromolecules. Cells are broken open using a detergent solution containing buffering compounds. To prevent degradation and contamination, macromolecules such as proteins and RNA are inactivated using enzymes. The DNA is then brought out of solution using alcohol. The resulting DNA, because it is made up of long polymers, forms a gelatinous mass.

DNA Extraction

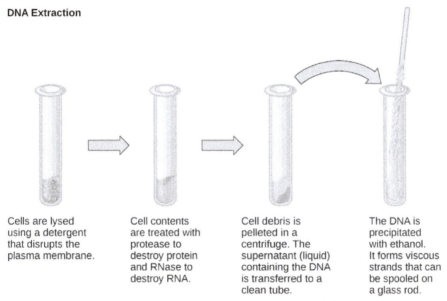

Cells are lysed using a detergent that disrupts the plasma membrane.

Cell contents are treated with protease to destroy protein and RNase to destroy RNA.

Cell debris is pelleted in a centrifuge. The supernatant (liquid) containing the DNA is transferred to a clean tube.

The DNA is precipitated with ethanol. It forms viscous strands that can be spooled on a glass rod.

Figure 10.2 This diagram shows the basic method used for the extraction of DNA.

RNA is studied to understand gene expression patterns in cells. RNA is naturally very unstable because enzymes that break down RNA are commonly present in nature. Some are even secreted by our own skin and are very difficult to inactivate. Similar to DNA extraction, RNA extraction involves the use of various buffers and enzymes to inactivate other macromolecules and preserve only the RNA.

Gel Electrophoresis

Because nucleic acids are negatively charged ions at neutral or alkaline pH in an aqueous environment, they can be moved by an electric field. **Gel electrophoresis** is a technique used to separate charged molecules on the basis of size and charge. The nucleic acids can be separated as whole chromosomes or as fragments. The nucleic acids are loaded into a slot at one end of a gel matrix, an electric current is applied, and negatively charged molecules are pulled toward the opposite end of the gel (the end with the positive electrode). Smaller molecules move through the pores in the gel faster than larger molecules; this difference in the rate of migration separates the fragments on the basis of size. The nucleic acids in a gel matrix are invisible until they are stained with a compound that allows them to be seen, such as a dye. Distinct fragments of nucleic acids appear as bands at specific distances from the top of the gel (the negative electrode end) that are based on their size (Figure 10.3). A mixture of many fragments of varying sizes appear as a long smear, whereas uncut genomic DNA is usually too large to run through the gel and forms a single large band at the top of the gel.

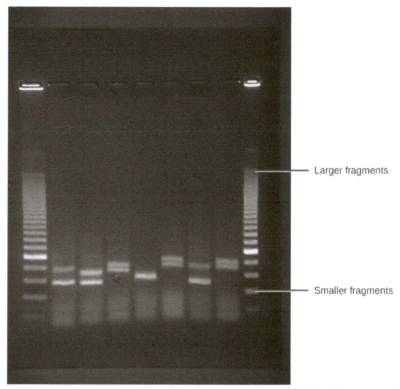

Figure 10.3 Shown are DNA fragments from six samples run on a gel, stained with a fluorescent dye and viewed under UV light. (credit: modification of work by James Jacob, Tompkins Cortland Community College)

Polymerase Chain Reaction

DNA analysis often requires focusing on one or more specific regions of the genome. It also frequently involves situations in which only one or a few copies of a DNA molecule are available for further analysis. These amounts are insufficient for most procedures, such as gel electrophoresis. **Polymerase chain reaction (PCR)** is a technique used to rapidly increase the number of copies of specific regions of DNA for further analyses (Figure 10.4). PCR uses a special form of DNA polymerase, the enzyme that replicates DNA, and other short nucleotide sequences called primers that base pair to a specific portion of the DNA being replicated. PCR is used for many purposes in laboratories. These include: 1) the identification of the owner of a DNA sample left at a crime scene; 2) paternity analysis; 3) the comparison of small amounts of ancient DNA with modern organisms; and 4) determining the sequence of nucleotides in a specific region.

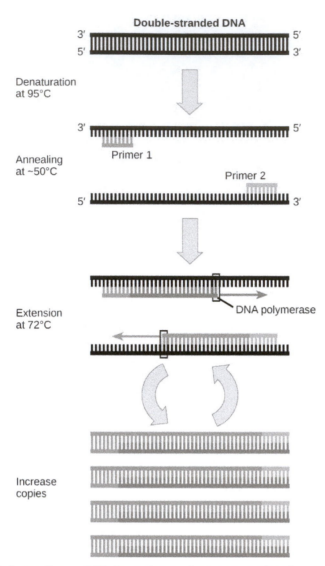

Figure 10.4 Polymerase chain reaction, or PCR, is used to produce many copies of a specific sequence of DNA using a special form of DNA polymerase.

Cloning

In general, **cloning** means the creation of a perfect replica. Typically, the word is used to describe the creation of a genetically identical copy. In biology, the re-creation of a whole organism is referred to as "reproductive cloning." Long before attempts were made to clone an entire organism, researchers learned how to copy short stretches of DNA—a process that is referred to as molecular cloning.

Molecular Cloning

Cloning allows for the creation of multiple copies of genes, expression of genes, and study of specific genes. To get the DNA fragment into a bacterial cell in a form that will be copied or expressed, the fragment is first inserted into a plasmid. A **plasmid** (also called a vector in this context) is a small circular DNA molecule that replicates independently of the chromosomal DNA in bacteria. In cloning, the plasmid molecules can be used to provide a "vehicle" in which to insert a desired DNA fragment. Modified plasmids are usually reintroduced into a bacterial host for replication. As the bacteria divide, they copy their own DNA (including the plasmids). The inserted DNA fragment is copied along with the rest of the bacterial DNA. In a bacterial cell, the fragment of DNA from the human genome (or another organism that is being studied) is referred to as foreign DNA to differentiate it from the DNA of the bacterium (the host DNA).

Plasmids occur naturally in bacterial populations (such as *Escherichia coli*) and have genes that can contribute favorable traits to the organism, such as antibiotic resistance (the ability to be unaffected by antibiotics). Plasmids have been highly engineered as vectors for molecular cloning and for the subsequent large-scale production of important molecules, such as insulin. A valuable characteristic of plasmid vectors is the ease with which a foreign DNA fragment can be introduced. These plasmid vectors contain many short DNA sequences that can be cut with different commonly available **restriction enzymes**. Restriction enzymes (also called restriction endonucleases) recognize specific DNA sequences and cut them in a predictable manner; they are naturally produced by bacteria as a defense mechanism against foreign DNA. Many restriction enzymes make staggered cuts in the two strands of DNA, such that the cut ends have a 2- to 4-nucleotide single-stranded overhang. The sequence that is recognized by the restriction enzyme is a four- to eight-nucleotide sequence that is a palindrome. Like with a word palindrome, this means the sequence reads the same forward and backward. In most cases, the sequence reads the same forward on one strand and backward on the complementary strand. When a staggered cut is made in a sequence like this, the overhangs are complementary (Figure 10.5).

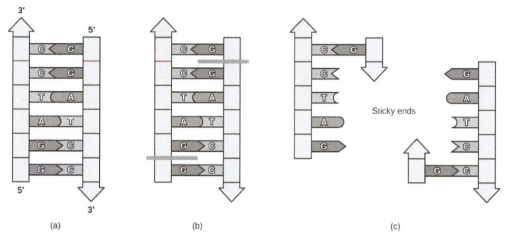

Figure 10.5 In this (a) six-nucleotide restriction enzyme recognition site, notice that the sequence of six nucleotides reads the same in the 5' to 3' direction on one strand as it does in the 5' to 3' direction on the complementary strand. This is known as a palindrome. (b) The restriction enzyme makes breaks in the DNA strands, and (c) the cut in the DNA results in "sticky ends". Another piece of DNA cut on either end by the same restriction enzyme could attach to these sticky ends and be inserted into the gap made by this cut.

Because these overhangs are capable of coming back together by hydrogen bonding with complementary overhangs on a piece of DNA cut with the same restriction enzyme, these are called "sticky ends." The process of forming hydrogen bonds between complementary sequences on single strands to form double-stranded DNA is called **annealing**. Addition of an enzyme called DNA ligase, which takes part in DNA replication in cells, permanently joins the DNA fragments when the sticky ends come together. In this way, any DNA fragment can be spliced between the two ends of a plasmid DNA that has been cut with the same restriction enzyme (Figure 10.6).

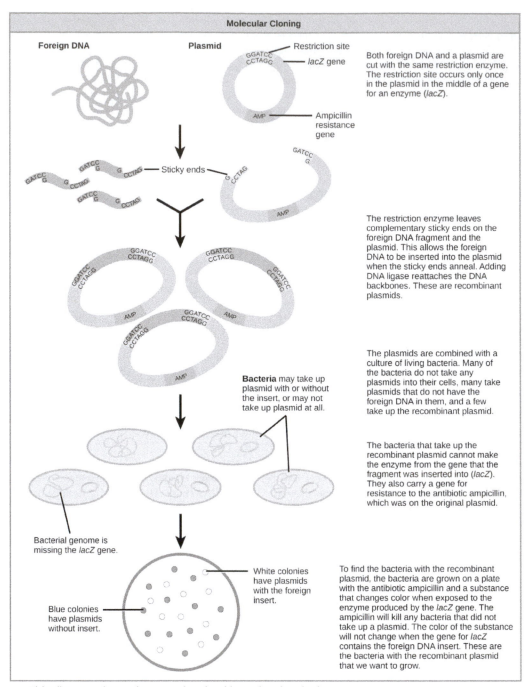

Molecular Cloning

Both foreign DNA and a plasmid are cut with the same restriction enzyme. The restriction site occurs only once in the plasmid in the middle of a gene for an enzyme (*lacZ*).

The restriction enzyme leaves complementary sticky ends on the foreign DNA fragment and the plasmid. This allows the foreign DNA to be inserted into the plasmid when the sticky ends anneal. Adding DNA ligase reattaches the DNA backbones. These are recombinant plasmids.

The plasmids are combined with a culture of living bacteria. Many of the bacteria do not take any plasmids into their cells, many take plasmids that do not have the foreign DNA in them, and a few take up the recombinant plasmid.

The bacteria that take up the recombinant plasmid cannot make the enzyme from the gene that the fragment was inserted into (*lacZ*). They also carry a gene for resistance to the antibiotic ampicillin, which was on the original plasmid.

To find the bacteria with the recombinant plasmid, the bacteria are grown on a plate with the antibiotic ampicillin and a substance that changes color when exposed to the enzyme produced by the *lacZ* gene. The ampicillin will kill any bacteria that did not take up a plasmid. The color of the substance will not change when the gene for *lacZ* contains the foreign DNA insert. These are the bacteria with the recombinant plasmid that we want to grow.

Figure 10.6 This diagram shows the steps involved in molecular cloning.

Plasmids with foreign DNA inserted into them are called **recombinant DNA** molecules because they contain new combinations of genetic material. Proteins that are produced from recombinant DNA molecules are called **recombinant proteins**. Not all recombinant plasmids are capable of expressing genes. Plasmids may also be engineered to express proteins only when stimulated by certain environmental factors, so that scientists can control the expression of the recombinant proteins.

Reproductive Cloning

Reproductive cloning is a method used to make a clone or an identical copy of an entire multicellular organism. Most multicellular organisms undergo reproduction by sexual means, which involves the contribution of DNA from two individuals (parents), making it impossible to generate an identical copy or a clone of either parent. Recent advances in biotechnology have made it possible to reproductively clone mammals in the laboratory.

Natural sexual reproduction involves the union, during fertilization, of a sperm and an egg. Each of these gametes is haploid, meaning they contain one set of chromosomes in their nuclei. The resulting cell, or zygote, is then diploid and contains two sets of chromosomes. This cell divides mitotically to produce a multicellular organism. However, the union of just any two cells cannot produce a viable zygote; there are components in the cytoplasm of the egg cell that are essential for the early development of the embryo during its first few cell divisions. Without these provisions, there would be no subsequent development. Therefore, to produce a new individual, both a diploid genetic complement and an egg cytoplasm are required. The approach to producing an artificially cloned individual is to take the egg cell of one individual and to remove the haploid nucleus. Then a diploid nucleus from a body cell of a second individual, the donor, is put into the egg cell. The egg is then stimulated to divide so that development proceeds. This sounds simple, but in fact it takes many attempts before each of the steps is completed successfully.

The first cloned agricultural animal was Dolly, a sheep who was born in 1996. The success rate of reproductive cloning at the time was very low. Dolly lived for six years and died of a lung tumor (Figure 10.7). There was speculation that because the cell DNA that gave rise to Dolly came from an older individual, the age of the DNA may have affected her life expectancy. Since Dolly, several species of animals (such as horses, bulls, and goats) have been successfully cloned.

There have been attempts at producing cloned human embryos as sources of embryonic stem cells. In the procedure, the DNA from an adult human is introduced into a human egg cell, which is then stimulated to divide. The technology is similar to the technology that was used to produce Dolly, but the embryo is never implanted into a surrogate mother. The cells produced are called embryonic stem cells because they have the capacity to develop into many different kinds of cells, such as muscle or nerve cells. The stem cells could be used to research and ultimately provide therapeutic applications, such as replacing damaged tissues. The benefit of cloning in this instance is that the cells used to regenerate new tissues would be a perfect match to the donor of the original DNA. For example, a leukemia patient would not require a sibling with a tissue match for a bone-marrow transplant.

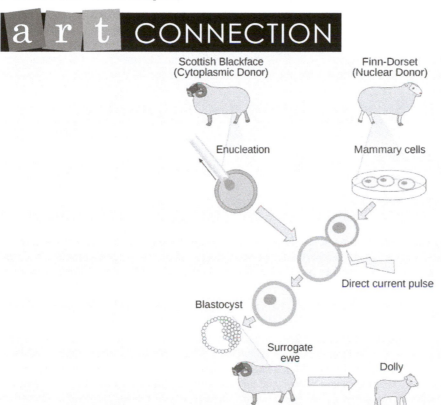

Figure 10.7 Dolly the sheep was the first agricultural animal to be cloned. To create Dolly, the nucleus was removed from a donor egg cell. The enucleated egg was placed next to the other cell, then they were shocked to fuse. They were shocked again to start division. The cells were allowed to divide for several days until an early embryonic stage was reached, before being implanted in a surrogate mother.

Why was Dolly a Finn-Dorset and not a Scottish Blackface sheep?

Genetic Engineering

Using recombinant DNA technology to modify an organism's DNA to achieve desirable traits is called **genetic engineering**. Addition of foreign DNA in the form of recombinant DNA vectors that are generated by molecular cloning is the most common method of genetic engineering. An organism that receives the recombinant DNA is called a **genetically modified organism** (GMO). If the foreign DNA that is introduced comes from a different species, the host organism is called **transgenic**. Bacteria, plants, and animals have been genetically modified since the early 1970s for academic, medical, agricultural, and industrial purposes. These applications will be examined in more detail in the next module.

Watch this short video (http://openstaxcollege.org/l/transgenic) explaining how scientists create a transgenic animal.

Although the classic methods of studying the function of genes began with a given phenotype and determined the genetic basis of that phenotype, modern techniques allow researchers to start at the DNA sequence level and ask: "What does this gene or DNA element do?" This technique, called **reverse genetics**, has resulted in reversing the classical genetic methodology. One example of this method is analogous to damaging a body part to determine its function. An insect that loses a wing cannot fly, which means that the wing's function is flight. The classic genetic method compares insects that cannot fly with insects that can fly, and observes that the non-flying insects have lost wings. Similarly in a reverse genetics approach, mutating or deleting genes provides researchers with clues about gene function. Alternately, reverse genetics can be used to cause a gene to overexpress itself to determine what phenotypic effects may occur.

10.2 | Biotechnology in Medicine and Agriculture

By the end of this section, you will be able to:

- Describe uses of biotechnology in medicine
- Describe uses of biotechnology in agriculture

It is easy to see how biotechnology can be used for medicinal purposes. Knowledge of the genetic makeup of our species, the genetic basis of heritable diseases, and the invention of technology to manipulate and fix mutant genes provides methods to treat diseases. Biotechnology in agriculture can enhance resistance to disease, pests, and environmental stress to improve both crop yield and quality.

Genetic Diagnosis and Gene Therapy

The process of testing for suspected genetic defects before administering treatment is called genetic diagnosis by genetic testing. In some cases in which a genetic disease is present in an individual's family, family members may be advised to undergo genetic testing. For example, mutations in the *BRCA* genes may increase the likelihood of developing breast and ovarian cancers in women and some other cancers in women and men. A woman with breast cancer can be screened for these mutations. If one of the high-risk mutations is found, her female relatives may also wish to be screened for that particular mutation, or simply be more vigilant for the occurrence of cancers. Genetic testing is also offered for fetuses (or embryos with in vitro fertilization) to determine the presence or absence of disease-causing genes in families with specific debilitating diseases.

See how human DNA is extracted (http://openstaxcollege.org/l/DNA_extraction) for uses such as genetic testing.

Gene therapy is a genetic engineering technique that may one day be used to cure certain genetic diseases. In its simplest form, it involves the introduction of a non-mutated gene at a random location in the genome to cure a disease by replacing a protein that may be absent in these individuals because of a genetic mutation. The non-mutated gene is usually introduced into diseased cells as part of a vector transmitted by a virus, such as an adenovirus, that can infect the host cell and deliver the foreign DNA into the genome of the targeted cell (Figure 10.8). To date, gene therapies have been primarily experimental procedures in humans. A few of these experimental treatments have been successful, but the methods may be important in the future as the factors limiting its success are resolved.

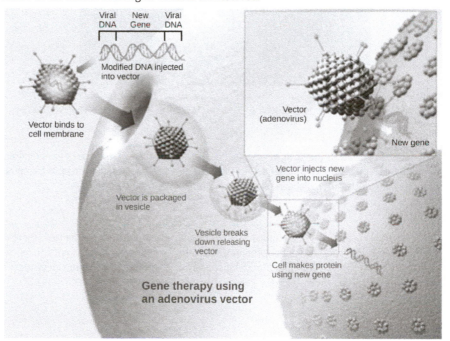

Figure 10.8 This diagram shows the steps involved in curing disease with gene therapy using an adenovirus vector. (credit: modification of work by NIH)

Production of Vaccines, Antibiotics, and Hormones

Traditional vaccination strategies use weakened or inactive forms of microorganisms or viruses to stimulate the immune system. Modern techniques use specific genes of microorganisms cloned into vectors and mass-produced in bacteria to make large quantities of specific substances to stimulate the immune system. The substance is then used as a vaccine. In some cases, such as the H1N1 flu vaccine, genes cloned from the virus have been used to combat the constantly changing strains of this virus.

Antibiotics kill bacteria and are naturally produced by microorganisms such as fungi; penicillin is perhaps the most well-known example. Antibiotics are produced on a large scale by cultivating and manipulating fungal cells. The fungal cells have typically been genetically modified to improve the yields of the antibiotic compound.

Recombinant DNA technology was used to produce large-scale quantities of the human hormone insulin in *E. coli* as early as 1978. Previously, it was only possible to treat diabetes with pig insulin, which caused allergic reactions in many humans because of differences in the insulin molecule. In addition, human growth hormone (HGH) is used to treat growth disorders in children. The HGH gene was cloned from a cDNA (complementary DNA) library and inserted into *E. coli* cells by cloning it into a bacterial vector.

Transgenic Animals

Although several recombinant proteins used in medicine are successfully produced in bacteria, some proteins need a eukaryotic animal host for proper processing. For this reason, genes have been cloned and expressed in animals such as sheep, goats, chickens, and mice. Animals that have been modified to express recombinant DNA are called transgenic animals (Figure 10.9).

Figure 10.9 It can be seen that two of these mice are transgenic because they have a gene that causes them to fluoresce under a UV light. The non-transgenic mouse does not have the gene that causes fluorescence. (credit: Ingrid Moen et al.)

Several human proteins are expressed in the milk of transgenic sheep and goats. In one commercial example, the FDA has approved a blood anticoagulant protein that is produced in the milk of transgenic goats for use in humans. Mice have been used extensively for expressing and studying the effects of recombinant genes and mutations.

Transgenic Plants

Manipulating the DNA of plants (creating genetically modified organisms, or GMOs) has helped to create desirable traits such as disease resistance, herbicide, and pest resistance, better nutritional value, and better shelf life (Figure 10.10). Plants are the most important source of food for the human population. Farmers developed ways to select for plant varieties with desirable traits long before modern-day biotechnology practices were established.

Figure 10.10 Corn, a major agricultural crop used to create products for a variety of industries, is often modified through plant biotechnology. (credit: Keith Weller, USDA)

Transgenic plants have received DNA from other species. Because they contain unique combinations of genes and are not restricted to the laboratory, transgenic plants and other GMOs are closely monitored by government agencies to ensure that they are fit for human consumption and do not endanger other plant and animal life. Because foreign genes can spread to other species in the environment, particularly in the pollen and seeds of plants, extensive testing is required to ensure ecological stability. Staples like corn, potatoes, and tomatoes were the first crop plants to be genetically engineered.

Transformation of Plants Using Agrobacterium tumefaciens

In plants, tumors caused by the bacterium *Agrobacterium tumefaciens* occur by transfer of DNA from the bacterium to the plant. The artificial introduction of DNA into plant cells is more challenging than in animal cells because of the thick plant cell wall. Researchers used the natural transfer of DNA from *Agrobacterium* to a plant host to introduce DNA fragments of their choice into plant hosts. In nature, the disease-causing *A. tumefaciens* have a set of plasmids that contain genes that integrate into the infected plant cell's genome. Researchers manipulate the plasmids to carry the desired DNA fragment and insert it into the plant genome.

The Organic Insecticide Bacillus thuringiensis

Bacillus thuringiensis (Bt) is a bacterium that produces protein crystals that are toxic to many insect species that feed on plants. Insects that have eaten Bt toxin stop feeding on the plants within a few hours. After the toxin is activated in the intestines of the insects, death occurs within a couple of days. The crystal toxin genes have been cloned from the bacterium and introduced into plants, therefore allowing plants to produce their own crystal Bt toxin that acts against insects. Bt toxin is safe for the environment and non-toxic to mammals (including humans). As a result, it has been approved for use by organic farmers as a natural insecticide. There is some concern, however, that insects may evolve resistance to the Bt toxin in the same way that bacteria evolve resistance to antibiotics.

FlavrSavr Tomato

The first GM crop to be introduced into the market was the FlavrSavr Tomato produced in 1994. Molecular genetic technology was used to slow down the process of softening and rotting caused by fungal infections, which led to increased shelf life of the GM tomatoes. Additional genetic modification improved the flavor of this tomato. The FlavrSavr tomato did not successfully stay in the market because of problems maintaining and shipping the crop.

10.3 | Genomics and Proteomics

By the end of this section, you will be able to:

- Define genomics and proteomics
- Define whole genome sequencing
- Explain different applications of genomics and proteomics

The study of nucleic acids began with the discovery of DNA, progressed to the study of genes and small fragments, and has now exploded to the field of **genomics**. Genomics is the study of entire genomes, including the complete set of genes, their nucleotide sequence and organization, and their interactions within a species and with other species. The advances in genomics have been made possible by DNA sequencing technology. Just as information technology has led to Google Maps that enable us to get detailed information about locations around the globe, genomic information is used to create similar maps of the DNA of different organisms.

Mapping Genomes

Genome mapping is the process of finding the location of genes on each chromosome. The maps that are created are comparable to the maps that we use to navigate streets. A **genetic map** is an illustration that lists genes and their location on a chromosome. Genetic maps provide the big picture (similar to a map of interstate highways) and use genetic markers (similar to landmarks). A genetic marker is a gene or sequence on a chromosome that shows genetic linkage with a trait of interest. The genetic marker tends to be inherited with the gene of interest, and one measure of distance between them is the recombination frequency during meiosis. Early geneticists called this linkage analysis.

Physical maps get into the intimate details of smaller regions of the chromosomes (similar to a detailed road map) (Figure 10.11). A physical map is a representation of the physical distance, in nucleotides, between genes or genetic markers. Both genetic linkage maps and physical maps are required to build a complete picture of the genome. Having a complete map of the genome makes it easier for researchers to study individual genes. Human genome maps help researchers in their efforts to identify human disease-causing genes related to illnesses such as cancer, heart disease, and cystic fibrosis, to name a few. In addition, genome mapping can be used to help identify organisms with beneficial traits, such as microbes with the ability to clean up pollutants or even prevent pollution. Research involving plant genome mapping may lead to methods that produce higher crop yields or to the development of plants that adapt better to climate change.

Download for free at https://openstax.org/details/books/concepts-biology

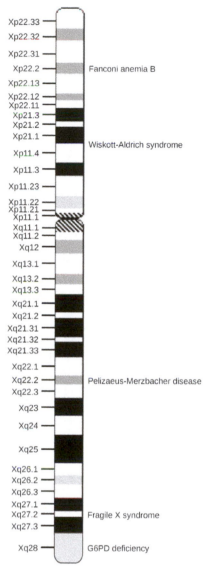

Figure 10.11 This is a physical map of the human X chromosome. (credit: modification of work by NCBI, NIH)

Genetic maps provide the outline, and physical maps provide the details. It is easy to understand why both types of genome-mapping techniques are important to show the big picture. Information obtained from each technique is used in combination to study the genome. Genomic mapping is used with different model organisms that are used for research. Genome mapping is still an ongoing process, and as more advanced techniques are developed, more advances are expected. Genome mapping is similar to completing a complicated puzzle using every piece of available data. Mapping information generated in laboratories all over the world is entered into central databases, such as the National Center for Biotechnology Information (NCBI). Efforts are made to make the information more easily accessible to researchers and the general public. Just as we use global positioning systems instead of paper maps to navigate through roadways, NCBI allows us to use a genome viewer tool to simplify the data mining process.

Whole Genome Sequencing

Although there have been significant advances in the medical sciences in recent years, doctors are still confounded by many diseases and researchers are using whole genome sequencing to get to the bottom of the problem. **Whole genome sequencing** is a process that determines the DNA sequence of an entire genome. Whole genome sequencing is a brute-force approach to problem solving when there is a genetic basis at the core of a disease. Several laboratories now provide services to sequence, analyze, and interpret entire genomes.

In 2010, whole genome sequencing was used to save a young boy whose intestines had multiple mysterious abscesses. The child had several colon operations with no relief. Finally, a whole genome sequence revealed a defect in a pathway that controls apoptosis (programmed cell death). A bone marrow transplant was used to overcome this genetic disorder, leading to a cure for the boy. He was the first person to be successfully diagnosed using whole genome sequencing.

The first genomes to be sequenced, such as those belonging to viruses, bacteria, and yeast, were smaller in terms of the number of nucleotides than the genomes of multicellular organisms. The genomes of other model organisms, such as the mouse (*Mus musculus*), the fruit fly (*Drosophila melanogaster*), and the nematode (*Caenorhabditis elegans*) are now known. A great deal of basic research is performed in **model organisms** because the information can be applied to other organisms. A model organism is a species that is studied as a model to understand the biological processes in other species that can be represented by the model organism. For example, fruit flies are able to metabolize alcohol like humans, so the genes affecting sensitivity to alcohol have been studied in fruit flies in an effort to understand the variation in sensitivity to alcohol in humans. Having entire genomes sequenced helps with the research efforts in these model organisms (Figure 10.12).

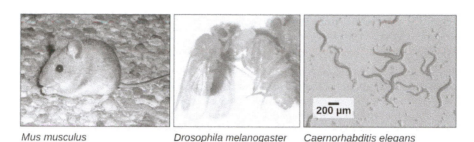

Mus musculus   *Drosophila melanogaster*   *Caernorhabditis elegans*

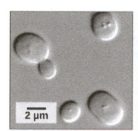

Saccharomyces cerevisiae   *Arabidopsis thaliana*

Figure 10.12 Much basic research is done with model organisms, such as the mouse, *Mus musculus*; the fruit fly, *Drosophila melanogaster*; the nematode *Caenorhabditis elegans*; the yeast *Saccharomyces cerevisiae*; and the common weed, *Arabidopsis thaliana*. (credit "mouse": modification of work by Florean Fortescue; credit "nematodes": modification of work by "snickclunk"/Flickr; credit "common weed": modification of work by Peggy Greb, USDA; scale-bar data from Matt Russell)

The first human genome sequence was published in 2003. The number of whole genomes that have been sequenced steadily increases and now includes hundreds of species and thousands of individual human genomes.

Applying Genomics

The introduction of DNA sequencing and whole genome sequencing projects, particularly the Human Genome Project, has expanded the applicability of DNA sequence information. Genomics is now being used in a wide variety of fields, such as metagenomics, pharmacogenomics, and mitochondrial genomics. The most commonly known application of genomics is to understand and find cures for diseases.

Predicting Disease Risk at the Individual Level

Predicting the risk of disease involves screening and identifying currently healthy individuals by genome analysis at the individual level. Intervention with lifestyle changes and drugs can be recommended before disease onset. However, this approach is most applicable when the problem arises from a single gene mutation. Such defects only account for about 5 percent of diseases found in developed countries. Most of the common diseases, such as heart disease, are multifactorial or polygenic, which refers to a phenotypic characteristic that is determined by two or more genes, and also environmental factors such as diet. In April 2010, scientists at Stanford University published the genome analysis of a healthy individual (Stephen Quake, a scientist at Stanford University, who had his genome sequenced); the analysis predicted his propensity to acquire various diseases. A risk assessment was done to analyze Quake's percentage of risk for 55 different medical conditions. A rare genetic mutation was found that showed him to be at risk for sudden heart attack. He was also predicted to have a 23 percent risk of developing prostate cancer and a 1.4 percent risk of developing Alzheimer's disease. The scientists used databases and several publications to analyze the genomic data. Even though genomic sequencing is becoming more affordable and analytical tools are becoming more reliable, ethical issues surrounding genomic analysis at a population level remain to be addressed. For example, could such data be legitimately used to charge more or less for insurance or to affect credit ratings?

Genome-wide Association Studies

Since 2005, it has been possible to conduct a type of study called a genome-wide association study, or GWAS. A GWAS is a method that identifies differences between individuals in single nucleotide polymorphisms (SNPs) that may be involved in causing diseases. The method is particularly suited to diseases that may be affected by one or many genetic changes throughout the genome. It is very difficult to identify the genes involved in such a disease using family history information. The GWAS method relies on a genetic database that has been in development since 2002 called the International HapMap Project. The HapMap Project sequenced the genomes of several hundred individuals from around the world and identified groups of SNPs. The groups include SNPs that are located near to each other on chromosomes so they tend to stay together through recombination. The fact that the group stays together means that identifying one marker SNP is all that is needed to identify all the SNPs in the group. There are several million SNPs identified, but identifying them in other individuals who have not had their complete genome sequenced is much easier because only the marker SNPs need to be identified.

In a common design for a GWAS, two groups of individuals are chosen; one group has the disease, and the other group does not. The individuals in each group are matched in other characteristics to reduce the effect of confounding variables causing differences between the two groups. For example, the genotypes may differ because the two groups are mostly taken from different parts of the world. Once the individuals are chosen, and typically their numbers are a thousand or more for the study to work, samples of their DNA are obtained. The DNA is analyzed using automated systems to identify large differences in the percentage of particular SNPs between the two groups. Often the study examines a million or more SNPs in the DNA. The results of GWAS can be used in two ways: the genetic differences may be used as markers for susceptibility to the disease in undiagnosed individuals, and the particular genes identified can be targets for research into the molecular pathway of the disease and potential therapies. An offshoot of the discovery of gene associations with disease has been the formation of companies that provide so-called "personal genomics" that will identify risk levels for various diseases based on an individual's SNP complement. The science behind these services is controversial.

Because GWAS looks for associations between genes and disease, these studies provide data for other research into causes, rather than answering specific questions themselves. An association between a gene difference and a disease does not necessarily mean there is a cause-and-effect relationship. However, some studies have provided useful information about the genetic causes of diseases. For example, three different studies in 2005 identified a gene for a protein involved in regulating inflammation in the body that is associated with a disease-causing blindness called age-related macular degeneration. This opened up new possibilities for research into the cause of this disease. A large number of genes have been identified to be associated with Crohn's disease using GWAS, and some of these have suggested new hypothetical mechanisms for the cause of the disease.

Pharmacogenomics

Pharmacogenomics involves evaluating the effectiveness and safety of drugs on the basis of information from an individual's genomic sequence. Personal genome sequence information can be used to prescribe medications that will be most effective and least toxic on the basis of the individual patient's genotype. Studying changes in gene expression could provide information about the gene transcription profile in the presence of the drug, which can be used as an early indicator of the potential for toxic effects. For example, genes involved in cellular growth and controlled cell death, when disturbed, could lead to the growth of cancerous cells. Genome-wide studies can also help to find new genes involved in drug toxicity. The gene signatures may not be completely accurate, but can be tested further before pathologic symptoms arise.

Metagenomics

Traditionally, microbiology has been taught with the view that microorganisms are best studied under pure culture conditions, which involves isolating a single type of cell and culturing it in the laboratory. Because microorganisms can go through several generations in a matter of hours, their gene expression profiles adapt to the new laboratory environment very quickly. On the other hand, many species resist being cultured in isolation. Most microorganisms do not live as isolated entities, but in microbial communities known as biofilms. For all of these reasons, pure culture is not always the best way to study microorganisms. **Metagenomics** is the study of the collective genomes of multiple species that grow and interact in an environmental niche. Metagenomics can be used to identify new species more rapidly and to analyze the effect of pollutants on the environment (Figure 10.13). Metagenomics techniques can now also be applied to communities of higher eukaryotes, such as fish.

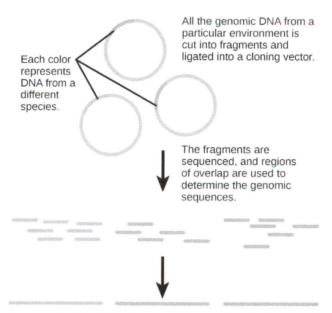

Each color represents DNA from a different species.

All the genomic DNA from a particular environment is cut into fragments and ligated into a cloning vector.

The fragments are sequenced, and regions of overlap are used to determine the genomic sequences.

Figure 10.13 Metagenomics involves isolating DNA from multiple species within an environmental niche. The DNA is cut up and sequenced, allowing entire genome sequences of multiple species to be reconstructed from the sequences of overlapping pieces.

Creation of New Biofuels

Knowledge of the genomics of microorganisms is being used to find better ways to harness biofuels from algae and cyanobacteria. The primary sources of fuel today are coal, oil, wood, and other plant products such as ethanol. Although plants are renewable resources, there is still a need to find more alternative renewable sources of energy to meet our population's energy demands. The microbial world is one of the largest resources for genes that encode new enzymes and produce new organic compounds, and it remains largely untapped. This vast genetic resource holds the potential to provide new sources of biofuels (Figure 10.14).

Figure 10.14 Renewable fuels were tested in Navy ships and aircraft at the first Naval Energy Forum. (credit: modification of work by John F. Williams, US Navy)

Mitochondrial Genomics

Mitochondria are intracellular organelles that contain their own DNA. Mitochondrial DNA mutates at a rapid rate and is often used to study evolutionary relationships. Another feature that makes studying the mitochondrial genome interesting is that in most multicellular organisms, the mitochondrial DNA is passed on from the mother during the process of fertilization. For this reason, mitochondrial genomics is often used to trace genealogy.

Genomics in Forensic Analysis

Information and clues obtained from DNA samples found at crime scenes have been used as evidence in court cases, and genetic markers have been used in forensic analysis. Genomic analysis has also become useful in this field. In 2001, the first use of genomics in forensics was published. It was a collaborative effort between academic research institutions and the FBI to solve the mysterious cases of anthrax (Figure 10.15) that was transported by the US Postal Service. Anthrax bacteria were made into an infectious powder and mailed to news media and two U.S. Senators. The powder infected the administrative staff and postal workers who opened or handled the letters. Five people died, and 17 were sickened from the bacteria. Using microbial genomics, researchers determined that a specific strain of anthrax was used in all the mailings; eventually, the source was traced to a scientist at a national biodefense laboratory in Maryland.

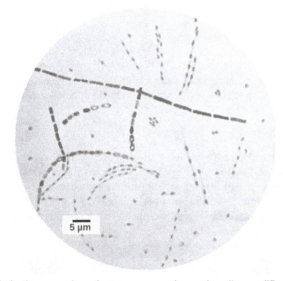

Figure 10.15 *Bacillus anthracis* is the organism that causes anthrax. (credit: modification of work by CDC; scale-bar data from Matt Russell)

Genomics in Agriculture

Genomics can reduce the trials and failures involved in scientific research to a certain extent, which could improve the quality and quantity of crop yields in agriculture (Figure 10.16). Linking traits to genes or gene signatures helps to improve crop breeding to generate hybrids with the most desirable qualities. Scientists use genomic data to identify desirable traits, and then transfer those traits to a different organism to create a new genetically modified organism, as described in the previous module. Scientists are discovering how genomics can improve the quality and quantity of agricultural production. For example, scientists could use desirable traits to create a useful product or enhance an existing product, such as making a drought-sensitive crop more tolerant of the dry season.

Figure 10.16 Transgenic agricultural plants can be made to resist disease. These transgenic plums are resistant to the plum pox virus. (credit: Scott Bauer, USDA ARS)

Proteomics

Proteins are the final products of genes that perform the function encoded by the gene. Proteins are composed of amino acids and play important roles in the cell. All enzymes (except ribozymes) are proteins and act as catalysts that affect the rate of reactions. Proteins are also regulatory molecules, and some are hormones. Transport proteins, such as hemoglobin, help transport oxygen to various organs. Antibodies that defend against foreign particles are also proteins. In the diseased state, protein function can be impaired because of changes at the genetic level or because of direct impact on a specific protein.

A proteome is the entire set of proteins produced by a cell type. Proteomes can be studied using the knowledge of genomes because genes code for mRNAs, and the mRNAs encode proteins. The study of the function of proteomes is called **proteomics**. Proteomics complements genomics and is useful when scientists want to test their hypotheses that were based on genes. Even though all cells in a multicellular organism have the same set of genes, the set of proteins produced in different tissues is different and dependent on gene expression. Thus, the genome is constant, but the proteome varies and is dynamic within an organism. In addition, RNAs can be alternatively spliced (cut and pasted to create novel combinations and novel proteins), and many proteins are modified after translation. Although the genome provides a blueprint, the final architecture depends on several factors that can change the progression of events that generate the proteome.

Genomes and proteomes of patients suffering from specific diseases are being studied to understand the genetic basis of the disease. The most prominent disease being studied with proteomic approaches is cancer (Figure 10.17). Proteomic approaches are being used to improve the screening and early detection of cancer; this is achieved by identifying proteins whose expression is affected by the disease process. An individual protein is called a **biomarker**, whereas a set of proteins with altered expression levels is called a **protein signature**. For a biomarker or protein signature to be useful as a candidate for early screening and detection of a cancer, it must be secreted in body fluids such as sweat, blood, or urine, so that large-scale screenings can be performed in a noninvasive fashion. The current problem with using biomarkers for the early detection of cancer is the high rate of false-negative results. A false-negative result is a negative test result that should have been positive. In other words, many cases of cancer go undetected, which makes biomarkers unreliable. Some examples of protein biomarkers used in cancer detection are CA-125 for ovarian cancer and PSA for prostate cancer. Protein signatures may be more reliable than biomarkers to detect cancer cells. Proteomics is also being used to develop individualized treatment plans, which involves the prediction of whether or not an individual will respond to specific drugs and the side effects that the individual may have. Proteomics is also being used to predict the possibility of disease recurrence.

Figure 10.17 This machine is preparing to do a proteomic pattern analysis to identify specific cancers so that an accurate cancer prognosis can be made. (credit: Dorie Hightower, NCI, NIH)

The National Cancer Institute has developed programs to improve the detection and treatment of cancer. The Clinical Proteomic Technologies for Cancer and the Early Detection Research Network are efforts to identify protein signatures specific to different types of cancers. The Biomedical Proteomics Program is designed to identify protein signatures and design effective therapies for cancer patients.

KEY TERMS

anneal in molecular biology, the process by which two single strands of DNA hydrogen bond at complementary nucleotides to form a double-stranded molecule

biomarker an individual protein that is uniquely produced in a diseased state

biotechnology the use of artificial methods to modify the genetic material of living organisms or cells to produce novel compounds or to perform new functions

cloning the production of an exact copy—specifically, an exact genetic copy—of a gene, cell, or organism

gel electrophoresis a technique used to separate molecules on the basis of their ability to migrate through a semisolid gel in response to an electric current

gene therapy the technique used to cure heritable diseases by replacing mutant genes with good genes

genetic engineering alteration of the genetic makeup of an organism using the molecular methods of biotechnology

genetic map an outline of genes and their location on a chromosome that is based on recombination frequencies between markers

genetic testing identifying gene variants in an individual that may lead to a genetic disease in that individual

genetically modified organism (GMO) an organism whose genome has been artificially changed

genomics the study of entire genomes, including the complete set of genes, their nucleotide sequence and organization, and their interactions within a species and with other species

metagenomics the study of the collective genomes of multiple species that grow and interact in an environmental niche

model organism a species that is studied and used as a model to understand the biological processes in other species represented by the model organism

pharmacogenomics the study of drug interactions with the genome or proteome; also called toxicogenomics

physical map a representation of the physical distance between genes or genetic markers

plasmid a small circular molecule of DNA found in bacteria that replicates independently of the main bacterial chromosome; plasmids code for some important traits for bacteria and can be used as vectors to transport DNA into bacteria in genetic engineering applications

polymerase chain reaction (PCR) a technique used to make multiple copies of DNA

protein signature a set of over- or under-expressed proteins characteristic of cells in a particular diseased tissue

proteomics study of the function of proteomes

recombinant DNA a combination of DNA fragments generated by molecular cloning that does not exist in nature

recombinant protein a protein that is expressed from recombinant DNA molecules

reproductive cloning cloning of entire organisms

restriction enzyme an enzyme that recognizes a specific nucleotide sequence in DNA and cuts the DNA double strand at that recognition site, often with a staggered cut leaving short single strands or "sticky" ends

reverse genetics a form of genetic analysis that manipulates DNA to disrupt or affect the product of a gene to analyze the gene's function

transgenic describing an organism that receives DNA from a different species

whole genome sequencing a process that determines the nucleotide sequence of an entire genome

CHAPTER SUMMARY

10.1 Cloning and Genetic Engineering

Nucleic acids can be isolated from cells for the purposes of further analysis by breaking open the cells and enzymatically destroying all other major macromolecules. Fragmented or whole chromosomes can be separated on the basis of size by gel electrophoresis. Short stretches of DNA can be amplified by PCR. DNA can be cut (and subsequently re-spliced together) using restriction enzymes. The molecular and cellular techniques of biotechnology allow researchers to genetically engineer organisms, modifying them to achieve desirable traits.

Cloning may involve cloning small DNA fragments (molecular cloning), or cloning entire organisms (reproductive cloning). In molecular cloning with bacteria, a desired DNA fragment is inserted into a bacterial plasmid using restriction enzymes and the plasmid is taken up by a bacterium, which will then express the foreign DNA. Using other techniques, foreign genes can be inserted into eukaryotic organisms. In each case, the organisms are called transgenic organisms. In reproductive cloning, a donor nucleus is put into an enucleated egg cell, which is then stimulated to divide and develop into an organism.

In reverse genetics methods, a gene is mutated or removed in some way to identify its effect on the phenotype of the whole organism as a way to determine its function.

10.2 Biotechnology in Medicine and Agriculture

Genetic testing is performed to identify disease-causing genes, and can be used to benefit affected individuals and their relatives who have not developed disease symptoms yet. Gene therapy—by which functioning genes are incorporated into the genomes of individuals with a non-functioning mutant gene—has the potential to cure heritable diseases. Transgenic organisms possess DNA from a different species, usually generated by molecular cloning techniques. Vaccines, antibiotics, and hormones are examples of products obtained by recombinant DNA technology. Transgenic animals have been created for experimental purposes and some are used to produce some human proteins.

Genes are inserted into plants, using plasmids in the bacterium *Agrobacterium tumefaciens*, which infects plants. Transgenic plants have been created to improve the characteristics of crop plants—for example, by giving them insect resistance by inserting a gene for a bacterial toxin.

10.3 Genomics and Proteomics

Genome mapping is similar to solving a big, complicated puzzle with pieces of information coming from laboratories all over the world. Genetic maps provide an outline for the location of genes within a genome, and they estimate the distance between genes and genetic markers on the basis of the recombination frequency during meiosis. Physical maps provide detailed information about the physical distance between the genes. The most detailed information is available through sequence mapping. Information from all mapping and sequencing sources is combined to study an entire genome.

Whole genome sequencing is the latest available resource to treat genetic diseases. Some doctors are using whole genome sequencing to save lives. Genomics has many industrial applications, including biofuel development, agriculture, pharmaceuticals, and pollution control.

Imagination is the only barrier to the applicability of genomics. Genomics is being applied to most fields of biology; it can be used for personalized medicine, prediction of disease risks at an individual level, the study of drug interactions before the conduction of clinical trials, and the study of microorganisms in the environment as opposed to the laboratory. It is also being applied to the generation of new biofuels, genealogical assessment using mitochondria, advances in forensic science, and improvements in agriculture.

Proteomics is the study of the entire set of proteins expressed by a given type of cell under certain environmental conditions. In a multicellular organism, different cell types will have different proteomes, and these will vary with changes in the environment. Unlike a genome, a proteome is dynamic and under constant flux, which makes it more complicated and more useful than the knowledge of genomes alone.

ART CONNECTION QUESTIONS

1. Figure 10.7 Why was Dolly a Finn-Dorset and not a Scottish Blackface sheep?

REVIEW QUESTIONS

2. In gel electrophoresis of DNA, the different bands in the final gel form because the DNA molecules _____.

    a. are from different organisms
    b. have different lengths
    c. have different nucleotide compositions
    d. have different genes

3. In the reproductive cloning of an animal, the genome of the cloned individual comes from _____.
    a. a sperm cell
    b. an egg cell
    c. any gamete cell
    d. a body cell

4. What carries a gene from one organism into a bacteria cell?
    a. a plasmid
    b. an electrophoresis gel
    c. a restriction enzyme
    d. polymerase chain reaction

5. What is a genetically modified organism (GMO)?

    a. a plant with certain genes removed
    b. an organism with an artificially altered genome
    c. a hybrid organism
    d. any agricultural organism produced by breeding or biotechnology

6. What is the role of *Agrobacterium tumefaciens* in the production of transgenic plants?

    a. Genes from *A. tumefaciens* are inserted into plant DNA to give the plant different traits.
    b. Transgenic plants have been given resistance to the pest *A. tumefaciens*.
    c. *A. tumefaciens* is used as a vector to move genes into plant cells.
    d. Plant genes are incorporated into the genome of *Agrobacterium tumefaciens*.

7. What is the most challenging issue facing genome sequencing?
    a. the inability to develop fast and accurate sequencing techniques
    b. the ethics of using information from genomes at the individual level
    c. the availability and stability of DNA
    d. all of the above

8. Genomics can be used in agriculture to:
    a. generate new hybrid strains
    b. improve disease resistance
    c. improve yield
    d. all of the above

9. What kind of diseases are studied using genome-wide association studies?
    a. viral diseases
    b. single-gene inherited diseases
    c. diseases caused by multiple genes
    d. diseases caused by environmental factors

CRITICAL THINKING QUESTIONS

10. What is the purpose and benefit of the polymerase chain reaction?

11. Today, it is possible for a diabetic patient to purchase human insulin from a pharmacist. What technology makes this possible and why is it a benefit over how things used to be?

12. Describe two of the applications for genome mapping.

13. Identify a possible advantage and a possible disadvantage of a genetic test that would identify genes in individuals that increase their probability of having Alzheimer's disease later in life.

11 | EVOLUTION AND ITS PROCESSES

Figure 11.1 The diversity of life on Earth is the result of evolution, a continuous process that is still occurring. (credit "wolf": modification of work by Gary Kramer, USFWS; credit "coral": modification of work by William Harrigan, NOAA; credit "river": modification of work by Vojtěch Dostál; credit "protozoa": modification of work by Sharon Franklin, Stephen Ausmus, USDA ARS; credit "fish" modification of work by Christian Mehlführer; credit "mushroom", "bee": modification of work by Cory Zanker; credit "tree": modification of work by Joseph Kranak)

Chapter Outline

11.1: Discovering How Populations Change

11.2: Mechanisms of Evolution

11.3: Evidence of Evolution

11.4: Speciation

11.5: Common Misconceptions about Evolution

Introduction

All species of living organisms—from the bacteria on our skin, to the trees in our yards, to the birds outside—evolved at some point from a different species. Although it may seem that living things today stay much the same from generation to generation, that is not the case: evolution is ongoing. Evolution is the process through which the characteristics of species change and through which new species arise.

The theory of evolution is the unifying theory of biology, meaning it is the framework within which biologists ask questions about the living world. Its power is that it provides direction for predictions about living things that are borne out in experiment after experiment. The Ukrainian-born American geneticist Theodosius Dobzhansky famously wrote that "nothing makes sense in biology except in the light of evolution."[1] He meant that the principle that all life has evolved

---

1. Theodosius Dobzhansky. "Biology, Molecular and Organismic." *American Zoologist* 4, no. 4 (1964): 449.

and diversified from a common ancestor is the foundation from which we understand all other questions in biology. This chapter will explain some of the mechanisms for evolutionary change and the kinds of questions that biologists can and have answered using evolutionary theory.

11.1 | Discovering How Populations Change

By the end of this section, you will be able to:

- Explain how Darwin's theory of evolution differed from the current view at the time
- Describe how the present-day theory of evolution was developed
- Describe how population genetics is used to study the evolution of populations

The theory of evolution by natural selection describes a mechanism for species change over time. That species change had been suggested and debated well before Darwin. The view that species were static and unchanging was grounded in the writings of Plato, yet there were also ancient Greeks that expressed evolutionary ideas.

In the eighteenth century, ideas about the evolution of animals were reintroduced by the naturalist Georges-Louis Leclerc, Comte de Buffon and even by Charles Darwin's grandfather, Erasmus Darwin. During this time, it was also accepted that there were extinct species. At the same time, James Hutton, the Scottish naturalist, proposed that geological change occurred gradually by the accumulation of small changes from processes (over long periods of time) just like those happening today. This contrasted with the predominant view that the geology of the planet was a consequence of catastrophic events occurring during a relatively brief past. Hutton's view was later popularized by the geologist Charles Lyell in the nineteenth century. Lyell became a friend to Darwin and his ideas were very influential on Darwin's thinking. Lyell argued that the greater age of Earth gave more time for gradual change in species, and the process provided an analogy for gradual change in species.

In the early nineteenth century, Jean-Baptiste Lamarck published a book that detailed a mechanism for evolutionary change that is now referred to as **inheritance of acquired characteristics**. In Lamarck's theory, modifications in an individual caused by its environment, or the use or disuse of a structure during its lifetime, could be inherited by its offspring and, thus, bring about change in a species. While this mechanism for evolutionary change as described by Lamarck was discredited, Lamarck's ideas were an important influence on evolutionary thought. The inscription on the statue of Lamarck that stands at the gates of the Jardin des Plantes in Paris describes him as the "founder of the doctrine of evolution."

Charles Darwin and Natural Selection

The actual mechanism for evolution was independently conceived of and described by two naturalists, Charles Darwin and Alfred Russell Wallace, in the mid-nineteenth century. Importantly, each spent time exploring the natural world on expeditions to the tropics. From 1831 to 1836, Darwin traveled around the world on *H.M.S. Beagle*, visiting South America, Australia, and the southern tip of Africa. Wallace traveled to Brazil to collect insects in the Amazon rainforest from 1848 to 1852 and to the Malay Archipelago from 1854 to 1862. Darwin's journey, like Wallace's later journeys in the Malay Archipelago, included stops at several island chains, the last being the Galápagos Islands (west of Ecuador). On these islands, Darwin observed species of organisms on different islands that were clearly similar, yet had distinct differences. For example, the ground finches inhabiting the Galápagos Islands comprised several species that each had a unique beak shape (Figure 11.2). He observed both that these finches closely resembled another finch species on the mainland of South America and that the group of species in the Galápagos formed a graded series of beak sizes and shapes, with very small differences between the most similar. Darwin imagined that the island species might be all species modified from one original mainland species. In 1860, he wrote, "Seeing this gradation and diversity of structure in one small, intimately related group of birds, one might really fancy that from an original paucity of birds in this archipelago, one species had been taken and modified for different ends."[2]

2. Charles Darwin, *Journal of Researches into the Natural History and Geology of the Countries Visited during the Voyage of H.M.S. Beagle Round the World, under the Command of Capt. Fitz Roy, R.N*, 2nd. ed. (London: John Murray, 1860), http://www.archive.org/details/journalofresea00darw.

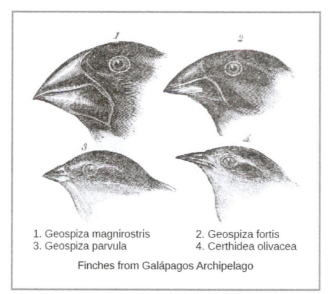

1. Geospiza magnirostris   2. Geospiza fortis
3. Geospiza parvula   4. Certhidea olivacea

Finches from Galápagos Archipelago

Figure 11.2 Darwin observed that beak shape varies among finch species. He postulated that the beak of an ancestral species had adapted over time to equip the finches to acquire different food sources. This illustration shows the beak shapes for four species of ground finch: 1. *Geospiza magnirostris* (the large ground finch), 2. *G. fortis* (the medium ground finch), 3. *G. parvula* (the small tree finch), and 4. *Certhidea olivacea* (the green-warbler finch).

Wallace and Darwin both observed similar patterns in other organisms and independently conceived a mechanism to explain how and why such changes could take place. Darwin called this mechanism natural selection. **Natural selection**, Darwin argued, was an inevitable outcome of three principles that operated in nature. First, the characteristics of organisms are inherited, or passed from parent to offspring. Second, more offspring are produced than are able to survive; in other words, resources for survival and reproduction are limited. The capacity for reproduction in all organisms outstrips the availability of resources to support their numbers. Thus, there is a competition for those resources in each generation. Both Darwin and Wallace's understanding of this principle came from reading an essay by the economist Thomas Malthus, who discussed this principle in relation to human populations. Third, offspring vary among each other in regard to their characteristics and those variations are inherited. Out of these three principles, Darwin and Wallace reasoned that offspring with inherited characteristics that allow them to best compete for limited resources will survive and have more offspring than those individuals with variations that are less able to compete. Because characteristics are inherited, these traits will be better represented in the next generation. This will lead to change in populations over generations in a process that Darwin called "descent with modification."

Papers by Darwin and Wallace (Figure 11.3) presenting the idea of natural selection were read together in 1858 before the Linnaean Society in London. The following year Darwin's book, *On the Origin of Species,* was published, which outlined in considerable detail his arguments for evolution by natural selection.

(a)   (b)

Figure 11.3 (a) Charles Darwin and (b) Alfred Wallace wrote scientific papers on natural selection that were presented together before the Linnean Society in 1858.

Demonstrations of evolution by natural selection can be time consuming. One of the best demonstrations has been in the very birds that helped to inspire the theory, the Galápagos finches. Peter and Rosemary Grant and their colleagues have studied Galápagos finch populations every year since 1976 and have provided important demonstrations of the operation of natural selection. The Grants found changes from one generation to the next in the beak shapes of the medium ground finches on the Galápagos island of Daphne Major. The medium ground finch feeds on seeds. The birds have inherited variation in the bill shape with some individuals having wide, deep bills and others having thinner bills. Large-billed birds feed more efficiently on large, hard seeds, whereas smaller billed birds feed more efficiently on small, soft seeds. During 1977, a drought period altered vegetation on the island. After this period, the number of seeds declined dramatically: the decline in small, soft seeds was greater than the decline in large, hard seeds. The large-billed birds were able to survive better than the small-billed birds the following year. The year following the drought when the Grants measured beak sizes in the much-reduced population, they found that the average bill size was larger (Figure 11.4). This was clear evidence for natural selection (differences in survival) of bill size caused by the availability of seeds. The Grants had studied the inheritance of bill sizes and knew that the surviving large-billed birds would tend to produce offspring with larger bills, so the selection would lead to evolution of bill size. Subsequent studies by the Grants have demonstrated selection on and evolution of bill size in this species in response to changing conditions on the island. The evolution has occurred both to larger bills, as in this case, and to smaller bills when large seeds became rare.

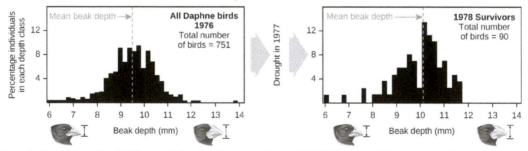

Figure 11.4 A drought on the Galápagos island of Daphne Major in 1977 reduced the number of small seeds available to finches, causing many of the small-beaked finches to die. This caused an increase in the finches' average beak size between 1976 and 1978.

Variation and Adaptation

Natural selection can only take place if there is **variation**, or differences, among individuals in a population. Importantly, these differences must have some genetic basis; otherwise, selection will not lead to change in the next generation. This is critical because variation among individuals can be caused by non-genetic reasons, such as an individual being taller because of better nutrition rather than different genes.

Genetic diversity in a population comes from two main sources: mutation and sexual reproduction. Mutation, a change in DNA, is the ultimate source of new alleles or new genetic variation in any population. An individual that has a mutated gene might have a different trait than other individuals in the population. However, this is not always the case. A mutation can have one of three outcomes on the organisms' appearance (or phenotype):

- A mutation may affect the phenotype of the organism in a way that gives it reduced fitness—lower likelihood of survival, resulting in fewer offspring.

- A mutation may produce a phenotype with a beneficial effect on fitness.

- Many mutations, called neutral mutations, will have no effect on fitness.

Mutations may also have a whole range of effect sizes on the fitness of the organism that expresses them in their phenotype, from a small effect to a great effect. Sexual reproduction and crossing over in meiosis also lead to genetic diversity: when two parents reproduce, unique combinations of alleles assemble to produce unique genotypes and, thus, phenotypes in each of the offspring.

A heritable trait that aids the survival and reproduction of an organism in its present environment is called an **adaptation**. An adaptation is a "match" of the organism to the environment. Adaptation to an environment comes about when a change in the range of genetic variation occurs over time that increases or maintains the match of the population with its environment. The variations in finch beaks shifted from generation to generation providing adaptation to food availability.

Whether or not a trait is favorable depends on the environment at the time. The same traits do not always have the same relative benefit or disadvantage because environmental conditions can change. For example, finches with large bills were benefited in one climate, while small bills were a disadvantage; in a different climate, the relationship reversed.

Patterns of Evolution

The evolution of species has resulted in enormous variation in form and function. When two species evolve in different directions from a common point, it is called **divergent evolution**. Such divergent evolution can be seen in the forms of the reproductive organs of flowering plants, which share the same basic anatomies; however, they can look very different as a result of selection in different physical environments, and adaptation to different kinds of pollinators (Figure 11.5).

(a)   (b)

Figure 11.5 Flowering plants evolved from a common ancestor. Notice that the (a) dense blazing star and (b) purple coneflower vary in appearance, yet both share a similar basic morphology. (credit a, b: modification of work by Cory Zanker)

In other cases, similar phenotypes evolve independently in distantly related species. For example, flight has evolved in both bats and insects, and they both have structures we refer to as wings, which are adaptations to flight. The wings of bats and insects, however, evolved from very different original structures. When similar structures arise through evolution independently in different species it is called **convergent evolution**. The wings of bats and insects are called **analogous structures**; they are similar in function and appearance, but do not share an origin in a common ancestor. Instead they evolved independently in the two lineages. The wings of a hummingbird and an ostrich are **homologous structures**, meaning they share similarities (despite their differences resulting from evolutionary divergence). The wings of hummingbirds and ostriches did not evolve independently in the hummingbird lineage and the ostrich lineage—they descended from a common ancestor with wings.

The Modern Synthesis

The mechanisms of inheritance, genetics, were not understood at the time Darwin and Wallace were developing their idea of natural selection. This lack of understanding was a stumbling block to comprehending many aspects of evolution. In fact, blending inheritance was the predominant (and incorrect) genetic theory of the time, which made it difficult to understand how natural selection might operate. Darwin and Wallace were unaware of the genetics work by Austrian monk Gregor Mendel, which was published in 1866, not long after publication of *On the Origin of Species*. Mendel's work was rediscovered in the early twentieth century at which time geneticists were rapidly coming to an understanding of the basics of inheritance. Initially, the newly discovered particulate nature of genes made it difficult for biologists to understand how gradual evolution could occur. But over the next few decades genetics and evolution were integrated in what became known as the **modern synthesis**—the coherent understanding of the relationship between natural selection and genetics that took shape by the 1940s and is generally accepted today. In sum, the modern synthesis describes how evolutionary pressures, such as natural selection, can affect a population's genetic makeup, and, in turn, how this can result in the gradual evolution of populations and species. The theory also connects this gradual change of a population over time, called **microevolution**, with the processes that gave rise to new species and higher taxonomic groups with widely divergent characters, called **macroevolution**.

Population Genetics

Recall that a gene for a particular character may have several variants, or alleles, that code for different traits associated with that character. For example, in the ABO blood type system in humans, three alleles determine the particular blood-type protein on the surface of red blood cells. Each individual in a population of diploid organisms can only carry two alleles for a particular gene, but more than two may be present in the individuals that make up the population. Mendel followed alleles as they were inherited from parent to offspring. In the early twentieth century, biologists began to study what happens to all the alleles in a population in a field of study known as **population genetics**.

Until now, we have defined evolution as a change in the characteristics of a population of organisms, but behind that phenotypic change is genetic change. In population genetic terms, evolution is defined as a change in the frequency of an allele in a population. Using the ABO system as an example, the frequency of one of the alleles, I^A, is the number of copies of that allele divided by all the copies of the ABO gene in the population. For example, a study in Jordan found a frequency of I^A to be 26.1 percent.[3] The I^B, I^0 alleles made up 13.4 percent and 60.5 percent of the alleles respectively, and all of the frequencies add up to 100 percent. A change in this frequency over time would constitute evolution in the population.

There are several ways the allele frequencies of a population can change. One of those ways is natural selection. If a given allele confers a phenotype that allows an individual to have more offspring that survive and reproduce, that allele, by virtue of being inherited by those offspring, will be in greater frequency in the next generation. Since allele frequencies always add up to 100 percent, an increase in the frequency of one allele always means a corresponding decrease in one or more of the other alleles. Highly beneficial alleles may, over a very few generations, become "fixed" in this way, meaning that every individual of the population will carry the allele. Similarly, detrimental alleles may be swiftly eliminated from the **gene pool**, the sum of all the alleles in a population. Part of the study of population genetics is tracking how selective forces change the allele frequencies in a population over time, which can give scientists clues regarding the selective forces that may be operating on a given population. The studies of changes in wing coloration in the peppered moth from mottled white to dark in response to soot-covered tree trunks and then back to mottled white when factories stopped producing so much soot is a classic example of studying evolution in natural populations (Figure 11.6).

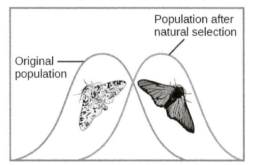

Light-colored peppered moths are better camouflaged against a pristine environment; likewise, dark-colored peppered moths are better camouflaged against a sooty environment. Thus, as the Industrial Revolution progressed in nineteenth-century England, the color of the moth population shifted from light to dark.

Figure 11.6 As the Industrial Revolution caused trees to darken from soot, darker colored peppered moths were better camouflaged than the lighter colored ones, which caused there to be more of the darker colored moths in the population.

In the early twentieth century, English mathematician Godfrey Hardy and German physician Wilhelm Weinberg independently provided an explanation for a somewhat counterintuitive concept. Hardy's original explanation was in response to a misunderstanding as to why a "dominant" allele, one that masks a recessive allele, should not increase in frequency in a population until it eliminated all the other alleles. The question resulted from a common confusion about what "dominant" means, but it forced Hardy, who was not even a biologist, to point out that if there are no factors that affect an allele frequency those frequencies will remain constant from one generation to the next. This principle is now known as the Hardy-Weinberg equilibrium. The theory states that a population's allele and genotype frequencies are inherently stable—unless some kind of evolutionary force is acting on the population, the population would carry the same alleles in the same proportions generation after generation. Individuals would, as a whole, look essentially the same and this would

3. Sahar S. Hanania, Dhia S. Hassawi, and Nidal M. Irshaid, "Allele Frequency and Molecular Genotypes of ABO Blood Group System in a Jordanian Population," *Journal of Medical Sciences* 7 (2007): 51-58, doi:10.3923/jms.2007.51.58

be unrelated to whether the alleles were dominant or recessive. The four most important evolutionary forces, which will disrupt the equilibrium, are natural selection, mutation, **genetic drift**, and **migration** into or out of a population. A fifth factor, nonrandom mating, will also disrupt the Hardy-Weinberg equilibrium but only by shifting genotype frequencies, not allele frequencies. In nonrandom mating, individuals are more likely to mate with like individuals (or unlike individuals) rather than at random. Since nonrandom mating does not change allele frequencies, it does not cause evolution directly. Natural selection has been described. Mutation creates one allele out of another one and changes an allele's frequency by a small, but continuous amount each generation. Each allele is generated by a low, constant mutation rate that will slowly increase the allele's frequency in a population if no other forces act on the allele. If natural selection acts against the allele, it will be removed from the population at a low rate leading to a frequency that results from a balance between selection and mutation. This is one reason that genetic diseases remain in the human population at very low frequencies. If the allele is favored by selection, it will increase in frequency. Genetic drift causes random changes in allele frequencies when populations are small. Genetic drift can often be important in evolution, as discussed in the next section. Finally, if two populations of a species have different allele frequencies, migration of individuals between them will cause frequency changes in both populations. As it happens, there is no population in which one or more of these processes are not operating, so populations are always evolving, and the Hardy-Weinberg equilibrium will never be exactly observed. However, the Hardy-Weinberg principle gives scientists a baseline expectation for allele frequencies in a non-evolving population to which they can compare evolving populations and thereby infer what evolutionary forces might be at play. The population is evolving if the frequencies of alleles or genotypes deviate from the value expected from the Hardy-Weinberg principle.

Darwin identified a special case of natural selection that he called sexual selection. Sexual selection affects an individual's ability to mate and thus produce offspring, and it leads to the evolution of dramatic traits that often appear maladaptive in terms of survival but persist because they give their owners greater reproductive success. Sexual selection occurs in two ways: through male–male competition for mates and through female selection of mates. Male–male competition takes the form of conflicts between males, which are often ritualized, but may also pose significant threats to a male's survival. Sometimes the competition is for territory, with females more likely to mate with males with higher quality territories. Female choice occurs when females choose a male based on a particular trait, such as feather colors, the performance of a mating dance, or the building of an elaborate structure. In some cases male–male competition and female choice combine in the mating process. In each of these cases, the traits selected for, such as fighting ability or feather color and length, become enhanced in the males. In general, it is thought that sexual selection can proceed to a point at which natural selection against a character's further enhancement prevents its further evolution because it negatively impacts the male's ability to survive. For example, colorful feathers or an elaborate display make the male more obvious to predators.

11.2 | Mechanisms of Evolution

By the end of this section, you will be able to:

- Describe the four basic causes of evolution: natural selection, mutation, genetic drift, and gene flow
- Explain how each evolutionary force can influence the allele frequencies of a population

The Hardy-Weinberg equilibrium principle says that allele frequencies in a population will remain constant in the absence of the four factors that could change them. Those factors are natural selection, mutation, genetic drift, and migration (gene flow). In fact, we know they are probably always affecting populations.

Natural Selection

Natural selection has already been discussed. Alleles are expressed in a phenotype. Depending on the environmental conditions, the phenotype confers an advantage or disadvantage to the individual with the phenotype relative to the other phenotypes in the population. If it is an advantage, then that individual will likely have more offspring than individuals with the other phenotypes, and this will mean that the allele behind the phenotype will have greater representation in the next generation. If conditions remain the same, those offspring, which are carrying the same allele, will also benefit. Over time, the allele will increase in frequency in the population.

Mutation

Mutation is a source of new alleles in a population. Mutation is a change in the DNA sequence of the gene. A mutation can change one allele into another, but the net effect is a change in frequency. The change in frequency resulting from mutation is small, so its effect on evolution is small unless it interacts with one of the other factors, such as selection. A mutation may produce an allele that is selected against, selected for, or selectively neutral. Harmful mutations are removed from the population by selection and will generally only be found in very low frequencies equal to the mutation rate. Beneficial mutations will spread through the population through selection, although that initial spread is slow. Whether or not a mutation is beneficial or harmful is determined by whether it helps an organism survive to sexual maturity and reproduce. It should be noted that mutation is the ultimate source of genetic variation in all populations—new alleles, and, therefore, new genetic variations arise through mutation.

Genetic Drift

Another way a population's allele frequencies can change is genetic drift (Figure 11.7), which is simply the effect of chance. Genetic drift is most important in small populations. Drift would be completely absent in a population with infinite individuals, but, of course, no population is this large. Genetic drift occurs because the alleles in an offspring generation are a random sample of the alleles in the parent generation. Alleles may or may not make it into the next generation due to chance events including mortality of an individual, events affecting finding a mate, and even the events affecting which gametes end up in fertilizations. If one individual in a population of ten individuals happens to die before it leaves any offspring to the next generation, all of its genes—a tenth of the population's gene pool—will be suddenly lost. In a population of 100, that 1 individual represents only 1 percent of the overall gene pool; therefore, it has much less impact on the population's genetic structure and is unlikely to remove all copies of even a relatively rare allele.

Imagine a population of ten individuals, half with allele A and half with allele a (the individuals are haploid). In a stable population, the next generation will also have ten individuals. Choose that generation randomly by flipping a coin ten times and let heads be A and tails be a. It is unlikely that the next generation will have exactly half of each allele. There might be six of one and four of the other, or some different set of frequencies. Thus, the allele frequencies have changed and evolution has occurred. A coin will no longer work to choose the next generation (because the odds are no longer one half for each allele). The frequency in each generation will drift up and down on what is known as a random walk until at one point either all A or all a are chosen and that allele is fixed from that point on. This could take a very long time for a large population. This simplification is not very biological, but it can be shown that real populations behave this way. The effect of drift on frequencies is greater the smaller a population is. Its effect is also greater on an allele with a frequency far from one half. Drift will influence every allele, even those that are being naturally selected.

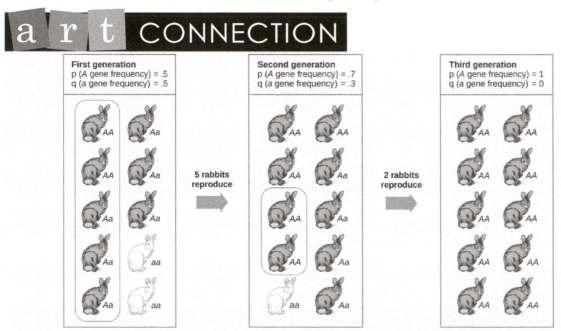

Figure 11.7 Genetic drift in a population can lead to the elimination of an allele from a population by chance. In each generation, a random set of individuals reproduces to produce the next generation. The frequency of alleles in the next generation is equal to the frequency of alleles among the individuals reproducing.

Do you think genetic drift would happen more quickly on an island or on the mainland?

Genetic drift can also be magnified by natural or human-caused events, such as a disaster that randomly kills a large portion of the population, which is known as the **bottleneck effect** that results in a large portion of the genome suddenly being wiped out (Figure 11.8). In one fell swoop, the genetic structure of the survivors becomes the genetic structure of the entire population, which may be very different from the pre-disaster population. The disaster must be one that kills for reasons unrelated to the organism's traits, such as a hurricane or lava flow. A mass killing caused by unusually cold temperatures at night, is likely to affect individuals differently depending on the alleles they possess that confer cold hardiness.

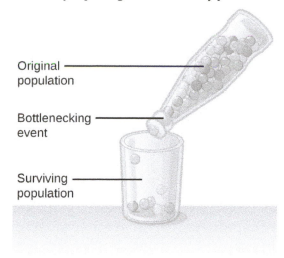

Figure 11.8 A chance event or catastrophe can reduce the genetic variability within a population.

Another scenario in which populations might experience a strong influence of genetic drift is if some portion of the population leaves to start a new population in a new location, or if a population gets divided by a physical barrier of some kind. In this situation, those individuals are unlikely to be representative of the entire population which results in the **founder effect**. The founder effect occurs when the genetic structure matches that of the new population's founding fathers and mothers. The founder effect is believed to have been a key factor in the genetic history of the Afrikaner population of Dutch settlers in South Africa, as evidenced by mutations that are common in Afrikaners but rare in most other populations. This is likely due to a higher-than-normal proportion of the founding colonists, which were a small sample of the original population, carried these mutations. As a result, the population expresses unusually high incidences of Huntington's disease (HD) and Fanconi anemia (FA), a genetic disorder known to cause bone marrow and congenital abnormalities, and even cancer.[4]

Visit this site (http://openstaxcollege.org/l/genetic_drift2) to learn more about genetic drift and to run simulations of allele changes caused by drift.

Gene Flow

Another important evolutionary force is **gene flow**, or the flow of alleles in and out of a population resulting from the migration of individuals or gametes (Figure 11.9). While some populations are fairly stable, others experience more flux. Many plants, for example, send their seeds far and wide, by wind or in the guts of animals; these seeds may introduce alleles common in the source population to a new population in which they are rare.

---

4. A. J. Tipping et al., "Molecular and Genealogical Evidence for a Founder Effect in Fanconi Anemia Families of the Afrikaner Population of South Africa," *PNAS* 98, no. 10 (2001): 5734-5739, doi: 10.1073/pnas.091402398.

Figure 11.9 Gene flow can occur when an individual travels from one geographic location to another and joins a different population of the species. In the example shown here, the brown allele is introduced into the green population.

11.3 | Evidence of Evolution

By the end of this section, you will be able to:

- Explain sources of evidence for evolution
- Define homologous and vestigial structures

The evidence for evolution is compelling and extensive. Looking at every level of organization in living systems, biologists see the signature of past and present evolution. Darwin dedicated a large portion of his book, *On the Origin of Species*, identifying patterns in nature that were consistent with evolution and since Darwin our understanding has become clearer and broader.

Fossils

Fossils provide solid evidence that organisms from the past are not the same as those found today; fossils show a progression of evolution. Scientists determine the age of fossils and categorize them all over the world to determine when the organisms lived relative to each other. The resulting fossil record tells the story of the past, and shows the evolution of form over millions of years (Figure 11.10). For example, highly detailed fossil records have been recovered for sequences of species in the evolution of whales and modern horses. The fossil record of horses in North America is especially rich and many contain transition fossils: those showing intermediate anatomy between earlier and later forms. The fossil record extends back to a dog-like ancestor some 55 million years ago that gave rise to the first horse-like species 55 to 42 million years ago in the genus *Eohippus*. The series of fossils tracks the change in anatomy resulting from a gradual drying trend that changed the landscape from a forested one to a prairie. Successive fossils show the evolution of teeth shapes and foot and leg anatomy to a grazing habit, with adaptations for escaping predators, for example in species of *Mesohippus* found from 40 to 30 million years ago. Later species showed gains in size, such as those of *Hipparion*, which existed from about 23 to 2 million years ago. The fossil record shows several adaptive radiations in the horse lineage, which is now much reduced to only one genus, *Equus*, with several species.

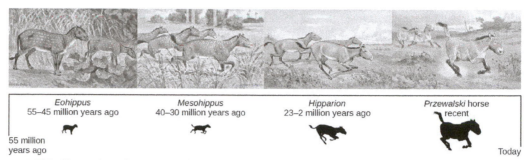

Eohippus	Mesohippus	Hipparion	*Przewalski* horse
55–45 million years ago	40–30 million years ago	23–2 million years ago	recent

55 million years ago   Today

Figure 11.10 This illustration shows an artist's renderings of these species derived from fossils of the evolutionary history of the horse and its ancestors. The species depicted are only four from a very diverse lineage that contains many branches, dead ends, and adaptive radiations. One of the trends, depicted here is the evolutionary tracking of a drying climate and increase in prairie versus forest habitat reflected in forms that are more adapted to grazing and predator escape through running. Przewalski's horse is one of a few living species of horse.

Anatomy and Embryology

Another type of evidence for evolution is the presence of structures in organisms that share the same basic form. For example, the bones in the appendages of a human, dog, bird, and whale all share the same overall construction (Figure 11.11). That similarity results from their origin in the appendages of a common ancestor. Over time, evolution led to changes in the shapes and sizes of these bones in different species, but they have maintained the same overall layout, evidence of descent from a common ancestor. Scientists call these synonymous parts homologous structures. Some structures exist in organisms that have no apparent function at all, and appear to be residual parts from a past ancestor. For example, some snakes have pelvic bones despite having no legs because they descended from reptiles that did have legs. These unused structures without function are called **vestigial structures**. Other examples of vestigial structures are wings on flightless birds (which may have other functions), leaves on some cacti, traces of pelvic bones in whales, and the sightless eyes of cave animals.

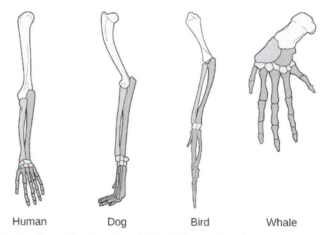

Human   Dog   Bird   Whale

Figure 11.11 The similar construction of these appendages indicates that these organisms share a common ancestor.

Click through the activities at this interactive site (http://openstaxcollege.org/l/bone_structure2) to guess which bone structures are homologous and which are analogous, and to see examples of all kinds of evolutionary adaptations that illustrate these concepts.

Another evidence of evolution is the convergence of form in organisms that share similar environments. For example, species of unrelated animals, such as the arctic fox and ptarmigan (a bird), living in the arctic region have temporary white coverings during winter to blend with the snow and ice (Figure 11.12). The similarity occurs not because of common ancestry, indeed one covering is of fur and the other of feathers, but because of similar selection pressures—the benefits of not being seen by predators.

(a)   (b)

Figure 11.12 The white winter coat of (a) the arctic fox and (b) the ptarmigan's plumage are adaptations to their environments. (credit a: modification of work by Keith Morehouse)

Embryology, the study of the development of the anatomy of an organism to its adult form also provides evidence of relatedness between now widely divergent groups of organisms. Structures that are absent in some groups often appear in their embryonic forms and disappear by the time the adult or juvenile form is reached. For example, all vertebrate embryos, including humans, exhibit gill slits at some point in their early development. These disappear in the adults of terrestrial groups, but are maintained in adult forms of aquatic groups such as fish and some amphibians. Great ape embryos, including humans, have a tail structure during their development that is lost by the time of birth. The reason embryos of unrelated species are often similar is that mutational changes that affect the organism during embryonic development can cause amplified differences in the adult, even while the embryonic similarities are preserved.

Biogeography

The geographic distribution of organisms on the planet follows patterns that are best explained by evolution in conjunction with the movement of tectonic plates over geological time. Broad groups that evolved before the breakup of the supercontinent Pangaea (about 200 million years ago) are distributed worldwide. Groups that evolved since the breakup appear uniquely in regions of the planet, for example the unique flora and fauna of northern continents that formed from the supercontinent Laurasia and of the southern continents that formed from the supercontinent Gondwana. The presence of Proteaceae in Australia, southern Africa, and South America is best explained by the plant family's presence there prior to the southern supercontinent Gondwana breaking up (Figure 11.13).

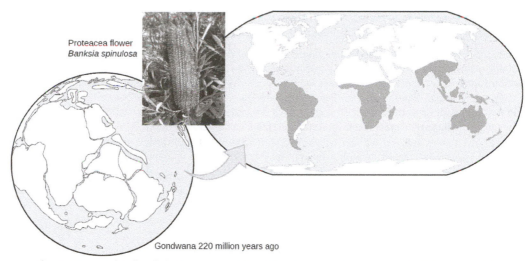

Proteacea flower
Banksia spinulosa

Gondwana 220 million years ago

Figure 11.13 The Proteacea family of plants evolved before the supercontinent Gondwana broke up. Today, members of this plant family are found throughout the southern hemisphere (shown in red). (credit "Proteacea flower": modification of work by "dorofofoto"/Flickr)

The great diversification of the marsupials in Australia and the absence of other mammals reflects that island continent's long isolation. Australia has an abundance of endemic species—species found nowhere else—which is typical of islands whose isolation by expanses of water prevents migration of species to other regions. Over time, these species diverge evolutionarily into new species that look very different from their ancestors that may exist on the mainland. The marsupials of Australia, the finches on the Galápagos, and many species on the Hawaiian Islands are all found nowhere else but on their island, yet display distant relationships to ancestral species on mainlands.

Molecular Biology

Like anatomical structures, the structures of the molecules of life reflect descent with modification. Evidence of a common ancestor for all of life is reflected in the universality of DNA as the genetic material and of the near universality of the genetic code and the machinery of DNA replication and expression. Fundamental divisions in life between the three domains are reflected in major structural differences in otherwise conservative structures such as the components of ribosomes and the structures of membranes. In general, the relatedness of groups of organisms is reflected in the similarity of their DNA sequences—exactly the pattern that would be expected from descent and diversification from a common ancestor.

DNA sequences have also shed light on some of the mechanisms of evolution. For example, it is clear that the evolution of new functions for proteins commonly occurs after gene duplication events. These duplications are a kind of mutation in which an entire gene is added as an extra copy (or many copies) in the genome. These duplications allow the free modification of one copy by mutation, selection, and drift, while the second copy continues to produce a functional protein. This allows the original function for the protein to be kept, while evolutionary forces tweak the copy until it functions in a new way.

11.4 | Speciation

By the end of this section, you will be able to:

- Describe the definition of species and how species are identified as different
- Explain allopatric and sympatric speciation
- Describe adaptive radiation

The biological definition of species, which works for sexually reproducing organisms, is a group of actually or potentially interbreeding individuals. According to this definition, one species is distinguished from another by the possibility of matings between individuals from each species to produce fertile offspring. There are exceptions to this rule. Many species are similar enough that hybrid offspring are possible and may often occur in nature, but for the majority of species this rule generally holds. In fact, the presence of hybrids between similar species suggests that they may have descended from a single interbreeding species and that the speciation process may not yet be completed.

Given the extraordinary diversity of life on the planet there must be mechanisms for **speciation**: the formation of two species from one original species. Darwin envisioned this process as a branching event and diagrammed the process in the only illustration found in *On the Origin of Species* (Figure 11.14a). For speciation to occur, two new populations must be formed from one original population, and they must evolve in such a way that it becomes impossible for individuals from the two new populations to interbreed. Biologists have proposed mechanisms by which this could occur that fall into two broad categories. **Allopatric speciation**, meaning speciation in "other homelands," involves a geographic separation of populations from a parent species and subsequent evolution. **Sympatric speciation**, meaning speciation in the "same homeland," involves speciation occurring within a parent species while remaining in one location.

Biologists think of speciation events as the splitting of one ancestral species into two descendant species. There is no reason why there might not be more than two species formed at one time except that it is less likely and such multiple events can also be conceptualized as single splits occurring close in time.

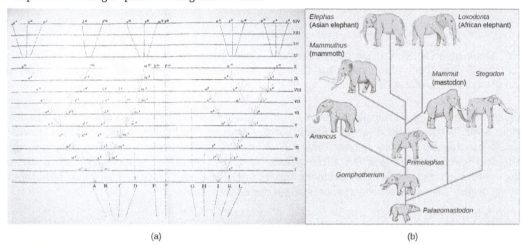

(a)   (b)

Figure 11.14 The only illustration in Darwin's *On the Origin of Species* is (a) a diagram showing speciation events leading to biological diversity. The diagram shows similarities to phylogenetic charts that are drawn today to illustrate the relationships of species. (b) Modern elephants evolved from the *Palaeomastodon*, a species that lived in Egypt 35–50 million years ago.

Speciation through Geographic Separation

A geographically continuous population has a gene pool that is relatively homogeneous. Gene flow, the movement of alleles across the range of the species, is relatively free because individuals can move and then mate with individuals in their new location. Thus, the frequency of an allele at one end of a distribution will be similar to the frequency of the allele at the other end. When populations become geographically discontinuous that free-flow of alleles is prevented. When that separation lasts for a period of time, the two populations are able to evolve along different trajectories. Thus, their allele frequencies at numerous genetic loci gradually become more and more different as new alleles independently arise by mutation in each population. Typically, environmental conditions, such as climate, resources, predators, and competitors, for the two populations will differ causing natural selection to favor divergent adaptations in each group. Different histories of genetic drift, enhanced because the populations are smaller than the parent population, will also lead to divergence.

Given enough time, the genetic and phenotypic divergence between populations will likely affect characters that influence reproduction enough that were individuals of the two populations brought together, mating would be less likely, or if a mating occurred, offspring would be non-viable or infertile. Many types of diverging characters may affect the reproductive isolation (inability to interbreed) of the two populations. These mechanisms of reproductive isolation can be divided into prezygotic mechanisms (those that operate before fertilization) and postzygotic mechanisms (those that operate after fertilization). Prezygotic mechanisms include traits that allow the individuals to find each other, such as the timing of mating, sensitivity to pheromones, or choice of mating sites. If individuals are able to encounter each other, character divergence may prevent courtship rituals from leading to a mating either because female preferences have changed or male behaviors have changed. Physiological changes may interfere with successful fertilization if mating is able to occur. Postzygotic mechanisms include genetic incompatibilities that prevent proper development of the offspring, or if the offspring live, they may be unable to produce viable gametes themselves as in the example of the mule, the infertile offspring of a female horse and a male donkey.

If the two isolated populations are brought back together and the hybrid offspring that formed from matings between individuals of the two populations have lower survivorship or reduced fertility, then selection will favor individuals that are able to discriminate between potential mates of their own population and the other population. This selection will enhance the reproductive isolation.

Isolation of populations leading to allopatric speciation can occur in a variety of ways: from a river forming a new branch, erosion forming a new valley, or a group of organisms traveling to a new location without the ability to return, such as seeds floating over the ocean to an island. The nature of the geographic separation necessary to isolate populations depends entirely on the biology of the organism and its potential for dispersal. If two flying insect populations took up residence in separate nearby valleys, chances are that individuals from each population would fly back and forth, continuing gene flow. However, if two rodent populations became divided by the formation of a new lake, continued gene flow would be unlikely; therefore, speciation would be more likely.

Biologists group allopatric processes into two categories. If a few members of a species move to a new geographical area, this is called **dispersal**. If a natural situation arises to physically divide organisms, this is called **vicariance**.

Scientists have documented numerous cases of allopatric speciation taking place. For example, along the west coast of the United States, two separate subspecies of spotted owls exist. The northern spotted owl has genetic and phenotypic differences from its close relative, the Mexican spotted owl, which lives in the south (Figure 11.15). The cause of their initial separation is not clear, but it may have been caused by the glaciers of the ice age dividing an initial population into two.[5]

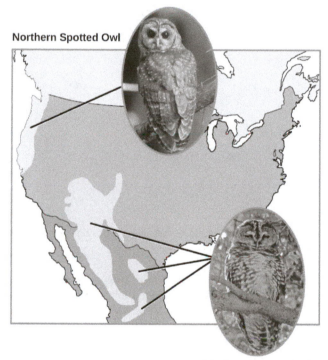

Figure 11.15 The northern spotted owl and the Mexican spotted owl inhabit geographically separate locations with different climates and ecosystems. The owl is an example of incipient speciation. (credit "northern spotted owl": modification of work by John and Karen Hollingsworth, USFWS; credit "Mexican spotted owl": modification of work by Bill Radke, USFWS)

Additionally, scientists have found that the further the distance between two groups that once were the same species, the more likely for speciation to take place. This seems logical because as the distance increases, the various environmental factors would likely have less in common than locations in close proximity. Consider the two owls; in the north, the climate is cooler than in the south; the other types of organisms in each ecosystem differ, as do their behaviors and habits; also, the hunting habits and prey choices of the owls in the south vary from the northern ones. These variances can lead to evolved differences in the owls, and over time speciation will likely occur unless gene flow between the populations is restored.

---

5. Courtney, S.P., et al, "Scientific Evaluation of the Status of the Northern Spotted Owl," Sustainable Ecosystems Institute (2004), Portland, OR.

In some cases, a population of one species disperses throughout an area, and each finds a distinct niche or isolated habitat. Over time, the varied demands of their new lifestyles lead to multiple speciation events originating from a single species, which is called **adaptive radiation**. From one point of origin, many adaptations evolve causing the species to radiate into several new ones. Island archipelagos like the Hawaiian Islands provide an ideal context for adaptive radiation events because water surrounds each island, which leads to geographical isolation for many organisms (Figure 11.16). The Hawaiian honeycreeper illustrates one example of adaptive radiation. From a single species, called the founder species, numerous species have evolved, including the eight shown in Figure 11.16.

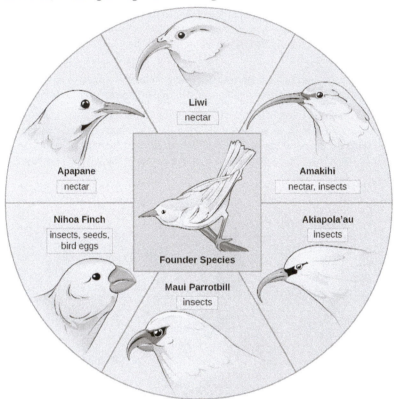

Figure 11.16 The honeycreeper birds illustrate adaptive radiation. From one original species of bird, multiple others evolved, each with its own distinctive characteristics.

Notice the differences in the species' beaks in Figure 11.16. Change in the genetic variation for beaks in response to natural selection based on specific food sources in each new habitat led to evolution of a different beak suited to the specific food source. The fruit and seed-eating birds have thicker, stronger beaks which are suited to break hard nuts. The nectar-eating birds have long beaks to dip into flowers to reach their nectar. The insect-eating birds have beaks like swords, appropriate for stabbing and impaling insects. Darwin's finches are another well-studied example of adaptive radiation in an archipelago.

Click through this interactive site (http://openstaxcollege.org/l/bird_evolution) to see how island birds evolved; click to see images of each species in evolutionary increments from five million years ago to today.

Speciation without Geographic Separation

Can divergence occur if no physical barriers are in place to separate individuals who continue to live and reproduce in the same habitat? A number of mechanisms for sympatric speciation have been proposed and studied.

One form of sympatric speciation can begin with a chromosomal error during meiosis or the formation of a hybrid individual with too many chromosomes. Polyploidy is a condition in which a cell, or organism, has an extra set, or sets, of chromosomes. Scientists have identified two main types of polyploidy that can lead to reproductive isolation of an individual in the polyploid state. In some cases a polyploid individual will have two or more complete sets of chromosomes from its own species in a condition called autopolyploidy (Figure 11.17). The prefix "auto" means self, so the term means multiple chromosomes from one's own species. Polyploidy results from an error in meiosis in which all of the chromosomes move into one cell instead of separating.

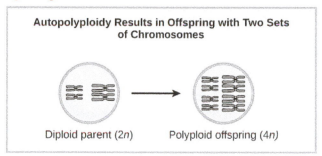

Figure 11.17 Autopolyploidy results when mitosis is not followed by cytokinesis.

For example, if a plant species with $2n = 6$ produces autopolyploid gametes that are also diploid ($2n = 6$, when they should be $n = 3$), the gametes now have twice as many chromosomes as they should have. These new gametes will be incompatible with the normal gametes produced by this plant species. But they could either self-pollinate or reproduce with other autopolyploid plants with gametes having the same diploid number. In this way, sympatric speciation can occur quickly by forming offspring with $4n$ called a tetraploid. These individuals would immediately be able to reproduce only with those of this new kind and not those of the ancestral species. The other form of polyploidy occurs when individuals of two different species reproduce to form a viable offspring called an allopolyploid. The prefix "allo" means "other" (recall from allopatric); therefore, an allopolyploid occurs when gametes from two different species combine. Figure 11.18 illustrates one possible way an allopolyploidy can form. Notice how it takes two generations, or two reproductive acts, before the viable fertile hybrid results.

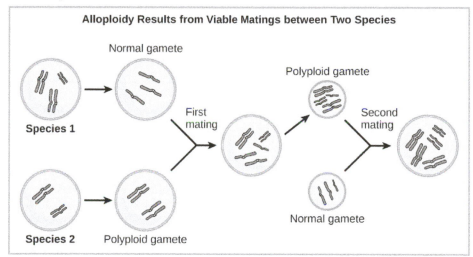

Figure 11.18 Alloploidy results when two species mate to produce viable offspring. In the example shown, a normal gamete from one species fuses with a polyploid gamete from another. Two matings are necessary to produce viable offspring.

The cultivated forms of wheat, cotton, and tobacco plants are all allopolyploids. Although polyploidy occurs occasionally in animals, most chromosomal abnormalities in animals are lethal; it takes place most commonly in plants. Scientists have discovered more than 1/2 of all plant species studied relate back to a species evolved through polyploidy.

Sympatric speciation may also take place in ways other than polyploidy. For example, imagine a species of fish that lived in a lake. As the population grew, competition for food also grew. Under pressure to find food, suppose that a group of these fish had the genetic flexibility to discover and feed off another resource that was unused by the other fish. What if this new food source was found at a different depth of the lake? Over time, those feeding on the second food source would interact more with each other than the other fish; therefore they would breed together as well. Offspring of these fish would likely behave as their parents and feed and live in the same area, keeping them separate from the original population. If this group of fish continued to remain separate from the first population, eventually sympatric speciation might occur as more genetic differences accumulated between them.

This scenario does play out in nature, as do others that lead to reproductive isolation. One such place is Lake Victoria in Africa, famous for its sympatric speciation of cichlid fish. Researchers have found hundreds of sympatric speciation events in these fish, which have not only happened in great number, but also over a short period of time. Figure 11.19 shows this type of speciation among a cichlid fish population in Nicaragua. In this locale, two types of cichlids live in the same geographic location; however, they have come to have different morphologies that allow them to eat various food sources.

Thin-lipped cichlid   Thick-lipped cichlid

Figure 11.19 Cichlid fish from Lake Apoyeque, Nicaragua, show evidence of sympatric speciation. Lake Apoyeque, a crater lake, is 1800 years old, but genetic evidence indicates that the lake was populated only 100 years ago by a single population of cichlid fish. Nevertheless, two populations with distinct morphologies and diets now exist in the lake, and scientists believe these populations may be in an early stage of speciation.

Finally, a well-documented example of ongoing sympatric speciation occurred in the apple maggot fly, *Rhagoletis pomonella*, which arose as an isolated population sometime after the introduction of the apple into North America. The native population of flies fed on hawthorn species and is host-specific: it only infests hawthorn trees. Importantly, it also uses the trees as a location to meet for mating. It is hypothesized that either through mutation or a behavioral mistake, flies jumped hosts and met and mated in apple trees, subsequently laying their eggs in apple fruit. The offspring matured and kept their preference for the apple trees effectively dividing the original population into two new populations separated by host species, not by geography. The host jump took place in the nineteenth century, but there are now measureable differences between the two populations of fly. It seems likely that host specificity of parasites in general is a common cause of sympatric speciation.

11.5 | Common Misconceptions about Evolution

By the end of this section, you will be able to:

- Identify common misconceptions about evolution
- Identify common criticisms of evolution

Although the theory of evolution initially generated some controversy, by 20 years after the publication of *On the Origin of Species* it was almost universally accepted by biologists, particularly younger biologists. Nevertheless, the theory of evolution is a difficult concept and misconceptions about how it works abound. In addition, there are those that reject it as an explanation for the diversity of life.

This website (http://openstaxcollege.org/l/misconception2) addresses some of the main misconceptions associated with the theory of evolution.

Evolution Is Just a Theory

Critics of the theory of evolution dismiss its importance by purposefully confounding the everyday usage of the word "theory" with the way scientists use the word. In science, a "theory" is understood to be a concept that has been extensively tested and supported over time. We have a theory of the atom, a theory of gravity, and the theory of relativity, each of which describes what scientists understand to be facts about the world. In the same way, the theory of evolution describes facts about the living world. As such, a theory in science has survived significant efforts to discredit it by scientists, who are naturally skeptical. While theories can sometimes be overturned or revised, this does not lessen their weight but simply reflects the constantly evolving state of scientific knowledge. In contrast, a "theory" in common vernacular means a guess or suggested explanation for something. This meaning is more akin to the concept of a "hypothesis" used by scientists, which is a tentative explanation for something that is proposed to either be supported or disproved. When critics of evolution say evolution is "just a theory," they are implying that there is little evidence supporting it and that it is still in the process of being rigorously tested. This is a mischaracterization. If this were the case, geneticist Theodosius Dobzhansky would not have said that "nothing in biology makes sense, except in the light of evolution."[6]

Individuals Evolve

An individual is born with the genes it has—these do not change as the individual ages. Therefore, an individual cannot evolve or adapt through natural selection. Evolution is the change in genetic composition of a population over time, specifically over generations, resulting from differential reproduction of individuals with certain alleles. Individuals do change over their lifetime, but this is called development; it involves changes programmed by the set of genes the individual acquired at birth in coordination with the individual's environment. When thinking about the evolution of a characteristic, it is probably best to think about the change of the average value of the characteristic in the population over time. For example, when natural selection leads to bill-size change in medium ground finches in the Galápagos, this does not mean that individual bills on the finches are changing. If one measures the average bill size among all individuals in the population at one time, and then measures the average bill size in the population several years later after there has been a strong selective pressure, this average value may be different as a result of evolution. Although some individuals may survive from the first time to the second, those individuals will still have the same bill size. However, there may be enough new individuals with different bill sizes to change the average bill size.

Evolution Explains the Origin of Life

It is a common misunderstanding that evolution includes an explanation of life's origins. Conversely, some of the theory's critics complain that it cannot explain the origin of life. The theory does not try to explain the origin of life. The theory of evolution explains how populations change over time and how life diversifies—the origin of species. It does not shed light on the beginnings of life including the origins of the first cells, which is how life is defined. The mechanisms of the origin of life on Earth are a particularly difficult problem because it occurred a very long time ago, over a very long time, and presumably just occurred once. Importantly, biologists believe that the presence of life on Earth precludes the possibility that the events that led to life on Earth can be repeated because the intermediate stages would immediately become food for existing living things. The early stages of life included the formation of organic molecules such as carbohydrates, amino acids, or nucleotides. If these were formed from inorganic precursors today, they would simply be broken down by living things. The early stages of life also probably included more complex aggregations of molecules into enclosed structures with an internal environment, a boundary layer of some form, and the external environment. Such structures, if they were formed now, would be quickly consumed or broken down by living organisms.

---

6. Theodosius Dobzhansky. "Biology, Molecular and Organismic." *American Zoologist* 4, no. 4 (1964): 449.

However, once a mechanism of inheritance was in place in the form of a molecule like DNA or RNA, either within a cell or within a pre-cell, these entities would be subject to the principle of natural selection. More effective reproducers would increase in frequency at the expense of inefficient reproducers. So while evolution does not explain the origin of life, it may have something to say about some of the processes operating once pre-living entities acquired certain properties.

Organisms Evolve on Purpose

Statements such as "organisms evolve in response to a change in an environment," are quite common. There are two easy misunderstandings possible with such a statement. First of all, the statement must not be understood to mean that individual organisms evolve, as was discussed above. The statement is shorthand for "a population evolves in response to a changing environment." However, a second misunderstanding may arise by interpreting the statement to mean that the evolution is somehow intentional. A changed environment results in some individuals in the population, those with particular phenotypes, benefiting and, therefore, producing proportionately more offspring than other phenotypes. This results in change in the population if the characters are genetically determined.

It is also important to understand that the variation that natural selection works on is already in a population and does not arise in response to an environmental change. For example, applying antibiotics to a population of bacteria will, over time, select for a population of bacteria that are resistant to antibiotics. The resistance, which is caused by a gene, did not arise by mutation because of the application of the antibiotic. The gene for resistance was already present in the gene pool of the bacteria, likely at a low frequency. The antibiotic, which kills the bacterial cells without the resistance gene, strongly selects for individuals that are resistant, since these would be the only ones that survived and divided. Experiments have demonstrated that mutations for antibiotic resistance do not arise as a result of antibiotic application.

In a larger sense, evolution is also not goal directed. Species do not become "better" over time; they simply track their changing environment with adaptations that maximize their reproduction in a particular environment at a particular time. Evolution has no goal of making faster, bigger, more complex, or even smarter species. This kind of language is common in popular literature. Certain organisms, ourselves included, are described as the "pinnacle" of evolution, or "perfected" by evolution. What characteristics evolve in a species are a function of the variation present and the environment, both of which are constantly changing in a non-directional way. What trait is fit in one environment at one time may well be fatal at some point in the future. This holds equally well for a species of insect as it does the human species.

Evolution Is Controversial among Scientists

The theory of evolution was controversial when it was first proposed in 1859, yet within 20 years virtually every working biologist had accepted evolution as the explanation for the diversity of life. The rate of acceptance was extraordinarily rapid, partly because Darwin had amassed an impressive body of evidence. The early controversies involved both scientific arguments against the theory and the arguments of religious leaders. It was the arguments of the biologists that were resolved after a short time, while the arguments of religious leaders have persisted to this day.

The theory of evolution replaced the predominant theory at the time that species had all been specially created within relatively recent history. Despite the prevalence of this theory, it was becoming increasingly clear to naturalists during the nineteenth century that it could no longer explain many observations of geology and the living world. The persuasiveness of the theory of evolution to these naturalists lay in its ability to explain these phenomena, and it continues to hold extraordinary explanatory power to this day. Its continued rejection by some religious leaders results from its replacement of special creation, a tenet of their religious belief. These leaders cannot accept the replacement of special creation by a mechanistic process that excludes the actions of a deity as an explanation for the diversity of life including the origins of the human species. It should be noted, however, that most of the major denominations in the United States have statements supporting the acceptance of evidence for evolution as compatible with their theologies.

The nature of the arguments against evolution by religious leaders has evolved over time. One current argument is that the theory is still controversial among biologists. This claim is simply not true. The number of working scientists who reject the theory of evolution, or question its validity and say so, is small. A Pew Research poll in 2009 found that 97 percent of the 2500 scientists polled believe species evolve.[7] The support for the theory is reflected in signed statements from many scientific societies such as the American Association for the Advancement of Science, which includes working scientists as members. Many of the scientists that reject or question the theory of evolution are non-biologists, such as engineers, physicians, and chemists. There are no experimental results or research programs that contradict the theory. There are no papers published in peer-reviewed scientific journals that appear to refute the theory. The latter observation might be considered a consequence of suppression of dissent, but it must be remembered that scientists are skeptics and that there is a long history of published reports that challenged scientific orthodoxy in unpopular ways. Examples include the endosymbiotic theory of eukaryotic origins, the theory of group selection, the microbial cause of stomach ulcers, the

7. Pew Research Center for the People & the Press, *Public Praises Science; Scientists Fault Public, Media* (Washington, DC, 2009), 37.

asteroid-impact theory of the Cretaceous extinction, and the theory of plate tectonics. Research with evidence and ideas with scientific merit are considered by the scientific community. Research that does not meet these standards is rejected.

Other Theories Should Be Taught

A common argument from some religious leaders is that alternative theories to evolution should be taught in public schools. Critics of evolution use this strategy to create uncertainty about the validity of the theory without offering actual evidence. In fact, there are no viable alternative scientific theories to evolution. The last such theory, proposed by Lamarck in the nineteenth century, was replaced by the theory of natural selection. A single exception was a research program in the Soviet Union based on Lamarck's theory during the early twentieth century that set that country's agricultural research back decades. Special creation is not a viable alternative scientific theory because it is not a scientific theory, since it relies on an untestable explanation. Intelligent design, despite the claims of its proponents, is also not a scientific explanation. This is because intelligent design posits the existence of an unknown designer of living organisms and their systems. Whether the designer is unknown or supernatural, it is a cause that cannot be measured; therefore, it is not a scientific explanation. There are two reasons not to teach nonscientific theories. First, these explanations for the diversity of life lack scientific usefulness because they do not, and cannot, give rise to research programs that promote our understanding of the natural world. Experiments cannot test non-material explanations for natural phenomena. For this reason, teaching these explanations as science in public schools is not in the public interest. Second, in the United States, it is illegal to teach them as science because the U.S. Supreme Court and lower courts have ruled that the teaching of religious belief, such as special creation or intelligent design, violates the establishment clause of the First Amendment of the U.S. Constitution, which prohibits government sponsorship of a particular religion.

The theory of evolution and science in general is, by definition, silent on the existence or non-existence of the spiritual world. Science is only able to study and know the material world. Individual biologists have sometimes been vocal atheists, but it is equally true that there are many deeply religious biologists. Nothing in biology precludes the existence of a god, indeed biology as a science has nothing to say about it. The individual biologist is free to reconcile her or his personal and scientific knowledge as they see fit. The Voices for Evolution project (http://ncse.com/voices), developed through the National Center for Science Education, works to gather the diversity of perspectives on evolution to advocate it being taught in public schools.

KEY TERMS

adaptation a heritable trait or behavior in an organism that aids in its survival in its present environment

adaptive radiation a speciation when one species radiates out to form several other species

allopatric speciation a speciation that occurs via a geographic separation

analogous structure a structure that is similar because of evolution in response to similar selection pressures resulting in convergent evolution, not similar because of descent from a common ancestor

bottleneck effect the magnification of genetic drift as a result of natural events or catastrophes

convergent evolution an evolution that results in similar forms on different species

dispersal an allopatric speciation that occurs when a few members of a species move to a new geographical area

divergent evolution an evolution that results in different forms in two species with a common ancestor

founder effect a magnification of genetic drift in a small population that migrates away from a large parent population carrying with it an unrepresentative set of alleles

gene flow the flow of alleles in and out of a population due to the migration of individuals or gametes

gene pool all of the alleles carried by all of the individuals in the population

genetic drift the effect of chance on a population's gene pool

homologous structure a structure that is similar because of descent from a common ancestor

inheritance of acquired characteristics a phrase that describes the mechanism of evolution proposed by Lamarck in which traits acquired by individuals through use or disuse could be passed on to their offspring thus leading to evolutionary change in the population

macroevolution a broader scale of evolutionary changes seen over paleontological time

microevolution the changes in a population's genetic structure (i.e., allele frequency)

migration the movement of individuals of a population to a new location; in population genetics it refers to the movement of individuals and their alleles from one population to another, potentially changing allele frequencies in both the old and the new population

modern synthesis the overarching evolutionary paradigm that took shape by the 1940s and is generally accepted today

natural selection the greater relative survival and reproduction of individuals in a population that have favorable heritable traits, leading to evolutionary change

population genetics the study of how selective forces change the allele frequencies in a population over time

speciation a formation of a new species

sympatric speciation a speciation that occurs in the same geographic space

variation the variety of alleles in a population

vestigial structure a physical structure present in an organism but that has no apparent function and appears to be from a functional structure in a distant ancestor

vicariance an allopatric speciation that occurs when something in the environment separates organisms of the same species into separate groups

CHAPTER SUMMARY

11.1 Discovering How Populations Change

Evolution by natural selection arises from three conditions: individuals within a species vary, some of those variations are heritable, and organisms have more offspring than resources can support. The consequence is that individuals with relatively advantageous variations will be more likely to survive and have higher reproductive rates than those individuals with different traits. The advantageous traits will be passed on to offspring in greater proportion. Thus, the trait will have higher representation in the next and subsequent generations leading to genetic change in the population.

The modern synthesis of evolutionary theory grew out of the reconciliation of Darwin's, Wallace's, and Mendel's thoughts on evolution and heredity. Population genetics is a theoretical framework for describing evolutionary change in populations through the change in allele frequencies. Population genetics defines evolution as a change in allele frequency over generations. In the absence of evolutionary forces allele frequencies will not change in a population; this is known as Hardy-Weinberg equilibrium principle. However, in all populations, mutation, natural selection, genetic drift, and migration act to change allele frequencies.

11.2 Mechanisms of Evolution

There are four factors that can change the allele frequencies of a population. Natural selection works by selecting for alleles that confer beneficial traits or behaviors, while selecting against those for deleterious qualities. Mutations introduce new alleles into a population. Genetic drift stems from the chance occurrence that some individuals have more offspring than others and results in changes in allele frequencies that are random in direction. When individuals leave or join the population, allele frequencies can change as a result of gene flow.

11.3 Evidence of Evolution

The evidence for evolution is found at all levels of organization in living things and in the extinct species we know about through fossils. Fossils provide evidence for the evolutionary change through now extinct forms that led to modern species. For example, there is a rich fossil record that shows the evolutionary transitions from horse ancestors to modern horses that document intermediate forms and a gradual adaptation o changing ecosystems. The anatomy of species and the embryological development of that anatomy reveal common structures in divergent lineages that have been modified over time by evolution. The geographical distribution of living species reflects the origins of species in particular geographic locations and the history of continental movements. The structures of molecules, like anatomical structures, reflect the relationships of living species and match patterns of similarity expected from descent with modification.

11.4 Speciation

Speciation occurs along two main pathways: geographic separation (allopatric speciation) and through mechanisms that occur within a shared habitat (sympatric speciation). Both pathways force reproductive isolation between populations. Sympatric speciation can occur through errors in meiosis that form gametes with extra chromosomes, called polyploidy. Autopolyploidy occurs within a single species, whereas allopolyploidy occurs because of a mating between closely related species. Once the populations are isolated, evolutionary divergence can take place leading to the evolution of reproductive isolating traits that prevent interbreeding should the two populations come together again. The reduced viability of hybrid offspring after a period of isolation is expected to select for stronger inherent isolating mechanisms.

11.5 Common Misconceptions about Evolution

The theory of evolution is a difficult concept and misconceptions abound. The factual nature of evolution is often challenged by wrongly associating the scientific meaning of a theory with the vernacular meaning. Evolution is sometimes mistakenly interpreted to mean that individuals evolve, when in fact only populations can evolve as their gene frequencies change over time. Evolution is often assumed to explain the origin of life, which it does not speak to. It is often spoken in goal-directed terms by which organisms change through intention, and selection operates on mutations present in a population that have not arisen in response to a particular environmental stress. Evolution is often characterized as being controversial among scientists; however, it is accepted by the vast majority of working scientists. Critics of evolution often argue that alternative theories to evolution should be taught in public schools; however, there are no viable alternative scientific theories to evolution. The alternative religious beliefs should not be taught as science because it cannot be proven, and in the United States it is unconstitutional. Science is silent on the question of the existence of a god while scientists are able to reconcile religious belief and scientific knowledge.

ART CONNECTION QUESTIONS

1. Figure 11.7 Do you think genetic drift would happen more quickly on an island or on the mainland?

REVIEW QUESTIONS

2. Which scientific concept did Charles Darwin and Alfred Wallace independently discover?
   a. mutation
   b. natural selection
   c. overbreeding
   d. sexual reproduction

3. Which of the following situations will lead to natural selection?
   a. The seeds of two plants land near each other and one grows larger than the other.
   b. Two types of fish eat the same kind of food, and one is better able to gather food than the other.
   c. Male lions compete for the right to mate with females, with only one possible winner.
   d. all of the above

4. What is the difference between micro- and macroevolution?
   a. Microevolution describes the evolution of small organisms, such as insects, while macroevolution describes the evolution of large organisms, like people and elephants.
   b. Microevolution describes the evolution of microscopic entities, such as molecules and proteins, while macroevolution describes the evolution of whole organisms.
   c. Microevolution describes the evolution of populations, while macroevolution describes the emergence of new species over long periods of time.
   d. Microevolution describes the evolution of organisms over their lifetimes, while macroevolution describes the evolution of organisms over multiple generations.

5. Population genetics is the study of _____.
   a. how allele frequencies in a population change over time
   b. populations of cells in an individual
   c. the rate of population growth
   d. how genes affect embryological development

6. Galápagos medium ground finches are found on Santa Cruz and San Cristóbal islands, which are separated by about 100 km of ocean. Occasionally, individuals from either island fly to the other island to stay. This can alter the allele frequencies of the population through which of the following mechanisms?
   a. natural selection
   b. genetic drift
   c. gene flow
   d. mutation

7. In which of the following pairs do both evolutionary processes introduce new genetic variation into a population?
   a. natural selection and genetic drift
   b. mutation and gene flow
   c. natural selection and gene flow
   d. gene flow and genetic drift

8. The wing of a bird and the arm of a human are examples of _____.
   a. vestigial structures
   b. molecular structures
   c. homologous structures
   d. analogous structures

9. The fact that DNA sequences are more similar in more closely related organisms is evidence of what?
   a. optimal design in organisms
   b. adaptation
   c. mutation
   d. descent with modification

10. Which situation would most likely lead to allopatric speciation?
   a. A flood causes the formation of a new lake.
   b. A storm causes several large trees to fall down.
   c. A mutation causes a new trait to develop.
   d. An injury causes an organism to seek out a new food source.

11. What is the main difference between dispersal and vicariance?
   a. One leads to allopatric speciation, whereas the other leads to sympatric speciation.
   b. One involves the movement of the organism, whereas the other involves a change in the environment.
   c. One depends on a genetic mutation occurring, whereas the other does not.
   d. One involves closely related organisms, whereas the other involves only individuals of the same species.

12. Which variable increases the likelihood of allopatric speciation taking place more quickly?
   a. lower rate of mutation
   b. longer distance between divided groups
   c. increased instances of hybrid formation
   d. equivalent numbers of individuals in each population

13. The word "theory" in theory of evolution is best replaced by _____.
   a. fact
   b. hypothesis

    c. idea
    d. alternate explanation

14. Why are alternative scientific theories to evolution not taught in public school?
    a. more theories would confuse students

    b. there are no viable scientific alternatives
    c. it is against the law
    d. alternative scientific theories are suppressed by the science establishment

CRITICAL THINKING QUESTIONS

15. If a person scatters a handful of plant seeds from one species in an area, how would natural selection work in this situation?

16. Explain the Hardy-Weinberg principle of equilibrium.

17. Describe natural selection and give an example of natural selection at work in a population.

18. Why do scientists consider vestigial structures evidence for evolution?

19. Why do island chains provide ideal conditions for adaptive radiation to occur?

20. Two species of fish had recently undergone sympatric speciation. The males of each species had a different coloring through which females could identify and choose a partner from her own species. After some time, pollution made the lake so cloudy it was hard for females to distinguish colors. What might take place in this situation?

21. How does the scientific meaning of "theory" differ from the common, everyday meaning of the word?

22. Explain why the statement that a monkey is more evolved than a mouse is incorrect.

12 | DIVERSITY OF LIFE

Figure 12.1 Although they look different, this bee and flower are distantly related. (credit: modification of work by John Beetham)

Chapter Outline
12.1: Organizing Life on Earth
12.2: Determining Evolutionary Relationships

Introduction

This bee and *Echinacea* flower could not look more different, yet they are related, as are all living organisms on Earth. By following pathways of similarities and differences—both visible and genetic—scientists seek to map the history of evolution from single-celled organisms to the tremendous diversity of creatures that have crawled, germinated, floated, swam, flown, and walked on this planet.

12.1 | Organizing Life on Earth

By the end of this section, you will be able to:

- Discuss the need for a comprehensive classification system
- List the different levels of the taxonomic classification system
- Describe how systematics and taxonomy relate to phylogeny

All life on Earth evolved from a common ancestor. Biologists map how organisms are related by constructing phylogenetic trees. In other words, a "tree of life" can be constructed to illustrate when different organisms evolved and to show the relationships among different organisms, as shown in Figure 12.2. Notice that from a single point, the three domains of Archaea, Bacteria, and Eukarya diverge and then branch repeatedly. The small branch that plants and animals (including humans) occupy in this diagram shows how recently these groups had their origin compared with other groups.

Phylogenetic Tree of Life

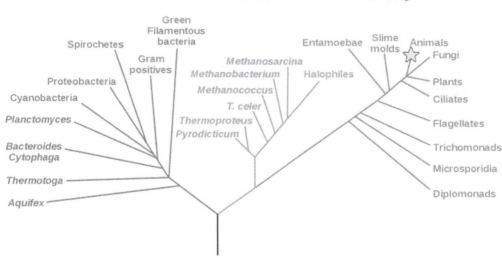

Figure 12.2 In the evolution of life on Earth, the three domains of life—Archaea, Bacteria, and Eukarya—branch from a single point. (credit: modification of work by Eric Gaba)

The phylogenetic tree in Figure 12.2 illustrates the pathway of evolutionary history. The pathway can be traced from the origin of life to any individual species by navigating through the evolutionary branches between the two points. Also, by starting with a single species and tracing backward to any branch point, the organisms related to it by various degrees of closeness can be identified.

A **phylogeny** is the evolutionary history and the relationships among a species or group of species. The study of organisms with the purpose of deriving their relationships is called **systematics**.

Many disciplines within the study of biology contribute to understanding how past and present life evolved over time, and together they contribute to building, updating, and maintaining the "tree of life." Information gathered may include data collected from fossils, from studying morphology, from the structure of body parts, or from molecular structure, such as the sequence of amino acids in proteins or DNA nucleotides. By considering the trees generated by different sets of data scientists can put together the phylogeny of a species.

Scientists continue to discover new species of life on Earth as well as new character information, thus trees change as new data arrive.

The Levels of Classification

Taxonomy (which literally means "arrangement law") is the science of naming and grouping species to construct an internationally shared classification system. The taxonomic classification system (also called the Linnaean system after its inventor, Carl Linnaeus, a Swedish naturalist) uses a hierarchical model. A hierarchical system has levels and each group at one of the levels includes groups at the next lowest level, so that at the lowest level each member belongs to a series of nested groups. An analogy is the nested series of directories on the main disk drive of a computer. For example, in the most inclusive grouping, scientists divide organisms into three **domains**: Bacteria, Archaea, and Eukarya. Within each domain is a second level called a **kingdom**. Each domain contains several kingdoms. Within kingdoms, the subsequent categories of increasing specificity are: **phylum**, **class**, **order**, **family**, **genus**, and **species**.

As an example, the classification levels for the domestic dog are shown in Figure 12.3. The group at each level is called a **taxon** (plural: taxa). In other words, for the dog, Carnivora is the taxon at the order level, Canidae is the taxon at the family level, and so forth. Organisms also have a common name that people typically use, such as domestic dog, or wolf. Each taxon name is capitalized except for species, and the genus and species names are italicized. Scientists refer to an organism by its genus and species names together, commonly called a scientific name, or Latin name. This two-name system is called **binomial nomenclature**. The scientific name of the wolf is therefore *Canis lupus*. Recent study of the DNA of domestic dogs and wolves suggest that the domestic dog is a subspecies of the wolf, not its own species, thus it is given an extra name to indicate its subspecies status, *Canis lupus familiaris*.

Figure 12.3 also shows how taxonomic levels move toward specificity. Notice how within the domain we find the dog grouped with the widest diversity of organisms. These include plants and other organisms not pictured, such as fungi and protists. At each sublevel, the organisms become more similar because they are more closely related. Before Darwin's theory of evolution was developed, naturalists sometimes classified organisms using arbitrary similarities, but since the theory of evolution was proposed in the 19th century, biologists work to make the classification system reflect evolutionary relationships. This means that all of the members of a taxon should have a common ancestor and be more closely related to each other than to members of other taxa.

Recent genetic analysis and other advancements have found that some earlier taxonomic classifications do not reflect actual evolutionary relationships, and therefore, changes and updates must be made as new discoveries take place. One dramatic and recent example was the breaking apart of prokaryotic species, which until the 1970s were all classified as bacteria. Their division into Archaea and Bacteria came about after the recognition that their large genetic differences warranted their separation into two of three fundamental branches of life.

art CONNECTION

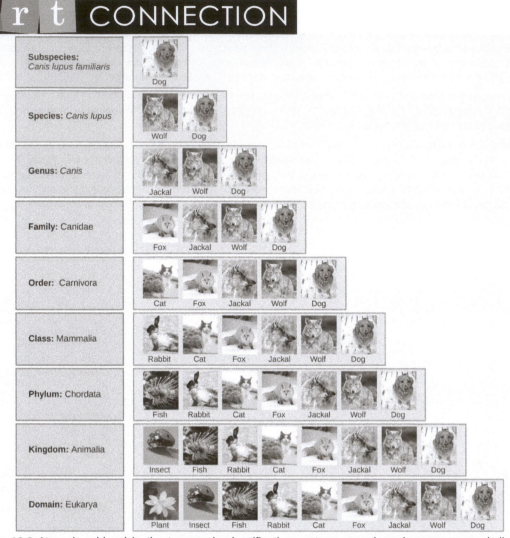

Figure 12.3 At each sublevel in the taxonomic classification system, organisms become more similar. Dogs and wolves are the same species because they can breed and produce viable offspring, but they are different enough to be classified as different subspecies. (credit "plant": modification of work by "berduchwal"/Flickr; credit "insect": modification of work by Jon Sullivan; credit "fish": modification of work by Christian Mehlführer; credit "rabbit": modification of work by Aidan Wojtas; credit "cat": modification of work by Jonathan Lidbeck; credit "fox": modification of work by Kevin Bacher, NPS; credit "jackal": modification of work by Thomas A. Hermann, NBII, USGS; credit "wolf" modification of work by Robert Dewar; credit "dog": modification of work by "digital_image_fan"/Flickr)

In what levels are cats and dogs considered to be part of the same group?

Visit this PBS site (http://openstaxcollege.org/l/classify_life2) to learn more about taxonomy. Under Classifying Life, click Launch Interactive.

Classification and Phylogeny

Scientists use a tool called a phylogenetic tree to show the evolutionary pathways and relationships between organisms. A **phylogenetic tree** is a diagram used to reflect evolutionary relationships among organisms or groups of organisms. The hierarchical classification of groups nested within more inclusive groups is reflected in diagrams. Scientists consider phylogenetic trees to be a hypothesis of the evolutionary past because one cannot go back through time to confirm the proposed relationships.

Unlike with a taxonomic classification, a phylogenetic tree can be read like a map of evolutionary history, as shown in Figure 12.4. Shared characteristics are used to construct phylogenetic trees. The point where a split occurs in a tree, called a **branch point**, represents where a single lineage evolved into distinct new ones. Many phylogenetic trees have a single branch point at the base representing a common ancestor of all the branches in the tree. Scientists call such trees **rooted**, which means there is a single ancestral taxon at the base of a phylogenetic tree to which all organisms represented in the diagram descend from. When two lineages stem from the same branch point, they are called **sister taxa**, for example the two species of orangutans. A branch point with more than two groups illustrates a situation for which scientists have not definitively determined relationships. An example is illustrated by the three branches leading to the gorilla subspecies; their exact relationships are not yet understood. It is important to note that sister taxa share an ancestor, which does not mean that one taxon evolved from the other. The branch point, or split, represents a common ancestor that existed in the past, but that no longer exists. Humans did not evolve from chimpanzees (nor did chimpanzees evolve from humans) although they are our closest living relatives. Both humans and chimpanzees evolved from a common ancestor that lived, scientists believe, six million years ago and looked different from both modern chimpanzees and modern humans.

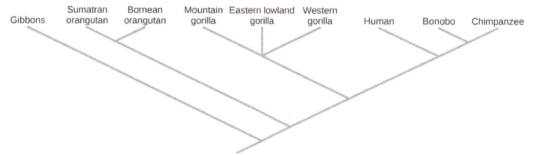

Figure 12.4 A phylogenetic tree is rooted and shows how different organisms, in this case the species and subspecies of living apes, evolved from a common ancestor.

The branch points and the branches in phylogenetic tree structure also imply evolutionary change. Sometimes the significant character changes are identified on a branch or branch point. For example, in Figure 12.5, the branch point that gives rise to the mammal and reptile lineage from the frog lineage shows the origin of the amniotic egg character. Also the branch point that gives rise to organisms with legs is indicated at the common ancestor of mammals, reptiles, amphibians, and jawed fishes.

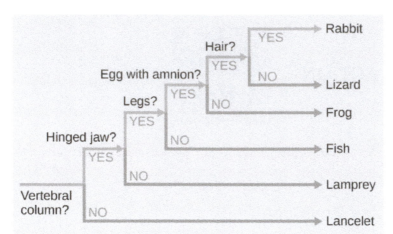

Figure 12.5 This phylogenetic tree is rooted by an organism that lacked a vertebral column. At each branch point, organisms with different characters are placed in different groups.

This interactive exercise (http://openstaxcollege.org/l/tree_of_life3) allows you to explore the evolutionary relationships among species.

Limitations of Phylogenetic Trees

It is easy to assume that more closely related organisms look more alike, and while this is often the case, it is not always true. If two closely related lineages evolved under significantly different surroundings or after the evolution of a major new adaptation, they may look quite different from each other, even more so than other groups that are not as closely related. For example, the phylogenetic tree in Figure 12.5 shows that lizards and rabbits both have amniotic eggs, whereas salamanders (within the frog lineage) do not; yet on the surface, lizards and salamanders appear more similar than the lizards and rabbits.

Another aspect of phylogenetic trees is that, unless otherwise indicated, the branches do not show length of time, they show only the order in time of evolutionary events. In other words, a long branch does not necessarily mean more time passed, nor does a short branch mean less time passed— unless specified on the diagram. For example, in Figure 12.5, the tree does not indicate how much time passed between the evolution of amniotic eggs and hair. What the tree does show is the order in which things took place. Again using Figure 12.5, the tree shows that the oldest trait is the vertebral column, followed by hinged jaws, and so forth. Remember that any phylogenetic tree is a part of the greater whole, and similar to a real tree, it does not grow in only one direction after a new branch develops. So, for the organisms in Figure 12.5, just because a vertebral column evolved does not mean that invertebrate evolution ceased, it only means that a new branch formed. Also, groups that are not closely related, but evolve under similar conditions, may appear more similar to each other than to a close relative.

12.2 | Determining Evolutionary Relationships

By the end of this section, you will be able to:

- Compare homologous and analogous traits
- Discuss the purpose of cladistics

Scientists collect information that allows them to make evolutionary connections between organisms. Similar to detective work, scientists must use evidence to uncover the facts. In the case of phylogeny, evolutionary investigations focus on two types of evidence: morphologic (form and function) and genetic.

Two Measures of Similarity

Organisms that share similar physical features *and* genetic sequences tend to be more closely related than those that do not. Features that overlap both morphologically and genetically are referred to as homologous structures; the similarities stem from common evolutionary paths. For example, as shown in Figure 12.6, the bones in the wings of bats and birds, the arms of humans, and the foreleg of a horse are homologous structures. Notice the structure is not simply a single bone, but rather a grouping of several bones arranged in a similar way in each organism even though the elements of the structure may have changed shape and size.

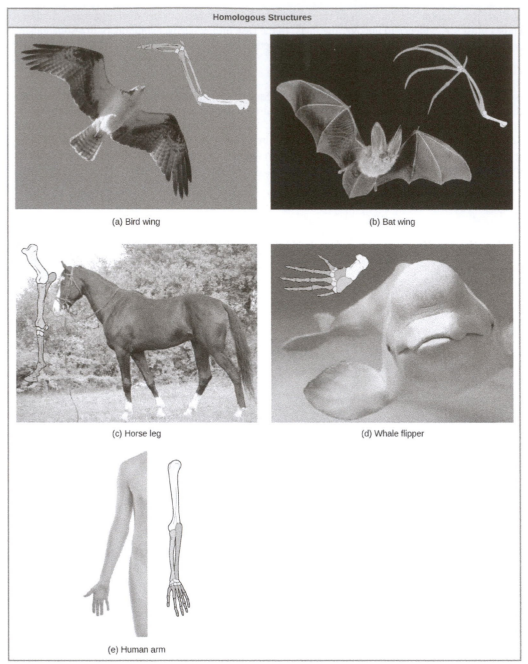

Figure 12.6 Bat and bird wings, the foreleg of a horse, the flipper of a whale, and the arm of a human are homologous structures, indicating that bats, birds, horses, whales, and humans share a common evolutionary past. (credit a photo: modification of work by Steve Hillebrand, USFWS; credit b photo: modification of work by U.S. BLM; credit c photo: modification of work by Virendra Kankariya; credit d photo: modification of work by Russian Gov./Wikimedia Commons)

Misleading Appearances

Some organisms may be very closely related, even though a minor genetic change caused a major morphological difference to make them look quite different. For example, chimpanzees and humans, the skulls of which are shown in Figure 12.7 are very similar genetically, sharing 99 percent[1] of their genes. However, chimpanzees and humans show considerable anatomical differences, including the degree to which the jaw protrudes in the adult and the relative lengths of our arms and legs.

---

1. Gibbons, A. (2012, June 13). *Science Now*. Retrieved from http://news.sciencemag.org/sciencenow/2012/06/bonobo-genome-sequenced.html

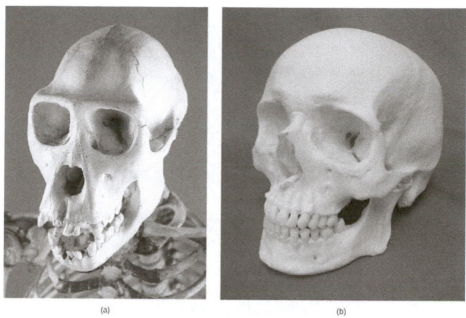

(a)   (b)

Figure 12.7 (a) The chimpanzee jaw protrudes to a much greater degree than (b) the human jaw. (credit a: modification of work by "Pastorius"/Wikimedia Commons)

However, unrelated organisms may be distantly related yet appear very much alike, usually because common adaptations to similar environmental conditions evolved in both. An example is the streamlined body shapes, the shapes of fins and appendages, and the shape of the tails in fishes and whales, which are mammals. These structures bear superficial similarity because they are adaptations to moving and maneuvering in the same environment—water. When a characteristic that is similar occurs by adaptive convergence (convergent evolution), and not because of a close evolutionary relationship, it is called an **analogous structure**. In another example, insects use wings to fly like bats and birds. We call them both wings because they perform the same function and have a superficially similar form, but the embryonic origin of the two wings is completely different. The difference in the development, or embryogenesis, of the wings in each case is a signal that insects and bats or birds do not share a common ancestor that had a wing. The wing structures, shown in Figure 12.8 evolved independently in the two lineages.

Similar traits can be either homologous or analogous. Homologous traits share an evolutionary path that led to the development of that trait, and analogous traits do not. Scientists must determine which type of similarity a feature exhibits to decipher the phylogeny of the organisms being studied.

(a) Bat wing   (b) Bird wing

(c) Insect wing

Figure 12.8 The wing of a honey bee is similar in shape to a bird wing and a bat wing and serves the same function (flight). The bird and bat wings are homologous structures. However, the honey bee wing has a different structure (it is made of a chitinous exoskeleton, not a boney endoskeleton) and embryonic origin. The bee and bird or bat wing types illustrate an analogy—similar structures that do not share an evolutionary history. (credit a photo: modification of work by U.S. BLM; credit b: modification of work by Steve Hillebrand, USFWS; credit c: modification of work by Jon Sullivan)

This website (http://openstaxcollege.org/l/relationships2) has several examples to show how appearances can be misleading in understanding the phylogenetic relationships of organisms.

Molecular Comparisons

With the advancement of DNA technology, the area of **molecular systematics**, which describes the use of information on the molecular level including DNA sequencing, has blossomed. New analysis of molecular characters not only confirms many earlier classifications, but also uncovers previously made errors. Molecular characters can include differences in the amino-acid sequence of a protein, differences in the individual nucleotide sequence of a gene, or differences in the arrangements of genes. Phylogenies based on molecular characters assume that the more similar the sequences are in two organisms, the more closely related they are. Different genes change evolutionarily at different rates and this affects the level at which they are useful at identifying relationships. Rapidly evolving sequences are useful for determining the relationships among closely related species. More slowly evolving sequences are useful for determining the relationships between distantly related species. To determine the relationships between very different species such as Eukarya and Archaea, the genes used must be very ancient, slowly evolving genes that are present in both groups, such as the genes for ribosomal RNA. Comparing phylogenetic trees using different sequences and finding them similar helps to build confidence in the inferred relationships.

Sometimes two segments of DNA in distantly related organisms randomly share a high percentage of bases in the same locations, causing these organisms to appear closely related when they are not. For example, the fruit fly shares 60 percent of its DNA with humans.[2] In this situation, computer-based statistical algorithms have been developed to help identify the actual relationships, and ultimately, the coupled use of both morphologic and molecular information is more effective in determining phylogeny.

e*olution IN ACTION

Why Does Phylogeny Matter?

In addition to enhancing our understanding of the evolutionary history of species, our own included, phylogenetic analysis has numerous practical applications. Two of those applications include understanding the evolution and transmission of disease and making decisions about conservation efforts. A 2010 study[3] of MRSA (methicillin-resistant *Staphylococcus aureus*), an antibiotic resistant pathogenic bacterium, traced the origin and spread of the strain throughout the past 40 years. The study uncovered the timing and patterns in which the resistant strain moved from its point of origin in Europe to centers of infection and evolution in South America, Asia, North America, and Australasia. The study suggested that introductions of the bacteria to new populations occurred very few times, perhaps only once, and then spread from that limited number of individuals. This is in contrast to the possibility that many individuals had carried the bacteria from one place to another. This result suggests that public health officials should concentrate on quickly identifying the contacts of individuals infected with a new strain of bacteria to control its spread.

A second area of usefulness for phylogenetic analysis is in conservation. Biologists have argued that it is important to protect species throughout a phylogenetic tree rather than just those from one branch of the tree. Doing this will preserve more of the variation produced by evolution. For example, conservation efforts should focus on a single species without sister species rather than another species that has a cluster of close sister species that recently evolved. If the single evolutionarily distinct species goes extinct a disproportionate amount of variation from the tree will be lost compared to one species in the cluster of closely related species. A study published in 2007[4] made recommendations for conservation of mammal species worldwide based on how evolutionarily distinct and at risk of extinction they are. The study found that their recommendations differed from priorities based on simply the level of extinction threat to the species. The study recommended protecting some threatened and valued large mammals such as the orangutans, the giant and lesser pandas, and the African and Asian elephants. But they also found that some much lesser known species should be protected based on how evolutionary distinct they are. These include a number of rodents, bats, shrews and hedgehogs. In addition there are some critically endangered species that did not rate as very important in evolutionary distinctiveness including species of deer mice and gerbils. While many criteria affect conservation decisions, preserving phylogenetic diversity provides an objective way to protect the full range of diversity generated by evolution.

Building Phylogenetic Trees

How do scientists construct phylogenetic trees? Presently, the most accepted method for constructing phylogenetic trees is a method called **cladistics**. This method sorts organisms into **clades**, groups of organisms that are most closely related to each other and the ancestor from which they descended. For example, in Figure 12.9, all of the organisms in the shaded region evolved from a single ancestor that had amniotic eggs. Consequently, all of these organisms also have amniotic eggs and make a single clade, also called a **monophyletic group**. Clades must include the ancestral species and all of the descendants from a branch point.

2. *Background on comparative genomic analysis.* (2002, December). Retrieved from http://www.genome.gov/10005835
3. Harris, S.R. et al. 2010. Evolution of MRSA during hospital transmission and intercontinental spread. *Science* 327:469–474.
4. Isaac NJ, Turvey ST, Collen B, Waterman C, Baillie JE (2007) Mammals on the EDGE: Conservation Priorities Based on Threat and Phylogeny. PLoS ONE 2(3): e296. doi:10.1371/journal.pone.0000296

art CONNECTION

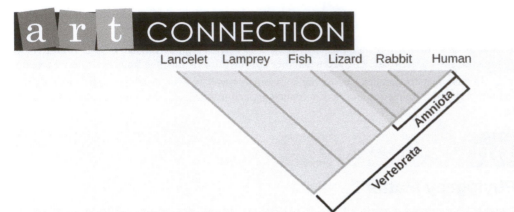

Figure 12.9 Lizards, rabbits, and humans all descend from a common ancestor in which the amniotic egg evolved. Thus, lizards, rabbits, and humans all belong to the clade Amniota. Vertebrata is a larger clade that also includes fish and lamprey.

Which animals in this figure belong to a clade that includes animals with hair? Which evolved first: hair or the amniotic egg?

Clades can vary in size depending on which branch point is being referenced. The important factor is that all of the organisms in the clade or monophyletic group stem from a single point on the tree. This can be remembered because monophyletic breaks down into "mono," meaning one, and "phyletic," meaning evolutionary relationship.

Shared Characteristics

Cladistics rests on three assumptions. The first is that living things are related by descent from a common ancestor, which is a general assumption of evolution. The second is that speciation occurs by splits of one species into two, never more than two at a time, and essentially at one point in time. This is somewhat controversial, but is acceptable to most biologists as a simplification. The third assumption is that traits change enough over time to be considered to be in a different state .It is also assumed that one can identify the actual direction of change for a state. In other words, we assume that an amniotic egg is a later character state than non-amniotic eggs. This is called the polarity of the character change. We know this by reference to a group outside the clade: for example, insects have non-amniotic eggs; therefore, this is the older or ancestral character state. Cladistics compares ingroups and outgroups. An ingroup (lizard, rabbit and human in our example) is the group of taxa being analyzed. An outgroup (lancelet, lamprey and fish in our example) is a species or group of species that diverged before the lineage containing the group(s) of interest. By comparing ingroup members to each other and to the outgroup members, we can determine which characteristics are evolutionary modifications determining the branch points of the ingroup's phylogeny.

If a characteristic is found in all of the members of a group, it is a **shared ancestral character** because there has been no change in the trait during the descent of each of the members of the clade. Although these traits appear interesting because they unify the clade, in cladistics they are considered not helpful when we are trying to determine the relationships of the members of the clade because every member is the same. In contrast, consider the amniotic egg characteristic of Figure 12.9. Only some of the organisms have this trait, and to those that do, it is called a **shared derived character** because this trait changed at some point during descent. This character does tell us about the relationships among the members of the clade; it tells us that lizards, rabbits, and humans group more closely together than any of these organisms do with fish, lampreys, and lancelets.

A sometimes confusing aspect of "ancestral" and "derived" characters is that these terms are relative. The same trait could be either ancestral or derived depending on the diagram being used and the organisms being compared. Scientists find these terms useful when distinguishing between clades during the building of phylogenetic trees, but it is important to remember that their meaning depends on context.

Choosing the Right Relationships

Constructing a phylogenetic tree, or cladogram, from the character data is a monumental task that is usually left up to a computer. The computer draws a tree such that all of the clades share the same list of derived characters. But there are other decisions to be made, for example, what if a species presence in a clade is supported by all of the shared derived characters for that clade except one? One conclusion is that the trait evolved in the ancestor, but then changed back in that one species. Also a character state that appears in two clades must be assumed to have evolved independently in those clades. These inconsistencies are common in trees drawn from character data and complicate the decision-making process about which tree most closely represents the real relationships among the taxa.

To aid in the tremendous task of choosing the best tree, scientists often use a concept called **maximum parsimony**, which means that events occurred in the simplest, most obvious way. This means that the "best" tree is the one with the fewest number of character reversals, the fewest number of independent character changes, and the fewest number of character changes throughout the tree. Computer programs search through all of the possible trees to find the small number of trees with the simplest evolutionary pathways. Starting with all of the homologous traits in a group of organisms, scientists can determine the order of evolutionary events of which those traits occurred that is the most obvious and simple.

Practice Parsimony: Go to this website (http://openstaxcollege.org/l/parsimony2) to learn how maximum parsimony is used to create phylogenetic trees (be sure to continue to the second page).

These tools and concepts are only a few of the strategies scientists use to tackle the task of revealing the evolutionary history of life on Earth. Recently, newer technologies have uncovered surprising discoveries with unexpected relationships, such as the fact that people seem to be more closely related to fungi than fungi are to plants. Sound unbelievable? As the information about DNA sequences grows, scientists will become closer to mapping the evolutionary history of all life on Earth.

KEY TERMS

analogous structure a character found in two taxa that looks similar because of convergent evolution, not because of descent from a common ancestor

binomial nomenclature a system of two-part scientific names for an organism, which includes genus and species names

branch point a point on a phylogenetic tree where a single lineage splits to distinct new ones

clade a group of taxa with the same set of shared derived characters, including an ancestral species and all its descendants

cladistics a method used to organize homologous traits to describe phylogenies using common descendent as the primary criterion used to classify organisms

class the category in the taxonomic classification system that falls within phylum and includes orders

domain the highest level category in the classification system and that includes all taxonomic classifications below it; it is the most inclusive taxon

family the category in the taxonomic classification system that falls within order and includes genera

genus the category in the taxonomic classification system that falls within family and includes species; the first part of the scientific name

kingdom the category in the taxonomic classification system that falls within domain and includes phyla

maximum parsimony applying the simplest, most obvious way with the least number of steps

molecular systematics the methods of using molecular evidence to identify phylogenetic relationships

monophyletic group (also, clade) organisms that share a single ancestor

order the category in the taxonomic classification system that falls within class and includes families

phylogenetic tree diagram used to reflect the evolutionary relationships between organisms or groups of organisms

phylogeny evolutionary history and relationship of an organism or group of organisms

phylum the category in the taxonomic classification system that falls within kingdom and includes classes

rooted describing a phylogenetic tree with a single ancestral lineage to which all organisms represented in the diagram relate

shared ancestral character a character on a phylogenetic branch that is shared by a particular clade

shared derived character a character on a phylogenetic tree that is shared only by a certain clade of organisms

sister taxa two lineages that diverged from the same branch point

species the most specific category of classification

systematics the science of determining the evolutionary relationships of organisms

taxon a single level in the taxonomic classification system

taxonomy the science of classifying organisms

CHAPTER SUMMARY

12.1 Organizing Life on Earth

Scientists continually obtain new information that helps to understand the evolutionary history of life on Earth. Each group of organisms went through its own evolutionary journey, called its phylogeny. Each organism shares relatedness with others, and based on morphologic and genetic evidence scientists attempt to map the evolutionary pathways of all life on Earth. Historically, organisms were organized into a taxonomic classification system. However, today many scientists build phylogenetic trees to illustrate evolutionary relationships and the taxonomic classification system is expected to reflect evolutionary relationships.

12.2 Determining Evolutionary Relationships

To build phylogenetic trees, scientists must collect character information that allows them to make evolutionary connections between organisms. Using morphologic and molecular data, scientists work to identify homologous characteristics and genes. Similarities between organisms can stem either from shared evolutionary history (homologies) or from separate evolutionary paths (analogies). After homologous information is identified, scientists use cladistics to organize these events as a means to determine an evolutionary timeline. Scientists apply the concept of maximum parsimony, which states that the likeliest order of events is probably the simplest shortest path. For evolutionary events, this would be the path with the least number of major divergences that correlate with the evidence.

ART CONNECTION QUESTIONS

1. Figure 12.3 In what levels are cats and dogs considered to be part of the same group?

2. Figure 12.8 Which animals in this figure belong to a clade that includes animals with hair? Which evolved first: hair or the amniotic egg?

REVIEW QUESTIONS

3. What is a phylogeny a description of?
   a. mutations
   b. DNA
   c. evolutionary history
   d. organisms on Earth

4. What do scientists in the field of systematics accomplish?
   a. discover new fossil sites
   b. organize and classify organisms
   c. name new species
   d. communicate between field biologists

5. Which statement about the taxonomic classification system is correct?
   a. There are more domains than kingdoms.
   b. Kingdoms are the top category of classification.
   c. A phylum may be represented in more than one kingdom.
   d. Species are the most specific category of classification.

6. Which best describes the relationship between chimpanzees and humans?
   a. chimpanzees evolved from humans
   b. humans evolved from chimpanzees
   c. chimpanzees and humans evolved from a common ancestor
   d. chimpanzees and humans belong to the same species

7. Which best describes a branch point in a phylogenetic tree?
   a. a hypothesis
   b. new lineage
   c. hybridization
   d. a mating

8. Which statement about analogies is correct?
   a. They occur only as errors.
   b. They are synonymous with homologous traits.
   c. They are derived by response to similar environmental pressures.
   d. They are a form of mutation.

9. What kind of trait is important to cladistics?
   a. shared derived traits
   b. shared ancestral traits
   c. analogous traits
   d. parsimonious traits

10. What is true about organisms that are a part of the same clade?
   a. They all share the same basic characteristics.
   b. They evolved from a shared ancestor.
   c. They all are on the same tree.
   d. They have identical phylogenies.

11. Which assumption of cladistics is stated incorrectly?

   a. Living things are related by descent from a common ancestor.
   b. Speciation can produce one, two, or three new species.
   c. Traits change from one state to another.
   d. The polarity of a character state change can be determined.

12. A monophyletic group is a _____.
   a. phylogenetic tree
   b. shared derived trait
   c. character state
   d. clade

CRITICAL THINKING QUESTIONS

13. How does a phylogenetic tree indicate major evolutionary events within a lineage?

14. List the different levels of the taxonomic classification system.

15. Dolphins and fish have similar body shapes. Is this feature more likely a homologous or analogous trait?

16. Describe maximum parsimony.

17. How does a biologist determine the polarity of a character change?

13 | DIVERSITY OF MICROBES, FUNGI, AND PROTISTS

Figure 13.1 Living things are very diverse, from simple, single-celled bacteria to complex, multicellular organisms. (credit "ringworm": modification of work by Dr. Lucille K. Georg, CDC; credit "Trypanosomes": modification of work by Dr. Myron G. Schultz, CDC; credit "tree mold": modification of work by Janice Haney Carr, Robert Simmons, CDC; credit "coral fungus": modification of work by Cory Zanker; credit "bacterium": modification of work by Dr. David Cox, CDC; credit "cup fungus": modification of work by "icelight"/Flickr; credit "MRSA": modification of work by Janice Haney Carr, CDC; credit "moldy grapefruit": modification of work by Joseph Smilanick)

Chapter Outline
13.1: Prokaryotic Diversity
13.2: Eukaryotic Origins
13.3: Protists
13.4: Fungi

Introduction

Until the late twentieth century, scientists most commonly grouped living things into five kingdoms—animals, plants, fungi, protists, and bacteria—based on several criteria, such as absence or presence of a nucleus and other membrane-bound organelles, absence or presence of cell walls, multicellularity, and mode of nutrition. In the late twentieth century, the pioneering work of Carl Woese and others compared nucleotide sequences of small-subunit ribosomal RNA (SSU rRNA), which resulted in a dramatically different way to group organisms on Earth. Based on differences in the structure of cell membranes and in rRNA, Woese and his colleagues proposed that all life on Earth evolved along three lineages, called domains. The three domains are called Bacteria, Archaea, and Eukarya.

Two of the three domains—Bacteria and Archaea—are prokaryotic, meaning that they lack both a nucleus and true membrane-bound organelles. However, they are now considered, on the basis of membrane structure and rRNA, to be as different from each other as they are from the third domain, the Eukarya. Prokaryotes were the first inhabitants on Earth, perhaps appearing approximately 3.9 billion years ago. Today they are ubiquitous—inhabiting the harshest environments on the planet, from boiling hot springs to permanently frozen environments in Antarctica, as well as more benign environments such as compost heaps, soils, ocean waters, and the guts of animals (including humans). The Eukarya include the familiar kingdoms of animals, plants, and fungi. They also include a diverse group of kingdoms formerly grouped together as protists.

13.1 | Prokaryotic Diversity

By the end of this section, you will be able to:

- Describe the evolutionary history of prokaryotes
- Describe the basic structure of a typical prokaryote
- Identify bacterial diseases that caused historically important plagues and epidemics
- Describe the uses of prokaryotes in food processing and bioremediation

Prokaryotes are present everywhere. They cover every imaginable surface where there is sufficient moisture, and they live on and inside of other living things. There are more prokaryotes inside and on the exterior of the human body than there are human cells in the body. Some prokaryotes thrive in environments that are inhospitable for most other living things. Prokaryotes recycle nutrients—essential substances (such as carbon and nitrogen)—and they drive the evolution of new ecosystems, some of which are natural while others are man-made. Prokaryotes have been on Earth since long before multicellular life appeared.

Prokaryotic Diversity

The advent of DNA sequencing provided immense insight into the relationships and origins of prokaryotes that were not possible using traditional methods of classification. A major insight identified two groups of prokaryotes that were found to be as different from each other as they were from eukaryotes. This recognition of prokaryotic diversity forced a new understanding of the classification of all life and brought us closer to understanding the fundamental relationships of all living things, including ourselves.

Early Life on Earth

When and where did life begin? What were the conditions on Earth when life began? Prokaryotes were the first forms of life on Earth, and they existed for billions of years before plants and animals appeared. Earth is about 4.54 billion years old. This estimate is based on evidence from the dating of meteorite material, since surface rocks on Earth are not as old as Earth itself. Most rocks available on Earth have undergone geological changes that make them younger than Earth itself. Some meteorites are made of the original material in the solar disk that formed the objects of the solar system, and they have not been altered by the processes that altered rocks on Earth. Thus, the age of meteorites is a good indicator of the age of the formation of Earth. The original estimate of 4.54 billion years was obtained by Clare Patterson in 1956. His meticulous work has since been corroborated by ages determined from other sources, all of which point to an Earth age of about 4.54 billion years.

Early Earth had a very different atmosphere than it does today. Evidence indicates that during the first 2 billion years of Earth's existence, the atmosphere was **anoxic**, meaning that there was no oxygen. Therefore, only those organisms that can grow without oxygen— **anaerobic** organisms—were able to live. Organisms that convert solar energy into chemical energy are called **phototrophs**. Phototrophic organisms that required an organic source of carbon appeared within one billion years of the formation of Earth. Then, **cyanobacteria**, also known as blue-green algae, evolved from these simple phototrophs one billion years later. Cyanobacteria are able to use carbon dioxide as a source of carbon. Cyanobacteria (Figure 13.2) began the oxygenation of the atmosphere. The increase in oxygen concentration allowed the evolution of other life forms.

Figure 13.2 This hot spring in Yellowstone National Park flows toward the foreground. Cyanobacteria in the spring are green, and as water flows down the heat gradient, the intensity of the color increases because cell density increases. The water is cooler at the edges of the stream than in the center, causing the edges to appear greener. (credit: Graciela Brelles-Mariño)

Before the atmosphere became oxygenated, the planet was subjected to strong radiation; thus, the first organisms would have flourished where they were more protected, such as in ocean depths or beneath the surface of Earth. At this time, too, strong volcanic activity was common on Earth, so it is likely that these first organisms—the first prokaryotes—were adapted to very high temperatures. These are not the typical temperate environments in which most life flourishes today; thus, we can conclude that the first organisms that appeared on Earth likely were able to withstand harsh conditions.

Microbial mats may represent the earliest forms of life on Earth, and there is fossil evidence of their presence, starting about 3.5 billion years ago. A **microbial mat** is a large biofilm, a multi-layered sheet of prokaryotes (Figure 13.3a), including mostly bacteria, but also archaea. Microbial mats are a few centimeters thick, and they typically grow on moist surfaces. Their various types of prokaryotes carry out different metabolic pathways, and for this reason, they reflect various colors. Prokaryotes in a microbial mat are held together by a gummy-like substance that they secrete.

The first microbial mats likely obtained their energy from hydrothermal vents. A **hydrothermal vent** is a fissure in Earth's surface that releases geothermally heated water. With the evolution of photosynthesis about 3 billion years ago, some prokaryotes in microbial mats came to use a more widely available energy source—sunlight—whereas others were still dependent on chemicals from hydrothermal vents for food.

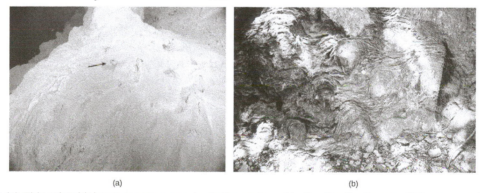

(a)   (b)

Figure 13.3 (a) This microbial mat grows over a hydrothermal vent in the Pacific Ocean. Chimneys such as the one indicated by the arrow allow gases to escape. (b) This photo shows stromatolites that are nearly 1.5 billion years old, found in Glacier National Park, Montana. (credit a: modification of work by Dr. Bob Embley, NOAA PMEL; credit b: modification of work by P. Carrara, NPS)

Fossilized microbial mats represent the earliest record of life on Earth. A **stromatolite** is a sedimentary structure formed when minerals are precipitated from water by prokaryotes in a microbial mat (Figure 13.3b). Stromatolites form layered rocks made of carbonate or silicate. Although most stromatolites are artifacts from the past, there are places on Earth where stromatolites are still forming. For example, living stromatolites have been found in the Anza-Borrego Desert State Park in San Diego County, California.

Some prokaryotes are able to thrive and grow under conditions that would kill a plant or animal. Bacteria and archaea that grow under extreme conditions are called **extremophiles**, meaning "lovers of extremes." Extremophiles have been found in extreme environments of all kinds, including the depths of the oceans, hot springs, the Arctic and the Antarctic, very dry places, deep inside Earth, harsh chemical environments, and high radiation environments. Extremophiles give us a better understanding of prokaryotic diversity and open up the possibility of the discovery of new therapeutic drugs or industrial applications. They have also opened up the possibility of finding life in other places in the solar system, which have harsher environments than those typically found on Earth. Many of these extremophiles cannot survive in moderate environments.

Watch a video (http://openstaxcollege.org/l/extremophiles) showing the Director of the Planetary Science Division of NASA discussing the implications that the existence extremophiles on Earth have on the possibility of finding life on other planets in our solar system, such as Mars.

Biofilms

Until a couple of decades ago, microbiologists thought of prokaryotes as isolated entities living apart. This model, however, does not reflect the true ecology of prokaryotes, most of which prefer to live in communities where they can interact. A **biofilm** is a microbial community held together in a gummy-textured matrix, consisting primarily of polysaccharides secreted by the organisms, together with some proteins and nucleic acids. Biofilms grow attached to surfaces. Some of the best-studied biofilms are composed of prokaryotes, although fungal biofilms have also been described.

Biofilms are present almost everywhere. They cause the clogging of pipes and readily colonize surfaces in industrial settings. They have played roles in recent, large-scale outbreaks of bacterial contamination of food. Biofilms also colonize household surfaces, such as kitchen counters, cutting boards, sinks, and toilets.

Interactions among the organisms that populate a biofilm, together with their protective environment, make these communities more robust than are free-living, or planktonic, prokaryotes. Overall, biofilms are very difficult to destroy, because they are resistant to many of the common forms of sterilization.

Characteristics of Prokaryotes

There are many differences between prokaryotic and eukaryotic cells. However, all cells have four common structures: a plasma membrane that functions as a barrier for the cell and separates the cell from its environment; cytoplasm, a jelly-like substance inside the cell; genetic material (DNA and RNA); and ribosomes, where protein synthesis takes place. Prokaryotes come in various shapes, but many fall into three categories: cocci (spherical), bacilli (rod-shaped), and spirilla (spiral-shaped) (Figure 13.4).

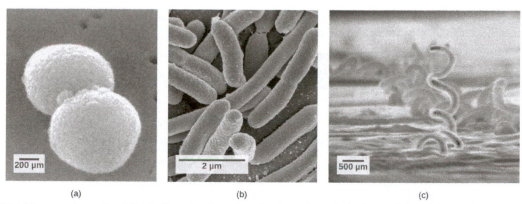

(a) (b) (c)

Figure 13.4 Many prokaryotes fall into three basic categories based on their shape: (a) cocci, or spherical; (b) bacilli, or rod-shaped; and (c) spirilla, or spiral-shaped. (credit a: modification of work by Janice Haney Carr, Dr. Richard Facklam, CDC; credit c: modification of work by Dr. David Cox, CDC; scale-bar data from Matt Russell)

The Prokaryotic Cell

Recall that prokaryotes (Figure 13.5) are unicellular organisms that lack organelles surrounded by membranes. Therefore, they do not have a nucleus but instead have a single chromosome—a piece of circular DNA located in an area of the cell called the nucleoid. Most prokaryotes have a cell wall lying outside the plasma membrane. The composition of the cell wall differs significantly between the domains Bacteria and Archaea (and their cell walls also differ from the eukaryotic cell walls found in plants and fungi.) The cell wall functions as a protective layer and is responsible for the organism's shape. Some other structures are present in some prokaryotic species, but not in others. For example, the **capsule** found in some species enables the organism to attach to surfaces and protects it from dehydration. Some species may also have flagella (singular, flagellum) used for locomotion, and pili (singular, pilus) used for attachment to surfaces and to other bacteria for conjugation. Plasmids, which consist of small, circular pieces of DNA outside of the main chromosome, are also present in many species of bacteria.

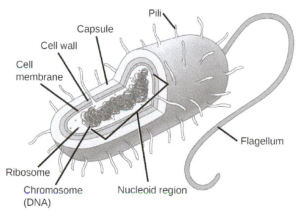

Figure 13.5 The features of a typical bacterium cell are shown.

Both Bacteria and Archaea are types of prokaryotic cells. They differ in the lipid composition of their cell membranes and in the characteristics of their cell walls. Both types of prokaryotes have the same basic structures, but these are built from different chemical components that are evidence of an ancient separation of their lineages. The archaeal plasma membrane is chemically different from the bacterial membrane; some archaeal membranes are lipid monolayers instead of phospholipid bilayers.

The Cell Wall

The cell wall is a protective layer that surrounds some prokaryotic cells and gives them shape and rigidity. It is located outside the cell membrane and prevents osmotic lysis (bursting caused by increasing volume). The chemical compositions of the cell walls vary between Archaea and Bacteria, as well as between bacterial species. Bacterial cell walls contain **peptidoglycan**, composed of polysaccharide chains cross-linked to peptides. Bacteria are divided into two major groups: **Gram-positive** and **Gram-negative**, based on their reaction to a procedure called Gram staining. The different bacterial responses to the staining procedure are caused by cell wall structure. Gram-positive organisms have a thick wall consisting of many layers of peptidoglycan. Gram-negative bacteria have a thinner cell wall composed of a few layers of peptidoglycan and additional structures, surrounded by an outer membrane (Figure 13.6).

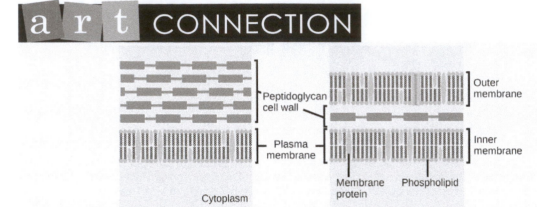

Figure 13.6 Bacteria are divided into two major groups: Gram-positive and Gram-negative. Both groups have a cell wall composed of peptidoglycans: In Gram-positive bacteria, the wall is thick, whereas in Gram-negative bacteria, the wall is thin. In Gram-negative bacteria, the cell wall is surrounded by an outer membrane.

Which of the following statements is true?

a. Gram-positive bacteria have a single cell wall formed from peptidoglycan.

b. Gram-positive bacteria have an outer membrane.

c. The cell wall of Gram-negative bacteria is thick, and the cell wall of Gram-positive bacteria is thin.

d. Gram-negative bacteria have a cell wall made of peptidoglycan, while Gram-positive bacteria have a cell wall made of phospholipids.

Archaeal cell walls do not contain peptidoglycan. There are four different types of archaeal cell walls. One type is composed of **pseudopeptidoglycan**. The other three types of cell walls contain polysaccharides, glycoproteins, and surface-layer proteins known as S-layers.

Reproduction

Reproduction in prokaryotes is primarily asexual and takes place by binary fission. Recall that the DNA of a prokaryote exists usually as a single, circular chromosome. Prokaryotes do not undergo mitosis. Rather, the chromosome loop is replicated, and the two resulting copies attached to the plasma membrane move apart as the cell grows in a process called binary fission. The prokaryote, now enlarged, is pinched inward at its equator, and the two resulting cells, which are clones, separate. Binary fission does not provide an opportunity for genetic recombination, but prokaryotes can alter their genetic makeup in three ways.

In a process called **transformation**, the cell takes in DNA found in its environment that is shed by other prokaryotes, alive or dead. A **pathogen** is an organism that causes a disease. If a nonpathogenic bacterium takes up DNA from a pathogen and incorporates the new DNA in its own chromosome, it too may become pathogenic. In **transduction**, bacteriophages, the viruses that infect bacteria, move DNA from one bacterium to another. Archaea have a different set of viruses that infect them and translocate genetic material from one individual to another. During **conjugation**, DNA is transferred from one prokaryote to another by means of a pilus that brings the organisms into contact with one another. The DNA transferred is usually a plasmid, but parts of the chromosome can also be moved.

Cycles of binary fission can be very rapid, on the order of minutes for some species. This short generation time coupled with mechanisms of genetic recombination result in the rapid evolution of prokaryotes, allowing them to respond to environmental changes (such as the introduction of an antibiotic) very quickly.

How Prokaryotes Obtain Energy and Carbon

Prokaryotes are metabolically diverse organisms. Prokaryotes fill many niches on Earth, including being involved in nutrient cycles such as the nitrogen and carbon cycles, decomposing dead organisms, and growing and multiplying inside living organisms, including humans. Different prokaryotes can use different sources of energy to assemble macromolecules from smaller molecules. Phototrophs obtain their energy from sunlight. Chemotrophs obtain their energy from chemical compounds.

Bacterial Diseases in Humans

Devastating pathogen-borne diseases and plagues, both viral and bacterial in nature, have affected and continue to affect humans. It is worth noting that all pathogenic prokaryotes are Bacteria; there are no known pathogenic Archaea in humans or any other organism. Pathogenic organisms evolved alongside humans. In the past, the true cause of these diseases was not understood, and some cultures thought that diseases were a spiritual punishment or were mistaken about material causes. Over time, people came to realize that staying apart from afflicted persons, improving sanitation, and properly disposing of the corpses and personal belongings of victims of illness reduced their own chances of getting sick.

Historical Perspective

There are records of infectious diseases as far back as 3,000 B.C. A number of significant **pandemics** caused by Bacteria have been documented over several hundred years. Some of the largest pandemics led to the decline of cities and cultures. Many were zoonoses that appeared with the domestication of animals, as in the case of tuberculosis. A zoonosis is a disease that infects animals but can be transmitted from animals to humans.

Infectious diseases remain among the leading causes of death worldwide. Their impact is less significant in many developed countries, but they are important determiners of mortality in developing countries. The development of antibiotics did much to lessen the mortality rates from bacterial infections, but access to antibiotics is not universal, and the overuse of antibiotics has led to the development of resistant strains of bacteria. Public sanitation efforts that dispose of sewage and provide clean drinking water have done as much or more than medical advances to prevent deaths caused by bacterial infections.

In 430 B.C., the plague of Athens killed one-quarter of the Athenian troops that were fighting in the Great Peloponnesian War. The disease killed a quarter of the population of Athens in over 4 years and weakened Athens' dominance and power. The source of the plague may have been identified recently when researchers from the University of Athens were able to analyze DNA from teeth recovered from a mass grave. The scientists identified nucleotide sequences from a pathogenic bacterium that causes typhoid fever.[1]

From 541 to 750 A.D., an outbreak called the plague of Justinian (likely a bubonic plague) eliminated, by some estimates, one-quarter to one-half of the human population. The population in Europe declined by 50 percent during this outbreak. Bubonic plague would decimate Europe more than once.

One of the most devastating pandemics was the **Black Death** (1346 to 1361), which is believed to have been another outbreak of bubonic plague caused by the bacterium *Yersinia pestis*. This bacterium is carried by fleas living on black rats. The Black Death reduced the world's population from an estimated 450 million to about 350 to 375 million. Bubonic plague struck London hard again in the mid-1600s. There are still approximately 1,000 to 3,000 cases of plague globally each year. Although contracting bubonic plague before antibiotics meant almost certain death, the bacterium responds to several types of modern antibiotics, and mortality rates from plague are now very low.

Watch a video (http://openstaxcollege.org/l/black_death2) on the modern understanding of the Black Death (bubonic plague) in Europe during the fourteenth century.

Over the centuries, Europeans developed resistance to many infectious diseases. However, European conquerors brought disease-causing bacteria and viruses with them when they reached the Western hemisphere, triggering **epidemics** that completely devastated populations of Native Americans (who had no natural resistance to many European diseases).

1. Papagrigorakis M. J., Synodinos P. N., Yapijakis C, "Ancient typhoid epidemic reveals possible ancestral strain of *Salmonella enterica* serovar Typhi, *Infect Genet Evol* 7 (2007): 126-7.

The Antibiotic Crisis

The word antibiotic comes from the Greek *anti*, meaning "against," and *bios*, meaning "life." An antibiotic is an organism-produced chemical that is hostile to the growth of other organisms. Today's news and media often address concerns about an antibiotic crisis. Are antibiotics that were used to treat bacterial infections easily treatable in the past becoming obsolete? Are there new "superbugs"—bacteria that have evolved to become more resistant to our arsenal of antibiotics? Is this the beginning of the end of antibiotics? All of these questions challenge the healthcare community.

One of the main reasons for resistant bacteria is the overuse and incorrect use of antibiotics, such as not completing a full course of prescribed antibiotics. The incorrect use of an antibiotic results in the natural selection of resistant forms of bacteria. The antibiotic kills most of the infecting bacteria, and therefore only the resistant forms remain. These resistant forms reproduce, resulting in an increase in the proportion of resistant forms over non-resistant ones.

Another problem is the excessive use of antibiotics in livestock. The routine use of antibiotics in animal feed promotes bacterial resistance as well. In the United States, 70 percent of the antibiotics produced are fed to animals. The antibiotics are not used to prevent disease, but to enhance production of their products.

Watch a recent news (http://openstaxcollege.org/l/antibiotics2) report on the problem of routine antibiotic administration to livestock and antibiotic-resistant bacteria.

Staphylococcus aureus, often called "staph," is a common bacterium that can live in and on the human body, which usually is easily treatable with antibiotics. A very dangerous strain, however, has made the news over the past few years (Figure 13.7). This strain, **methicillin-resistant *Staphylococcus aureus* (MRSA)**, is resistant to many commonly used antibiotics, including methicillin, amoxicillin, penicillin, and oxacillin. While MRSA infections have been common among people in healthcare facilities, it is appearing more commonly in healthy people who live or work in dense groups (like military personnel and prisoners). The *Journal of the American Medical Association* reported that, among MRSA-afflicted persons in healthcare facilities, the average age is 68 years, while people with "community-associated MRSA" (CA-MRSA) have an average age of 23 years.[2]

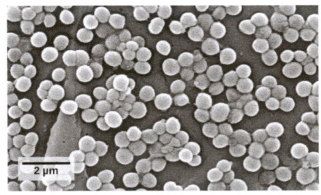

Figure 13.7 This scanning electron micrograph shows methicillin-resistant *Staphylococcus aureus* bacteria, commonly known as MRSA. (credit: modification of work by Janice Haney Carr, CDC; scale-bar data from Matt Russell)

In summary, society is facing an antibiotic crisis. Some scientists believe that after years of being protected from bacterial infections by antibiotics, we may be returning to a time in which a simple bacterial infection could again devastate the human population. Researchers are working on developing new antibiotics, but few are in the drug development pipeline, and it takes many years to generate an effective and approved drug.

---

2. Naimi, T. S., LeDell, K. H., Como-Sabetti, K., et al., "Comparison of community- and health care-associated methicillin-resistant *Staphylococcus aureus* infection," *JAMA* 290 (2003): 2976-2984, doi: 10.1001/jama.290.22.2976.

Download for free at https://openstax.org/details/books/concepts-biology

Foodborne Diseases

Prokaryotes are everywhere: They readily colonize the surface of any type of material, and food is not an exception. Outbreaks of bacterial infection related to food consumption are common. A **foodborne disease** (colloquially called "food poisoning") is an illness resulting from the consumption of food contaminated with pathogenic bacteria, viruses, or other parasites. Although the United States has one of the safest food supplies in the world, the Center for Disease Control and Prevention (CDC) has reported that "76 million people get sick, more than 300,000 are hospitalized, and 5,000 Americans die each year from foodborne illness."[3]

The characteristics of foodborne illnesses have changed over time. In the past, it was relatively common to hear about sporadic cases of **botulism**, the potentially fatal disease produced by a toxin from the anaerobic bacterium *Clostridium botulinum*. A can, jar, or package created a suitable anaerobic environment where *Clostridium* could grow. Proper sterilization and canning procedures have reduced the incidence of this disease.

Most cases of foodborne illnesses are now linked to produce contaminated by animal waste. For example, there have been serious, produce-related outbreaks associated with raw spinach in the United States and with vegetable sprouts in Germany (Figure 13.8). The raw spinach outbreak in 2006 was produced by the bacterium *E. coli* strain O157:H7. Most *E. coli* strains are not particularly dangerous to humans, (indeed, they live in our large intestine), but O157:H7 is potentially fatal.

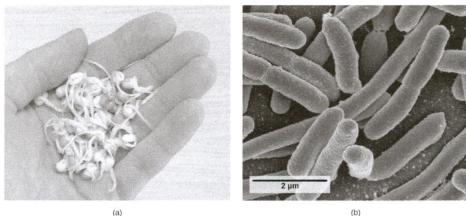

(a)   (b)

Figure 13.8 (a) Locally grown vegetable sprouts were the cause of a European *E. coli* outbreak that killed 31 people and sickened about 3,000 in 2010. (b) *Escherichia coli* are shown here in a scanning electron micrograph. The strain of *E. coli* that caused a deadly outbreak in Germany is a new one not involved in any previous *E. coli* outbreaks. It has acquired several antibiotic resistance genes and specific genetic sequences involved in aggregation ability and virulence. It has recently been sequenced. (credit b: Rocky Mountain Laboratories, NIAID, NIH; scale-bar data from Matt Russell)

All types of food can potentially be contaminated with harmful bacteria of different species. Recent outbreaks of *Salmonella* reported by the CDC occurred in foods as diverse as peanut butter, alfalfa sprouts, and eggs.

---

3. http://www.cdc.gov/ecoli/2006/september, Centers for Disease Control and Prevention, "Multi-state outbreak of *E. coli* O157:H7 infections from spinach," September-October (2006).

Epidemiologist

Epidemiology is the study of the occurrence, distribution, and determinants of health and disease in a population. It is, therefore, related to public health. An epidemiologist studies the frequency and distribution of diseases within human populations and environments.

Epidemiologists collect data about a particular disease and track its spread to identify the original mode of transmission. They sometimes work in close collaboration with historians to try to understand the way a disease evolved geographically and over time, tracking the natural history of pathogens. They gather information from clinical records, patient interviews, and any other available means. That information is used to develop strategies and design public health policies to reduce the incidence of a disease or to prevent its spread. Epidemiologists also conduct rapid investigations in case of an outbreak to recommend immediate measures to control it.

Epidemiologists typically have a graduate-level education. An epidemiologist often has a bachelor's degree in some field and a master's degree in public health (MPH). Many epidemiologists are also physicians (and have an MD) or they have a PhD in an associated field, such as biology or epidemiology.

Beneficial Prokaryotes

Not all prokaryotes are pathogenic. On the contrary, pathogens represent only a very small percentage of the diversity of the microbial world. In fact, our life and all life on this planet would not be possible without prokaryotes.

Prokaryotes, and Food and Beverages

According to the United Nations Convention on Biological Diversity, biotechnology is "any technological application that uses biological systems, living organisms, or derivatives thereof, to make or modify products or processes for specific use."[4] The concept of "specific use" involves some sort of commercial application. Genetic engineering, artificial selection, antibiotic production, and cell culture are current topics of study in biotechnology. However, humans have used prokaryotes to create products before the term biotechnology was even coined. And some of the goods and services are as simple as cheese, yogurt, sour cream, vinegar, cured sausage, sauerkraut, and fermented seafood that contains both bacteria and archaea (Figure 13.9).

---

4. http://www.cbd.int/convention/articles/?a=cbd-02http://www.cbd.int/convention/articles/?a=cbd-02, United Nations Convention on Biological Diversity, "Article 2: Use of Terms."

Figure 13.9 Some of the products derived from the use of prokaryotes in early biotechnology include (a) cheese, (b) salami, (c) yogurt, and (d) fish sauce. (credit b: modification of work by Alisdair McDiarmid; credit c: modification of work by Kris Miller; credit d: modification of work by Jane Whitney)

Cheese production began around 4,000 years ago when humans started to breed animals and process their milk. Evidence suggests that cultured milk products, like yogurt, have existed for at least 4,000 years.

Using Prokaryotes to Clean up Our Planet: Bioremediation

Microbial **bioremediation** is the use of prokaryotes (or microbial metabolism) to remove pollutants. Bioremediation has been used to remove agricultural chemicals (pesticides and fertilizers) that leach from soil into groundwater. Certain toxic metals, such as selenium and arsenic compounds, can also be removed from water by bioremediation. The reduction of SeO_4^{2-} to SeO_3^{2-} and to Se^0 (metallic selenium) is a method used to remove selenium ions from water. Mercury is an example of a toxic metal that can be removed from an environment by bioremediation. Mercury is an active ingredient of some pesticides; it is used in industry and is also a byproduct of certain industries, such as battery production. Mercury is usually present in very low concentrations in natural environments but it is highly toxic because it accumulates in living tissues. Several species of bacteria can carry out the biotransformation of toxic mercury into nontoxic forms. These bacteria, such as *Pseudomonas aeruginosa*, can convert Hg^{2+} to Hg^0, which is nontoxic to humans.

Probably one of the most useful and interesting examples of the use of prokaryotes for bioremediation purposes is the cleanup of oil spills. The importance of prokaryotes to petroleum bioremediation has been demonstrated in several oil spills in recent years, such as the Exxon Valdez spill in Alaska (1989) (Figure 13.10), the Prestige oil spill in Spain (2002), the spill into the Mediterranean from a Lebanon power plant (2006,) and more recently, the BP oil spill in the Gulf of Mexico (2010). To clean up these spills, bioremediation is promoted by adding inorganic nutrients that help bacteria already present in the environment to grow. Hydrocarbon-degrading bacteria feed on the hydrocarbons in the oil droplet, breaking them into inorganic compounds. Some species, such as *Alcanivorax borkumensis*, produce surfactants that solubilize the oil, while other bacteria degrade the oil into carbon dioxide. In the case of oil spills in the ocean, ongoing, natural bioremediation tends to occur, inasmuch as there are oil-consuming bacteria in the ocean prior to the spill. Under ideal conditions, it has been reported that up to 80 percent of the nonvolatile components in oil can be degraded within 1 year of the spill. Other oil fractions containing aromatic and highly branched hydrocarbon chains are more difficult to remove and remain in the environment for longer periods of time. Researchers have genetically engineered other bacteria to consume petroleum products; indeed, the first patent application for a bioremediation application in the U.S. was for a genetically modified oil-eating bacterium.

(a)   (b)

Figure 13.10 (a) Cleaning up oil after the Valdez spill in Alaska, the workers hosed oil from beaches and then used a floating boom to corral the oil, which was finally skimmed from the water surface. Some species of bacteria are able to solubilize and degrade the oil. (b) One of the most catastrophic consequences of oil spills is the damage to fauna. (credit a: modification of work by NOAA; credit b: modification of work by GOLUBENKOV, NGO: Saving Taman)

Prokaryotes in and on the Body

Humans are no exception when it comes to forming symbiotic relationships with prokaryotes. We are accustomed to thinking of ourselves as single organisms, but in reality, we are walking ecosystems. There are 10 to 100 times as many bacterial and archaeal cells inhabiting our bodies as we have cells in our bodies. Some of these are in mutually beneficial relationships with us, in which both the human host and the bacterium benefit, while some of the relationships are classified as **commensalism**, a type of relationship in which the bacterium benefits and the human host is neither benefited nor harmed.

Human gut flora lives in the large intestine and consists of hundreds of species of bacteria and archaea, with different individuals containing different species mixes. The term "flora," which is usually associated with plants, is traditionally used in this context because bacteria were once classified as plants. The primary functions of these prokaryotes for humans appear to be metabolism of food molecules that we cannot break down, assistance with the absorption of ions by the colon, synthesis of vitamin K, training of the infant immune system, maintenance of the adult immune system, maintenance of the epithelium of the large intestine, and formation of a protective barrier against pathogens.

The surface of the skin is also coated with prokaryotes. The different surfaces of the skin, such as the underarms, the head, and the hands, provide different habitats for different communities of prokaryotes. Unlike with gut flora, the possible beneficial roles of skin flora have not been well studied. However, the few studies conducted so far have identified bacteria that produce antimicrobial compounds as probably responsible for preventing infections by pathogenic bacteria.

Researchers are actively studying the relationships between various diseases and alterations to the composition of human microbial flora. Some of this work is being carried out by the Human Microbiome Project, funded in the United States by the National Institutes of Health.

13.2 | Eukaryotic Origins

By the end of this section, you will be able to:

- Describe the endosymbiotic theory
- Explain the origin of mitochondria and chloroplasts

The fossil record and genetic evidence suggest that prokaryotic cells were the first organisms on Earth. These cells originated approximately 3.5 billion years ago, which was about 1 billion years after Earth's formation, and were the only life forms on the planet until eukaryotic cells emerged approximately 2.1 billion years ago. During the prokaryotic reign, photosynthetic prokaryotes evolved that were capable of applying the energy from sunlight to synthesize organic materials (like carbohydrates) from carbon dioxide and an electron source (such as hydrogen, hydrogen sulfide, or water).

Photosynthesis using water as an electron donor consumes carbon dioxide and releases molecular oxygen (O_2) as a byproduct. The functioning of photosynthetic bacteria over millions of years progressively saturated Earth's water with oxygen and then oxygenated the atmosphere, which previously contained much greater concentrations of carbon dioxide and much lower concentrations of oxygen. Older anaerobic prokaryotes of the era could not function in their new, aerobic environment. Some species perished, while others survived in the remaining anaerobic environments left on Earth. Still other early prokaryotes evolved mechanisms, such as aerobic respiration, to exploit the oxygenated atmosphere by using oxygen to store energy contained within organic molecules. Aerobic respiration is a more efficient way of obtaining energy from organic molecules, which contributed to the success of these species (as evidenced by the number and diversity of aerobic organisms living on Earth today). The evolution of aerobic prokaryotes was an important step toward the evolution of the first eukaryote, but several other distinguishing features had to evolve as well.

Endosymbiosis

The origin of eukaryotic cells was largely a mystery until a revolutionary hypothesis was comprehensively examined in the 1960s by Lynn Margulis. The **endosymbiotic theory** states that eukaryotes are a product of one prokaryotic cell engulfing another, one living within another, and evolving together over time until the separate cells were no longer recognizable as such. This once-revolutionary hypothesis had immediate persuasiveness and is now widely accepted, with work progressing on uncovering the steps involved in this evolutionary process as well as the key players. It has become clear that many nuclear eukaryotic genes and the molecular machinery responsible for replicating and expressing those genes appear closely related to the Archaea. On the other hand, the metabolic organelles and the genes responsible for many energy-harvesting processes had their origins in bacteria. Much remains to be clarified about how this relationship occurred; this continues to be an exciting field of discovery in biology. Several endosymbiotic events likely contributed to the origin of the eukaryotic cell.

Mitochondria

Eukaryotic cells may contain anywhere from one to several thousand mitochondria, depending on the cell's level of energy consumption. Each mitochondrion measures 1 to 10 micrometers in length and exists in the cell as a moving, fusing, and dividing oblong spheroid (Figure 13.11). However, mitochondria cannot survive outside the cell. As the atmosphere was oxygenated by photosynthesis, and as successful aerobic prokaryotes evolved, evidence suggests that an ancestral cell engulfed and kept alive a free-living, aerobic prokaryote. This gave the host cell the ability to use oxygen to release energy stored in nutrients. Several lines of evidence support that mitochondria are derived from this endosymbiotic event. Mitochondria are shaped like a specific group of bacteria and are surrounded by two membranes, which would result when one membrane-bound organism was engulfed by another membrane-bound organism. The mitochondrial inner membrane involves substantial infoldings or cristae that resemble the textured outer surface of certain bacteria.

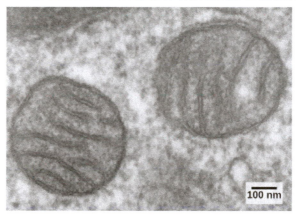

Figure 13.11 In this transmission electron micrograph of mitochondria in a mammalian lung cell, the cristae, infoldings of the mitochondrial inner membrane, can be seen in cross-section. (credit: modification of work by Louisa Howard; scale-bar data from Matt Russell)

Mitochondria divide on their own by a process that resembles binary fission in prokaryotes. Mitochondria have their own circular DNA chromosome that carries genes similar to those expressed by bacteria. Mitochondria also have special ribosomes and transfer RNAs that resemble these components in prokaryotes. These features all support that mitochondria were once free-living prokaryotes.

Chloroplasts

Chloroplasts are one type of **plastid**, a group of related organelles in plant cells that are involved in the storage of starches, fats, proteins, and pigments. Chloroplasts contain the green pigment chlorophyll and play a role in photosynthesis. Genetic and morphological studies suggest that plastids evolved from the endosymbiosis of an ancestral cell that engulfed a photosynthetic cyanobacterium. Plastids are similar in size and shape to cyanobacteria and are enveloped by two or more membranes, corresponding to the inner and outer membranes of cyanobacteria. Like mitochondria, plastids also contain circular genomes and divide by a process reminiscent of prokaryotic cell division. The chloroplasts of red and green algae exhibit DNA sequences that are closely related to photosynthetic cyanobacteria, suggesting that red and green algae are direct descendants of this endosymbiotic event.

Mitochondria likely evolved before plastids because all eukaryotes have either functional mitochondria or mitochondria-like organelles. In contrast, plastids are only found in a subset of eukaryotes, such as terrestrial plants and algae. One hypothesis of the evolutionary steps leading to the first eukaryote is summarized in Figure 13.12.

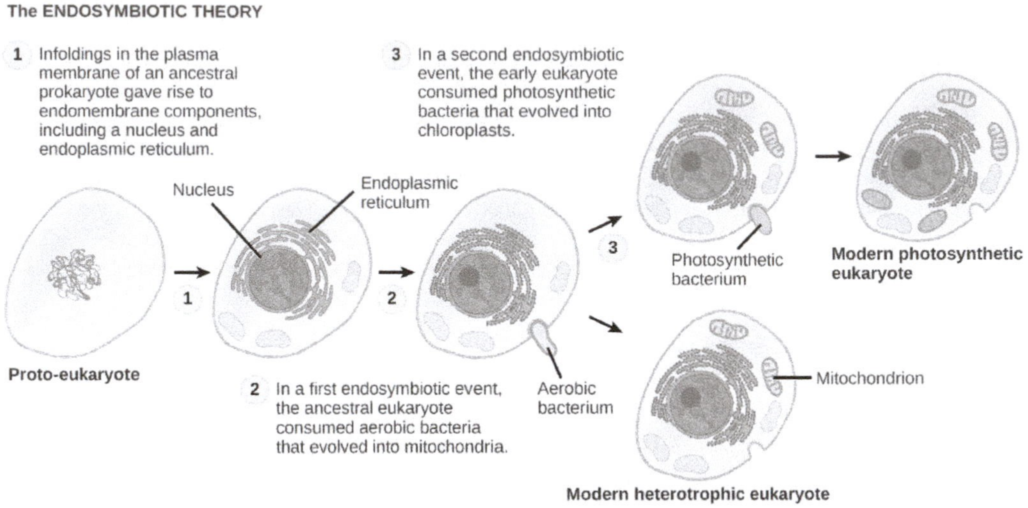

Figure 13.12 The first eukaryote may have originated from an ancestral prokaryote that had undergone membrane proliferation, compartmentalization of cellular function (into a nucleus, lysosomes, and an endoplasmic reticulum), and the establishment of endosymbiotic relationships with an aerobic prokaryote and, in some cases, a photosynthetic prokaryote to form mitochondria and chloroplasts, respectively.

The exact steps leading to the first eukaryotic cell can only be hypothesized, and some controversy exists regarding which events actually took place and in what order. Spirochete bacteria have been hypothesized to have given rise to microtubules, and a flagellated prokaryote may have contributed the raw materials for eukaryotic flagella and cilia. Other scientists suggest that membrane proliferation and compartmentalization, not endosymbiotic events, led to the development of mitochondria and plastids. However, the vast majority of studies support the endosymbiotic hypothesis of eukaryotic evolution.

The early eukaryotes were unicellular like most protists are today, but as eukaryotes became more complex, the evolution of multicellularity allowed cells to remain small while still exhibiting specialized functions. The ancestors of today's multicellular eukaryotes are thought to have evolved about 1.5 billion years ago.

13.3 | Protists

By the end of this section, you will be able to:

- Describe the main characteristics of protists
- Describe important pathogenic species of protists
- Describe the roles of protists as food sources and as decomposers

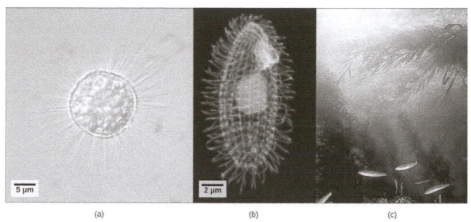

Figure 13.13 Protists range from the microscopic, single-celled (a) *Acanthocystis turfacea* and the (b) ciliate *Tetrahymena thermophila* to the enormous, multicellular (c) kelps (Chromalveolata) that extend for hundreds of feet in underwater "forests." (credit a: modification of work by Yuiuji Tsukii; credit b: modification of work by Richard Robinson, Public Library of Science; credit c: modification of work by Kip Evans, NOAA; scale-bar data from Matt Russell)

Eukaryotic organisms that did not fit the criteria for the kingdoms Animalia, Fungi, or Plantae historically were called protists and were classified into the kingdom Protista. Protists include the single-celled eukaryotes living in pond water (Figure 13.13), although protist species live in a variety of other aquatic and terrestrial environments, and occupy many different niches. Not all protists are microscopic and single-celled; there exist some very large multicellular species, such as the kelps. During the past two decades, the field of molecular genetics has demonstrated that some protists are more related to animals, plants, or fungi than they are to other protists. For this reason, protist lineages originally classified into the kingdom Protista have been reassigned into new kingdoms or other existing kingdoms. The evolutionary lineages of the protists continue to be examined and debated. In the meantime, the term "protist" still is used informally to describe this tremendously diverse group of eukaryotes. As a collective group, protists display an astounding diversity of morphologies, physiologies, and ecologies.

Characteristics of Protists

There are over 100,000 described living species of protists, and it is unclear how many undescribed species may exist. Since many protists live in symbiotic relationships with other organisms and these relationships are often species specific, there is a huge potential for undescribed protist diversity that matches the diversity of the hosts. As the catchall term for eukaryotic organisms that are not animals, plants, fungi, or any single phylogenetically related group, it is not surprising that few characteristics are common to all protists.

Nearly all protists exist in some type of aquatic environment, including freshwater and marine environments, damp soil, and even snow. Several protist species are **parasites** that infect animals or plants. A parasite is an organism that lives on or in another organism and feeds on it, often without killing it. A few protist species live on dead organisms or their wastes, and contribute to their decay.

Protist Structure

The cells of protists are among the most elaborate of all cells. Most protists are microscopic and unicellular, but some true multicellular forms exist. A few protists live as colonies that behave in some ways as a group of free-living cells and in other ways as a multicellular organism. Still other protists are composed of enormous, multinucleate, single cells that look like amorphous blobs of slime or, in other cases, like ferns. In fact, many protist cells are multinucleated; in some species, the nuclei are different sizes and have distinct roles in protist cell function.

Single protist cells range in size from less than a micrometer to the 3-meter lengths of the multinucleate cells of the seaweed *Caulerpa*. Protist cells may be enveloped by animal-like cell membranes or plant-like cell walls. Others are encased in glassy silica-based shells or wound with **pellicles** of interlocking protein strips. The pellicle functions like a flexible coat of armor, preventing the protist from being torn or pierced without compromising its range of motion.

The majority of protists are motile, but different types of protists have evolved varied modes of movement. Some protists have one or more flagella, which they rotate or whip. Others are covered in rows or tufts of tiny cilia that they beat in coordination to swim. Still others send out lobe-like pseudopodia from anywhere on the cell, anchor the pseudopodium to a substrate, and pull the rest of the cell toward the anchor point. Some protists can move toward light by coupling their locomotion strategy with a light-sensing organ.

How Protists Obtain Energy

Protists exhibit many forms of nutrition and may be aerobic or anaerobic. Photosynthetic protists (photoautotrophs) are characterized by the presence of chloroplasts. Other protists are heterotrophs and consume organic materials (such as other organisms) to obtain nutrition. Amoebas and some other heterotrophic protist species ingest particles by a process called phagocytosis, in which the cell membrane engulfs a food particle and brings it inward, pinching off an intracellular membranous sac, or vesicle, called a food vacuole (Figure 13.14). This vesicle then fuses with a lysosome, and the food particle is broken down into small molecules that can diffuse into the cytoplasm and be used in cellular metabolism. Undigested remains ultimately are expelled from the cell through exocytosis.

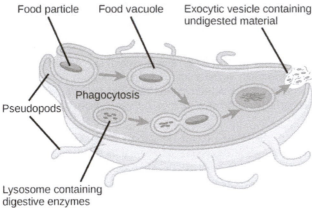

Figure 13.14 The stages of phagocytosis include the engulfment of a food particle, the digestion of the particle using hydrolytic enzymes contained within a lysosome, and the expulsion of undigested material from the cell.

Some heterotrophs absorb nutrients from dead organisms or their organic wastes, and others are able to use photosynthesis or feed on organic matter, depending on conditions.

Reproduction

Protists reproduce by a variety of mechanisms. Most are capable some form of asexual reproduction, such as binary fission to produce two daughter cells, or multiple fission to divide simultaneously into many daughter cells. Others produce tiny buds that go on to divide and grow to the size of the parental protist. Sexual reproduction, involving meiosis and fertilization, is common among protists, and many protist species can switch from asexual to sexual reproduction when necessary. Sexual reproduction is often associated with periods when nutrients are depleted or environmental changes occur. Sexual reproduction may allow the protist to recombine genes and produce new variations of progeny that may be better suited to surviving in the new environment. However, sexual reproduction is also often associated with cysts that are a protective, resting stage. Depending on their habitat, the cysts may be particularly resistant to temperature extremes, desiccation, or low pH. This strategy also allows certain protists to "wait out" stressors until their environment becomes more favorable for survival or until they are carried (such as by wind, water, or transport on a larger organism) to a different environment because cysts exhibit virtually no cellular metabolism.

Protist Diversity

With the advent of DNA sequencing, the relationships among protist groups and between protist groups and other eukaryotes are beginning to become clearer. Many relationships that were based on morphological similarities are being replaced by new relationships based on genetic similarities. Protists that exhibit similar morphological features may have evolved analogous structures because of similar selective pressures—rather than because of recent common ancestry. This phenomenon is called convergent evolution. It is one reason why protist classification is so challenging. The emerging classification scheme groups the entire domain Eukaryota into six "supergroups" that contain all of the protists as well as animals, plants, and fungi (Figure 13.15); these include the **Excavata**, **Chromalveolata**, **Rhizaria**, **Archaeplastida**, **Amoebozoa**, and **Opisthokonta**. The supergroups are believed to be monophyletic; all organisms within each supergroup are believed to have evolved from a single common ancestor, and thus all members are most closely related to each other than to organisms outside that group. There is still evidence lacking for the monophyly of some groups.

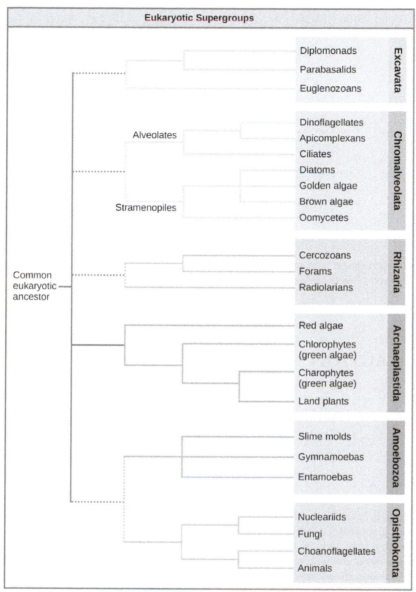

Figure 13.15 Protists appear in all six eukaryotic supergroups.

Human Pathogens

Many protists are pathogenic parasites that must infect other organisms to survive and propagate. Protist parasites include the causative agents of malaria, African sleeping sickness, and waterborne gastroenteritis in humans. Other protist pathogens prey on plants, effecting massive destruction of food crops.

Plasmodium Species

Members of the genus *Plasmodium* must infect a mosquito and a vertebrate to complete their life cycle. In vertebrates, the parasite develops in liver cells and goes on to infect red blood cells, bursting from and destroying the blood cells with each asexual replication cycle (Figure 13.16). Of the four *Plasmodium* species known to infect humans, *P. falciparum* accounts for 50 percent of all malaria cases and is the primary cause of disease-related fatalities in tropical regions of the world. In 2010, it was estimated that malaria caused between 0.5 and 1 million deaths, mostly in African children. During the course of malaria, *P. falciparum* can infect and destroy more than one-half of a human's circulating blood cells, leading to severe anemia. In response to waste products released as the parasites burst from infected blood cells, the host immune system mounts a massive inflammatory response with delirium-inducing fever episodes, as parasites destroy red blood cells, spilling parasite waste into the blood stream. *P. falciparum* is transmitted to humans by the African malaria mosquito, *Anopheles gambiae*. Techniques to kill, sterilize, or avoid exposure to this highly aggressive mosquito species are crucial to malaria control.

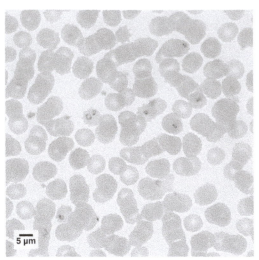

Figure 13.16 This light micrograph shows a 100× magnification of red blood cells infected with *P. falciparum* (seen as purple). (credit: modification of work by Michael Zahniser; scale-bar data from Matt Russell)

This movie (http://openstaxcollege.org/l/malaria2) depicts the pathogenesis of *Plasmodium falciparum*, the causative agent of malaria.

Trypanosomes

T. brucei, the parasite that is responsible for African sleeping sickness, confounds the human immune system by changing its thick layer of surface glycoproteins with each infectious cycle (Figure 13.17). The glycoproteins are identified by the immune system as foreign matter, and a specific antibody defense is mounted against the parasite. However, *T. brucei* has thousands of possible antigens, and with each subsequent generation, the protist switches to a glycoprotein coating with a different molecular structure. In this way, *T. brucei* is capable of replicating continuously without the immune system ever succeeding in clearing the parasite. Without treatment, African sleeping sickness leads invariably to death because of damage it does to the nervous system. During epidemic periods, mortality from the disease can be high. Greater surveillance and control measures have led to a reduction in reported cases; some of the lowest numbers reported in 50 years (fewer than 10,000 cases in all of sub-Saharan Africa) have happened since 2009.

In Latin America, another species in the genus, *T. cruzi*, is responsible for Chagas disease. *T. cruzi* infections are mainly caused by a blood-sucking bug. The parasite inhabits heart and digestive system tissues in the chronic phase of infection, leading to malnutrition and heart failure caused by abnormal heart rhythms. An estimated 10 million people are infected with Chagas disease, which caused 10,000 deaths in 2008.

Download for free at https://openstax.org/details/books/concepts-biology

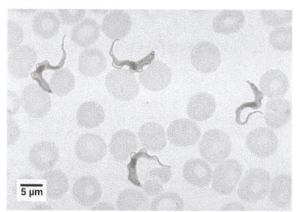

Figure 13.17 Trypanosomes are shown in this light micrograph among red blood cells. (credit: modification of work by Myron G. Schultz, CDC; scale-bar data from Matt Russell)

This movie (http://openstaxcollege.org/l/African_sleep2) discusses the pathogenesis of *Trypanosoma brucei*, the causative agent of African sleeping sickness.

Plant Parasites

Protist parasites of terrestrial plants include agents that destroy food crops. The oomycete *Plasmopara viticola* parasitizes grape plants, causing a disease called downy mildew (Figure 13.18a). Grape plants infected with *P. viticola* appear stunted and have discolored withered leaves. The spread of downy mildew caused the near collapse of the French wine industry in the nineteenth century.

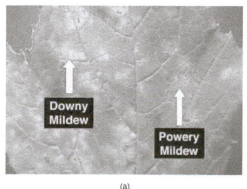 

(a)   (b)

Figure 13.18 (a) The downy and powdery mildews on this grape leaf are caused by an infection of *P. viticola*. (b) This potato exhibits the results of an infection with *P. infestans*, the potato late blight. (credit a: modification of work by David B. Langston, University of Georgia, USDA ARS; credit b: USDA ARS)

Phytophthora infestans is an oomycete responsible for potato late blight, which causes potato stalks and stems to decay into black slime (Figure 13.18b). Widespread potato blight caused by *P. infestans* precipitated the well-known Irish potato famine in the nineteenth century that claimed the lives of approximately 1 million people and led to the emigration from Ireland of at least 1 million more. Late blight continues to plague potato crops in certain parts of the United States and Russia, wiping out as much as 70 percent of crops when no pesticides are applied.

Beneficial Protists

Protists play critically important ecological roles as producers particularly in the world's oceans. They are equally important on the other end of food webs as decomposers.

Protists as Food Sources

Protists are essential sources of nutrition for many other organisms. In some cases, as in plankton, protists are consumed directly. Alternatively, photosynthetic protists serve as producers of nutrition for other organisms by carbon fixation. For instance, photosynthetic dinoflagellates called zooxanthellae pass on most of their energy to the coral polyps that house them (Figure 13.19). In this mutually beneficial relationship, the polyps provide a protective environment and nutrients for the zooxanthellae. The polyps secrete the calcium carbonate that builds coral reefs. Without dinoflagellate symbionts, corals lose algal pigments in a process called coral bleaching, and they eventually die. This explains why reef-building corals do not reside in waters deeper than 20 meters: Not enough light reaches those depths for dinoflagellates to photosynthesize.

Figure 13.19 Coral polyps obtain nutrition through a symbiotic relationship with dinoflagellates.

Protists themselves and their products of photosynthesis are essential—directly or indirectly—to the survival of organisms ranging from bacteria to mammals. As primary producers, protists feed a large proportion of the world's aquatic species. (On land, terrestrial plants serve as primary producers.) In fact, approximately one-quarter of the world's photosynthesis is conducted by protists, particularly dinoflagellates, diatoms, and multicellular algae.

Protists do not create food sources only for sea-dwelling organisms. For instance, certain anaerobic species exist in the digestive tracts of termites and wood-eating cockroaches, where they contribute to digesting cellulose ingested by these insects as they bore through wood. The actual enzyme used to digest the cellulose is actually produced by bacteria living within the protist cells. The termite provides the food source to the protist and its bacteria, and the protist and bacteria provide nutrients to the termite by breaking down the cellulose.

Agents of Decomposition

Many fungus-like protists are **saprobes**, organisms that feed on dead organisms or the waste matter produced by organisms (saprophyte is an equivalent term), and are specialized to absorb nutrients from nonliving organic matter. For instance, many types of oomycetes grow on dead animals or algae. Saprobic protists have the essential function of returning inorganic nutrients to the soil and water. This process allows for new plant growth, which in turn generates sustenance for other organisms along the food chain. Indeed, without saprobic species, such as protists, fungi, and bacteria, life would cease to exist as all organic carbon became "tied up" in dead organisms.

13.4 | Fungi

By the end of this section, you will be able to:

* List the characteristics of fungi
* Describe fungal parasites and pathogens of plants and infections in humans
* Describe the importance of fungi to the environment
* Summarize the beneficial role of fungi in food and beverage preparation and in the chemical and pharmaceutical industry

Figure 13.20 The (a) familiar mushroom is only one type of fungus. The brightly colored fruiting bodies of this (b) coral fungus are displayed. This (c) electron micrograph shows the spore-bearing structures of *Aspergillus*, a type of toxic fungi found mostly in soil and plants. (credit a: modification of work by Chris Wee; credit b: modification of work by Cory Zanker; credit c: modification of work by Janice Haney Carr, Robert Simmons, CDC; scale-bar data from Matt Russell)

The word *fungus* comes from the Latin word for mushroom. Indeed, the familiar mushrooms are fungi, but there are many other types of fungi as well (Figure 13.20). The kingdom Fungi includes an enormous variety of living organisms collectively referred to as Eumycota, or true fungi. While scientists have identified about 100,000 species of fungi, this is only a fraction of the over 1 million species likely present on Earth. Edible mushrooms, yeasts, black mold, and *Penicillium notatum* (the producer of the antibiotic penicillin) are all members of the kingdom Fungi, which belongs to the domain Eukarya. As eukaryotes, a typical fungal cell contains a true nucleus and many membrane-bound organelles.

Fungi were once considered plant-like organisms; however, DNA comparisons have shown that fungi are more closely related to animals than plants. Fungi are not capable of photosynthesis: They use complex organic compounds as sources of energy and carbon. Some fungal organisms multiply only asexually, whereas others undergo both asexual reproduction and sexual reproduction. Most fungi produce a large number of spores that are disseminated by the wind. Like bacteria, fungi play an essential role in ecosystems, because they are decomposers and participate in the cycling of nutrients by breaking down organic materials into simple molecules.

Fungi often interact with other organisms, forming mutually beneficial or mutualistic associations. Fungi also cause serious infections in plants and animals. For example, Dutch elm disease is a particularly devastating fungal infection that destroys many native species of elm (*Ulmus* spp.). The fungus infects the vascular system of the tree. It was accidentally introduced to North America in the 1900s and decimated elm trees across the continent. Dutch elm disease is caused by the fungus *Ophiostoma ulmi*. The elm bark beetle acts as a vector and transmits the disease from tree to tree. Many European and Asiatic elms are less susceptible than American elms.

In humans, fungal infections are generally considered challenging to treat because, unlike bacteria, they do not respond to traditional antibiotic therapy since they are also eukaryotes. These infections may prove deadly for individuals with a compromised immune system.

Fungi have many commercial applications. The food industry uses yeasts in baking, brewing, and wine making. Many industrial compounds are byproducts of fungal fermentation. Fungi are the source of many commercial enzymes and antibiotics.

Cell Structure and Function

Fungi are eukaryotes and as such have a complex cellular organization. As eukaryotes, fungal cells contain a membrane-bound nucleus. A few types of fungi have structures comparable to the plasmids (loops of DNA) seen in bacteria. Fungal cells also contain mitochondria and a complex system of internal membranes, including the endoplasmic reticulum and Golgi apparatus.

Fungal cells do not have chloroplasts. Although the photosynthetic pigment chlorophyll is absent, many fungi display bright colors, ranging from red to green to black. The poisonous *Amanita muscaria* (fly agaric) is recognizable by its bright red cap with white patches (Figure 13.21). Pigments in fungi are associated with the cell wall and play a protective role against ultraviolet radiation. Some pigments are toxic.

Figure 13.21 The poisonous *Amanita muscaria* is native to the temperate and boreal regions of North America. (credit: Christine Majul)

Like plant cells, fungal cells are surrounded by a thick cell wall; however, the rigid layers contain the complex polysaccharides chitin and glucan and not cellulose that is used by plants. Chitin, also found in the exoskeleton of insects, gives structural strength to the cell walls of fungi. The cell wall protects the cell from desiccation and predators. Fungi have plasma membranes similar to other eukaryotes, except that the structure is stabilized by ergosterol, a steroid molecule that functions like the cholesterol found in animal cell membranes. Most members of the kingdom Fungi are nonmotile. Flagella are produced only by the gametes in the primitive division Chytridiomycota.

Growth and Reproduction

The vegetative body of a fungus is called a **thallus** and can be unicellular or multicellular. Some fungi are dimorphic because they can go from being unicellular to multicellular depending on environmental conditions. Unicellular fungi are generally referred to as **yeasts**.*Saccharomyces cerevisiae* (baker's yeast) and *Candida* species (the agents of thrush, a common fungal infection) are examples of unicellular fungi.

Most fungi are multicellular organisms. They display two distinct morphological stages: vegetative and reproductive. The vegetative stage is characterized by a tangle of slender thread-like structures called hyphae (singular, **hypha**), whereas the reproductive stage can be more conspicuous. A mass of hyphae is called a **mycelium** (Figure 13.22). It can grow on a surface, in soil or decaying material, in a liquid, or even in or on living tissue. Although individual hypha must be observed under a microscope, the mycelium of a fungus can be very large with some species truly being "the fungus humongous." The giant *Armillaria ostoyae* (honey mushroom) is considered the largest organism on Earth, spreading across over 2,000 acres of underground soil in eastern Oregon; it is estimated to be at least 2,400 years old.

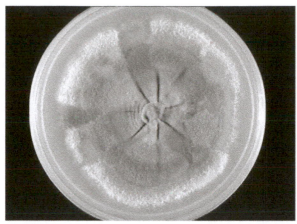

Figure 13.22 The mycelium of the fungus *Neotestudina rosati* can be pathogenic to humans. The fungus enters through a cut or scrape and develops into a mycetoma, a chronic subcutaneous infection. (credit: CDC)

Most fungal hyphae are divided into separate cells by end walls called septa (singular, **septum**). In most divisions (like plants, fungal phyla are called *divisions* by tradition) of fungi, tiny holes in the septa allow for the rapid flow of nutrients and small molecules from cell to cell along the hyphae. They are described as perforated septa. The hyphae in bread **molds** (which belong to the division Zygomycota) are not separated by septa. They are formed of large cells containing many nuclei, an arrangement described as coenocytic hyphae.

Fungi thrive in environments that are moist and slightly acidic, and can grow with or without light. They vary in their oxygen requirements. Most fungi are obligate aerobes, requiring oxygen to survive. Other species, such as the Chytridiomycota that reside in the rumen of cattle, are obligate anaerobes, meaning that they cannot grow and reproduce in an environment with oxygen. Yeasts are intermediate: They grow best in the presence of oxygen but can use fermentation in the absence of oxygen. The alcohol produced from yeast fermentation is used in wine and beer production, and the carbon dioxide they produce carbonates beer and sparkling wine, and makes bread rise.

Fungi can reproduce sexually or asexually. In both sexual and asexual reproduction, fungi produce spores that disperse from the parent organism by either floating in the wind or hitching a ride on an animal. Fungal spores are smaller and lighter than plant seeds, but they are not usually released as high in the air. The giant puffball mushroom bursts open and releases trillions of spores: The huge number of spores released increases the likelihood of spores landing in an environment that will support growth (Figure 13.23).

(a)   (b)

Figure 13.23 The (a) giant puffball mushroom releases (b) a cloud of spores when it reaches maturity. (credit a: modification of work by Roger Griffith; credit b: modification of work by Pearson Scott Foresman, donated to the Wikimedia Foundation)

How Fungi Obtain Nutrition

Like animals, fungi are heterotrophs: They use complex organic compounds as a source of carbon rather than fixing carbon dioxide from the atmosphere, as some bacteria and most plants do. In addition, fungi do not fix nitrogen from the atmosphere. Like animals, they must obtain it from their diet. However, unlike most animals that ingest food and then digest it internally in specialized organs, fungi perform these steps in the reverse order. Digestion precedes ingestion. First, exoenzymes, enzymes that catalyze reactions on compounds outside of the cell, are transported out of the hyphae where they break down nutrients in the environment. Then, the smaller molecules produced by the external digestion are absorbed through the large surface areas of the mycelium. As with animal cells, the fungal storage polysaccharide is glycogen rather than starch, as found in plants.

Fungi are mostly saprobes, organisms that derive nutrients from decaying organic matter. They obtain their nutrients from dead or decomposing organic matter, mainly plant material. Fungal exoenzymes are able to break down insoluble polysaccharides, such as the cellulose and lignin of dead wood, into readily absorbable glucose molecules. Decomposers are important components of ecosystems, because they return nutrients locked in dead bodies to a form that is usable for other organisms. This role is discussed in more detail later. Because of their varied metabolic pathways, fungi fulfill an important ecological role and are being investigated as potential tools in bioremediation. For example, some species of fungi can be used to break down diesel oil and polycyclic aromatic hydrocarbons. Other species take up heavy metals such as cadmium and lead.

Fungal Diversity

The kingdom Fungi contains four major divisions that were established according to their mode of sexual reproduction. Polyphyletic, unrelated fungi that reproduce without a sexual cycle, are placed for convenience in a fifth division, and a sixth major fungal group that does not fit well with any of the previous five has recently been described. Not all mycologists agree with this scheme. Rapid advances in molecular biology and the sequencing of 18S rRNA (a component of ribosomes) continue to reveal new and different relationships between the various categories of fungi.

The traditional divisions of Fungi are the **Chytridiomycota** (chytrids), the **Zygomycota** (conjugated fungi), the **Ascomycota** (sac fungi), and the **Basidiomycota** (club fungi). An older classification scheme grouped fungi that strictly use asexual reproduction into Deuteromycota, a group that is no longer in use. The **Glomeromycota** belong to a newly described group (Figure 13.24).

Figure 13.24 Divisions of fungi include (a) chytrids, (b) conjugated fungi, (c) sac fungi, and (d) club fungi. (credit a: modification of work by USDA APHIS PPQ; credit c: modification of work by "icelight"/Flickr; credit d: modification of work by Cory Zanker.)

Pathogenic Fungi

Many fungi have negative impacts on other species, including humans and the organisms they depend on for food. Fungi may be parasites, pathogens, and, in a very few cases, predators.

Plant Parasites and Pathogens

The production of enough good-quality crops is essential to our existence. Plant diseases have ruined crops, bringing widespread famine. Most plant pathogens are fungi that cause tissue decay and eventual death of the host (Figure 13.25). In addition to destroying plant tissue directly, some plant pathogens spoil crops by producing potent toxins. Fungi are also responsible for food spoilage and the rotting of stored crops. For example, the fungus *Claviceps purpurea* causes ergot, a disease of cereal crops (especially of rye). Although the fungus reduces the yield of cereals, the effects of the ergot's alkaloid toxins on humans and animals are of much greater significance: In animals, the disease is referred to as ergotism. The most common signs and symptoms are convulsions, hallucination, gangrene, and loss of milk in cattle. The active ingredient of ergot is lysergic acid, which is a precursor of the drug LSD. Smuts, rusts, and powdery or downy mildew are other examples of common fungal pathogens that affect crops.

Figure 13.25 Some fungal pathogens include (a) green mold on grapefruit, (b) fungus on grapes, (c) powdery mildew on a zinnia, and (d) stem rust on a sheaf of barley. Notice the brownish color of the fungus in (b) *Botrytis cinerea*, also referred to as the "noble rot," which grows on grapes and other fruit. Controlled infection of grapes by *Botrytis* is used to produce strong and much-prized dessert wines. (credit a: modification of work by Scott Bauer, USDA ARS; credit b: modification of work by Stephen Ausmus, USDA ARS; credit c: modification of work by David Marshall, USDA ARS; credit d: modification of work by Joseph Smilanick, USDA ARS)

Aflatoxins are toxic and carcinogenic compounds released by fungi of the genus *Aspergillus*. Periodically, harvests of nuts and grains are tainted by aflatoxins, leading to massive recall of produce, sometimes ruining producers, and causing food shortages in developing countries.

Animal and Human Parasites and Pathogens

Fungi can affect animals, including humans, in several ways. Fungi attack animals directly by colonizing and destroying tissues. Humans and other animals can be poisoned by eating toxic mushrooms or foods contaminated by fungi. In addition, individuals who display hypersensitivity to molds and spores develop strong and dangerous allergic reactions. Fungal infections are generally very difficult to treat because, unlike bacteria, fungi are eukaryotes. Antibiotics only target prokaryotic cells, whereas compounds that kill fungi also adversely affect the eukaryotic animal host.

Many fungal infections (**mycoses**) are superficial and termed cutaneous (meaning "skin") mycoses. They are usually visible on the skin of the animal. Fungi that cause the superficial mycoses of the epidermis, hair, and nails rarely spread to the underlying tissue (Figure 13.26). These fungi are often misnamed "dermatophytes" from the Greek *dermis* skin and *phyte* plant, but they are not plants. Dermatophytes are also called "ringworms" because of the red ring that they cause on skin (although the ring is caused by fungi, not a worm). These fungi secrete extracellular enzymes that break down keratin (a protein found in hair, skin, and nails), causing a number of conditions such as athlete's foot, jock itch, and other cutaneous fungal infections. These conditions are usually treated with over-the-counter topical creams and powders, and are easily cleared. More persistent, superficial mycoses may require prescription oral medications.

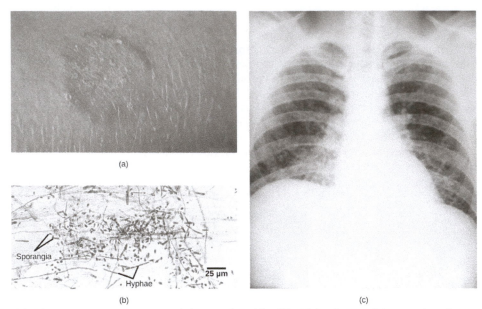

Figure 13.26 (a) Ringworm presents as a red ring on the skin. (b) *Trichophyton violaceum* is a fungus that causes superficial mycoses on the scalp. (c) *Histoplasma capsulatum*, seen in this X-ray as speckling of light areas in the lung, is a species of Ascomycota that infects airways and causes symptoms similar to the flu. (credit a, b: modification of work by Dr. Lucille K. Georg, CDC; credit c: modification of work by M Renz, CDC; scale-bar data from Matt Russell)

Systemic mycoses spread to internal organs, most commonly entering the body through the respiratory system. For example, coccidioidomycosis (valley fever) is commonly found in the southwestern United States, where the fungus resides in the dust. Once inhaled, the spores develop in the lungs and cause signs and symptoms similar to those of tuberculosis. Histoplasmosis (Figure 13.26c) is caused by the dimorphic fungus *Histoplasma capsulatum*; it causes pulmonary infections and, in rare cases, swelling of the membranes of the brain and spinal cord. Treatment of many fungal diseases requires the use of antifungal medications that have serious side effects.

Opportunistic mycoses are fungal infections that are either common in all environments or part of the normal biota. They affect mainly individuals who have a compromised immune system. Patients in the late stages of AIDS suffer from opportunistic mycoses, such as *Pneumocystis*, which can be life threatening. The yeast *Candida* spp., which is a common member of the natural biota, can grow unchecked if the pH, the immune defenses, or the normal population of bacteria is altered, causing yeast infections of the vagina or mouth (oral thrush).

Fungi may even take on a predatory lifestyle. In soil environments that are poor in nitrogen, some fungi resort to predation of nematodes (small roundworms). Species of *Arthrobotrys* fungi have a number of mechanisms to trap nematodes. For example, they have constricting rings within their network of hyphae. The rings swell when the nematode touches it and closes around the body of the nematode, thus trapping it. The fungus extends specialized hyphae that can penetrate the body of the worm and slowly digest the hapless prey.

Beneficial Fungi

Fungi play a crucial role in the balance of ecosystems. They colonize most habitats on Earth, preferring dark, moist conditions. They can thrive in seemingly hostile environments, such as the tundra, thanks to a most successful symbiosis with photosynthetic organisms, like lichens. Fungi are not obvious in the way that large animals or tall trees are. Yet, like bacteria, they are major decomposers of nature. With their versatile metabolism, fungi break down organic matter that is insoluble and would not be recycled otherwise.

Importance to Ecosystems

Food webs would be incomplete without organisms that decompose organic matter and fungi are key participants in this process. Decomposition allows for cycling of nutrients such as carbon, nitrogen, and phosphorus back into the environment so they are available to living things, rather than being trapped in dead organisms. Fungi are particularly important because they have evolved enzymes to break down cellulose and lignin, components of plant cell walls that few other organisms are able to digest, releasing their carbon content.

Fungi are also involved in ecologically important coevolved symbioses, both mutually beneficial and pathogenic with organisms from the other kingdoms. **Mycorrhiza**, a term combining the Greek roots *myco* meaning fungus and *rhizo* meaning root, refers to the association between vascular plant roots and their symbiotic fungi. Somewhere between 80–90 percent of all plant species have mycorrhizal partners. In a mycorrhizal association, the fungal mycelia use their extensive network of hyphae and large surface area in contact with the soil to channel water and minerals from the soil into the plant. In exchange, the plant supplies the products of photosynthesis to fuel the metabolism of the fungus. Ectomycorrhizae ("outside" mycorrhiza) depend on fungi enveloping the roots in a sheath (called a mantle) and a net of hyphae that extends into the roots between cells. In a second type, the Glomeromycota fungi form arbuscular mycorrhiza. In these mycorrhiza, the fungi form arbuscles, a specialized highly branched hypha, which penetrate root cells and are the sites of the metabolic exchanges between the fungus and the host plant. Orchids rely on a third type of mycorrhiza. Orchids form small seeds without much storage to sustain germination and growth. Their seeds will not germinate without a mycorrhizal partner (usually Basidiomycota). After nutrients in the seed are depleted, fungal symbionts support the growth of the orchid by providing necessary carbohydrates and minerals. Some orchids continue to be mycorrhizal throughout their lifecycle.

Lichens blanket many rocks and tree bark, displaying a range of colors and textures. Lichens are important pioneer organisms that colonize rock surfaces in otherwise lifeless environments such as are created by glacial recession. The lichen is able to leach nutrients from the rocks and break them down in the first step to creating soil. Lichens are also present in mature habitats on rock surfaces or the trunks of trees. They are an important food source for caribou. Lichens are not a single organism, but rather a fungus (usually an Ascomycota or Basidiomycota species) living in close contact with a photosynthetic organism (an alga or cyanobacterium). The body of a lichen, referred to as a thallus, is formed of hyphae wrapped around the green partner. The photosynthetic organism provides carbon and energy in the form of carbohydrates and receives protection from the elements by the thallus of the fungal partner. Some cyanobacteria fix nitrogen from the atmosphere, contributing nitrogenous compounds to the association. In return, the fungus supplies minerals and protection from dryness and excessive light by encasing the algae in its mycelium. The fungus also attaches the symbiotic organism to the substrate.

Fungi have evolved mutualistic associations with numerous arthropods. The association between species of Basidiomycota and scale insects is one example. The fungal mycelium covers and protects the insect colonies. The scale insects foster a flow of nutrients from the parasitized plant to the fungus. In a second example, leaf-cutting ants of Central and South America literally farm fungi. They cut disks of leaves from plants and pile them up in gardens. Fungi are cultivated in these gardens, digesting the cellulose that the ants cannot break down. Once smaller sugar molecules are produced and consumed by the fungi, they in turn become a meal for the ants. The insects also patrol their garden, preying on competing fungi. Both ants and fungi benefit from the association. The fungus receives a steady supply of leaves and freedom from competition, while the ants feed on the fungi they cultivate.

Importance to Humans

Although we often think of fungi as organisms that cause diseases and rot food, fungi are important to human life on many levels. As we have seen, they influence the well-being of human populations on a large scale because they help nutrients cycle in ecosystems. They have other ecosystem roles as well. For example, as animal pathogens, fungi help to control the population of damaging pests. These fungi are very specific to the insects they attack and do not infect other animals or plants. The potential to use fungi as microbial insecticides is being investigated, with several species already on the market. For example, the fungus *Beauveria bassiana* is a pesticide that is currently being tested as a possible biological control for the recent spread of emerald ash borer. It has been released in Michigan, Illinois, Indiana, Ohio, West Virginia, and Maryland.

The mycorrhizal relationship between fungi and plant roots is essential for the productivity of farmland. Without the fungal partner in the root systems, 80–90% of trees and grasses would not survive. Mycorrhizal fungal inoculants are available as soil amendments from gardening supply stores and promoted by supporters of organic agriculture.

We also eat some types of fungi. Mushrooms figure prominently in the human diet. Morels, shiitake mushrooms, chanterelles, and truffles are considered delicacies (Figure 13.27). The humble meadow mushroom, *Agaricus campestris*, appears in many dishes. Molds of the genus *Penicillium* ripen many cheeses. They originate in the natural environment such as the caves of Roquefort, France, where wheels of sheep milk cheese are stacked to capture the molds responsible for the blue veins and pungent taste of the cheese.

Figure 13.27 The morel mushroom is an ascomycete that is much appreciated for its delicate taste. (credit: Jason Hollinger)

Fermentation—of grains to produce beer, and of fruits to produce wine—is an ancient art that humans in most cultures have practiced for millennia. Wild yeasts are acquired from the environment and used to ferment sugars into CO_2 and ethyl alcohol under anaerobic conditions. It is now possible to purchase isolated strains of wild yeasts from different wine-making regions. Pasteur was instrumental in developing a reliable strain of brewer's yeast, *Saccharomyces cerevisiae*, for the French brewing industry in the late 1850s. It was one of the first examples of biotechnology patenting. Yeast is also used to make breads that rise. The carbon dioxide they produce is responsible for the bubbles produced in the dough that become the air pockets of the baked bread.

Many secondary metabolites of fungi are of great commercial importance. Antibiotics are naturally produced by fungi to kill or inhibit the growth of bacteria, and limit competition in the natural environment. Valuable drugs isolated from fungi include the immunosuppressant drug cyclosporine (which reduces the risk of rejection after organ transplant), the precursors of steroid hormones, and ergot alkaloids used to stop bleeding. In addition, as easily cultured eukaryotic organisms, some fungi are important model research organisms including the red bread mold *Neurospora crassa* and the yeast, *S. cerevisiae.*

KEY TERMS

Amoebozoa the eukaryotic supergroup that contains the amoebas and slime molds

anaerobic refers to organisms that grow without oxygen

anoxic without oxygen

Archaeplastida the eukaryotic supergroup that contains land plants, green algae, and red algae

Ascomycota (sac fungi) a division of fungi that store spores in a sac called ascus

basidiomycota (club fungi) a division of fungi that produce club shaped structures, basidia, which contain spores

biofilm a microbial community that is held together by a gummy-textured matrix

bioremediation the use of microbial metabolism to remove pollutants

Black Death a devastating pandemic that is believed to have been an outbreak of bubonic plague caused by the bacterium *Yersinia pestis*

botulism a disease produce by the toxin of the anaerobic bacterium *Clostridium botulinum*

capsule an external structure that enables a prokaryote to attach to surfaces and protects it from dehydration

Chromalveolata the eukaryotic supergroup that contains the dinoflagellates, ciliates, the brown algae, diatoms, and water molds

Chytridiomycota (chytrids) a primitive division of fungi that live in water and produce gametes with flagella

commensalism a symbiotic relationship in which one member benefits while the other member is not affected

conjugation the process by which prokaryotes move DNA from one individual to another using a pilus

cyanobacteria bacteria that evolved from early phototrophs and oxygenated the atmosphere; also known as blue-green algae

Deuteromycota a division of fungi that do not have a known sexual reproductive cycle (presently members of two phyla: Ascomycota and Basidiomycota)

endosymbiosis the engulfment of one cell by another such that the engulfed cell survives and both cells benefit; the process responsible for the evolution of mitochondria and chloroplasts in eukaryotes

epidemic a disease that occurs in an unusually high number of individuals in a population at the same time

Excavata the eukaryotic supergroup that contains flagellated single-celled organisms with a feeding groove

extremophile an organism that grows under extreme or harsh conditions

foodborne disease any illness resulting from the consumption of contaminated food, or of the pathogenic bacteria, viruses, or other parasites that contaminate food

Glomeromycota a group of fungi that form symbiotic relationships with the roots of trees

Gram-negative describes a bacterium whose cell wall contains little peptidoglycan but has an outer membrane

Gram-positive describes a bacterium that contains mainly peptidoglycan in its cell walls

hydrothermal vent a fissure in Earth's surface that releases geothermally heated water

hypha a fungal filament composed of one or more cells

lichen the close association of a fungus with a photosynthetic alga or bacterium that benefits both partners

microbial mat a multi-layered sheet of prokaryotes that may include bacteria and archaea

mold a tangle of visible mycelia with a fuzzy appearance

MRSA (methicillin-resistant *Staphylococcus aureus*) a very dangerous *Staphylococcus aureus* strain resistant to antibiotics

mycelium a mass of fungal hyphae

mycorrhiza a mutualistic association between fungi and vascular plant roots

mycosis a fungal infection

Opisthokonta the eukaryotic supergroup that contains the fungi, animals, and choanoflagellates

pandemic a widespread, usually worldwide, epidemic disease

parasite an organism that lives on or in another organism and feeds on it, often without killing it

pathogen an organism, or infectious agent, that causes a disease

pellicle an outer cell covering composed of interlocking protein strips that function like a flexible coat of armor, preventing cells from being torn or pierced without compromising their range of motion

peptidoglycan a material composed of polysaccharide chains cross-linked to unusual peptides

phototroph an organism that uses energy from sunlight

plastid one of a group of related organelles in plant cells that are involved in the storage of starches, fats, proteins, and pigments

pseudopeptidoglycan a component of some cell walls of Archaea

Rhizaria the eukaryotic supergroup that contains organisms that move by amoeboid movement

saprobe an organism that feeds on dead organic material

septum the cell wall division between hyphae

stromatolite a layered sedimentary structure formed by precipitation of minerals by prokaryotes in microbial mats

thallus a vegetative body of a fungus

transduction the process by which a bacteriophage moves DNA from one prokaryote to another

transformation a mechanism of genetic change in prokaryotes in which DNA present in the environment is taken into the cell and incorporated into the genome

yeast a general term used to describe unicellular fungi

Zygomycota (conjugated fungi) the division of fungi that form a zygote contained in a zygospore

CHAPTER SUMMARY

13.1 Prokaryotic Diversity

Prokaryotes existed for billions of years before plants and animals appeared. Microbial mats are thought to represent the earliest forms of life on Earth, and there is fossil evidence, called stromatolites, of their presence about 3.5 billion years ago. During the first 2 billion years, the atmosphere was anoxic and only anaerobic organisms were able to live. Cyanobacteria began the oxygenation of the atmosphere. The increase in oxygen concentration allowed the evolution of other life forms.

Prokaryotes (domains Archaea and Bacteria) are single-celled organisms lacking a nucleus. They have a single piece of circular DNA in the nucleoid area of the cell. Most prokaryotes have cell wall outside the plasma membrane. Bacteria and Archaea differ in the compositions of their cell membranes and the characteristics of their cell walls.

Bacterial cell walls contain peptidoglycan. Archaean cell walls do not have peptidoglycan. Bacteria can be divided into two major groups: Gram-positive and Gram-negative. Gram-positive organisms have a thick cell wall. Gram-negative organisms have a thin cell wall and an outer membrane. Prokaryotes use diverse sources of energy to assemble macromolecules from smaller molecules. Phototrophs obtain their energy from sunlight, whereas chemotrophs obtain it from chemical compounds.

Infectious diseases caused by bacteria remain among the leading causes of death worldwide. The excessive use of antibiotics to control bacterial infections has resulted in resistant forms of bacteria being selected. Foodborne diseases result from the consumption of contaminated food, pathogenic bacteria, viruses, or parasites that contaminate food. Prokaryotes are used in human food products. Microbial bioremediation is the use of microbial metabolism to remove pollutants. The human body contains a huge community of prokaryotes, many of which provide beneficial services such as the development and maintenance of the immune system, nutrition, and protection from pathogens.

13.2 Eukaryotic Origins

The first eukaryotes evolved from ancestral prokaryotes by a process that involved membrane proliferation, the loss of a cell wall, the evolution of a cytoskeleton, and the acquisition and evolution of organelles. Nuclear eukaryotic genes appear to have had an origin in the Archaea, whereas the energy machinery of eukaryotic cells appears to be bacterial in origin. The mitochondria and plastids originated from endosymbiotic events when ancestral cells engulfed an aerobic bacterium (in the case of mitochondria) and a photosynthetic bacterium (in the case of chloroplasts). The evolution of mitochondria likely preceded the evolution of chloroplasts. There is evidence of secondary endosymbiotic events in which plastids appear to be the result of endosymbiosis after a previous endosymbiotic event.

13.3 Protists

Protists are extremely diverse in terms of biological and ecological characteristics due in large part to the fact that they are an artificial assemblage of phylogenetically unrelated groups. Protists display highly varied cell structures, several types of reproductive strategies, virtually every possible type of nutrition, and varied habitats. Most single-celled protists are motile, but these organisms use diverse structures for transportation.

The process of classifying protists into meaningful groups is ongoing, but genetic data in the past 20 years have clarified many relationships that were previously unclear or mistaken. The majority view at present is to order all eukaryotes into six supergroups. The goal of this classification scheme is to create clusters of species that all are derived from a common ancestor.

13.4 Fungi

Fungi are eukaryotic organisms that appeared on land over 450 million years ago. They are heterotrophs and contain neither photosynthetic pigments such as chlorophylls nor organelles such as chloroplasts. Because they feed on decaying and dead matter, they are saprobes. Fungi are important decomposers and release essential elements into the environment. External enzymes digest nutrients that are absorbed by the body of the fungus called a thallus. A thick cell wall made of chitin surrounds the cell. Fungi can be unicellular as yeasts or develop a network of filaments called a mycelium, often described as mold. Most species multiply by asexual and sexual reproductive cycles, and display an alternation of generations.

The divisions of fungi are the Chytridiomycota, Zygomycota, Ascomycota, Basidiomycota, Glomeromycota, and the Deuteromycota, a polyphyletic group.

Fungi establish parasitic relationships with plants and animals. Fungal diseases can decimate crops and spoil food during storage. Compounds produced by fungi can be toxic to humans and other animals. Mycoses are infections caused by fungi. Superficial mycoses affect the skin, whereas systemic mycoses spread through the body. Fungal infections are difficult to cure.

Fungi have colonized all environments on Earth but are most often found in cool, dark, moist places with a supply of decaying material. Fungi are important decomposers because they are saprobes. Many successful mutualistic relationships involve a fungus and another organism. They establish complex mycorrhizal associations with the roots of plants. Lichens are a symbiotic relationship between a fungus and a photosynthetic organism, usually an alga or cyanobacterium.

Fungi are important to everyday human life. Fungi are important decomposers in most ecosystems. Mycorrhizal fungi are essential for the growth of most plants. Fungi, as food, play a role in human nutrition in the form of mushrooms and as agents of fermentation in the production of bread, cheeses, alcoholic beverages, and numerous other food preparations. Secondary metabolites of fungi are used in medicine as antibiotics and anticoagulants. Fungi are used in research as model organisms for the study of eukaryotic genetics and metabolism.

ART CONNECTION QUESTIONS

1. Figure 13.6 Which of the following statements is true?

a. Gram-positive bacteria have a single cell wall formed from peptidoglycan.
b. Gram-positive bacteria have an outer membrane.

c. The cell wall of Gram-negative bacteria is thick, and the cell wall of Gram-positive bacteria is thin.
d. Gram-negative bacteria have a cell wall made of peptidoglycan, while Gram-positive bacteria have a cell wall made of phospholipids.

REVIEW QUESTIONS

2. The first forms of life on Earth were thought to be_____.

a. single-celled plants
b. prokaryotes
c. insects
d. large animals such as dinosaurs

3. The first organisms that oxygenated the atmosphere were _____.

a. cyanobacteria
b. phototrophic organisms
c. anaerobic organisms
d. all of the above

4. Which of the following consist of prokaryotic cells?

a. bacteria and fungi
b. archaea and fungi
c. protists and animals
d. bacteria and archaea

5. Prokaryotes stain as Gram-positive or Gram-negative because of differences in the _____.

a. cell wall
b. cytoplasm
c. nucleus
d. chromosome

6. Prokaryotes that obtain their energy from chemical compounds are called _____.

a. phototrophs
b. auxotrophs
c. chemotrophs
d. lithotrophs

7. Bioremediation includes _____.

a. the use of prokaryotes that can fix nitrogen
b. the use of prokaryotes to clean up pollutants
c. the use of prokaryotes as natural fertilizers
d. All of the above

8. What event is thought to have contributed to the evolution of eukaryotes?

a. global warming
b. glaciation
c. volcanic activity
d. oxygenation of the atmosphere

9. Mitochondria most likely evolved from _____.

a. a photosynthetic cyanobacterium
b. cytoskeletal elements
c. aerobic bacteria
d. membrane proliferation

10. Protists with the capabilities to absorb nutrients from dead organisms are called_____.

a. photoautotrophs
b. autotrophs
c. saprobes
d. heterotrophs

11. Which parasitic protist evades the host immune system by altering its surface proteins with each generation?

a. *Paramecium caudatum*
b. *Trypanosoma brucei*
c. *Plasmodium falciparum*
d. *Phytophthora infestans*

12. Which polysaccharide is usually found in the cell walls of fungi?

a. starch
b. glycogen
c. chitin
d. cellulose

13. What term describes the close association of a fungus with the root of a tree?

a. a rhizoid
b. a lichen
c. a mycorrhiza
d. an endophyte

CRITICAL THINKING QUESTIONS

Download for free at https://openstax.org/details/books/concepts-biology

14. Explain the reason why the imprudent and excessive use of antibiotics has resulted in a major global problem.

15. Your friend believes that prokaryotes are always detrimental and pathogenic. How would you explain to them that they are wrong?

16. Describe the hypothesized steps in the origin of eukaryote cells.

17. How does killing *Anopheles* mosquitoes affect the *Plasmodium* protists?

18. Without treatment, why does African sleeping sickness invariably lead to death?

19. Why can superficial mycoses in humans lead to bacterial infections?

14 | DIVERSITY OF PLANTS

Figure 14.1 Plants dominate the landscape and play an integral role in human societies. (a) Palm trees grow in tropical or subtropical climates; (b) wheat is a crop in most of the world; the flower of (c) the cotton plant produces fibers that are woven into fabric; the potent alkaloids of (d) the beautiful opium poppy have influenced human life both as a medicinal remedy and as a dangerously addictive drug. (credit a: modification of work by "3BoysInSanDiego"/Wikimedia Commons"; credit b: modification of work by Stephen Ausmus, USDA ARS; credit c: modification of work by David Nance, USDA ARS; credit d: modification of work by Jolly Janner)

Chapter Outline
14.1: The Plant Kingdom
14.2: Seedless Plants
14.3: Seed Plants: Gymnosperms
14.4: Seed Plants: Angiosperms

Introduction

Plants play an integral role in all aspects of life on the planet, shaping the physical terrain, influencing the climate, and maintaining life as we know it. For millennia, human societies have depended on plants for nutrition and medicinal compounds, and for many industrial by-products, such as timber, paper, dyes, and textiles. Palms provide materials including rattans, oils, and dates. Wheat is grown to feed both human and animal populations. The cotton boll flower is harvested and its fibers transformed into clothing or pulp for paper. The showy opium poppy is valued both as an ornamental flower and as a source of potent opiate compounds.

Current evolutionary thought holds that all plants are monophyletic: that is, descendants of a single common ancestor. The evolutionary transition from water to land imposed severe constraints on the ancestors of contemporary plants. Plants had to evolve strategies to avoid drying out, to disperse reproductive cells in air, for structural support, and to filter sunlight. While seed plants developed adaptations that allowed them to populate even the most arid habitats on Earth, full independence from water did not happen in all plants, and most seedless plants still require a moist environment.

14.1 | The Plant Kingdom

By the end of this section, you will be able to:

- Describe the major characteristics of the plant kingdom
- Discuss the challenges to plant life on land
- Describe the adaptations that allowed plants to colonize land

Plants are a large and varied group of organisms. There are close to 300,000 species of catalogued plants.[1] Of these, about 260,000 are plants that produce seeds. Mosses, ferns, conifers, and flowering plants are all members of the plant kingdom. The plant kingdom contains mostly photosynthetic organisms; a few parasitic forms have lost the ability to photosynthesize. The process of photosynthesis uses chlorophyll, which is located in organelles called chloroplasts. Plants possess cell walls containing cellulose. Most plants reproduce sexually, but they also have diverse methods of asexual reproduction. Plants exhibit indeterminate growth, meaning they do not have a final body form, but continue to grow body mass until they die.

Plant Adaptations to Life on Land

As organisms adapt to life on land, they have to contend with several challenges in the terrestrial environment. Water has been described as "the stuff of life." The cell's interior—the medium in which most small molecules dissolve and diffuse, and in which the majority of the chemical reactions of metabolism take place—is a watery soup. Desiccation, or drying out, is a constant danger for an organism exposed to air. Even when parts of a plant are close to a source of water, their aerial structures are likely to dry out. Water provides buoyancy to organisms that live in aquatic habitats. On land, plants need to develop structural support in air—a medium that does not give the same lift. Additionally, the male gametes must reach the female gametes using new strategies because swimming is no longer possible. Finally, both gametes and zygotes must be protected from drying out. The successful land plants evolved strategies to deal with all of these challenges, although not all adaptations appeared at once. Some species did not move far from an aquatic environment, whereas others left the water and went on to conquer the driest environments on Earth.

To balance these survival challenges, life on land offers several advantages. First, sunlight is abundant. On land, the spectral quality of light absorbed by the photosynthetic pigment, chlorophyll, is not filtered out by water or competing photosynthetic species in the water column above. Second, carbon dioxide is more readily available because its concentration is higher in air than in water. Additionally, land plants evolved before land animals; therefore, until dry land was colonized by animals, no predators threatened the well-being of plants. This situation changed as animals emerged from the water and found abundant sources of nutrients in the established flora. In turn, plants evolved strategies to deter predation: from spines and thorns to toxic chemicals.

The early land plants, like the early land animals, did not live far from an abundant source of water and developed survival strategies to combat dryness. One of these strategies is drought tolerance. Mosses, for example, can dry out to a brown and brittle mat, but as soon as rain makes water available, mosses will soak it up and regain their healthy, green appearance. Another strategy is to colonize environments with high humidity where droughts are uncommon. Ferns, an early lineage of plants, thrive in damp and cool places, such as the understory of temperate forests. Later, plants moved away from aquatic environments using resistance to desiccation, rather than tolerance. These plants, like the cactus, minimize water loss to such an extent they can survive in the driest environments on Earth.

In addition to adaptations specific to life on land, land plants exhibit adaptations that were responsible for their diversity and predominance in terrestrial ecosystems. Four major adaptations are found in many terrestrial plants: the alternation of

1. A.D. Chapman (2009) *Numbers of Living Species in Australia and the World*. 2nd edition. A Report for the Australian Biological Resources Study. Australian Biodiversity Information Services, Toowoomba, Australia. Available online at http://www.environment.gov.au/biodiversity/abrs/publications/other/species-numbers/2009/04-03-groups-plants.html.

generations, a sporangium in which spores are formed, a gametangium that produces haploid cells, and in vascular plants, apical meristem tissue in roots and shoots.

Alternation of Generations

Alternation of generations describes a life cycle in which an organism has both haploid and diploid multicellular stages (Figure 14.2).

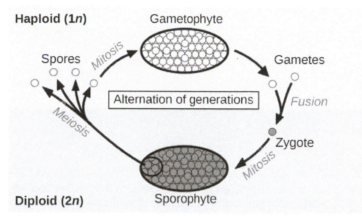

Figure 14.2 Alternation of generations between the haploid (1n) gametophyte and diploid (2n) sporophyte is shown. (credit: modification of work by Peter Coxhead)

Haplontic refers to a life cycle in which there is a dominant haploid stage. **Diplontic** refers to a life cycle in which the diploid stage is the dominant stage, and the haploid chromosome number is only seen for a brief time in the life cycle during sexual reproduction. Humans are diplontic, for example. Most plants exhibit alternation of generations, which is described as **haplodiplontic**: the haploid multicellular form known as a gametophyte is followed in the development sequence by a multicellular diploid organism, the **sporophyte**. The **gametophyte** gives rise to the gametes, or reproductive cells, by mitosis. It can be the most obvious phase of the life cycle of the plant, as in the mosses, or it can occur in a microscopic structure, such as a pollen grain in the higher plants (the collective term for the vascular plants). The sporophyte stage is barely noticeable in lower plants (the collective term for the plant groups of mosses, liverworts, and hornworts). Towering trees are the diplontic phase in the lifecycles of plants such as sequoias and pines.

Sporangia in the Seedless Plants

The sporophyte of seedless plants is diploid and results from **syngamy** or the fusion of two gametes (Figure 14.2). The sporophyte bears the **sporangia** (singular, sporangium), organs that first appeared in the land plants. The term "sporangia" literally means "spore in a vessel," as it is a reproductive sac that contains spores. Inside the multicellular sporangia, the diploid sporocytes, or mother cells, produce haploid spores by meiosis, which reduces the 2n chromosome number to 1n. The spores are later released by the sporangia and disperse in the environment. Two different types of spores are produced in land plants, resulting in the separation of sexes at different points in the life cycle. Seedless nonvascular plants (more appropriately referred to as "seedless nonvascular plants with a dominant gametophyte phase") produce only one kind of spore, and are called **homosporous**. After germinating from a spore, the gametophyte produces both male and female **gametangia**, usually on the same individual. In contrast, **heterosporous** plants produce two morphologically different types of spores. The male spores are called microspores because of their smaller size; the comparatively larger megaspores will develop into the female gametophyte. Heterospory is observed in a few seedless vascular plants and in all seed plants.

When the haploid spore germinates, it generates a multicellular gametophyte by mitosis. The gametophyte supports the zygote formed from the fusion of gametes and the resulting young sporophyte or vegetative form, and the cycle begins anew (Figure 14.3 and Figure 14.4).

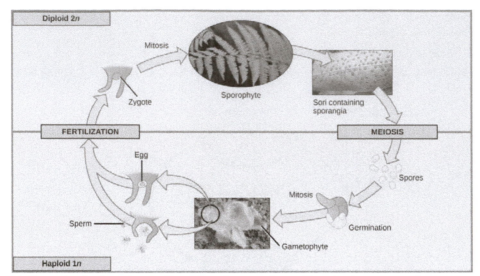

Figure 14.3 This life cycle of a fern shows alternation of generations with a dominant sporophyte stage. (credit "fern": modification of work by Cory Zanker; credit "gametophyte": modification of work by "Vlmastra"/Wikimedia Commons)

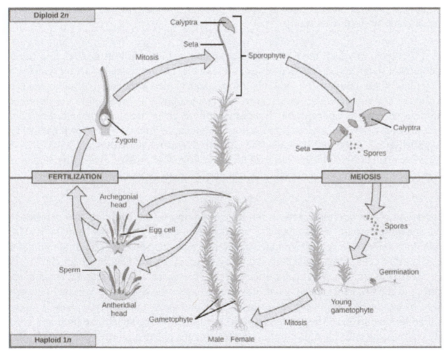

Figure 14.4 This life cycle of a moss shows alternation of generations with a dominant gametophyte stage. (credit: modification of work by Mariana Ruiz Villareal)

The spores of seedless plants and the pollen of seed plants are surrounded by thick cell walls containing a tough polymer known as sporopollenin. This substance is characterized by long chains of organic molecules related to fatty acids and carotenoids, and gives most pollen its yellow color. Sporopollenin is unusually resistant to chemical and biological degradation. Its toughness explains the existence of well-preserved fossils of pollen. Sporopollenin was once thought to be an innovation of land plants; however, the green algae *Coleochaetes* is now known to form spores that contain sporopollenin.

Protection of the embryo is a major requirement for land plants. The vulnerable embryo must be sheltered from desiccation and other environmental hazards. In both seedless and seed plants, the female gametophyte provides nutrition, and in seed plants, the embryo is also protected as it develops into the new generation of sporophyte.

Gametangia in the Seedless Plants

Gametangia (singular, gametangium) are structures on the gametophytes of seedless plants in which gametes are produced by mitosis. The male gametangium, the antheridium, releases sperm. Many seedless plants produce sperm equipped with flagella that enable them to swim in a moist environment to the archegonia, the female gametangium. The embryo develops inside the archegonium as the sporophyte.

Apical Meristems

The shoots and roots of plants increase in length through rapid cell division within a tissue called the **apical meristem** (Figure 14.5). The apical meristem is a cap of cells at the shoot tip or root tip made of undifferentiated cells that continue to proliferate throughout the life of the plant. Meristematic cells give rise to all the specialized tissues of the plant. Elongation of the shoots and roots allows a plant to access additional space and resources: light in the case of the shoot, and water and minerals in the case of roots. A separate meristem, called the lateral meristem, produces cells that increase the diameter of stems and tree trunks. Apical meristems are an adaptation to allow vascular plants to grow in directions essential to their survival: upward to greater availability of sunlight, and downward into the soil to obtain water and essential minerals.

Figure 14.5 This apple seedling is an example of a plant in which the apical meristem gives rise to new shoots and root growth.

Additional Land Plant Adaptations

As plants adapted to dry land and became independent of the constant presence of water in damp habitats, new organs and structures made their appearance. Early land plants did not grow above a few inches off the ground, and on these low mats, they competed for light. By evolving a shoot and growing taller, individual plants captured more light. Because air offers substantially less support than water, land plants incorporated more rigid molecules in their stems (and later, tree trunks). The evolution of vascular tissue for the distribution of water and solutes was a necessary prerequisite for plants to evolve larger bodies. The vascular system contains xylem and phloem tissues. Xylem conducts water and minerals taken from the soil up to the shoot; phloem transports food derived from photosynthesis throughout the entire plant. The root system that evolved to take up water and minerals also anchored the increasingly taller shoot in the soil.

In land plants, a waxy, waterproof cover called a cuticle coats the aerial parts of the plant: leaves and stems. The cuticle also prevents intake of carbon dioxide needed for the synthesis of carbohydrates through photosynthesis. Stomata, or pores, that open and close to regulate traffic of gases and water vapor therefore appeared in plants as they moved into drier habitats.

Plants cannot avoid predatory animals. Instead, they synthesize a large range of poisonous secondary metabolites: complex organic molecules such as alkaloids, whose noxious smells and unpleasant taste deter animals. These toxic compounds can cause severe diseases and even death.

Additionally, as plants coevolved with animals, sweet and nutritious metabolites were developed to lure animals into providing valuable assistance in dispersing pollen grains, fruit, or seeds. Plants have been coevolving with animal associates for hundreds of millions of years (Figure 14.6).

(a)   (b)

(c)   (d)

Figure 14.6 Plants have evolved various adaptations to life on land. (a) Early plants grew close to the ground, like this moss, to avoid desiccation. (b) Later plants developed a waxy cuticle to prevent desiccation. (c) To grow taller, like these maple trees, plants had to evolve new structural chemicals to strengthen their stems and vascular systems to transport water and minerals from the soil and nutrients from the leaves. (d) Plants developed physical and chemical defenses to avoid being eaten by animals. (credit a, b: modification of work by Cory Zanker; credit c: modification of work by Christine Cimala; credit d: modification of work by Jo Naylor)

e/olution IN ACTION

Paleobotany

How organisms acquired traits that allow them to colonize new environments, and how the contemporary ecosystem is shaped, are fundamental questions of evolution. Paleobotany addresses these questions by specializing in the study of extinct plants. Paleobotanists analyze specimens retrieved from field studies, reconstituting the morphology of organisms that have long disappeared. They trace the evolution of plants by following the modifications in plant morphology, and shed light on the connection between existing plants by identifying common ancestors that display the same traits. This field seeks to find transitional species that bridge gaps in the path to the development of modern organisms. Fossils are formed when organisms are trapped in sediments or environments where their shapes are preserved (Figure 14.7). Paleobotanists determine the geological age of specimens and the nature of their environment using the geological sediments and fossil organisms surrounding them. The activity requires great care to preserve the integrity of the delicate fossils and the layers in which they are found.

One of the most exciting recent developments in paleobotany is the use of analytical chemistry and molecular biology to study fossils. Preservation of molecular structures requires an environment free of oxygen, since oxidation and degradation of material through the activity of microorganisms depend on the presence of oxygen. One example of the use of analytical chemistry and molecular biology is in the identification of oleanane, a compound that deters pests and which, up to this point, appears to be unique to flowering plants. Oleanane was recovered from sediments dating from the Permian, much earlier than the current dates given for the appearance of the first flowering plants. Fossilized nucleic acids—DNA and RNA—yield the most information. Their sequences are analyzed and compared to those of living and related organisms. Through this analysis, evolutionary relationships can be built for plant lineages.

Some paleobotanists are skeptical of the conclusions drawn from the analysis of molecular fossils. For one, the chemical materials of interest degrade rapidly during initial isolation when exposed to air, as well as in further manipulations. There is always a high risk of contaminating the specimens with extraneous material, mostly from microorganisms. Nevertheless, as technology is refined, the analysis of DNA from fossilized plants will provide invaluable information on the evolution of plants and their adaptation to an ever-changing environment.

Figure 14.7 This fossil of a palm leaf (*Palmacites* sp.) discovered in Wyoming dates to about 40 million years ago.

The Major Divisions of Land Plants

Land plants are classified into two major groups according to the absence or presence of vascular tissue, as detailed in Figure 14.8. Plants that lack vascular tissue formed of specialized cells for the transport of water and nutrients are referred to as **nonvascular plants**. The bryophytes, liverworts, mosses, and hornworts are seedless and nonvascular, and likely appeared early in land plant evolution. **Vascular plants** developed a network of cells that conduct water and solutes through the plant body. The first vascular plants appeared in the late Ordovician (461–444 million years ago) and were probably similar to lycophytes, which include club mosses (not to be confused with the mosses) and the pterophytes (ferns, horsetails, and whisk ferns). Lycophytes and pterophytes are referred to as seedless vascular plants. They do not produce seeds, which are embryos with their stored food reserves protected by a hard casing. The seed plants form the largest group of all existing plants and, hence, dominate the landscape. Seed plants include gymnosperms, most notably conifers, which produce "naked seeds," and the most successful plants, the flowering plants, or angiosperms, which protect their seeds inside chambers at the center of a flower. The walls of these chambers later develop into fruits.

Embryophytes: The Land Plants						
Nonvascular Plants "Bryophytes"			Vascular Plants			
			Seedless Plants		Seed Plants	
			Lycophytes	Pterophytes	Gymno-sperms	Angio-sperms
Liverworts	Hornworts	Mosses	Club Mosses	Whisk Ferns		
			Quillworts	Horsetails		
			Spike Mosses	Ferns		

Figure 14.8 This table shows the major divisions of plants.

14.2 | Seedless Plants

By the end of this section, you will be able to:

- Describe the distinguishing traits of the three types of bryophytes
- Identify the new traits that first appear in seedless vascular plants
- Describe the major classes of seedless vascular plants

An incredible variety of seedless plants populates the terrestrial landscape. Mosses grow on tree trunks, and horsetails (Figure 14.9) display their jointed stems and spindly leaves on the forest floor. Yet, seedless plants represent only a small fraction of the plants in our environment. Three hundred million years ago, seedless plants dominated the landscape and grew in the enormous swampy forests of the Carboniferous period. Their decomposing bodies created large deposits of coal that we mine today.

Figure 14.9 Seedless plants like these horsetails (*Equisetum* sp.) thrive in damp, shaded environments under the tree canopy where dryness is a rare occurrence. (credit: Jerry Kirkhart)

Bryophytes

Bryophytes, an informal grouping of the nonvascular plants, are the closest extant relative of early terrestrial plants. The first bryophytes most probably appeared in the Ordovician period, about 490 million years ago. Because of the lack of lignin—the tough polymer in cell walls in the stems of vascular plants—and other resistant structures, the likelihood of bryophytes forming fossils is rather small, though some spores made up of sporopollenin have been discovered that have been attributed to early bryophytes. By the Silurian period (440 million years ago), however, vascular plants had spread throughout the continents. This fact is used as evidence that nonvascular plants must have preceded the Silurian period.

There are about 18,000 species of bryophytes, which thrive mostly in damp habitats, although some grow in deserts. They constitute the major flora of inhospitable environments like the tundra, where their small size and tolerance to desiccation offer distinct advantages. They do not have the specialized cells that conduct fluids found in the vascular plants, and generally lack lignin. In bryophytes, water and nutrients circulate inside specialized conducting cells. Although the name nontracheophyte is more accurate, bryophytes are commonly referred to as nonvascular plants.

In a bryophyte, all the conspicuous vegetative organs belong to the haploid organism, or gametophyte. The diploid sporophyte is barely noticeable. The gametes formed by bryophytes swim using flagella. The sporangium, the multicellular sexual reproductive structure, is present in bryophytes. The embryo also remains attached to the parent plant, which nourishes it. This is a characteristic of land plants.

The bryophytes are divided into three divisions (in plants, the taxonomic level "division" is used instead of phylum): the liverworts, or Marchantiophyta; the hornworts, or Anthocerotophyta; and the mosses, or true Bryophyta.

Liverworts

Liverworts (Marchantiophyta) may be viewed as the plants most closely related to the ancestor that moved to land. Liverworts have colonized many habitats on Earth and diversified to more than 6,000 existing species (Figure 14.10a). Some gametophytes form lobate green structures, as seen in Figure 14.10b. The shape is similar to the lobes of the liver and, hence, provides the origin of the common name given to the division.

(a)   (b)

Figure 14.10 (a) A 1904 drawing of liverworts shows the variety of their forms. (b) A liverwort, *Lunularia cruciata*, displays its lobate, flat thallus. The organism in the photograph is in the gametophyte stage.

Hornworts

The **hornworts** (Anthocerotophyta) have colonized a variety of habitats on land, although they are never far from a source of moisture. There are about 100 described species of hornworts. The dominant phase of the life cycle of hornworts is the short, blue-green gametophyte. The sporophyte is the defining characteristic of the group. It is a long and narrow pipe-like structure that emerges from the parent gametophyte and maintains growth throughout the life of the plant (Figure 14.11).

Figure 14.11 Hornworts grow a tall and slender sporophyte. (credit: modification of work by Jason Hollinger)

Mosses

More than 12,000 species of **mosses** have been catalogued. Their habitats vary from the tundra, where they are the main vegetation, to the understory of tropical forests. In the tundra, their shallow rhizoids allow them to fasten to a substrate without digging into the frozen soil. They slow down erosion, store moisture and soil nutrients, and provide shelter for small animals and food for larger herbivores, such as the musk ox. Mosses are very sensitive to air pollution and are used to monitor the quality of air. The sensitivity of mosses to copper salts makes these salts a common ingredient of compounds marketed to eliminate mosses in lawns (Figure 14.12).

Figure 14.12 This green feathery moss has reddish-brown sporophytes growing upward. (credit: "Lordgrunt"/Wikimedia Commons)

Vascular Plants

The vascular plants are the dominant and most conspicuous group of land plants. There are about 275,000 species of vascular plants, which represent more than 90 percent of Earth's vegetation. Several evolutionary innovations explain their success and their spread to so many habitats.

Vascular Tissue: Xylem and Phloem

The first fossils that show the presence of vascular tissue are dated to the Silurian period, about 430 million years ago. The simplest arrangement of conductive cells shows a pattern of xylem at the center surrounded by phloem. **Xylem** is the tissue responsible for long-distance transport of water and minerals, the transfer of water-soluble growth factors from the organs of synthesis to the target organs, and storage of water and nutrients.

A second type of vascular tissue is **phloem**, which transports sugars, proteins, and other solutes through the plant. Phloem cells are divided into sieve elements, or conducting cells, and supportive tissue. Together, xylem and phloem tissues form the vascular system of plants.

Roots: Support for the Plant

Roots are not well preserved in the fossil record; nevertheless, it seems that they did appear later in evolution than vascular tissue. The development of an extensive network of roots represented a significant new feature of vascular plants. Thin rhizoids attached the bryophytes to the substrate. Their rather flimsy filaments did not provide a strong anchor for the plant; neither did they absorb water and nutrients. In contrast, roots, with their prominent vascular tissue system, transfer water and minerals from the soil to the rest of the plant. The extensive network of roots that penetrates deep in the ground to reach sources of water also stabilizes trees by acting as ballast and an anchor. The majority of roots establish a symbiotic relationship with fungi, forming mycorrhizae. In the mycorrhizae, fungal hyphae grow around the root and within the root around the cells, and in some instances within the cells. This benefits the plant by greatly increasing the surface area for absorption.

Leaves, Sporophylls, and Strobili

A third adaptation marks seedless vascular plants. Accompanying the prominence of the sporophyte and the development of vascular tissue, the appearance of true leaves improved photosynthetic efficiency. Leaves capture more sunlight with their increased surface area.

In addition to photosynthesis, leaves play another role in the life of the plants. Pinecones, mature fronds of ferns, and flowers are all **sporophylls**—leaves that were modified structurally to bear sporangia. **Strobili** are structures that contain the sporangia. They are prominent in conifers and are known commonly as cones: for example, the pine cones of pine trees.

Seedless Vascular Plants

By the Late Devonian period (385 million years ago), plants had evolved vascular tissue, well-defined leaves, and root systems. With these advantages, plants increased in height and size. During the Carboniferous period (359–299 million years ago), swamp forests of club mosses and horsetails, with some specimens reaching more than 30 meters tall, covered most of the land. These forests gave rise to the extensive coal deposits that gave the Carboniferous its name. In seedless vascular plants, the sporophyte became the dominant phase of the lifecycle.

Water is still required for fertilization of seedless vascular plants, and most favor a moist environment. Modern-day seedless vascular plants include club mosses, horsetails, ferns, and whisk ferns.

Club Mosses

The **club mosses**, or Lycophyta, are the earliest group of seedless vascular plants. They dominated the landscape of the Carboniferous period, growing into tall trees and forming large swamp forests. Today's club mosses are diminutive, evergreen plants consisting of a stem (which may be branched) and small leaves called microphylls (Figure 14.13). The division Lycophyta consists of close to 1,000 species, including quillworts (*Isoetales*), club mosses (Lycopodiales), and spike mosses (Selaginellales): none of which is a true moss.

Figure 14.13 *Lycopodium clavatum* is a club moss. (credit: Cory Zanker)

Horsetails

Ferns and whisk ferns belong to the division Pterophyta. A third group of plants in the Pterophyta, the horsetails, is sometimes classified separately from ferns. **Horsetails** have a single genus, *Equisetum*. They are the survivors of a large group of plants, known as Arthrophyta, which produced large trees and entire swamp forests in the Carboniferous. The plants are usually found in damp environments and marshes (Figure 14.14).

Figure 14.14 Horsetails thrive in a marsh. (credit: Myriam Feldman)

The stem of a horsetail is characterized by the presence of joints, or nodes: hence the name Arthrophyta, which means "jointed plant". Leaves and branches come out as whorls from the evenly spaced rings. The needle-shaped leaves do not contribute greatly to photosynthesis, the majority of which takes place in the green stem (Figure 14.15).

Figure 14.15 Thin leaves originating at the joints are noticeable on the horsetail plant. (credit: Myriam Feldman)

Ferns and Whisk Ferns

Ferns are considered the most advanced seedless vascular plants and display characteristics commonly observed in seed plants. Ferns form large leaves and branching roots. In contrast, **whisk ferns**, the psilophytes, lack both roots and leaves, which were probably lost by evolutionary reduction. Evolutionary reduction is a process by which natural selection reduces the size of a structure that is no longer favorable in a particular environment. Photosynthesis takes place in the green stem of a whisk fern. Small yellow knobs form at the tip of the branch stem and contain the sporangia. Whisk ferns have been classified outside the true ferns; however, recent comparative analysis of DNA suggests that this group may have lost both vascular tissue and roots through evolution, and is actually closely related to ferns.

With their large fronds, **ferns** are the most readily recognizable seedless vascular plants (Figure 14.16). About 12,000 species of ferns live in environments ranging from tropics to temperate forests. Although some species survive in dry environments, most ferns are restricted to moist and shaded places. They made their appearance in the fossil record during the Devonian period (416–359 million years ago) and expanded during the Carboniferous period, 359–299 million years ago (Figure 14.17).

Figure 14.16 Some specimens of this short tree-fern species can grow very tall. (credit: Adrian Pingstone)

EON	ERA	PERIOD	MILLIONS OF YEARS AGO
Phanerozoic	Cenozoic	Quaternary	---- 1.6 -----
		Tertiary	---- 66 -----
	Mesozoic	Cretaceous	----138-----
		Jurassic	----205-----
		Triassic	----240-----
	Paleozoic	Permian	----290-----
		Carboniferous	----360-----
		Devonian	----410-----
		Silurian	----435-----
		Ordovician	----500-----
		Cambrian	----570-----
Proterozoic			----2500-----
Archean			---3800?----
Pre-Archean			

Figure 14.17 This chart shows the geological time scale, beginning with the Pre-Archean eon 3800 million years ago and ending with the Quaternary period in present time. (credit: modification of work by USGS)

Go to this website (http://openstaxcollege.org/l/fern_life_cycl2) to see an animation of the lifecycle of a fern and to test your knowledge.

careers IN ACTION

Landscape Designer

Looking at the well-laid gardens of flowers and fountains seen in royal castles and historic houses of Europe, it is clear that the creators of those gardens knew more than art and design. They were also familiar with the biology of the plants they chose. Landscape design also has strong roots in the United States' tradition. A prime example of early American classical design is Monticello, Thomas Jefferson's private estate; among his many other interests, Jefferson maintained a passion for botany. Landscape layout can encompass a small private space, like a backyard garden; public gathering places, like Central Park in New York City; or an entire city plan, like Pierre L'Enfant's design for Washington, DC.

A landscape designer will plan traditional public spaces—such as botanical gardens, parks, college campuses, gardens, and larger developments—as well as natural areas and private gardens (Figure 14.18). The restoration of natural places encroached upon by human intervention, such as wetlands, also requires the expertise of a landscape designer.

With such an array of required skills, a landscape designer's education includes a solid background in botany, soil science, plant pathology, entomology, and horticulture. Coursework in architecture and design software is also required for the completion of the degree. The successful design of a landscape rests on an extensive knowledge of plant growth requirements, such as light and shade, moisture levels, compatibility of different species, and susceptibility to pathogens and pests. For example, mosses and ferns will thrive in a shaded area where fountains provide moisture; cacti, on the other hand, would not fare well in that environment. The future growth of the individual plants must be taken into account to avoid crowding and competition for light and nutrients. The appearance of the space over time is also of concern. Shapes, colors, and biology must be balanced for a well-maintained and sustainable green space. Art, architecture, and biology blend in a beautifully designed and implemented landscape.

Figure 14.18 This campus garden was designed by students in the horticulture and landscaping department of the college. (credit: Myriam Feldman)

14.3 | Seed Plants: Gymnosperms

By the end of this section, you will be able to:

- Discuss the type of seeds produced by gymnosperms, as well as other characteristics of gymnosperms
- List the four groups of modern-day gymnosperms and provide examples of each

The first plants to colonize land were most likely closely related to modern-day mosses (bryophytes) and are thought to have appeared about 500 million years ago. They were followed by liverworts (also bryophytes) and primitive vascular plants, the pterophytes, from which modern ferns are derived. The life cycle of bryophytes and pterophytes is characterized by the alternation of generations. The completion of the life cycle requires water, as the male gametes must swim to the female gametes. The male gametophyte releases sperm, which must swim—propelled by their flagella—to reach and fertilize the female gamete or egg. After fertilization, the zygote matures and grows into a sporophyte, which in turn will form sporangia, or "spore vessels," in which mother cells undergo meiosis and produce haploid spores. The release of spores in a suitable environment will lead to germination and a new generation of gametophytes.

The Evolution of Seed Plants

In seed plants, the evolutionary trend led to a dominant sporophyte generation, in which the larger and more ecologically significant generation for a species is the diploid plant. At the same time, the trend led to a reduction in the size of the gametophyte, from a conspicuous structure to a microscopic cluster of cells enclosed in the tissues of the sporophyte. Lower vascular plants, such as club mosses and ferns, are mostly homosporous (produce only one type of spore). In contrast, all seed plants, or spermatophytes, are heterosporous, forming two types of spores: megaspores (female) and microspores (male). Megaspores develop into female gametophytes that produce eggs, and microspores mature into male gametophytes that generate sperm. Because the gametophytes mature within the spores, they are not free-living, as are the gametophytes of other seedless vascular plants. Heterosporous seedless plants are seen as the evolutionary forerunners of seed plants.

Seeds and pollen—two adaptations to drought—distinguish seed plants from other (seedless) vascular plants. Both adaptations were critical to the colonization of land. Fossils place the earliest distinct seed plants at about 350 million years ago. The earliest reliable record of gymnosperms dates their appearance to the Carboniferous period (359–299 million years ago). Gymnosperms were preceded by the progymnosperms ("first naked seed plants"). This was a transitional group of plants that superficially resembled conifers ("cone bearers") because they produced wood from the secondary growth of the vascular tissues; however, they still reproduced like ferns, releasing spores to the environment. In the Mesozoic era (251–65.5 million years ago), gymnosperms dominated the landscape. Angiosperms took over by the middle of the Cretaceous period (145.5–65.5 million years ago) in the late Mesozoic era, and have since become the most abundant plant group in most terrestrial biomes.

The two innovative structures of pollen and seed allowed seed plants to break their dependence on water for reproduction and development of the embryo, and to conquer dry land. The pollen grains carry the male gametes of the plant. The small haploid ($1n$) cells are encased in a protective coat that prevents desiccation (drying out) and mechanical damage. Pollen can travel far from the sporophyte that bore it, spreading the plant's genes and avoiding competition with other plants. The seed offers the embryo protection, nourishment and a mechanism to maintain dormancy for tens or even thousands of years, allowing it to survive in a harsh environment and ensuring germination when growth conditions are optimal. Seeds allow plants to disperse the next generation through both space and time. With such evolutionary advantages, seed plants have become the most successful and familiar group of plants.

Gymnosperms

Gymnosperms ("naked seed") are a diverse group of seed plants and are paraphyletic. Paraphyletic groups do not include descendants of a single common ancestor. Gymnosperm characteristics include naked seeds, separate female and male gametes, pollination by wind, and tracheids, which transport water and solutes in the vascular system.

Life Cycle of a Conifer

Pine trees are conifers and carry both male and female sporophylls on the same plant. Like all gymnosperms, pines are heterosporous and produce male microspores and female megaspores. In the male cones, or staminate cones, the **microsporocytes** give rise to microspores by meiosis. The microspores then develop into pollen grains. Each pollen grain contains two cells: one generative cell that will divide into two sperm, and a second cell that will become the pollen tube cell. In the spring, pine trees release large amounts of yellow pollen, which is carried by the wind. Some gametophytes will land on a female cone. The pollen tube grows from the pollen grain slowly, and the generative cell in the pollen grain divides into two sperm cells by mitosis. One of the sperm cells will finally unite its haploid nucleus with the haploid nucleus of an egg cell in the process of fertilization.

Female **cones**, or ovulate cones, contain two ovules per scale. One **megasporocyte** undergoes meiosis in each ovule. Only a single surviving haploid cell will develop into a female multicellular gametophyte that encloses an egg. On fertilization, the zygote will give rise to the embryo, which is enclosed in a seed coat of tissue from the parent plant. Fertilization and seed development is a long process in pine trees—it may take up to two years after pollination. The seed that is formed contains three generations of tissues: the seed coat that originates from the parent plant tissue, the female gametophyte that will provide nutrients, and the embryo itself. Figure 14.19 illustrates the life cycle of a conifer.

art CONNECTION

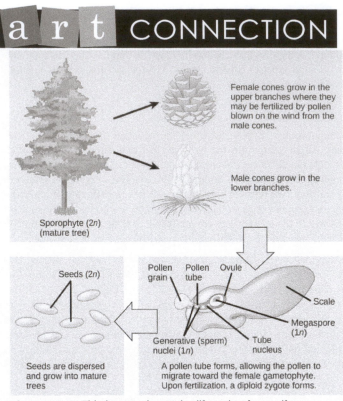

Figure 14.19 This image shows the lifecycle of a conifer.

At what stage does the diploid zygote form?

a. when the female cone begins to bud from the tree

b. when the sperm nucleus and the egg nucleus fuse

c. when the seeds drop from the tree

d. when the pollen tube begins to grow

Watch this video (http://openstaxcollege.org/l/gymnosperm) to see the process of seed production in gymnosperms.

Diversity of Gymnosperms

Modern gymnosperms are classified into four major divisions and comprise about 1,000 described species. Coniferophyta, Cycadophyta, and Ginkgophyta are similar in their production of secondary cambium (cells that generate the vascular system of the trunk or stem) and their pattern of seed development, but are not closely related phylogenetically to each other. Gnetophyta are considered the closest group to angiosperms because they produce true xylem tissue that contains both tracheids and vessel elements.

Conifers

Conifers are the dominant phylum of gymnosperms, with the most variety of species. Most are tall trees that usually bear scale-like or needle-like leaves. The thin shape of the needles and their waxy cuticle limits water loss through transpiration. Snow slides easily off needle-shaped leaves, keeping the load light and decreasing breaking of branches. These adaptations to cold and dry weather explain the predominance of conifers at high altitudes and in cold climates. Conifers include familiar evergreen trees, such as pines, spruces, firs, cedars, sequoias, and yews (Figure 14.20). A few species are deciduous and lose their leaves all at once in fall. The European larch and the tamarack are examples of deciduous conifers. Many coniferous trees are harvested for paper pulp and timber. The wood of conifers is more primitive than the wood of angiosperms; it contains tracheids, but no vessel elements, and is referred to as "soft wood."

Figure 14.20 Conifers are the dominant form of vegetation in cold or arid environments and at high altitudes. Shown here are the (a) evergreen spruce, (b) sequoia, (c) juniper, and (d) a deciduous gymnosperm: the tamarack *Larix larcinia*. Notice the yellow leaves of the tamarack. (credit b: modification of work by Alan Levine; credit c: modification of work by Wendy McCormac; credit d: modification of work by Micky Zlimen)

Cycads

Cycads thrive in mild climates and are often mistaken for palms because of the shape of their large, compound leaves. They bear large cones, and unusually for gymnosperms, may be pollinated by beetles, rather than wind. They dominated the landscape during the age of dinosaurs in the Mesozoic era (251–65.5 million years ago). Only a hundred or so cycad species persisted to modern times. They face possible extinction, and several species are protected through international conventions. Because of their attractive shape, they are often used as ornamental plants in gardens (Figure 14.21).

Figure 14.21 This *Encephalartos ferox* cycad exhibits large cones. (credit: Wendy Cutler)

Gingkophytes

The single surviving species of **ginkgophyte** is the *Ginkgo biloba* (Figure 14.22). Its fan-shaped leaves, unique among seed plants because they feature a dichotomous venation pattern, turn yellow in autumn and fall from the plant. For centuries, Buddhist monks cultivated *Ginkgo biloba,* ensuring its preservation. It is planted in public spaces because it is unusually resistant to pollution. Male and female organs are found on separate plants. Usually, only male trees are planted by gardeners because the seeds produced by the female plant have an off-putting smell of rancid butter.

Figure 14.22 This plate from the 1870 book *Flora Japonica, Sectio Prima (Tafelband)* depicts the leaves and fruit of *Gingko biloba*, as drawn by Philipp Franz von Siebold and Joseph Gerhard Zuccarini.

Gnetophytes

Gnetophytes are the closest relatives to modern angiosperms, and include three dissimilar genera of plants. Like angiosperms, they have broad leaves. *Gnetum* species are mostly vines in tropical and subtropical zones. The single species of *Welwitschia* is an unusual, low-growing plant found in the deserts of Namibia and Angola. It may live for up to 2000 years. The genus *Ephedra* is represented in North America in dry areas of the southwestern United States and Mexico (Figure 14.23). *Ephedra's* small, scale-like leaves are the source of the compound ephedrine, which is used in medicine as a potent decongestant. Because ephedrine is similar to amphetamines, both in chemical structure and neurological effects, its use is restricted to prescription drugs. Like angiosperms, but unlike other gymnosperms, all gnetophytes possess vessel elements in their xylem.

Figure 14.23 *Ephedra viridis*, known by the common name Mormon tea, grows in the western United States. (credit: US National Park Service, USDA-NRCS PLANTS Database)

Watch this BBC video (http://openstaxcollege.org/l/welwitschia) describing the amazing strangeness of Welwitschia.

14.4 | Seed Plants: Angiosperms

By the end of this section, you will be able to:

- Describe the main parts of a flower and their purpose
- Detail the life cycle of an angiosperm
- Discuss the two main groups into which flower plants are divided, as well as explain how basal angiosperms differ from others

From their humble and still obscure beginning during the early Jurassic period (202–145.5 MYA), the angiosperms, or flowering plants, have successfully evolved to dominate most terrestrial ecosystems. Angiosperms include a staggering number of genera and species; with more than 260,000 species, the division is second only to insects in terms of diversification (Figure 14.24).

Figure 14.24 These flowers grow in a botanical garden border in Bellevue, WA. Flowering plants dominate terrestrial landscapes. The vivid colors of flowers are an adaptation to pollination by insects and birds. (credit: Myriam Feldman)

Angiosperm success is a result of two novel structures that ensure reproductive success: flowers and fruit. Flowers allowed plants to form cooperative evolutionary relationships with animals, in particular insects, to disperse their pollen to female gametophytes in a highly targeted way. Fruit protect the developing embryo and serve as an agent of dispersal. Different structures on fruit reflect the dispersal strategies that help with the spreading of seeds.

Flowers

Flowers are modified leaves or sporophylls organized around a central stalk. Although they vary greatly in appearance, all flowers contain the same structures: sepals, petals, pistils, and stamens. A whorl of **sepals** (the **calyx**) is located at the base of the peduncle, or stem, and encloses the floral bud before it opens. Sepals are usually photosynthetic organs, although there are some exceptions. For example, the corolla in lilies and tulips consists of three sepals and three petals that look virtually identical—this led botanists to coin the word tepal. **Petals** (collectively the **corolla**) are located inside the whorl of sepals and usually display vivid colors to attract pollinators. Flowers pollinated by wind are usually small and dull. The sexual organs are located at the center of the flower.

As illustrated in Figure 14.25, the stigma, style, and ovary constitute the female organ, the **carpel** or **pistil**, which is also referred to as the **gynoecium**. A gynoecium may contain one or more carpels within a single flower. The megaspores and the female gametophytes are produced and protected by the thick tissues of the carpel. A long, thin structure called a **style** leads from the sticky **stigma**, where pollen is deposited, to the **ovary** enclosed in the carpel. The ovary houses one or more ovules that will each develop into a seed upon fertilization. The male reproductive organs, the androecium or **stamens**, surround the central carpel. Stamens are composed of a thin stalk called a **filament** and a sac-like structure, the **anther**, in which microspores are produced by meiosis and develop into pollen grains. The filament supports the anther.

Download for free at https://openstax.org/details/books/concepts-biology

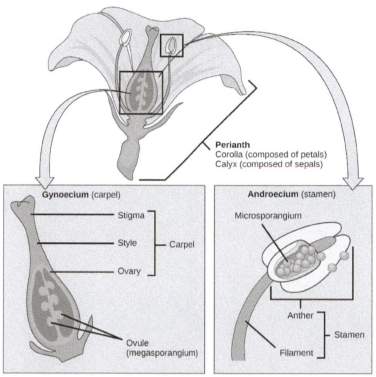

Figure 14.25 This image depicts the structure of a perfect and complete flower. Perfect flowers carry both male and female floral organs. (credit: modification of work by Mariana Ruiz Villareal)

Fruit

The seed forms in an ovary, which enlarges as the seeds grow. As the seed develops, the walls of the ovary also thicken and form the fruit. In botany, a fruit is a fertilized and fully grown, ripened ovary. Many foods commonly called vegetables are actually fruit. Eggplants, zucchini, string beans, and bell peppers are all technically fruit because they contain seeds and are derived from the thick ovary tissue. Acorns and winged maple keys, whose scientific name is a samara, are also fruit.

Mature fruit can be described as fleshy or dry. Fleshy fruit include the familiar berries, peaches, apples, grapes, and tomatoes. Rice, wheat, and nuts are examples of dry fruit. Another distinction is that not all fruits are derived from the ovary. Some fruits are derived from separate ovaries in a single flower, such as the raspberry. Other fruits, such as the pineapple, form from clusters of flowers. Additionally, some fruits, like watermelon and orange, have rinds. Regardless of how they are formed, fruits are an agent of dispersal. The variety of shapes and characteristics reflect the mode of dispersal. The light, dry fruits of trees and dandelions are carried by the wind. Floating coconuts are transported by water. Some fruits are colored, perfumed, sweet, and nutritious to attract herbivores, which eat the fruit and disperse the tough undigested seeds in their feces. Other fruits have burs and hooks that cling to fur and hitch rides on animals.

The Life Cycle of an Angiosperm

The adult, or sporophyte, phase is the main phase in an angiosperm's life cycle. Like gymnosperms, angiosperms are heterosporous. They produce microspores, which develop into pollen grains (the male gametophytes), and megaspores, which form an ovule containing the female gametophytes. Inside the anthers' microsporangia (Figure 14.26), male microsporocytes divide by meiosis, generating haploid microspores that undergo mitosis and give rise to pollen grains. Each pollen grain contains two cells: one generative cell that will divide into two sperm, and a second cell that will become the pollen tube cell.

art CONNECTION

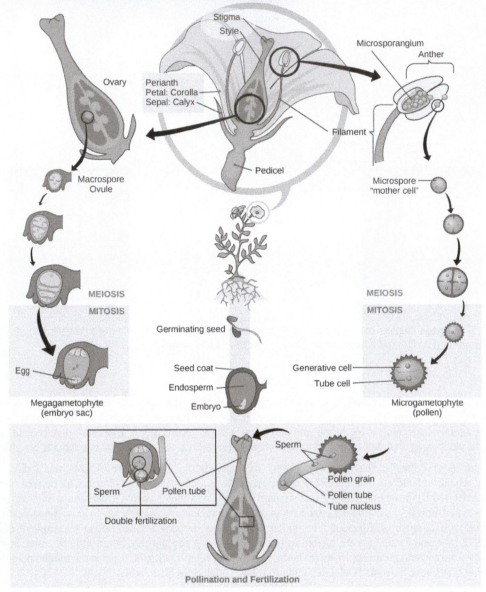

Figure 14.26 This diagram shows the lifecycle of an angiosperm. Anthers and ovaries are structures that shelter the actual gametophytes: the pollen grain and embryo sac. Double fertilization is a process unique to angiosperms. (credit: modification of work by Mariana Ruiz Villareal)

If a flower lacked a megasporangium, what type of gamete would it not be able to form? If it lacked a microsporangium, what type of gamete would not form?

In the ovules, the female gametophyte is produced when a megasporocyte undergoes meiosis to produce four haploid megaspores. One of these is larger than the others and undergoes mitosis to form the female gametophyte or embryo sac. Three mitotic divisions produce eight nuclei in seven cells. The egg and two cells move to one end of the embryo sac (gametophyte) and three cells move to the other end. Two of the nuclei remain in a single cell and fuse to form a 2*n* nucleus; this cell moves to the center of the embryo sac.

When a pollen grain reaches the stigma, a pollen tube extends from the grain, grows down the style, and enters through an opening in the integuments of the ovule. The two sperm cells are deposited in the embryo sac.

What occurs next is called a double fertilization event (Figure 14.27) and is unique to angiosperms. One sperm and the egg combine, forming a diploid zygote—the future embryo. The other sperm fuses with the diploid nucleus in the center of the embryo sac, forming a triploid cell that will develop into the endosperm: a tissue that serves as a food reserve. The zygote develops into an embryo with a radicle, or small root, and one or two leaf-like organs called **cotyledons**. Seed food reserves are stored outside the embryo, and the cotyledons serve as conduits to transmit the broken-down food reserves to the developing embryo. The seed consists of a toughened layer of integuments forming the coat, the endosperm with food reserves and, at the center, the well-protected embryo.

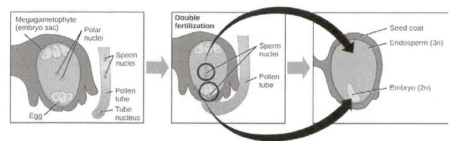

Figure 14.27 Double fertilization occurs only in angiosperms. (credit: modification of work by Mariana Ruiz Villareal)

Most flowers carry both stamens and carpels; however, a few species self-pollinate. These are known as "perfect" flowers because they contain both types of sex organs (Figure 14.25. Biochemical and anatomical barriers to self-pollination promote cross-pollination. Self-pollination is a severe form of inbreeding, and can increase the number of genetic defects in offspring.

A plant may have perfect flowers, and thus have both genders in each flower; or, it may have imperfect flowers of both kinds on one plant (Figure 14.28). In each case, such species are called monoecious plants, meaning "one house." Some botanists refer to plants with perfect flowers simply as hermaphroditic. Some plants are dioecious, meaning "two houses," and have male and female flowers ("imperfect flowers") on different plants. In these species, cross-pollination occurs all the time.

Figure 14.28 Monoecious plants have both male and female reproductive structures on the same flower or plant. In dioecious plants, males and females reproductive structures are on separate plants. (credit a: modification of work by Liz West; credit c: modification of work by Scott Zona)

Diversity of Angiosperms

Angiosperms are classified in a single division, the **Anthophyta**. Modern angiosperms appear to be a monophyletic group, which means that they originate from a single ancestor. Flowering plants are divided into two major groups, according to the structure of the cotyledons, the pollen grains, and other features: **monocots**, which include grasses and lilies, and **eudicots** or **dicots**, a polyphyletic group. **Basal angiosperms** are a group of plants that are believed to have branched off before the separation into monocots and eudicots because they exhibit traits from both groups. They are categorized separately in many classification schemes, and correspond to a grouping known as the Magnoliidae. The Magnoliidae group is comprised of magnolia trees, laurels, water lilies, and the pepper family.

Basal Angiosperms

The Magnoliidae are represented by the magnolias: tall trees that bear large, fragrant flowers with many parts, and are considered archaic (Figure 14.29**d**). Laurel trees produce fragrant leaves and small inconspicuous flowers. The Laurales are small trees and shrubs that grow mostly in warmer climates. Familiar plants in this group include the bay laurel, cinnamon, spice bush (Figure 14.29**a**), and the avocado tree. The Nymphaeales are comprised of the water lilies, lotus (Figure 14.29**c**), and similar plants. All species of the Nymphaeales thrive in freshwater biomes, and have leaves that float on the water surface or grow underwater. Water lilies are particularly prized by gardeners, and have graced ponds and pools since antiquity. The Piperales are a group of herbs, shrubs, and small trees that grow in tropical climates. They have small flowers without petals that are tightly arranged in long spikes. Many species are the source of prized fragrances or spices; for example, the berries of *Piper nigrum* (Figure 14.29**b**) are the familiar black pepper that is used to flavor many dishes.

(a)   (b)

(c)   (d)

Figure 14.29 The (a) southern spicebush belongs to the *Laurales*, the same family as cinnamon and bay laurel. The fruit of (b) the *Piper nigrum* plant is black pepper, the main product that was traded along spice routes. Notice the small, unobtrusive clustered flowers. (c) Lotus flowers, *Nelumbo nucifera*, have been cultivated since antiquity for their ornamental value; the root of the lotus flower is eaten as a vegetable. The (d) red berries of a magnolia tree, characteristic of the final stage, are just starting to appear. (credit a: modification of work by Cory Zanker; credit b: modification of work by Franz Eugen Köhler; credit c: modification of work by "berduchwal"/Flickr; credit d: modification of work by "Coastside2"/Wikimedia Commons)

Monocots

Plants in the monocot group have a single cotyledon in the seedling, and also share other anatomical features. Veins run parallel to the length of the leaves, and flower parts are arranged in a three- or six-fold symmetry. The pollen from the first angiosperms was monosulcate (containing a single furrow or pore through the outer layer). This feature is still seen in the modern monocots. True woody tissue is rarely found in monocots, and the vascular tissue of the stem is not arranged in any particular pattern. The root system is mostly adventitious (unusually positioned) with no major taproot. The monocots include familiar plants such as the true lilies (not to be confused with the water lilies), orchids, grasses, and palms. Many important crops, such as rice and other cereals (Figure 14.30a), corn, sugar cane, and tropical fruit, including bananas and pineapple, belong to the monocots.

Figure 14.30 The major crops in the world are flowering plants. One staple food, (a) rice, is a monocot, as are other cereals, while (b) beans are eudicots. Some popular flowers, such as this (c) lily are monocots; while others, such as this (d) daisy are eudicots. (credit a: modification of work by David Nance; credit b: modification of work by USDA, ARS; credit c: modification of work by "longhorndave"/Flickr; credit d: modification of work by "Cellulaer"/NinjaPhoto)

Eudicots

Eudicots, or true dicots, are characterized by the presence of two cotyledons. Veins form a network in leaves. Flower parts come in four, five, or many whorls. Vascular tissue forms a ring in the stem. (In monocots, vascular tissue is scattered in the stem.) Eudicots can be **herbaceous** (like dandelions or violets), or produce woody tissues. Most eudicots produce pollen that is trisulcate or triporate, with three furrows or pores. The root system is usually anchored by one main root developed from the embryonic radicle. Eudicots comprise two-thirds of all flowering plants. Many species seem to exhibit characteristics that belong to either group; therefore, the classification of a plant as a monocot or a eudicot is not always clearly evident (Table 14.1).

Comparison of Structural Characteristics of Monocots and Eudicots

Characteristic	Monocot	Eudicot
Cotyledon	One	Two
Veins in leaves	Parallel	Network (branched)
Vascular tissue	Scattered	Arranged in ring pattern
Roots	Network of adventitious roots	Tap root with many lateral roots
Pollen	Monosulcate	Trisulcate
Flower parts	Three or multiple of three	Four, five, multiple of four or five and whorls

Table 14.1

Explore this website (http://openstaxcollege.org/l/pollinators) for more information on poillinators.

KEY TERMS

anther a sac-like structure at the tip of the stamen in which pollen grains are produced

Anthophyta the division to which angiosperms belong

apical meristem the growing point in a vascular plant at the tip of a shoot or root where cell division occurs

basal angiosperms a group of plants that probably branched off before the separation of monocots and eudicots

calyx the whorl of sepals

carpel the female reproductive part of a flower consisting of the stigma, style, and ovary

club moss the earliest group of seedless vascular plants

cone the ovulate strobilus on gymnosperms that contains ovules

conifer the dominant division of gymnosperms with the most variety of species

corolla the collection of petals

cotyledon the one (monocot) or two (dicot) primitive leaves present in a seed

cycad a division of gymnosperms that grow in tropical climates and resemble palm trees

dicot a group of angiosperms whose embryos possess two cotyledons; also known as eudicot

diplontic describes a life cycle in which the diploid stage is the dominant stage

eudicots a group of angiosperms whose embryos possess two cotyledons; also known as dicot

fern a seedless vascular plant that produces large fronds; the most advanced group of seedless vascular plants

filament the thin stalk that links the anther to the base of the flower

gametangium (plural: gametangia) the structure within which gametes are produced

gametophyte the haploid plant that produces gametes

gingkophyte a division of gymnosperm with one living species, the *Gingko biloba*, a tree with fan-shaped leaves

gnetophyte a division of gymnosperms with varied morphological features that produce vessel elements in their woody tissues

gymnosperm a seed plant with naked seeds (seeds exposed on modified leaves or in cones)

gynoecium the group of structures that constitute the female reproductive organ; also called the pistil

haplodiplontic describes a life cycle in which the haploid and diploid stages alternate; also known as an alternation of generations life cycle

haplontic describes a life cycle in which the haploid stage is the dominant stage

herbaceous describes a plant without woody tissue

heterosporous having two kinds of spores that give rise to male and female gametophytes

homosporous having one kind of spore that gives rise to gametophytes that give rise to both male and female gametes

hornwort a group of non-vascular plants in which stomata appear

horsetail a seedless vascular plant characterized by a jointed stem

liverwort the most primitive group of non-vascular plants

megasporocyte a megaspore mother cell; larger spore that germinates into a female gametophyte in a heterosporous plant

microsporocyte smaller spore that produces a male gametophyte in a heterosporous plant

monocot a related group of angiosperms that produce embryos with one cotyledon and pollen with a single ridge

moss a group of plants in which a primitive conductive system appears

nonvascular plant a plant that lacks vascular tissue formed of specialized cells for the transport of water and nutrients

ovary the chamber that contains and protects the ovule or female megasporangium

petal a modified leaf interior to the sepal; colorful petals attract animal pollinator

phloem the vascular tissue responsible for transport of sugars, proteins, and other solutes

pistil the group of structures that constitute the female reproductive organ; also called the carpel

sepal a modified leaf that encloses the bud; outermost structure of a flower

sporangium (plural: sporangia) the organ within which spores are produced

sporophyll a leaf modified structurally to bear sporangia

sporophyte the diploid plant that produces spores

stamen the group of structures that contain the male reproductive organs

stigma uppermost structure of the carpel where pollen is deposited

strobili cone-like structures that contain the sporangia

style the long thin structure that links the stigma to the ovary

syngamy the union of two gametes in fertilization

vascular plant a plant in which there is a network of cells that conduct water and solutes through the organism

whisk fern a seedless vascular plant that lost roots and leaves by evolutionary reduction

xylem the vascular tissue responsible for long-distance transport of water and nutrients

CHAPTER SUMMARY

14.1 The Plant Kingdom

Land plants evolved traits that made it possible to colonize land and survive out of water. Adaptations to life on land include vascular tissues, roots, leaves, waxy cuticles, and a tough outer layer that protects the spores. Land plants include nonvascular plants and vascular plants. Vascular plants, which include seedless plants and plants with seeds, have apical meristems, and embryos with nutritional stores. All land plants share the following characteristics: alternation of generations, with the haploid plant called a gametophyte and the diploid plant called a sporophyte; formation of haploid spores in a sporangium; and formation of gametes in a gametangium.

14.2 Seedless Plants

Seedless nonvascular plants are small. The dominant stage of the life cycle is the gametophyte. Without a vascular system and roots, they absorb water and nutrients through all of their exposed surfaces. There are three main groups: the liverworts, the hornworts, and the mosses. They are collectively known as bryophytes.

Vascular systems consist of xylem tissue, which transports water and minerals, and phloem tissue, which transports sugars and proteins. With the vascular system, there appeared leaves—large photosynthetic organs—and roots to absorb water from the ground. The seedless vascular plants include club mosses, which are the most primitive; whisk ferns, which lost leaves and roots by reductive evolution; horsetails, and ferns.

14.3 Seed Plants: Gymnosperms

Gymnosperms are heterosporous seed plants that produce naked seeds. They appeared in the Carboniferous period (359–299 million years ago) and were the dominant plant life during the Mesozoic era (251–65.5 million years ago). Modern-day gymnosperms belong to four divisions. The division Coniferophyta—the conifers—are the predominant woody plants at high altitudes and latitudes. Cycads resemble palm trees and grow in tropical climates. *Gingko biloba* is the only species of the division Gingkophyta. The last division, the Gnetophytes, is a diverse group of species that produce vessel elements in their wood.

14.4 Seed Plants: Angiosperms

Angiosperms are the dominant form of plant life in most terrestrial ecosystems, comprising about 90 percent of all plant species. Most crop and ornamental plants are angiosperms. Their success results, in part, from two innovative structures: the flower and the fruit. Flowers are derived evolutionarily from modified leaves. The main parts of a flower are the sepals and petals, which protect the reproductive parts: the stamens and the carpels. The stamens produce the male gametes, which are pollen grains. The carpels contain the female gametes, which are the eggs inside ovaries. The walls of the ovary thicken after fertilization, ripening into fruit that can facilitate seed dispersal.

Angiosperms' life cycles are dominated by the sporophyte stage. Double fertilization is an event unique to angiosperms. The flowering plants are divided into two main groups—the monocots and eudicots—according to the number of cotyledons in the seedlings. Basal angiosperms belong to a lineage older than monocots and eudicots.

ART CONNECTION QUESTIONS

1. Figure 14.19 At what stage does the diploid zygote form?
   a. When the female cone begins to bud from the tree
   b. When the sperm nucleus and the egg nucleus fuse
   c. When the seeds drop from the tree
   d. When the pollen tube begins to grow

2. Figure 14.26 If a flower lacked a megasporangium, what type of gamete would it not be able to form? If it lacked a microsporangium, what type of gamete would not form?

REVIEW QUESTIONS

3. The land plants are probably descendants of which of these groups?
   a. green algae
   b. red algae
   c. brown algae
   d. angiosperms

4. The event that leads from the haploid stage to the diploid stage in alternation of generations is _____.

   a. meiosis
   b. mitosis
   c. fertilization
   d. germination

5. Moss is an example of which type of plant?
   a. haplontic plant
   b. vascular plant
   c. diplontic plant
   d. seed plant

6. Why do mosses grow well in the Arctic tundra?

   a. They grow better at cold temperatures.
   b. They do not require moisture.
   c. They do not have true roots and can grow on hard surfaces.
   d. There are no herbivores in the tundra.

7. Which is the most diverse group of seedless vascular plants?
   a. the liverworts
   b. the horsetails
   c. the club mosses
   d. the ferns

8. Which group are vascular plants?
   a. liverworts
   b. mosses
   c. hornworts
   d. ferns

9. Which of the following traits characterizes gymnosperms?
   a.   The plants carry exposed seeds on modified leaves.
   b.   Reproductive structures are located in a flower.
   c.   After fertilization, the ovary thickens and forms a fruit.
   d.   The gametophyte is longest phase of the life cycle.

10. What adaptation do seed plants have in addition to the seed that is not found in seedless plants?
   a.   gametophytes
   b.   vascular tissue
   c.   pollen

   d.   chlorophyll

11. Pollen grains develop in which structure?
   a.   the anther
   b.   the stigma
   c.   the filament
   d.   the carpel

12. Corn develops from a seedling with a single cotyledon, displays parallel veins on its leaves, and produces monosulcate pollen. It is most likely:
   a.   a gymnosperm
   b.   a monocot
   c.   a eudicot
   d.   a basal angiosperm

CRITICAL THINKING QUESTIONS

13. What adaptations do plants have that allow them to survive on land?

14. What are the three classes of bryophytes?

15. How did the development of a vascular system contribute to the increase in size of plants?

16. What are the four modern-day groups of gymnosperms?

17. Cycads are considered endangered species and their trade is severely restricted. Customs officials stop suspected smugglers, who claim that the plants in their possession are palm trees and not cycads. How would a botanist distinguish between the two types of plants?

18. What are the two structures that allow angiosperms to be the dominant form of plant life in most terrestrial ecosystems?

15 | DIVERSITY OF ANIMALS

Figure 15.1 The leaf chameleon (*Brookesia micra*) was discovered in northern Madagascar in 2012. At just over one inch long, it is the smallest known chameleon. (credit: modification of work by Frank Glaw, et al., PLOS)

Chapter Outline

15.1: Features of the Animal Kingdom

15.2: Sponges and Cnidarians

15.3: Flatworms, Nematodes, and Arthropods

15.4: Mollusks and Annelids

15.5: Echinoderms and Chordates

15.6: Vertebrates

Introduction

While we can easily identify dogs, lizards, fish, spiders, and worms as animals, other animals, such as corals and sponges, might be easily mistaken as plants or some other form of life. Yet scientists have recognized a set of common characteristics shared by all animals, including sponges, jellyfish, sea urchins, and humans.

The kingdom Animalia is a group of multicellular Eukarya. Animal evolution began in the ocean over 600 million years ago, with tiny creatures that probably do not resemble any living organism today. Since then, animals have evolved into a highly diverse kingdom. Although over one million currently living species of animals have been identified, scientists are continually discovering more species. The number of described living animal species is estimated to be about 1.4 million,[1] and there may be as many as 6.8 million.

Understanding and classifying the variety of living species helps us to better understand how to conserve and benefit from this diversity. The animal classification system characterizes animals based on their anatomy, features of embryological development, and genetic makeup. Scientists are faced with the task of classifying animals within a system of taxonomy that reflects their evolutionary history. Additionally, they must identify traits that are common to all animals as well as traits that can be used to distinguish among related groups of animals. However, animals vary in the complexity of their organization

---

1. "Number of Living Species in Australia and the World," A.D. Chapman, Australia Biodiversity Information Services, last modified August 26, 2010, http://www.environment.gov.au/biodiversity/abrs/publications/other/species-numbers/2009/03-exec-summary.html.

and exhibit a huge diversity of body forms, so the classification scheme is constantly changing as new information about species is learned.

15.1 | Features of the Animal Kingdom

By the end of this section, you will be able to:

- List the features that distinguish the animal kingdom from other kingdoms

- Explain the processes of animal reproduction and embryonic development

- Describe the hierarchy of basic animal classification

- Compare and contrast the embryonic development of protostomes and deuterostomes

Even though members of the animal kingdom are incredibly diverse, animals share common features that distinguish them from organisms in other kingdoms. All animals are eukaryotic, multicellular organisms, and almost all animals have specialized tissues. Most animals are motile, at least during certain life stages. Animals require a source of food to grow and develop. All animals are heterotrophic, ingesting living or dead organic matter. This form of obtaining energy distinguishes them from autotrophic organisms, such as most plants, which make their own nutrients through photosynthesis and from fungi that digest their food externally. Animals may be carnivores, herbivores, omnivores, or parasites (Figure 15.2). Most animals reproduce sexually: The offspring pass through a series of developmental stages that establish a determined body plan, unlike plants, for example, in which the exact shape of the body is indeterminate. The **body plan** refers to the shape of an animal.

(a)   (b)

Figure 15.2 All animals that derive energy from food are heterotrophs. The (a) black bear is an omnivore, eating both plants and animals. The (b) heartworm *Dirofilaria immitis* is a parasite that derives energy from its hosts. It spends its larval stage in mosquitos and its adult stage infesting the hearts of dogs and other mammals, as shown here. (credit a: modification of work by USDA Forest Service; credit b: modification of work by Clyde Robinson)

Complex Tissue Structure

A hallmark trait of animals is specialized structures that are differentiated to perform unique functions. As multicellular organisms, most animals develop specialized cells that group together into tissues with specialized functions. A tissue is a collection of similar cells that had a common embryonic origin. There are four main types of animal tissues: nervous, muscle, connective, and epithelial. Nervous tissue contains neurons, or nerve cells, which transmit nerve impulses. Muscle tissue contracts to cause all types of body movement from locomotion of the organism to movements within the body itself. Animals also have specialized connective tissues that provide many functions, including transport and structural support. Examples of connective tissues include blood and bone. Connective tissue is comprised of cells separated by extracellular material made of organic and inorganic materials, such as the protein and mineral deposits of bone. Epithelial tissue covers the internal and external surfaces of organs inside the animal body and the external surface of the body of the organism.

View this video (http://openstaxcollege.org/l/saving_life2) to watch a presentation by biologist E.O. Wilson on the importance of animal diversity.

Animal Reproduction and Development

Most animals have diploid body (somatic) cells and a small number of haploid reproductive (gamete) cells produced through meiosis. Some exceptions exist: For example, in bees, wasps, and ants, the male is haploid because it develops from an unfertilized egg. Most animals undergo sexual reproduction, while many also have mechanisms of asexual reproduction.

Sexual Reproduction and Embryonic Development

Almost all animal species are capable of reproducing sexually; for many, this is the only mode of reproduction possible. This distinguishes animals from fungi, protists, and bacteria, where asexual reproduction is common or exclusive. During sexual reproduction, the male and female gametes of a species combine in a process called fertilization. Typically, the small, motile male sperm travels to the much larger, sessile female egg. Sperm form is diverse and includes cells with flagella or amoeboid cells to facilitate motility. Fertilization and fusion of the gamete nuclei produce a zygote. Fertilization may be internal, especially in land animals, or external, as is common in many aquatic species.

After fertilization, a developmental sequence ensues as cells divide and differentiate. Many of the events in development are shared in groups of related animal species, and these events are one of the main ways scientists classify high-level groups of animals. During development, animal cells specialize and form tissues, determining their future morphology and physiology. In many animals, such as mammals, the young resemble the adult. Other animals, such as some insects and amphibians, undergo complete metamorphosis in which individuals enter one or more larval stages. For these animals, the young and the adult have different diets and sometimes habitats. In other species, a process of incomplete metamorphosis occurs in which the young somewhat resemble the adults and go through a series of stages separated by molts (shedding of the skin) until they reach the final adult form.

Asexual Reproduction

Asexual reproduction, unlike sexual reproduction, produces offspring genetically identical to each other and to the parent. A number of animal species—especially those without backbones, but even some fish, amphibians, and reptiles—are capable of asexual reproduction. Asexual reproduction, except for occasional identical twinning, is absent in birds and mammals. The most common forms of asexual reproduction for stationary aquatic animals include budding and fragmentation, in which part of a parent individual can separate and grow into a new individual. In contrast, a form of asexual reproduction found in certain invertebrates and rare vertebrates is called parthenogenesis (or "virgin beginning"), in which unfertilized eggs develop into new offspring.

Classification Features of Animals

Animals are classified according to morphological and developmental characteristics, such as a body plan. With the exception of sponges, the animal body plan is symmetrical. This means that their distribution of body parts is balanced along an axis. Additional characteristics that contribute to animal classification include the number of tissue layers formed during development, the presence or absence of an internal body cavity, and other features of embryological development.

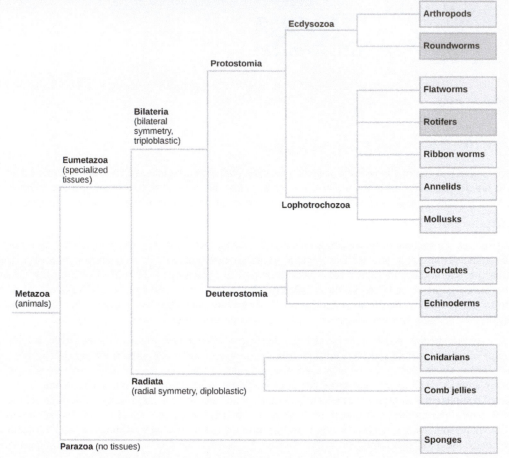

Figure 15.3 The phylogenetic tree of animals is based on morphological, fossil, and genetic evidence.

Which of the following statements is false?

a. Eumetazoa have specialized tissues and Parazoa do not.

b. Both acoelomates and pseudocoelomates have a body cavity.

c. Chordates are more closely related to echinoderms than to rotifers according to the figure.

d. Some animals have radial symmetry, and some animals have bilateral symmetry.

Body Symmetry

Animals may be asymmetrical, radial, or bilateral in form (Figure 15.4). **Asymmetrical** animals are animals with no pattern or symmetry; an example of an asymmetrical animal is a sponge (Figure 15.4a). An organism with **radial symmetry** (Figure 15.4b) has a longitudinal (up-and-down) orientation: Any plane cut along this up–down axis produces roughly mirror-image halves. An example of an organism with radial symmetry is a sea anemone.

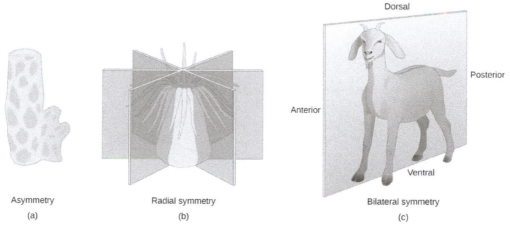

Figure 15.4 Animals exhibit different types of body symmetry. The (a) sponge is asymmetrical and has no planes of symmetry, the (b) sea anemone has radial symmetry with multiple planes of symmetry, and the (c) goat has bilateral symmetry with one plane of symmetry.

Bilateral symmetry is illustrated in Figure 15.4c using a goat. The goat also has upper and lower sides to it, but they are not symmetrical. A vertical plane cut from front to back separates the animal into roughly mirror-image right and left sides. Animals with bilateral symmetry also have a "head" and "tail" (anterior versus posterior) and a back and underside (dorsal versus ventral).

Watch this video (http://openstaxcollege.org/l/symmetry2) to see a quick sketch of the different types of body symmetry.

Layers of Tissues

Most animal species undergo a layering of early tissues during embryonic development. These layers are called **germ layers**. Each layer develops into a specific set of tissues and organs. Animals develop either two or three embryonic germs layers (Figure 15.5). The animals that display radial symmetry develop two germ layers, an inner layer (endoderm) and an outer layer (ectoderm). These animals are called **diploblasts**. Animals with bilateral symmetry develop three germ layers: an inner layer (endoderm), an outer layer (ectoderm), and a middle layer (mesoderm). Animals with three germ layers are called **triploblasts**.

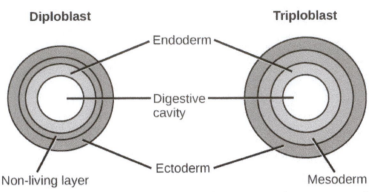

Figure 15.5 During embryogenesis, diploblasts develop two embryonic germ layers: an ectoderm and an endoderm. Triploblasts develop a third layer—the mesoderm—between the endoderm and ectoderm.

Presence or Absence of a Coelom

Triploblasts may develop an internal body cavity derived from mesoderm, called a **coelom** (pr. see-LŌM). This epithelial-lined cavity is a space, usually filled with fluid, which lies between the digestive system and the body wall. It houses organs such as the kidneys and spleen, and contains the circulatory system. Triploblasts that do not develop a coelom are called **acoelomates**, and their mesoderm region is completely filled with tissue, although they have a gut cavity. Examples of acoelomates include the flatworms. Animals with a true coelom are called **eucoelomates** (or coelomates) (Figure 15.6). A true coelom arises entirely within the mesoderm germ layer. Animals such as earthworms, snails, insects, starfish, and vertebrates are all eucoelomates. A third group of triploblasts has a body cavity that is derived partly from mesoderm and partly from endoderm tissue. These animals are called **pseudocoelomates**. Roundworms are examples of pseudocoelomates. New data on the relationships of pseudocoelomates suggest that these phyla are not closely related and so the evolution of the pseudocoelom must have occurred more than once (Figure 15.3). True coelomates can be further characterized based on features of their early embryological development.

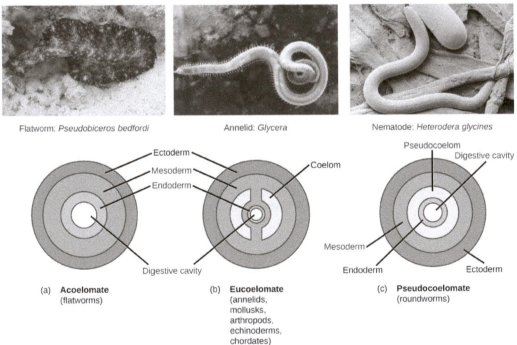

Figure 15.6 Triploblasts may be acoelomates, eucoelomates, or pseudocoelomates. Eucoelomates have a body cavity within the mesoderm, called a coelom, which is lined with mesoderm tissue. Pseudocoelomates have a similar body cavity, but it is lined with mesoderm and endoderm tissue. (credit a: modification of work by Jan Derk; credit b: modification of work by NOAA; credit c: modification of work by USDA, ARS)

Protostomes and Deuterostomes

Bilaterally symmetrical, triploblastic eucoelomates can be divided into two groups based on differences in their early embryonic development. **Protostomes** include phyla such as arthropods, mollusks, and annelids. **Deuterostomes** include the chordates and echinoderms. These two groups are named from which opening of the digestive cavity develops first: mouth or anus. The word *protostome* comes from Greek words meaning "mouth first," and *deuterostome* originates from words meaning "mouth second" (in this case, the anus develops first). This difference reflects the fate of a structure called the blastopore (Figure 15.7), which becomes the mouth in protostomes and the anus in deuterostomes. Other developmental characteristics differ between protostomes and deuterostomes, including the mode of formation of the coelom and the early cell division of the embryo.

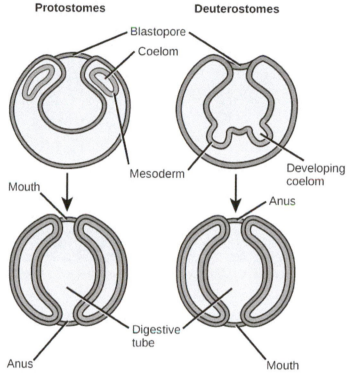

Figure 15.7 Eucoelomates can be divided into two groups, protostomes and deuterostomes, based on their early embryonic development. Two of these differences include the origin of the mouth opening and the way in which the coelom is formed.

15.2 | Sponges and Cnidarians

By the end of this section, you will be able to:

- Describe the organizational features of the simplest animals
- Describe the organizational features of cnidarians

The kingdom of animals is informally divided into invertebrate animals, those without a backbone, and vertebrate animals, those with a backbone. Although in general we are most familiar with vertebrate animals, the vast majority of animal species, about 95 percent, are invertebrates. Invertebrates include a huge diversity of animals, millions of species in about 32 phyla, which we can just begin to touch on here.

The sponges and the cnidarians represent the simplest of animals. Sponges appear to represent an early stage of multicellularity in the animal clade. Although they have specialized cells for particular functions, they lack true tissues in which specialized cells are organized into functional groups. Sponges are similar to what might have been the ancestor of animals: colonial, flagellated protists. The cnidarians, or the jellyfish and their kin, are the simplest animal group that displays true tissues, although they possess only two tissue layers.

Sponges

Animals in subkingdom Parazoa represent the simplest animals and include the sponges, or phylum **Porifera** (Figure 15.8). All sponges are aquatic and the majority of species are marine. Sponges live in intimate contact with water, which plays a role in their feeding, gas exchange, and excretion. Much of the body structure of the sponge is dedicated to moving water through the body so it can filter out food, absorb dissolved oxygen, and eliminate wastes.

Figure 15.8 Sponges are members of the phylum Porifera, which contains the simplest animals. (credit: Andrew Turner)

The body of the simplest sponges takes the shape of a cylinder with a large central cavity, the **spongocoel**. Water enters the spongocoel from numerous pores in the body wall. Water flows out through a large opening called the **osculum** (Figure 15.9). However, sponges exhibit a diversity of body forms, which vary in the size and branching of the spongocoel, the number of osculi, and where the cells that filter food from the water are located.

Sponges consist of an outer layer of flattened cells and an inner layer of cells called choanocytes separated by a jelly-like substance called **mesohyl**. The mesohyl contains embedded amoeboid cells that secrete tiny needles called **spicules** or protein fibers that help give the sponge its structural strength. The cell body of the **choanocyte** is embedded in mesohyl but protruding into the spongocoel is a mesh-like collar surrounding a single flagellum. The beating of flagella from all choanocytes moves water through the sponge. Food particles are trapped in mucus produced by the sieve-like collar of the choanocytes and are ingested by phagocytosis. This process is called **intracellular digestion**. **Amoebocytes** take up nutrients repackaged in food vacuoles of the choanocytes and deliver them to other cells within the sponge.

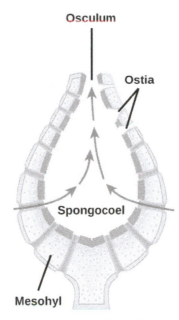

Basic sponge body plan

Figure 15.9 The sponge's basic body plan is shown.

Physiological Processes in Sponges

Despite their lack of complexity, sponges are clearly successful organisms, having persisted on Earth for more than half a billion years. Lacking a true digestive system, sponges depend on the intracellular digestive processes of their choanocytes for their energy intake. The limit of this type of digestion is that food particles must be smaller than individual cells. Gas exchange, circulation, and excretion occur by diffusion between cells and the water.

Sponges reproduce both sexually and asexually. Asexual reproduction is either by **fragmentation** (in which a piece of the sponge breaks off and develops into a new individual), or **budding** (an outgrowth from the parent that eventually detaches). A type of asexual reproduction found only in freshwater sponges occurs through the formation of **gemmules**, clusters of cells surrounded by a tough outer layer. Gemmules survive hostile environments and can attach to a substrate and grow into a new sponge.

Sponges are **monoecious** (or hermaphroditic), meaning one individual can produce both eggs and sperm. Sponges may be sequentially hermaphroditic, producing eggs first and sperm later. Eggs arise from amoebocytes and are retained within the spongocoel, whereas sperm arise from choanocytes and are ejected through the osculum. Sperm carried by water currents fertilize the eggs of other sponges. Early larval development occurs within the sponge, and free-swimming larvae are then released through the osculum. This is the only time that sponges exhibit mobility. Sponges are sessile as adults and spend their lives attached to a fixed substrate.

Watch this video (http://openstaxcollege.org/l/sponge_feed) that demonstrates the feeding of sponges.

Cnidarians

The phylum **Cnidaria** includes animals that show radial or biradial symmetry and are diploblastic. Nearly all (about 99 percent) cnidarians are marine species. Cnidarians have specialized cells known as **cnidocytes** ("stinging cells") containing organelles called **nematocysts**. These cells are concentrated around the mouth and tentacles of the animal and can immobilize prey with toxins. Nematocysts contain coiled threads that may bear barbs. The outer wall of the cell has a hairlike projection that is sensitive to touch. When touched, the cells fire the toxin-containing coiled threads that can penetrate and stun the predator or prey (see Figure 15.10).

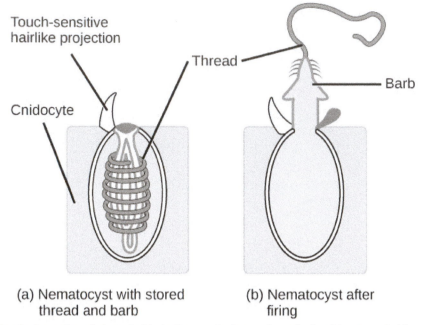

(a) Nematocyst with stored thread and barb

(b) Nematocyst after firing

Figure 15.10 Animals from the phylum Cnidaria have stinging cells called cnidocytes. Cnidocytes contain large organelles called (a) nematocysts that store a coiled thread and barb. When hairlike projections on the cell surface are touched, (b) the thread, barb, and a toxin are fired from the organelle.

Cnidarians display two distinct body plans: **polyp** or "stalk" and **medusa** or "bell" (Figure 15.11). Examples of the polyp form are freshwater species of the genus *Hydra*; perhaps the best-known medusoid animals are the jellies (jellyfish). Polyps are sessile as adults, with a single opening to the digestive system (the mouth) facing up with tentacles surrounding it. Medusae are motile, with the mouth and tentacles hanging from the bell-shaped body. In other cnidarians, both a polyp and medusa form exist, and the life cycle alternates between these forms.

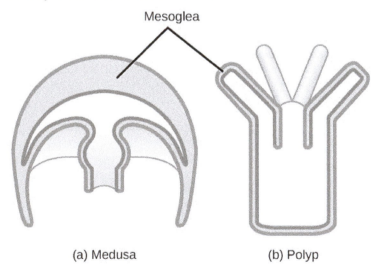

(a) Medusa   **(b) Polyp**

Figure 15.11 Cnidarians have two distinct body plans, the (a) medusa and the (b) polyp. All cnidarians have two tissue layers, with a jelly-like mesoglea between them.

Physiological Processes of Cnidarians

All cnidarians have two tissue layers. The outer layer is called the **epidermis**, whereas the inner layer is called the **gastrodermis** and lines the digestive cavity. Between these two layers is a non-living, jelly-like **mesoglea**. There are differentiated cell types in each tissue layer, such as nerve cells, enzyme-secreting cells, and nutrient-absorbing cells, as well as intercellular connections between the cells. However, organs and organ systems are not present in this phylum.

The nervous system is primitive, with nerve cells scattered across the body in a network. The function of the nerve cells is to carry signals from sensory cells and to contractile cells. Groups of cells in the nerve net form nerve cords that may be essential for more rapid transmission. Cnidarians perform **extracellular digestion**, with digestion completed by intracellular digestive processes. Food is taken into the **gastrovascular cavity**, enzymes are secreted into the cavity, and the cells lining the cavity absorb the nutrient products of the extracellular digestive process. The gastrovascular cavity has only one opening that serves as both a mouth and an anus (an incomplete digestive system). Like the sponges, Cnidarian cells exchange oxygen, carbon dioxide, and nitrogenous wastes by diffusion between cells in the epidermis and gastrodermis with water.

Cnidarian Diversity

The phylum Cnidaria contains about 10,000 described species divided into four classes: Anthozoa, Scyphozoa, Cubozoa, and Hydrozoa.

The class Anthozoa includes all cnidarians that exhibit a sessile polyp body plan only; in other words, there is no medusa stage within their life cycle. Examples include sea anemones, sea pens, and corals, with an estimated number of 6,100 described species. Sea anemones are usually brightly colored and can attain a size of 1.8 to 10 cm in diameter. These animals are usually cylindrical in shape and are attached to a substrate. A mouth opening is surrounded by tentacles bearing cnidocytes (Figure 15.12).

Figure 15.12 Sea anemones are cnidarians of class Anthozoa. (credit: "Dancing With Ghosts"/Flickr)

Scyphozoans include all the jellies and are motile and exclusively marine with about 200 described species. The medusa is the dominant stage in the life cycle, although there is also a polyp stage. Species range from 2 cm in length to the largest scyphozoan species, *Cyanea capillata*, at 2 m across. Jellies display a characteristic bell-like body shape (Figure 15.13).

Figure 15.13 Scyphozoans include the jellies. (credit: "Jimg944"/Flickr)

Identify the life cycle stages of jellies using this video animation game (http://openstaxcollege.org/l/amazing_jelly2) from the New England Aquarium.

The class Cubozoa includes jellies that are square in cross-section and so are known as "box jellyfish." These species may achieve sizes of 15–25 cm. Cubozoans are anatomically similar to the jellyfish. A prominent difference between the two classes is the arrangement of tentacles. Cubozoans have muscular pads called pedalia at the corners of the square bell canopy, with one or more tentacles attached to each pedalium. In some cases, the digestive system may extend into the pedalia. Cubozoans typically exist in a polyp form that develops from a larva. The polyps may bud to form more polyps and then transform into the medusoid forms.

Watch this video (http://openstaxcollege.org/l/box_jellyfish) to learn more about the deadly toxins of the box jellyfish.

Hydrozoa includes nearly 3,500 species,[2] most of which are marine. Most species in this class have both polyp and medusa forms in their life cycle. Many hydrozoans form colonies composed of branches of specialized polyps that share a gastrovascular cavity. Colonies may also be free-floating and contain both medusa and polyp individuals in the colony, as in the Portuguese Man O'War (*Physalia*) or By-the-Wind Sailor (*Velella*). Other species are solitary polyps or solitary medusae. The characteristic shared by all of these species is that their gonads are derived from epidermal tissue, whereas in all other cnidarians, they are derived from gastrodermal tissue (Figure 15.14ab).

---

2. "The Hydrozoa Directory," Peter Schuchert, Muséum Genève, last updated November 2012, http://www.ville-ge.ch/mhng/hydrozoa/hydrozoa-directory.htm.

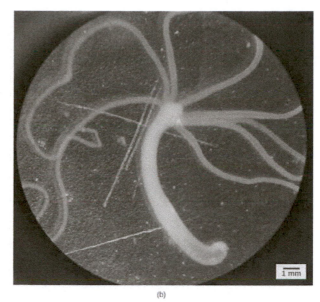

(a)   (b)

Figure 15.14 A (a) box jelly is an example from class Cubozoa. The (b) hydra is from class Hydrozoa. (credit b: scale-bar data from Matt Russell)

15.3 | Flatworms, Nematodes, and Arthropods

By the end of this section, you will be able to:

- Describe the structure and systems of flatworms
- Describe the structural organization of nematodes
- Compare the internal systems and the appendage specialization of arthropods

The animal phyla of this and subsequent modules are triploblastic and have an embryonic mesoderm sandwiched between the ectoderm and endoderm. These phyla are also bilaterally symmetrical, meaning that a longitudinal section will divide them into right and left sides that are mirror images of each other. Associated with bilateralism is the beginning of cephalization, the evolution of a concentration of nervous tissues and sensory organs in the head of the organism, which is where the organism first encounters its environment.

The flatworms are acoelomate organisms that include free-living and parasitic forms. The nematodes, or roundworms, possess a pseudocoelom and consist of both free-living and parasitic forms. Finally, the arthropods, one of the most successful taxonomic groups on the planet, are coelomate organisms with a hard exoskeleton and jointed appendages. The nematodes and the arthropods belong to a clade with a common ancestor, called Ecdysozoa. The name comes from the word *ecdysis*, which refers to the periodic shedding, or molting, of the exoskeleton. The ecdysozoan phyla have a hard cuticle covering their bodies that must be periodically shed and replaced for them to increase in size.

Flatworms

The relationships among flatworms, or phylum Platyhelminthes, is being revised and the description here will follow the traditional groupings. Most flatworms are parasitic, including important parasites of humans. Flatworms have three embryonic germ layers that give rise to surfaces covering tissues, internal tissues, and the lining of the digestive system. The epidermal tissue is a single layer of cells or a layer of fused cells covering a layer of circular muscle above a layer of longitudinal muscle. The mesodermal tissues include support cells and secretory cells that secrete mucus and other materials to the surface. The flatworms are acoelomate, so their bodies contain no cavities or spaces between the outer surface and the inner digestive tract.

Physiological Processes of Flatworms

Free-living species of flatworms are predators or scavengers, whereas parasitic forms feed from the tissues of their hosts. Most flatworms have an incomplete digestive system with an opening, the "mouth," that is also used to expel digestive system wastes. Some species also have an anal opening. The gut may be a simple sac or highly branched. Digestion is extracellular, with enzymes secreted into the space by cells lining the tract, and digested materials taken into the same cells by phagocytosis. One group, the cestodes, does not have a digestive system, because their parasitic lifestyle and the environment in which they live (suspended within the digestive cavity of their host) allows them to absorb nutrients directly across their body wall. Flatworms have an excretory system with a network of tubules throughout the body that open to the environment and nearby flame cells, whose cilia beat to direct waste fluids concentrated in the tubules out of the body. The system is responsible for regulation of dissolved salts and excretion of nitrogenous wastes. The nervous system consists of a pair of nerve cords running the length of the body with connections between them and a large ganglion or concentration of nerve cells at the anterior end of the worm; here, there may also be a concentration of photosensory and chemosensory cells (Figure 15.15).

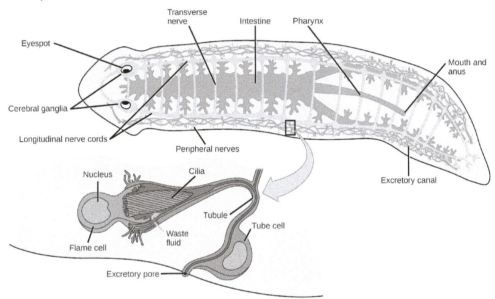

Figure 15.15 This planarian is a free-living flatworm that has an incomplete digestive system, an excretory system with a network of tubules throughout the body, and a nervous system made up of nerve cords running the length of the body with a concentration of nerves and photosensory and chemosensory cells at the anterior end.

Since there is no circulatory or respiratory system, gas and nutrient exchange is dependent on diffusion and intercellular junctions. This necessarily limits the thickness of the body in these organisms, constraining them to be "flat" worms. Most flatworm species are monoecious (hermaphroditic, possessing both sets of sex organs), and fertilization is typically internal. Asexual reproduction is common in some groups in which an entire organism can be regenerated from just a part of itself.

Diversity of Flatworms

Flatworms are traditionally divided into four classes: Turbellaria, Monogenea, Trematoda, and Cestoda (Figure 15.16). The turbellarians include mainly free-living marine species, although some species live in freshwater or moist terrestrial environments. The simple planarians found in freshwater ponds and aquaria are examples. The epidermal layer of the underside of turbellarians is ciliated, and this helps them move. Some turbellarians are capable of remarkable feats of regeneration in which they may regrow the body, even from a small fragment.

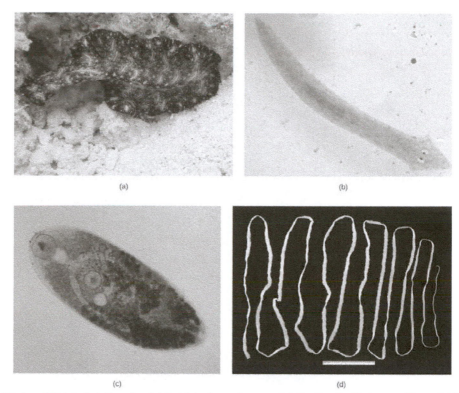

Figure 15.16 Phylum Platyhelminthes is divided into four classes: (a) Bedford's Flatworm (*Pseudobiceros bedfordi*) and the (b) planarian belong to class Turbellaria; (c) the Trematoda class includes about 20,000 species, most of which are parasitic; (d) class Cestoda includes tapeworms such as this *Taenia saginata*; and the parasitic class Monogenea (not shown). (credit a: modification of work by Jan Derk; credit c: modification of work by "Sahaquiel9102"/Wikimedia Commons; credit d: modification of work by CDC)

The monogeneans are external parasites mostly of fish with life cycles consisting of a free-swimming larva that attaches to a fish to begin transformation to the parasitic adult form. They have only one host during their life, typically of just one species. The worms may produce enzymes that digest the host tissues or graze on surface mucus and skin particles. Most monogeneans are hermaphroditic, but the sperm develop first, and it is typical for them to mate between individuals and not to self-fertilize.

The trematodes, or flukes, are internal parasites of mollusks and many other groups, including humans. Trematodes have complex life cycles that involve a primary host in which sexual reproduction occurs and one or more secondary hosts in which asexual reproduction occurs. The primary host is almost always a mollusk. Trematodes are responsible for serious human diseases including schistosomiasis, caused by a blood fluke (*Schistosoma*). The disease infects an estimated 200 million people in the tropics and leads to organ damage and chronic symptoms including fatigue. Infection occurs when a human enters the water, and a larva, released from the primary snail host, locates and penetrates the skin. The parasite infects various organs in the body and feeds on red blood cells before reproducing. Many of the eggs are released in feces and find their way into a waterway where they are able to reinfect the primary snail host.

The cestodes, or tapeworms, are also internal parasites, mainly of vertebrates. Tapeworms live in the intestinal tract of the primary host and remain fixed using a sucker on the anterior end, or scolex, of the tapeworm body. The remaining body of the tapeworm is made up of a long series of units called proglottids, each of which may contain an excretory system with flame cells, but will contain reproductive structures, both male and female. Tapeworms do not have a digestive system, they absorb nutrients from the food matter passing them in the host's intestine. Proglottids are produced at the scolex and are pushed to the end of the tapeworm as new proglottids form, at which point, they are "mature" and all structures except fertilized eggs have degenerated. Most reproduction occurs by cross-fertilization. The proglottid detaches and is released in the feces of the host. The fertilized eggs are eaten by an intermediate host. The juvenile worms emerge and infect the intermediate host, taking up residence, usually in muscle tissue. When the muscle tissue is eaten by the primary host, the cycle is completed. There are several tapeworm parasites of humans that are acquired by eating uncooked or poorly cooked pork, beef, and fish.

Nematodes

The phylum **Nematoda**, or roundworms, includes more than 28,000 species with an estimated 16,000 parasitic species. The name Nematoda is derived from the Greek word "nemos," which means "thread." Nematodes are present in all habitats and are extremely common, although they are usually not visible (Figure 15.17).

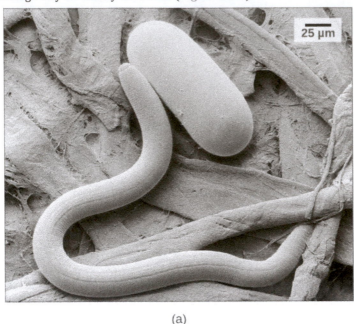

(a)

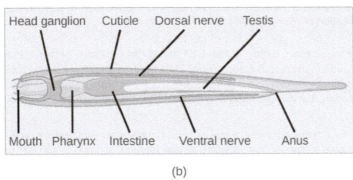

(b)

Figure 15.17 (a) An scanning electron micrograph of the nematode *Heterodera glycines* and (b) a schematic representation of the anatomy of a nematode are shown. (credit a: modification of work by USDA, ARS; scale-bar data from Matt Russell)

Most nematodes look similar to each other: slender tubes, tapered at each end (Figure 15.17). Nematodes are pseudocoelomates and have a **complete digestive system** with a distinct mouth and anus.

The nematode body is encased in a cuticle, a flexible but tough exoskeleton, or external skeleton, which offers protection and support. The cuticle contains a carbohydrate-protein polymer called **chitin**. The cuticle also lines the pharynx and rectum. Although the exoskeleton provides protection, it restricts growth, and therefore must be continually shed and replaced as the animal increases in size.

A nematode's mouth opens at the anterior end with three or six lips and, in some species, teeth in the form of cuticular extensions. There may also be a sharp stylet that can protrude from the mouth to stab prey or pierce plant or animal cells. The mouth leads to a muscular pharynx and intestine, leading to the rectum and anal opening at the posterior end.

Physiological Processes of Nematodes

In nematodes, the excretory system is not specialized. Nitrogenous wastes are removed by diffusion. In marine nematodes, regulation of water and salt is achieved by specialized glands that remove unwanted ions while maintaining internal body fluid concentrations.

Most nematodes have four nerve cords that run along the length of the body on the top, bottom, and sides. The nerve cords fuse in a ring around the pharynx, to form a head ganglion or "brain" of the worm, as well as at the posterior end to form the tail ganglion. Beneath the epidermis lies a layer of longitudinal muscles that permits only side-to-side, wave-like undulation of the body.

View this video (http://openstaxcollege.org/l/nematode) to see nematodes move about and feed on bacteria.

Nematodes employ a diversity of sexual reproductive strategies depending on the species; they may be monoecious, **dioecious** (separate sexes), or may reproduce asexually by parthenogenesis. *Caenorhabditis elegans* is nearly unique among animals in having both self-fertilizing hermaphrodites and a male sex that can mate with the hermaphrodite.

Arthropoda

The name "arthropoda" means "jointed legs," which aptly describes each of the enormous number of species belonging to this phylum. **Arthropoda** dominate the animal kingdom with an estimated 85 percent of known species, with many still undiscovered or undescribed. The principal characteristics of all the animals in this phylum are functional segmentation of the body and the presence of jointed appendages (Figure 15.18). As members of Ecdysozoa, arthropods also have an exoskeleton made principally of chitin. Arthropoda is the largest phylum in the animal world in terms of numbers of species, and insects form the single largest group within this phylum. Arthropods are true coelomate animals and exhibit prostostomic development.

Figure 15.18 Trilobites, like the one in this fossil, are an extinct group of arthropods. (credit: Kevin Walsh)

Physiological Processes of Arthropods

A unique feature of arthropods is the presence of a segmented body with fusion of certain sets of segments to give rise to functional segments. Fused segments may form a head, thorax, and abdomen, or a cephalothorax and abdomen, or a head and trunk. The coelom takes the form of a **hemocoel** (or blood cavity). The open circulatory system, in which blood bathes the internal organs rather than circulating in vessels, is regulated by a two-chambered heart. Respiratory systems vary, depending on the group of arthropod: Insects and myriapods use a series of tubes (**tracheae**) that branch throughout the body, open to the outside through openings called **spiracles**, and perform gas exchange directly between the cells and air in the tracheae. Aquatic crustaceans use gills, arachnids employ "book lungs," and aquatic chelicerates use "book gills." The book lungs of arachnids are internal stacks of alternating air pockets and hemocoel tissue shaped like the pages of a book. The book gills of crustaceans are external structures similar to book lungs with stacks of leaf-like structures that exchange gases with the surrounding water (Figure 15.19).

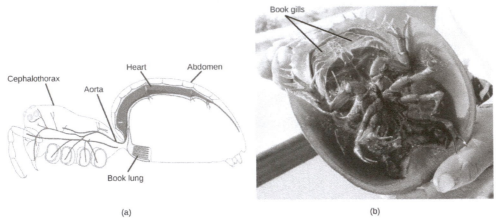

(a)   (b)

Figure 15.19 The book lungs of (a) arachnids are made up of alternating air pockets and hemocoel tissue shaped like a stack of books. The book gills of (b) crustaceans are similar to book lungs but are external so that gas exchange can occur with the surrounding water. (credit a: modification of work by Ryan Wilson based on original work by John Henry Comstock; credit b: modification of work by Angel Schatz)

Arthropod Diversity

Phylum Arthropoda includes animals that have been successful in colonizing terrestrial, aquatic, and aerial habitats. The phylum is further classified into five subphyla: Trilobitomorpha (trilobites), Hexapoda (insects and relatives), Myriapoda (millipedes, centipedes, and relatives), Crustacea (crabs, lobsters, crayfish, isopods, barnacles, and some zooplankton), and Chelicerata (horseshoe crabs, arachnids, scorpions, and daddy longlegs). Trilobites are an extinct group of arthropods found from the Cambrian period (540–490 million years ago) until they became extinct in the Permian (300–251 million years ago) that are probably most closely related to the Chelicerata. The 17,000 described species have been identified from fossils (Figure 15.18).

The Hexapoda have six legs (three pairs) as their name suggests. Hexapod segments are fused into a head, thorax, and abdomen (Figure 15.20). The thorax bears the wings and three pairs of legs. The insects we encounter on a daily basis—such as ants, cockroaches, butterflies, and bees—are examples of Hexapoda.

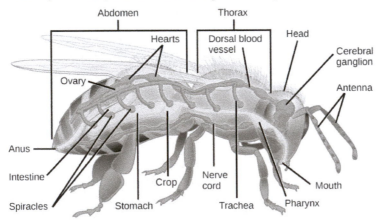

Figure 15.20 In this basic anatomy of a hexapod, note that insects have a developed digestive system (yellow), a respiratory system (blue), a circulatory system (red), and a nervous system (purple).

Subphylum Myriapoda includes arthropods with legs that may vary in number from 10 to 750. This subphylum includes 13,000 species; the most commonly found examples are millipedes and centipedes. All myriapods are terrestrial animals and prefer a humid environment (Figure 15.21).

(a)   (b)

Figure 15.21 (a) The centipede *Scutigera coleoptrata* has up to 15 pairs of legs. (b) This North American millipede (*Narceus americanus)* bears many legs, although not one thousand, as its name might suggest. (credit a: modification of work by Bruce Marlin; credit b: modification of work by Cory Zanker)

Crustaceans, such as shrimp, lobsters, crabs, and crayfish, are the dominant aquatic arthropods. A few crustaceans are terrestrial species like the pill bugs or sow bugs. The number of described crustacean species stands at about 47,000.[3]

Although the basic body plan in crustaceans is similar to the Hexapoda—head, thorax, and abdomen—the head and thorax may be fused in some species to form a **cephalothorax**, which is covered by a plate called the carapace (Figure 15.22). The exoskeleton of many species is also infused with calcium carbonate, which makes it even stronger than in other arthropods. Crustaceans have an open circulatory system in which blood is pumped into the hemocoel by the dorsal heart. Most crustaceans typically have separate sexes, but some, like barnacles, may be hermaphroditic. Serial hermaphroditism, in which the gonad can switch from producing sperm to ova, is also found in some crustacean species. Larval stages are seen in the early development of many crustaceans. Most crustaceans are carnivorous, but detritivores and filter feeders are also common.

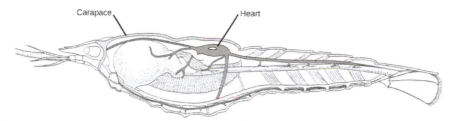

Figure 15.22 The crayfish is an example of a crustacean. It has a carapace around the cephalothorax and the heart in the dorsal thorax area. (credit: Jane Whitney)

Subphylum Chelicerata includes animals such as spiders, scorpions, horseshoe crabs, and sea spiders. This subphylum is predominantly terrestrial, although some marine species also exist. An estimated 103,000[4] described species are included in subphylum Chelicerata.

The body of chelicerates may be divided into two parts and a distinct "head" is not always discernible. The phylum derives its name from the first pair of appendages: the **chelicerae** (Figure 15.23a), which are specialized mouthparts. The chelicerae are mostly used for feeding, but in spiders, they are typically modified to inject venom into their prey (Figure 15.23b). As in other members of Arthropoda, chelicerates also utilize an open circulatory system, with a tube-like heart that pumps blood into the large hemocoel that bathes the internal organs. Aquatic chelicerates utilize gill respiration, whereas terrestrial species use either tracheae or book lungs for gaseous exchange.

3. "Number of Living Species in Australia and the World," A.D. Chapman, Australia Biodiversity Information Services, last modified August 26, 2010, http://www.environment.gov.au/biodiversity/abrs/publications/other/species-numbers/2009/03-exec-summary.html.
4. "Number of Living Species in Australia and the World," A.D. Chapman, Australia Biodiversity Information Services, last modified August 26, 2010, http://www.environment.gov.au/biodiversity/abrs/publications/other/species-numbers/2009/03-exec-summary.html.

(a)   (b)

Figure 15.23 (a) The chelicerae (first set of appendages) are well developed in the Chelicerata, which includes scorpions (a) and spiders (b). (credit a: modification of work by Kevin Walsh; credit b: modification of work by Marshal Hedin)

Click through (http://openstaxcollege.org/l/arthropod2) this lesson on arthropods to explore interactive habitat maps and more.

15.4 | Mollusks and Annelids

By the end of this section, you will be able to:

- Describe the unique anatomical features of mollusks
- Describe the features of an animal classified in phylum Annelida

The mollusks are a diverse group (85,000 described species) of mostly marine species. They have a variety of forms, ranging from large predatory squid and octopus, some of which show a high degree of intelligence, to small grazing forms with elaborately sculpted and colored shells. The annelids traditionally include the oligochaetes, which include the earthworms and leeches, the polychaetes, which are a marine group, and two other smaller classes.

The phyla Mollusca and Annelida belong to a clade called the **Lophotrochozoa**, which also includes the phylum Nemertea, or ribbon worms (Figure 15.3). They are distinct from the Ecdysozoa (nematodes and arthropods) based on evidence from analysis of their DNA, which has changed our views of the relationships among invertebrates.

Phylum Mollusca

Mollusca is the predominant phylum in marine environments, where it is estimated that 23 percent of all known marine species belong to this phylum. It is the second most diverse phylum of animals with over 75,000 described species. The name "mollusca" signifies a soft body, as the earliest descriptions of mollusks came from observations of unshelled, soft-bodied cuttlefish (squid relatives). Although mollusk body forms vary, they share key characteristics, such as a ventral, muscular foot that is typically used for locomotion; the visceral mass, which contains most of the internal organs of the animal; and a dorsal mantle, which is a flap of tissue over the visceral mass that creates a space called the mantle cavity. The mantle may or may not secrete a shell of calcium carbonate. In addition, many mollusks have a scraping structure at the mouth, called a **radula** (Figure 15.24).

The muscular foot varies in shape and function, depending on the type of mollusk (described below in the section on mollusk diversity). It is a retractable as well as extendable organ, used for locomotion and anchorage. Mollusks are eucoelomates, but the coelomic cavity is restricted to a cavity around the heart in adult animals. The mantle cavity, formed inside the **mantle**, develops independently of the coelomic cavity. It is a multi-purpose space, housing the gills, the anus, organs for sensing food particles in the water, and an outlet for gametes. Most mollusks have an open circulatory system with a heart that circulates the hemolymph in open spaces around the organs. The octopuses and squid are an exception to this and have a closed circulatory system with two hearts that move blood through the gills and a third, systemic heart that pumps blood through the rest of the body.

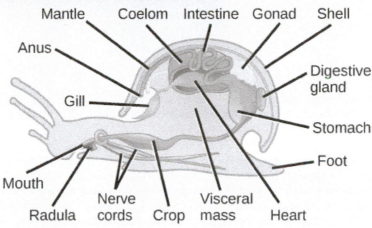

Figure 15.24 There are many species and variations of mollusks; the gastropod mollusk anatomy is shown here, which shares many characteristics common with other groups.

Which of the following statements about the anatomy of a mollusk is false?

a. Mollusks have a radula for scraping food.
b. Mollusks have ventral nerve cords.
c. The tissue beneath the shell is called the mantle.
d. The mantle cavity contains hemolymph.

Mollusk Diversity

This phylum is comprised of seven classes: Aplacophora, Monoplacophora, Polyplacophora, Bivalvia, Gastropoda, Cephalopoda, and Scaphopoda.

Class Aplacophora ("bearing no plates") includes worm-like animals living mostly on deep ocean bottoms. These animals lack a shell but have aragonite spicules on their skin. Members of class Monoplacophora ("bearing one plate") have a single, cap-like shell enclosing the body. The monoplacophorans were believed extinct and only known as fossils until the discovery of *Neopilina galatheae* in 1952. Today, scientists have identified nearly two dozen living species.

Animals in the class Polyplacophora ("bearing many plates") are commonly known as "chitons" and bear an armor-like, eight-plated shell (Figure 15.25). These animals have a broad, ventral foot that is adapted for attachment to rocks and a mantle that extends beyond the shell in the form of a girdle. They breathe with **ctenidia** (gills) present ventrally. These animals have a radula modified for scraping. A single pair of nephridia for excretion is present.

Figure 15.25 This chiton from the class Polyplacophora has the eight-plated shell indicative of its class. (credit: Jerry Kirkhart)

Class Bivalvia ("two shells") includes clams, oysters, mussels, scallops, and geoducks. They are found in marine and freshwater habitats. As the name suggests, bivalves are enclosed in a pair of shells (or valves) that are hinged at the dorsal side. The body is flattened on the sides. They feed by filtering particles from water and a radula is absent. They exchange gases using a pair of ctenidia, and excretion and osmoregulation are carried out by a pair of nephridia. In some species, the posterior edges of the mantle may fuse to form two siphons that inhale and exhale water. Some bivalves like oysters and mussels have the unique ability to secrete and deposit a calcareous **nacre** or "mother of pearl" around foreign particles that enter the mantle cavity. This property is commercially exploited to produce pearls.

Watch animations of clams (http://openstaxcollege.org/l/clams2) and mussels (http://openstaxcollege.org/l/mussels2) feeding to understand more about bivalves.

Gastropods ("stomach foot") include well-known mollusks like snails, slugs, conchs, sea hares, and sea butterflies. Gastropods include shell-bearing species as well as species with a reduced shell. These animals are asymmetrical and usually present a coiled shell (Figure 15.26).

(a)   (b)

Figure 15.26 (a) Like many gastropods, this snail has a stomach foot and a coiled shell. (b) This slug, which is also a gastropod, lacks a shell. (credit a: modification of work by Murray Stevenson; credit b: modification of work by Rosendahl)

The visceral mass in the shelled species is characteristically twisted and the foot is modified for crawling. Most gastropods bear a head with tentacles that support eyes. A complex radula is used to scrape food particles from the substrate. The mantle cavity encloses the ctenidia as well as a pair of nephridia.

The class Cephalopoda ("head foot" animals) includes octopuses, squids, cuttlefish, and nautilus. Cephalopods include shelled and reduced-shell groups. They display vivid coloration, typically seen in squids and octopuses, which is used for camouflage. The ability of some octopuses to rapidly adjust their colors to mimic a background pattern or to startle a predator is one of the more awe-inspiring feats of these animals. All animals in this class are predators and have beak-like jaws. All cephalopods have a well-developed nervous system, complex eyes, and a closed circulatory system. The foot is lobed and developed into tentacles and a funnel, which is used for locomotion. Suckers are present on the tentacles in octopuses and squid. Ctenidia are enclosed in a large mantle cavity and are serviced by large blood vessels, each with its own heart.

Cephalopods (Figure 15.27) are able to move quickly via jet propulsion by contracting the mantle cavity to forcefully eject a stream of water. Cephalopods have separate sexes, and the females of some species care for the eggs for an extended period of time. Although the shell is much reduced and internal in squid and cuttlefish, and absent altogether in octopus, nautilus live inside a spiral, multi-chambered shell that is filled with gas or water to regulate buoyancy.

Figure 15.27 The (a) nautilus, (b) giant cuttlefish, (c) reef squid, and (d) blue-ring octopus are all members of the class Cephalopoda. (credit a: modification of work by J. Baecker; credit b: modification of work by Adrian Mohedano; credit c: modification of work by Silke Baron; credit d: modification of work by Angell Williams)

Members of the class Scaphopoda ("boat feet") are known colloquially as "tusk shells" or "tooth shells." Tooth shells are open at both ends and usually lie buried in sand with the front opening exposed to water and the reduced head end projecting from the back of the shell. Tooth shells have a radula and a foot modified into tentacles, each with a bulbous end that catches and manipulates prey (Figure 15.28).

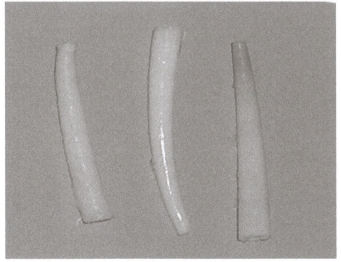

Figure 15.28 *Antalis vulgaris* shows the classic Dentaliidae shape that gives these animals their common name of "tusk shell." (credit: Georges Jansoone)

Annelida

Phylum **Annelida** are segmented worms found in marine, terrestrial, and freshwater habitats, but the presence of water or humidity is a critical factor for their survival in terrestrial habitats. The name of the phylum is derived from the Latin word *annellus*, which means a small ring. Approximately 16,500 species have been described. The phylum includes earthworms, polychaete worms, and leeches. Like mollusks, annelids exhibit protostomic development.

Annelids are bilaterally symmetrical and have a worm-like appearance. Their particular segmented body plan results in repetition of internal and external features in each body segment. This type of body plan is called **metamerism**. The evolutionary benefit of such a body plan is thought to be the capacity it allows for the evolution of independent modifications in different segments that perform different functions. The overall body can then be divided into head, body, and tail.

Physiological Processes of Annelida

The skin of annelids is protected by a cuticle that is thinner than the cuticle of the ecdysozoans and does not need to be molted for growth. Chitinous hairlike extensions, anchored in the skin and projecting from the cuticle, called **chaetae**, are present in every segment in most groups. The chaetae are a defining character of annelids. Polychaete worms have paired, unjointed limbs called parapodia on each segment used for locomotion and breathing. Beneath the cuticle there are two layers of muscle, one running around its circumference (circular) and one running the length of the worm (longitudinal). Annelids have a true coelom in which organs are distributed and bathed in coelomic fluid. Annelids possess a well-developed complete digestive system with specialized organs: mouth, muscular pharynx, esophagus, and crop. A cross-sectional view of a body segment of an earthworm is shown in Figure 15.29; each segment is limited by a membrane that divides the body cavity into compartments.

Annelids have a closed circulatory system with muscular pumping "hearts" in the anterior segments, dorsal and ventral blood vessels that run the length of the body with connections in each segment, and capillaries that service individual tissues. Gas exchange occurs across the moist body surface. Excretion is carried out by pairs of primitive "kidneys" called metanephridia that consist of a convoluted tubule and an open, ciliated funnel present in every segment. Annelids have a well-developed nervous system with two ventral nerve cords and a nerve ring of fused ganglia present around the pharynx.

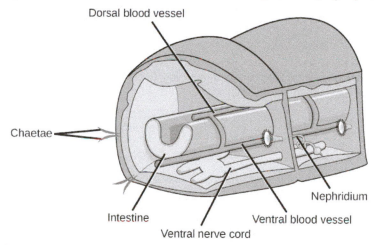

Figure 15.29 In this schematic showing the basic anatomy of annelids, the digestive system is indicated in green, the nervous system is indicated in yellow, and the circulatory system is indicated in red.

Annelids may be either monoecious with permanent gonads (as in earthworms and leeches) or dioecious with temporary or seasonal gonads (as in polychaetes).

This video and animation (http://openstaxcollege.org/l/annelid2) provides a close-up look at annelid anatomy.

Annelid Diversity

Phylum Annelida includes the classes Polychaeta and Clitellata (Figure 15.30); the latter contains subclasses Oligochaeta, Hirudinoidea, and Branchiobdellida.

Earthworms are the most abundant members of the subclass Oligochaeta, distinguished by the presence of the **clitellum**, a ring structure in the skin that secretes mucus to bind mating individuals and forms a protective cocoon for the eggs. They also have a few, reduced chaetae (oligo- = "few"; -chaetae = "hairs"). The number and size of chaetae is greatly diminished in oligochaetes as compared to the polychaetes (poly- = "many"; -chaetae = "hairs"). The chaetae of polychaetes are also arranged within fleshy, flat, paired appendages on each segment called parapodia.

The subclass Hirudinoidea includes leeches. Significant differences between leeches and other annelids include the development of suckers at the anterior and posterior ends, and the absence of chaetae. Additionally, the segmentation of the body wall may not correspond to internal segmentation of the coelomic cavity. This adaptation may allow leeches to swell when ingesting blood from host vertebrates. The subclass Branchiobdellida includes about 150 species that show similarity to leeches as well as oligochaetes. All species are obligate symbionts, meaning that they can only survive associated with their host, mainly with freshwater crayfish. They feed on the algae that grows on the carapace of the crayfish.

(a)   (b)

Figure 15.30 The (a) earthworm and (b) leech are both annelids. (credit a: modification of work by "schizoform"/Flickr; credit b: modification of work by "Sarah G..."/Flickr)

15.5 | Echinoderms and Chordates

By the end of this section, you will be able to:
- Describe the distinguishing characteristics of echinoderms
- Describe the distinguishing characteristics of chordates

Deuterostomes include the phyla Echinodermata and Chordata (which includes the vertebrates) and two smaller phyla. Deuterostomes share similar patterns of early development.

Echinoderms

Echinodermata are named for their spiny skin (from the Greek "echinos" meaning "spiny" and "dermos" meaning "skin"). The phylum includes about 7,000[5] described living species, such as sea stars, sea cucumbers, sea urchins, sand dollars, and brittle stars. **Echinodermata** are exclusively marine.

Adult echinoderms exhibit pentaradial symmetry and have a calcareous endoskeleton made of ossicles (Figure 15.31), although the early larval stages of all echinoderms have bilateral symmetry. The endoskeleton is developed by epidermal cells, which may also possess pigment cells, giving vivid colors to these animals, as well as cells laden with toxins. These animals have a true coelom, a portion of which is modified into a unique circulatory system called a **water vascular system**. An interesting feature of these animals is their power to regenerate, even when over 75 percent of their body mass is lost.

---

5. "Number of Living Species in Australia and the World," A.D. Chapman, Australia Biodiversity Information Services, last modified August 26, 2010, http://www.environment.gov.au/biodiversity/abrs/publications/other/species-numbers/2009/03-exec-summary.html.

Physiological Processes of Echinoderms

Echinoderms have a unique system for gas exchange, nutrient circulation, and locomotion called the water vascular system. The system consists of a central ring canal and radial canals extending along each arm. Water circulates through these structures allowing for gas, nutrient, and waste exchange. A structure on top of the body, called the **madreporite**, regulates the amount of water in the water vascular system. "Tube feet," which protrude through openings in the endoskeleton, may be expanded or contracted using the hydrostatic pressure in the system. The system allows for slow movement, but a great deal of power, as witnessed when the tube feet latch on to opposite halves of a bivalve mollusk, like a clam, and slowly, but surely pull the shells apart, exposing the flesh within.

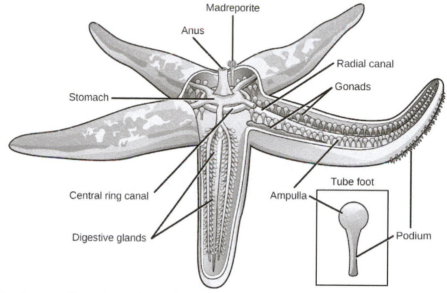

Figure 15.31 This diagram shows the anatomy of a sea star.

The echinoderm nervous system has a nerve ring at the center and five radial nerves extending outward along the arms. There is no centralized nervous control. Echinoderms have separate sexes and release their gametes into the water where fertilization takes place. Echinoderms may also reproduce asexually through regeneration from body parts.

Echinoderm Diversity

This phylum is divided into five classes: Asteroidea (sea stars), Ophiuroidea (brittle stars), Echinoidea (sea urchins and sand dollars), Crinoidea (sea lilies or feather stars), and Holothuroidea (sea cucumbers) (Figure 15.32).

Perhaps the best-known echinoderms are members of the class Asteroidea, or sea stars. They come in a large variety of shapes, colors, and sizes, with more than 1,800 species known. The characteristics of sea stars that set them apart from other echinoderm classes include thick arms that extend from a central disk where organs penetrate into the arms. Sea stars use their tube feet not only for gripping surfaces but also for grasping prey. Sea stars have two stomachs, one of which they can evert through their mouths to secrete digestive juices into or onto prey before ingestion. This process can essentially liquefy the prey and make digestion easier.

View this video (http://openstaxcollege.org/l/echinoderm2) to explore a sea star's body plan up close, watch one move across the sea floor, and see it devour a mussel.

Brittle stars have long, thin arms that do not contain any organs. Sea urchins and sand dollars do not have arms but are hemispherical or flattened with five rows of tube feet, which help them in slow movement. Sea lilies and feather stars are stalked suspension feeders. Sea cucumbers are soft-bodied and elongate with five rows of tube feet and a series of tube feet around the mouth that are modified into tentacles used in feeding.

Figure 15.32 Different members of Echinodermata include the (a) sea star in class Asteroidea, (b) the brittle star in class Ophiuroidea, (c) the sea urchins of class Echinoidea, (d) the sea lilies belonging to class Crinoidea, and (e) sea cucumbers representing class Holothuroidea. (credit a: modification of work by Adrian Pingstone; credit b: modification of work by Joshua Ganderson; credit c: modification of work by Samuel Chow; credit d: modification of work by Sarah Depper; credit e: modification of work by Ed Bierman)

Chordates

The majority of species in the phylum Chordata are found in the subphylum Vertebrata, which include many species with which we are familiar. The vertebrates contain more than 60,000 described species, divided into major groupings of the lampreys, fishes, amphibians, reptiles, birds, and mammals.

Animals in the phylum **Chordata** share four key features that appear at some stage of their development: a notochord, a dorsal hollow nerve cord, pharyngeal slits, and a post-anal tail (Figure 15.33). In certain groups, some of these traits are present only during embryonic development.

The chordates are named for the **notochord**, which is a flexible, rod-shaped structure that is found in the embryonic stage of all chordates and in the adult stage of some chordate species. It is located between the digestive tube and the nerve cord, and provides skeletal support through the length of the body. In some chordates, the notochord acts as the primary axial support of the body throughout the animal's lifetime. In vertebrates, the notochord is present during embryonic development, at which time it induces the development of the neural tube and serves as a support for the developing embryonic body. The notochord, however, is not found in the postnatal stage of vertebrates; at this point, it has been replaced by the **vertebral column** (the spine).

The **dorsal hollow nerve cord** is derived from ectoderm that sinks below the surface of the skin and rolls into a hollow tube during development. In chordates, it is located dorsally to the notochord. In contrast, other animal phyla possess solid nerve cords that are located either ventrally or laterally. The nerve cord found in most chordate embryos develops into the brain and spinal cord, which compose the central nervous system.

Pharyngeal slits are openings in the pharynx, the region just posterior to the mouth, that extend to the outside environment. In organisms that live in aquatic environments, pharyngeal slits allow for the exit of water that enters the mouth during feeding. Some invertebrate chordates use the pharyngeal slits to filter food from the water that enters the mouth. In fishes, the pharyngeal slits are modified into gill supports, and in jawed fishes, jaw supports. In tetrapods, the slits are further modified into components of the ear and tonsils, since there is no longer any need for gill supports in these air-breathing animals. **Tetrapod** means "four-footed," and this group includes amphibians, reptiles, birds, and mammals. (Birds are considered tetrapods because they evolved from tetrapod ancestors.)

The **post-anal tail** is a posterior elongation of the body extending beyond the anus. The tail contains skeletal elements and muscles, which provide a source of locomotion in aquatic species, such as fishes. In some terrestrial vertebrates, the tail may also function in balance, locomotion, courting, and signaling when danger is near. In many species, the tail is absent or reduced; for example, in apes, including humans, it is present in the embryo, but reduced in size and nonfunctional in adults.

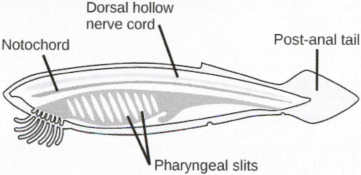

Figure 15.33 In chordates, four common features appear at some point in development: a notochord, a dorsal hollow nerve cord, pharyngeal slits, and a post-anal tail. The anatomy of a cephalochordate shown here illustrates all of these features.

Which of the following statements about common features of chordates is true?

a. The dorsal hollow nerve cord is part of the chordate central nervous system.

b. In vertebrate fishes, the pharyngeal slits become the gills.

c. Humans are not chordates because humans do not have a tail.

d. Vertebrates do not have a notochord at any point in their development; instead, they have a vertebral column.

Invertebrate Chordates

In addition to the vertebrates, the phylum Chordata contains two clades of invertebrates: **Urochordata** (tunicates) and **Cephalochordata** (lancelets). Members of these groups possess the four distinctive features of chordates at some point during their development.

The **tunicates** (Figure 15.34) are also called sea squirts. The name tunicate derives from the cellulose-like carbohydrate material, called the tunic, which covers the outer body. Although tunicates are classified as chordates, the adult forms are much modified in body plan and do not have a notochord, a dorsal hollow nerve cord, or a post-anal tail, although they do have pharyngeal slits. The larval form possesses all four structures. Most tunicates are hermaphrodites. Tunicate larvae hatch from eggs inside the adult tunicate's body. After hatching, a tunicate larva swims for a few days until it finds a suitable surface on which it can attach, usually in a dark or shaded location. It then attaches by the head to the substrate and undergoes metamorphosis into the adult form, at which point the notochord, nerve cord, and tail disappear.

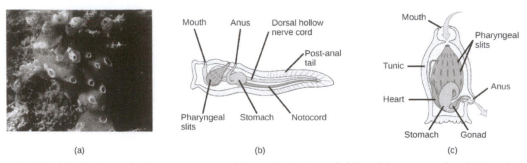

Figure 15.34 (a) This photograph shows a colony of the tunicate *Botrylloides violaceus*. In the (b) larval stage, the tunicate can swim freely until it attaches to a substrate to become (c) an adult. (credit a: modification of work by Dr. Dwayne Meadows, NOAA/NMFS/OPR)

Most tunicates live a sessile existence in shallow ocean waters and are suspension feeders. The primary foods of tunicates are plankton and detritus. Seawater enters the tunicate's body through its incurrent siphon. Suspended material is filtered out of this water by a mucus net (pharyngeal slits) and is passed into the intestine through the action of cilia. The anus empties into the excurrent siphon, which expels wastes and water.

Lancelets possess a notochord, dorsal hollow nerve cord, pharyngeal slits, and a post-anal tail in the adult stage (Figure 15.35). The notochord extends into the head, which gives the subphylum its name (Cephalochordata). Extinct fossils of this subphylum date to the middle of the Cambrian period (540–488 mya).The living forms, the lancelets, are named for their blade-like shape. Lancelets are only a few centimeters long and are usually found buried in sand at the bottom of warm temperate and tropical seas. Like tunicates, they are suspension feeders.

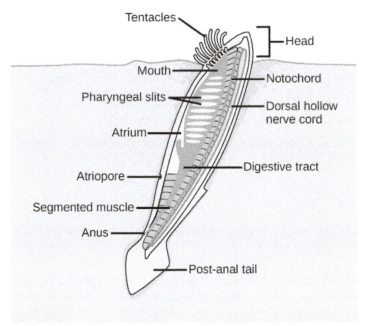

Figure 15.35 Adult lancelets retain the four key features of chordates: a notochord, a dorsal hollow nerve cord, pharyngeal slits, and a post-anal tail.

15.6 | Vertebrates

By the end of this section, you will be able to:

- Describe the difference between jawless and jawed fishes
- Explain the main characteristics of amphibians, reptiles, and birds
- Describe the derived characteristics in birds that facilitate flight
- Name and describe the distinguishing features of the three main groups of mammals
- Describe the derived features that distinguish primates from other animals

Vertebrates are among the most recognizable organisms of the animal kingdom (Figure 15.36). More than 62,000 vertebrate species have been identified. The vertebrate species now living represent only a small portion of the vertebrates that have existed. The best-known extinct vertebrates are the dinosaurs, a unique group of reptiles, reaching sizes not seen before or since in terrestrial animals. They were the dominant terrestrial animals for 150 million years, until they died out near the end of the Cretaceous period in a mass extinction. A great deal is known about the anatomy of the dinosaurs, given the preservation of their skeletal elements in the fossil record.

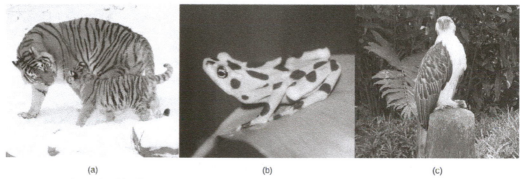

(a)   (b)   (c)

Figure 15.36 Examples of critically endangered vertebrate species include (a) the Siberian tiger (*Panthera tigris altaica*), (b) the Panamanian golden frog (*Atelopus zeteki*), and (c) the Philippine eagle (*Pithecophaga jefferyi*). (credit a: modification of work by Dave Pape; credit b: modification of work by Brian Gratwicke; credit c: modification of work by "cuatrok77"/Flickr)

Fishes

Modern fishes include an estimated 31,000 species. Fishes were the earliest vertebrates, and jawless fishes were the earliest of these. Jawless fishes—the present day hagfishes and lampreys—have a distinct cranium and complex sense organs including eyes, distinguishing them from the invertebrate chordates. The jawed fishes evolved later and are extraordinarily diverse today. Fishes are active feeders, rather than sessile, suspension feeders.

Jawless Fishes

Jawless fishes are **craniates** (which includes all the chordate groups except the tunicates and lancelets) that represent an ancient vertebrate lineage that arose over one half-billion years ago. Some of the earliest jawless fishes were the **ostracoderms** (which translates as "shell-skin"). Ostracoderms, now extinct, were vertebrate fishes encased in bony armor, unlike present-day jawless fishes, which lack bone in their scales.

The clade **Myxini** includes 67 species of hagfishes. **Hagfishes** are eel-like scavengers that live on the ocean floor and feed on dead invertebrates, other fishes, and marine mammals (Figure 15.37a). Hagfishes are entirely marine and are found in oceans around the world except for the polar regions. A unique feature of these animals is the slime glands beneath the skin that are able to release an extraordinary amount of mucus through surface pores. This mucus may allow the hagfish to escape from the grip of predators. Hagfish are known to enter the bodies of dead or dying organisms to devour them from the inside.

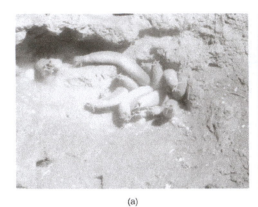

(a)   (b)

Figure 15.37 (a) Pacific hagfishes are scavengers that live on the ocean floor. (b) These parasitic sea lampreys attach to their lake trout host by suction and use their rough tongues to rasp away flesh in order to feed on the trout's blood. (credit a: modification of work by Linda Snook, NOAA/CBNMS; credit b: modification of work by USGS)

The skeleton of a hagfish is composed of cartilage, which includes a cartilaginous notochord, which runs the length of the body, and a skull. This notochord provides support to the fish's body. Although they are craniates, hagfishes are not vertebrates, since they do not replace the notochord with a vertebral column during development, as do the vertebrates.

The clade **Petromyzontidae** includes approximately 40 species of lampreys. **Lampreys** are similar to hagfishes in size and shape; however, lampreys have a brain case and incomplete vertebrae. Lampreys lack paired appendages and bone, as do the hagfishes. As adults, lampreys are characterized by a toothed, funnel-like sucking mouth. Some species are parasitic as adults, attaching to and feeding on the body fluids of fish (Figure 15.37b). Most species are free-living.

Lampreys live primarily in coastal and fresh waters and have a worldwide temperate region distribution. All species spawn in fresh waters. Eggs are fertilized externally, and the larvae are distinctly different from the adult form, spending 3 to 15 years as suspension feeders. Once they attain sexual maturity, the adults reproduce and die within days. Lampreys have a notochord as adults.

Jawed Fishes

Gnathostomes or "jaw-mouths" are vertebrates that have jaws and include both cartilaginous and bony fishes. One of the most significant developments in early vertebrate evolution was the origin of the jaw, which is a hinged structure attached to the cranium that allows an animal to grasp and tear its food. The evolution of jaws allowed early gnathostomes to exploit food resources that were unavailable to jawless fishes.

The clade **Chondrichthyes**, the cartilaginous fishes, is diverse, consisting of sharks (Figure 15.38a), rays, and skates, together with sawfishes and a few dozen species of fishes called *chimaeras,* or ghost sharks. Chondrichthyes have paired fins and a skeleton made of cartilage. This clade arose approximately 370 million years ago in the middle Devonian. They are thought to have descended from an extinct group that had a skeleton made of bone; thus, the cartilaginous skeleton of Chondrichthyes is a later development. Parts of the shark skeleton are strengthened by granules of calcium carbonate, but this is not the same as bone.

Most cartilaginous fishes live in marine habitats, with a few species living in fresh water for some or all of their lives. Most sharks are carnivores that feed on live prey, either swallowing it whole or using their jaws and teeth to tear it into smaller pieces. Shark teeth likely evolved from the jagged scales that cover their skin. Some species of sharks and rays are suspension feeders that feed on plankton.

(a)   (b)

Figure 15.38 (a) This hammerhead shark is an example of a predatory cartilaginous fish. (b) This stingray blends into the sandy bottom of the ocean floor when it is feeding or awaiting prey. (credit a: modification of work by Masashi Sugawara; credit b: modification of work by "Sailn1"/Flickr)

Sharks have well-developed sense organs that aid them in locating prey, including a keen sense of smell and electroreception, the latter being perhaps the most sensitive of any animal. Organs called **ampullae of Lorenzini** allow sharks to detect the electromagnetic fields that are produced by all living things, including their prey. Electroreception has only been observed in aquatic or amphibious animals. Sharks, together with most fishes, also have a sense organ called the **lateral line**, which is used to detect movement and vibration in the surrounding water, and a sense that is often considered homologous to "hearing" in terrestrial vertebrates. The lateral line is visible as a darker stripe that runs along the length of the fish's body.

Sharks reproduce sexually and eggs are fertilized internally. Most species are ovoviviparous, that is, the fertilized egg is retained in the oviduct of the mother's body, and the embryo is nourished by the egg yolk. The eggs hatch in the uterus and young are born alive and fully functional. Some species of sharks are oviparous: They lay eggs that hatch outside of the mother's body. Embryos are protected by a shark egg case or "mermaid's purse" that has the consistency of leather. The shark egg case has tentacles that snag in seaweed and give the newborn shark cover. A few species of sharks are viviparous, that is, the young develop within the mother's body, and she gives live birth.

Rays and skates include more than 500 species and are closely related to sharks. They can be distinguished from sharks by their flattened bodies, pectoral fins that are enlarged and fused to the head, and gill slits on their ventral surface (Figure 15.38b). Like sharks, rays and skates have a cartilaginous skeleton. Most species are marine and live on the sea floor, with nearly a worldwide distribution.

Bony Fishes

Members of the clade **Osteichthyes**, or bony fishes, are characterized by a bony skeleton. The vast majority of present-day fishes belong to this group, which consists of approximately 30,000 species, making it the largest class of vertebrates in existence today.

Nearly all bony fishes have an ossified skeleton with specialized bone cells (osteocytes) that produce and maintain a calcium phosphate matrix. This characteristic has only reverted in a few groups of Osteichthyes, such as sturgeons and paddlefish, which have primarily cartilaginous skeletons. The skin of bony fishes is often covered in overlapping scales, and glands in the skin secrete mucus that reduces drag when swimming and aids the fish in osmoregulation. Like sharks, bony fishes have a lateral line system that detects vibrations in water. Unlike sharks, some bony fish depend on their eyesight to locate prey. Bony fish are also unusual in possessing taste cells in the head and trunk region of the body that allow them to detect extremely small concentrations of molecules in the water.

All bony fishes, like the cartilaginous fishes, use gills to breathe. Water is drawn over gills that are located in chambers covered and ventilated by a protective, muscular flap called the operculum. Unlike sharks, bony fishes have a **swim bladder**, a gas-filled organ that helps to control the buoyancy of the fish. Bony fishes are further divided into two clades with living members: **Actinopterygii** (ray-finned fishes) and **Sarcopterygii** (lobe-finned fishes).

The ray-finned fishes include many familiar fishes—tuna, bass, trout, and salmon (Figure 15.39a), among others. Ray-finned fishes are named for the form of their fins—webs of skin supported by bony spines called rays. In contrast, the fins of lobe-finned fishes are fleshy and supported by bone (Figure 15.39b). Living members of lobe-finned fishes include the less familiar lungfishes and coelacanth.

(a)   (b)

Figure 15.39 The (a) sockeye salmon and (b) coelacanth are both bony fishes of the Osteichthyes clade. The coelacanth, sometimes called a lobe-finned fish, was thought to have gone extinct in the Late Cretaceous period 100 million years ago until one was discovered in 1938 between Africa and Madagascar. (credit a: modification of work by Timothy Knepp, USFWS; credit b: modification of work by Robbie Cada)

Amphibians

Amphibians are vertebrate tetrapods. **Amphibia** includes frogs, salamanders, and caecilians. The term amphibian means "dual life," which is a reference to the metamorphosis that many frogs undergo from a tadpole to an adult and the mixture of aquatic and terrestrial environments in their life cycle. Amphibians evolved in the Devonian period and were the earliest terrestrial tetrapods.

As tetrapods, most amphibians are characterized by four well-developed limbs, although some species of salamanders and all caecilians possess only vestigial limbs. An important characteristic of extant amphibians is a moist, permeable skin, achieved by mucus glands. The moist skin allows oxygen and carbon dioxide exchange with the environment, a process called **cutaneous respiration**. All living adult amphibian species are carnivorous, and some terrestrial amphibians have a sticky tongue that is used to capture prey.

Amphibian Diversity

Amphibia comprise an estimated 6,500 extant species that inhabit tropical and temperate regions around the world. Amphibians can be divided into three clades: **Urodela** ("tailed-ones"), the salamanders and newts; **Anura** ("tail-less ones"), the frogs and toads; and **Apoda** ("legless ones"), the caecilians.

Living **salamanders** (Figure 15.40a) include approximately 500 species, some of which are aquatic, others terrestrial, and some that live on land only as adults. Adult salamanders usually have a generalized tetrapod body plan with four limbs and a tail. Some salamanders are lungless, and respiration occurs through the skin or external gills. Some terrestrial salamanders have primitive lungs; a few species have both gills and lungs.

(a)   (b)

Figure 15.40 (a) Most salamanders have legs and a tail, but respiration varies among species. (b) The Australian green tree frog is a nocturnal predator that lives in the canopies of trees near a water source. (credit a: modification of work by Valentina Storti; credit b: modification of work by Evan Pickett)

Watch this video (http://openstaxcollege.org/l/river_monster2) about an unusually large salamander species.

Download for free at https://openstax.org/details/books/concepts-biology

Frogs (Figure 15.40b) are the most diverse group of amphibians, with approximately 5,000 species that live on all continents except Antarctica. Frogs have a body plan that is more specialized than the salamander body plan for movement on land. Adult frogs use their hind limbs to jump many times their body length on land. Frogs have a number of modifications that allow them to avoid predators, including skin that acts as camouflage and defensive chemicals that are poisonous to predators secreted from glands in the skin.

Frog eggs are fertilized externally, as they are laid in moist environments. Frogs demonstrate a range of parental behaviors, with some species exhibiting little care, to species that carry eggs and tadpoles on their hind legs or backs. The life cycle consists of two stages: the larval stage followed by metamorphosis to an adult stage. The larval stage of a frog, the **tadpole**, is often a filter-feeding herbivore. Tadpoles usually have gills, a lateral line system, long-finned tails, but no limbs. At the end of the tadpole stage, frogs undergo a gradual metamorphosis into the adult form. During this stage, the gills and lateral line system disappear, and four limbs develop. The jaws become larger and are suited for carnivorous feeding, and the digestive system transforms into the typical short gut of a predator. An eardrum and air-breathing lungs also develop. These changes during metamorphosis allow the larvae to move onto land in the adult stage (Figure 15.41).

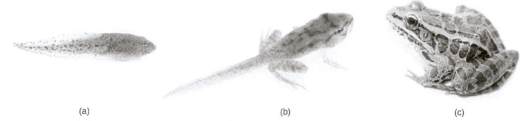

(a)　　　　　　　　　　　　　　(b)　　　　　　　　　　　　　　(c)

Figure 15.41 A frog begins as a (a) tadpole and undergoes metamorphosis to become (b) a juvenile and finally (c) an adult. (credit: modification of work by Brian Gratwicke)

Caecilians comprise an estimated 185 species. They lack external limbs and resemble giant earthworms. They inhabit soil and are found primarily in the tropics of South America, Africa, and southern Asia where they are adapted for a soil-burrowing lifestyle and are nearly blind. Unlike most of the other amphibians that breed in or near water, reproduction in a drier soil habitat means that caecilians must utilize internal fertilization, and most species give birth to live young (Figure 15.42).

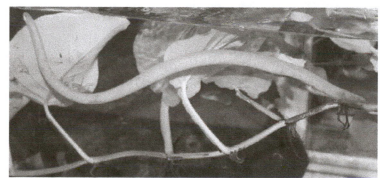

Figure 15.42 Caecilians lack external limbs and are well adapted for a soil-burrowing lifestyle. (credit: modification of work by "cliff1066"/Flickr)

Reptiles and Birds

The **amniotes**—reptiles, birds, and mammals—are distinguished from amphibians by their terrestrially adapted (shelled) egg and an embryo protected by amniotic membranes. The evolution of amniotic membranes meant that the embryos of amniotes could develop within an aquatic environment inside the egg. This led to less dependence on a water environment for development and allowed the amniotes to invade drier areas. This was a significant evolutionary change that distinguished them from amphibians, which were restricted to moist environments due to their shell-less eggs. Although the shells of various amniotic species vary significantly, they all allow retention of water. The membranes of the amniotic egg also allowed gas exchange and sequestering of wastes within the enclosure of an eggshell. The shells of bird eggs are composed of calcium carbonate and are hard and brittle, but possess pores for gas and water exchange. The shells of reptile eggs are more leathery and pliable. Most mammals do not lay eggs; however, even with internal gestation, amniotic membranes are still present.

In the past, the most common division of amniotes has been into classes Mammalia, Reptilia, and Aves. Birds are descended, however, from dinosaurs, so this classical scheme results in groups that are not true clades. We will discuss birds as a group distinct from reptiles with the understanding that this does not reflect evolutionary history.

Reptiles

Reptiles are tetrapods. Limbless reptiles—snakes—may have vestigial limbs and, like caecilians, are classified as tetrapods because they are descended from four-limbed ancestors. Reptiles lay shelled eggs on land. Even aquatic reptiles, like sea turtles, return to the land to lay eggs. They usually reproduce sexually with internal fertilization. Some species display ovoviviparity, with the eggs remaining in the mother's body until they are ready to hatch. Other species are viviparous, with the offspring born alive.

One of the key adaptations that permitted reptiles to live on land was the development of their scaly skin, containing the protein keratin and waxy lipids, which prevented water loss from the skin. This occlusive skin means that reptiles cannot use their skin for respiration, like amphibians, and thus all must breathe with lungs. In addition, reptiles conserve valuable body water by excreting nitrogen in the form of uric acid paste. These characteristics, along with the shelled, amniotic egg, were the major reasons why reptiles became so successful in colonizing a variety of terrestrial habitats far from water.

Reptiles are ectotherms, that is, animals whose main source of body heat comes from the environment. Behavioral maneuvers, like basking to heat themselves, or seeking shade or burrows to cool off, help them regulate their body temperature,

Class Reptilia includes diverse species classified into four living clades. These are the Crocodilia, Sphenodontia, Squamata, and Testudines.

The **Crocodilia** ("small lizard") arose approximately 84 million years ago, and living species include alligators, crocodiles, and caimans. Crocodilians (Figure 15.43a) live throughout the tropics of Africa, South America, the southeastern United States, Asia, and Australia. They are found in freshwater habitats, such as rivers and lakes, and spend most of their time in water. Some species are able to move on land due to their semi-erect posture.

(a)   (b)

(c)   (d)

Figure 15.43 (a) Crocodilians, such as this Siamese crocodile, provide parental care for their offspring. (b) This Jackson's chameleon blends in with its surroundings. (c) The garter snake belongs to the genus *Thamnophis*, the most widely distributed reptile genus in North America. (d) The African spurred tortoise lives at the southern edge of the Sahara Desert. It is the third largest tortoise in the world. (credit a: modification of work by Keshav Mukund Kandhadai; credit c: modification of work by Steve Jurvetson; credit d: modification of work by Jim Bowen)

The **Sphenodontia** ("wedge tooth") arose in the Mesozoic Era and includes only one living genus, *Tuatara*, with two species that are found in New Zealand. There are many fossil species extending back to the Triassic period (250–200 million years ago). Although the tuataras resemble lizards, they are anatomically distinct and share characteristics that are found in birds and turtles.

Squamata ("scaly") arose in the late Permian; living species include lizards and snakes, which are the largest extant clade of reptiles (Figure 15.43b). Lizards differ from snakes by having four limbs, eyelids, and external ears, which are lacking in snakes. Lizard species range in size from chameleons and geckos that are a few centimeters in length to the Komodo dragon, which is about 3 meters in length.

Snakes are thought to have descended from either burrowing lizards or aquatic lizards over 100 million years ago (Figure 15.43c). Snakes comprise about 3,000 species and are found on every continent except Antarctica. They range in size from 10 centimeter-long thread snakes to 7.5 meter-long pythons and anacondas. All snakes are carnivorous and eat small animals, birds, eggs, fish, and insects.

Turtles are members of the clade **Testudines** ("having a shell") (Figure 15.43d). Turtles are characterized by a bony or cartilaginous shell, made up of the carapace on the back and the plastron on the ventral surface, which develops from the ribs. Turtles arose approximately 200 million years ago, predating crocodiles, lizards, and snakes. Turtles lay eggs on land, although many species live in or near water. Turtles range in size from the speckled padloper tortoise at 8 centimeters (3.1 inches) to the leatherback sea turtle at 200 centimeters (over 6 feet). The term "turtle" is sometimes used to describe only those species of Testudines that live in the sea, with the terms "tortoise" and "terrapin" used to refer to species that live on land and in fresh water, respectively.

Birds

Data now suggest that birds belong within the reptile clade, but they display a number of unique adaptations that set them apart. Unlike the reptiles, birds are endothermic, meaning they generate their own body heat through metabolic processes. The most distinctive characteristic of birds is their feathers, which are modified reptilian scales. Birds have several different types of feathers that are specialized for specific functions, like contour feathers that streamline the bird's exterior and loosely structured **down feathers** that insulate (Figure 15.44a).

Feathers not only permitted the earliest birds to glide, and ultimately engage in flapping flight, but they insulated the bird's body, assisting the maintenance of endothermy, even in cooler temperatures. Powering a flying animal requires economizing on the amount of weight carried. As body weight increases, the muscle output and energetic cost required for flying increase. Birds have made several modifications to reduce body weight, including hollow or **pneumatic bones** (Figure 15.44b) with air spaces that may be connected to air sacs and cross-linked struts within their bones to provide structural reinforcement. Parts of the vertebral skeleton and braincase are fused to increase its strength while lightening its weight. Most species of bird only possess one ovary rather than two, and no living birds have teeth in their jaw, further reducing body mass.

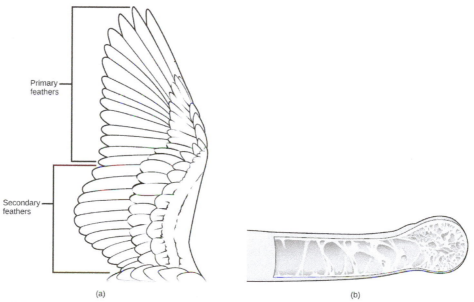

Figure 15.44 (a) Primary feathers are located at the wing tip and provide thrust; secondary feathers are located close to the body and provide lift. (b) Many birds have hollow pneumatic bones, which make flight easier.

Birds possess a system of air sacs branching from their primary airway that divert the path of air so that it passes unidirectionally through the lung, during both inspiration and expiration. Unlike mammalian lungs in which air flows in two directions as it is breathed in and out, air flows continuously through the bird's lung to provide a more efficient system of gas exchange.

Mammals

Mammals are vertebrates that have hair and mammary glands used to provide nutrition for their young. Certain features of the jaw, skeleton, skin, and internal anatomy are also unique to mammals. The presence of hair is one of the key characteristics of a mammal. Although it is not very extensive in some groups, such as whales, hair has many important functions for mammals. Mammals are endothermic, and hair provides insulation by trapping a layer of air close to the body to retain metabolic heat. Hair also serves as a sensory mechanism through specialized hairs called vibrissae, better known as whiskers. These attach to nerves that transmit touch information, which is particularly useful to nocturnal or burrowing mammals. Hair can also provide protective coloration.

Mammalian skin includes secretory glands with various functions. **Sebaceous glands** produce a lipid mixture called sebum that is secreted onto the hair and skin for water resistance and lubrication. Sebaceous glands are located over most of the body. **Sudoriferous glands** produce sweat and scent, which function in thermoregulation and communication, respectively. **Mammary glands** produce milk that is used to feed newborns. While male monotremes and eutherians possess mammary glands, male marsupials do not.

The skeletal system of mammals possesses unique features that differentiate them from other vertebrates. Most mammals have **heterodont teeth**, meaning they have different types and shapes of teeth that allow them to feed on different kinds of foods. These different types of teeth include the incisors, the canines, premolars, and molars. The first two types are for cutting and tearing, whereas the latter two types are for crushing and grinding. Different groups have different proportions of each type, depending on their diet. Most mammals are also **diphyodonts**, meaning they have two sets of teeth in their lifetime: deciduous or "baby" teeth, and permanent teeth. In other vertebrates, the teeth can be replaced throughout life.

Modern mammals are divided into three broad groups: monotremes, marsupials, and eutherians (or placental mammals). The eutherians, or placental mammals, and the marsupials collectively are called therian mammals, whereas monotremes are called metatherians.

There are three living species of **monotremes**: the platypus and two species of echidnas, or spiny anteaters (Figure 15.45). The platypus and one species of echidna are found in Australia, whereas the other species of echidna is found in New Guinea. Monotremes are unique among mammals, as they lay leathery eggs, similar to those of reptiles, rather than giving birth to live young. However, the eggs are retained within the mother's reproductive tract until they are almost ready to hatch. Once the young hatch, the female begins to secrete milk from pores in a ridge of mammary tissue along the ventral side of her body. Like other mammals, monotremes are endothermic but regulate body temperatures somewhat lower (90 °F, 32 °C) than placental mammals do (98 °F, 37 °C). Like reptiles, monotremes have one posterior opening for urinary, fecal, and reproductive products, rather than three separate openings like placental mammals do. Adult monotremes lack teeth.

Figure 15.45 The platypus (left), a monotreme, possesses a leathery beak and lays eggs rather than giving birth to live young. An echidna, another monotreme, is shown in the right photo. (credit "echidna": modification of work by Barry Thomas)

Marsupials are found primarily in Australia and nearby islands, although about 100 species of opossums and a few species of two other families are found in the Americas. Australian marsupials number over 230 species and include the kangaroo, koala, bandicoot, and Tasmanian devil (Figure 15.46). Most species of marsupials possess a pouch in which the young reside after birth, receiving milk and continuing to develop. Before birth, marsupials have a less complex placental connection, and the young are born much less developed than in placental mammals.

Figure 15.46 The Tasmanian devil is one of several marsupials native to Australia. (credit: Wayne McLean)

Eutherians are the most widespread of the mammals, occurring throughout the world. There are several groups of eutherians, including Insectivora, the insect eaters; Edentata, the toothless anteaters; Rodentia, the rodents; Chiroptera, the bats; Cetacea, the aquatic mammals including whales; Carnivora, carnivorous mammals including dogs, cats, and bears; and Primates, which includes humans. **Eutherian mammals** are sometimes called placental mammals, because all species have a complex placenta that connects a fetus to the mother, allowing for gas, fluid, waste, and nutrient exchange. While other mammals may possess a less complex placenta or briefly have a placenta, all eutherians have a complex placenta during gestation.

Primates

Order **Primates** of class Mammalia includes lemurs, tarsiers, monkeys, and the apes, which include humans. Non-human primates live primarily in tropical or subtropical regions of South America, Africa, and Asia. They range in size from the mouse lemur at 30 grams (1 ounce) to the mountain gorilla at 200 kilograms (441 pounds). The characteristics and evolution of primates are of particular interest to us as they allow us to understand the evolution of our own species.

All primate species have adaptations for climbing trees, as they all descended from tree-dwellers, although not all species are arboreal. This arboreal heritage of primates resulted in hands and feet that are adapted for **brachiation**, or climbing and swinging through trees. These adaptations include, but are not limited to 1) a rotating shoulder joint, 2) a big toe that is widely separated from the other toes and thumbs that are widely separated from fingers (except humans), which allow for gripping branches, and 3) **stereoscopic vision**, two overlapping visual fields, which allows for the depth perception necessary to gauge distance. Other characteristics of primates are brains that are larger than those of many other mammals, claws that have been modified into flattened nails, typically only one offspring per pregnancy, and a trend toward holding the body upright.

Order Primates is divided into two groups: prosimians and anthropoids. **Prosimians** include the bush babies of Africa, the lemurs of Madagascar, and the lorises, pottos, and tarsiers of Southeast Asia. **Anthropoids** include monkeys, lesser apes, and great apes (Figure 15.47). In general, prosimians tend to be nocturnal, smaller in size than anthropoids, and have relatively smaller brains compared to anthropoids.

Figure 15.47 Primates can be divided into prosimians, such as the (a) lemur, and anthropoids. Anthropoids include monkeys, such as the (b) howler monkey; lesser apes, such as the (c) gibbon; and great apes, such as the (d) chimpanzee, (e) bonobo, (f) gorilla, and (g) orangutan. (credit a: modification of work by Frank Vassen; credit b: modification of work by Xavi Talleda; credit d: modification of work by Aaron Logan; credit e: modification of work by Trisha Shears; credit f: modification of work by Dave Proffer; credit g: modification of work by Julie Langford)

KEY TERMS

acoelomate without a body cavity

Actinopterygii ray-finned fishes

amniote a clade of animals that possesses an amniotic egg; includes reptiles (including birds) and mammals

amoebocyte an amoeba-like cell of sponges whose functions include distribution of nutrients to other cells in the sponge

Amphibia frogs, salamanders, and caecilians

ampulla of Lorenzini a sensory organ that allows sharks to detect electromagnetic fields produced by living things

Annelida a phylum of worm-like animals with metamerism

anthropoids a clade consisting of monkeys, apes, and humans

Anura frogs

Apoda caecilians

Arthropoda a phylum of Ecdysozoa with jointed appendages and segmented bodies

asymmetrical having no plane of symmetry

bilateral symmetry a type of symmetry in which there is only one plane of symmetry that creates two mirror-image sides

body plan the shape and symmetry of an organism

brachiation swinging through trees

budding a form of asexual reproduction that occurs through the growth of a new organism as a branch on an adult organism that breaks off and becomes independent; found in plants, sponges, cnidarians, and some other invertebrates

caecilian a legless amphibian that belongs to clade Apoda

Cephalochordata a chordate clade whose members possess a notochord, dorsal hollow nerve cord, pharyngeal slits, and a post-anal tail in the adult stage

cephalothorax a fused head and thorax

chaeta a chitinous projection from the cuticle found in annelids

chelicerae a modified first pair of appendages in subphylum Chelicerata

chitin a tough nitrogen-containing polysaccharide found in the cuticles of arthropods and the cell walls of fungi

choanocyte a cell type unique to sponges with a flagellum surrounded by a collar used to maintain water flow through the sponge, and capture and digest food particles

Chondrichthyes jawed fishes with paired fins and a skeleton made of cartilage

Chordata a phylum of animals distinguished by their possession of a notochord, a dorsal hollow nerve cord, pharyngeal slits, and a post-anal tail at some point during their development

clitellum a specialized band of fused segments in some annelids, which aids in reproduction

Cnidaria a phylum of animals that are diploblastic and have radial symmetry and stinging cells

cnidocyte a specialized stinging cell found in Cnidaria

coelom a lined body cavity derived from mesodermal embryonic tissue

complete digestive system a digestive system that opens at one end, the mouth, and exits at the other end, the anus, and through which food normally moves in one direction

craniate a proposed clade of chordates that includes all groups except the tunicates and lancelets

Crocodilia crocodiles and alligators

ctenidia specialized gills in mollusks

cutaneous respiration gas exchange through the skin

deuterostome describing an animal in which the blastopore develops into the anus, with the second opening developing into the mouth

dioecious having separate male and female sexes

diphyodont refers to the possession of two sets of teeth in a lifetime

diploblast an animal that develops from two embryonic germ layers

dorsal hollow nerve cord a hollow, tubular structure derived from ectoderm, which is located dorsal to the notochord in chordates

down feather feather specialized for insulation

Echinodermata a phylum of deuterostomes with spiny skin; exclusively marine organisms

epidermis the layer of cells that lines the outer surface of an animal

eucoelomate describing animals with a body cavity completely lined with mesodermal tissue

eutherian mammal a mammal with a complex placenta, which connects a fetus to the mother; sometimes called placental mammals

extracellular digestion a form of digestion, the breakdown of food, which occurs outside of cells with the aid of enzymes released by cells

fragmentation a form of asexual reproduction in which a portion of the body of an organism breaks off and develops into a living independent organism; found in plants, sponges, and some other invertebrates

frog a tail-less amphibian that belongs to clade Anura

gastrodermis the layer of cells that lines the gastrovascular cavity of cnidarians

gastrovascular cavity the central cavity bounded by the gastrodermis in cnidarians

gemmule a structure produced by asexual reproduction in freshwater sponges that is able to survive harsh conditions

germ layer a collection of cells formed during embryogenesis that will give rise to future body tissues

gnathostome a jawed fish

hagfish an eel-like jawless fish that lives on the ocean floor and is a scavenger

hemocoel the internal body cavity seen in arthropods

heterodont teeth different types of teeth modified by different purposes

intracellular digestion the digestion of matter brought into a cell by phagocytosis

lamprey a jawless fish characterized by a toothed, funnel-like, sucking mouth

lancelet a member of Cephalochordata; named for its blade-like shape

lateral line the sense organ that runs the length of a fish's body, used to detect vibration in the water

Lophotrochozoa a clade of invertebrate organisms that is a sister group to the Ecdysozoa

madreporite a pore for regulating entry and exit of water into the water vascular system

mammal one of the groups of endothermic vertebrates that possess hair and mammary glands

mammary gland in female mammals, a gland that produces milk for newborns

mantle a specialized epidermis that encloses all visceral organs and secretes shells in mollusks

marsupial one of the groups of mammals that includes the kangaroo, koala, bandicoot, Tasmanian devil, and several other species; young develop within a pouch

medusa a free-floating cnidarian body plan with a mouth on the underside and tentacles hanging down from a bell

mesoglea the non-living, gel-like matrix present in between ectoderm and endoderm in cnidarians

mesohyl the collagen-like gel containing suspended cells that perform various functions in sponges

metamerism having a series of body structures that are similar internally and externally, such as segments

Mollusca a phylum of protostomes with soft bodies and no segmentation

monoecious having both sexes in one body, hermaphroditic

monotreme an egg-laying mammal

Myxini hagfishes

nacre a calcareous secretion produced by bivalve mollusks to line the inner side of shells as well as to coat foreign particulate matter

nematocyst the harpoon-like organelle within a cnidocyte with a pointed projectile and poison to stun and entangle prey

Nematoda a phylum of worms in Ecdysozoa commonly called roundworms containing both free-living and parasitic forms

notochord a flexible, rod-shaped structure that is found in the embryonic stage of all chordates and in the adult stage of some chordates

osculum the large opening in a sponge body through which water leaves

Osteichthyes bony fishes

ostracoderm one of the earliest jawless fishes covered in bone

Petromyzontidae the clade of lampreys

pharyngeal slit an opening in the pharynx

pneumatic bone an air-filled bone

polyp the stalk-like, sessile life form of a cnidarians with mouth and tentacles facing upward, usually sessile but may be able to glide along a surface

Porifera a phylum of animals with no true tissues, but a porous body with a rudimentary endoskeleton

post-anal tail a muscular, posterior elongation of the body extending beyond the anus in chordates

Primates includes lemurs, tarsiers, monkeys, apes, and humans

prosimians a group of primates that includes bush babies of Africa, lemurs of Madagascar, and lorises, pottos, and tarsiers of southeast Asia

protostome describing an animal in which the mouth develops first during embryogenesis and a second opening developing into the anus

pseudocoelomate an animal with a coelom that is not completely lined with tissues derived from the mesoderm as in eucoelomate animals

radial symmetry a type of symmetry with multiple planes of symmetry all cross at an axis through the center of the organism

radula a tongue-like scraping organ with chitinous ornamentation found in most mollusks

salamander a tailed amphibian that belongs to the clade Urodela

Sarcopterygii lobe-finned fishes

sebaceous gland in mammals, a skin gland that produce a lipid mixture called sebum

Sphenodontia the reptilian clade that includes the tuataras

spicule a short sliver or spike-like structure, in sponges, they are formed of silicon dioxide, calcium carbonate, or protein, and are found in the mesohyl

spiracle a respiratory openings in insects that allow air into the tracheae

spongocoel the central cavity within the body of some sponges

Squamata the reptilian clade of lizards and snakes

stereoscopic vision two overlapping fields of vision from the eyes that produces depth perception

sudoriferous gland a gland in mammals that produces sweat and scent molecules

swim bladder in fishes, a gas filled organ that helps to control the buoyancy of the fish

tadpole the larval stage of a frog

Testudines turtles

tetrapod a four-footed animal; includes amphibians, reptiles, birds, and mammals

trachea in some arthropods, such as insects, a respiratory tube that conducts air from the spiracles to the tissues

triploblast an animal that develops from three germ layers

tunicate a sessile chordate that is a member of Urochordata

Urochordata the clade composed of the tunicates

Urodela salamanders

vertebral column a series of separate bones that surround the spinal cord in vertebrates

water vascular system a system in echinoderms in which water is the circulatory fluid

CHAPTER SUMMARY

15.1 Features of the Animal Kingdom

Animals constitute a diverse kingdom of organisms. Although animals range in complexity from simple sea sponges to human beings, most members share certain features. Animals are eukaryotic, multicellular, heterotrophic organisms that ingest their food and usually develop into motile creatures with a fixed body plan. Most members of the animal kingdom have differentiated tissues of four main classes—nervous, muscular, connective, and epithelial—that are specialized to perform different functions. Most animals reproduce sexually, leading to a developmental sequence that is relatively similar across the animal kingdom.

Organisms in the animal kingdom are classified based on their body morphology and development. True animals are divided into those with radial versus bilateral symmetry. Animals with three germ layers, called triploblasts, are further characterized by the presence or absence of an internal body cavity called a coelom. Animals with a body cavity may be either coelomates or pseudocoelomates, depending on which tissue gives rise to the coelom. Coelomates are further divided into two groups called protostomes and deuterostomes, based on a number of developmental characteristics.

15.2 Sponges and Cnidarians

Animals included in phylum Porifera are parazoans and do not possess true tissues. These organisms show a simple organization. Sponges have multiple cell types that are geared toward executing various metabolic functions.

Cnidarians have outer and inner tissue layers sandwiching a noncellular mesoglea. Cnidarians possess a well-formed digestive system and carry out extracellular digestion. The cnidocyte is a specialized cell for delivering toxins to prey and predators. Cnidarians have separate sexes. They have a life cycle that involves morphologically distinct forms—medusoid and polypoid—at various stages in their life cycle.

15.3 Flatworms, Nematodes, and Arthropods

Flatworms are acoelomate, triploblastic animals. They lack circulatory and respiratory systems, and have a rudimentary excretory system. The digestive system is incomplete in most species. There are four traditional classes of flatworms, the largely free-living turbellarians, the ectoparasitic monogeneans, and the endoparasitic trematodes and cestodes. Trematodes have complex life cycles involving a secondary mollusk host and a primary host in which sexual reproduction takes place. Cestodes, or tapeworms, infect the digestive systems of primary vertebrate hosts.

Nematodes are pseudocoelomate members of the clade Ecdysozoa. They have a complete digestive system and a pseudocoelomic body cavity. This phylum includes free-living as well as parasitic organisms. They include dioecious and hermaphroditic species. Nematodes have a poorly developed excretory system. Embryonic development is external and proceeds through larval stages separated by molts.

Arthropods represent the most successful phylum of animals on Earth, in terms of number of species as well as the number of individuals. They are characterized by a segmented body and jointed appendages. In the basic body plan, a pair of appendages is present per body segment. Within the phylum, classification is based on mouthparts, number of appendages, and modifications of appendages. Arthropods bear a chitinous exoskeleton. Gills, tracheae, and book lungs facilitate respiration. Embryonic development may include multiple larval stages.

15.4 Mollusks and Annelids

The phylum Mollusca is a large, mainly marine group of invertebrates. Mollusks show a variety of morphologies. Many mollusks secrete a calcareous shell for protection, but in other species, the shell is reduced or absent. Mollusks are protostomes. The dorsal epidermis in mollusks is modified to form the mantle, which encloses the mantle cavity and visceral organs. This cavity is distinct from the coelomic cavity, which the adult animal retains, surrounding the heart. Respiration is facilitated by gills known as ctenidia. A chitinous scraper called the radula is present in most mollusks. Mollusks are mostly dioecious and are divided into seven classes.

The phylum Annelida includes worm-like, segmented animals. Segmentation is both external and internal, which is called metamerism. Annelids are protostomes. The presence of chitinous hairs called chaetae is characteristic of most members. These animals have well-developed nervous and digestive systems. Polychaete annelids have parapodia that participate in locomotion and respiration. Suckers are seen in the order Hirudinea. Breeding systems include separate sexes and hermaphroditism.

15.5 Echinoderms and Chordates

Echinoderms are deuterostome marine organisms. This phylum of animals bear a calcareous endoskeleton composed of ossicles covered by a spiny skin. Echinoderms possess a water-based circulatory system. The madreporite is the point of entry and exit for water for the water vascular system.

The characteristic features of Chordata are a notochord, a dorsal hollow nerve cord, pharyngeal slits, and a post-anal tail. Chordata contains two clades of invertebrates: Urochordata (tunicates) and Cephalochordata (lancelets), together with the vertebrates. Most tunicates live on the ocean floor and are suspension feeders. Lancelets are suspension feeders that feed on phytoplankton and other microorganisms.

15.6 Vertebrates

The earliest vertebrates that diverged from the invertebrate chordates were the jawless fishes. Hagfishes are eel-like scavengers that feed on dead invertebrates and other fishes. Lampreys are characterized by a toothed, funnel-like sucking mouth, and some species are parasitic on other fishes. Gnathostomes include the jawed fishes (cartilaginous and bony fishes) as well as all other tetrapods. Cartilaginous fishes include sharks, rays, skates, and ghost sharks. Bony fishes can be further divided into ray-finned and lobe-finned fishes.

As tetrapods, most amphibians are characterized by four well-developed limbs, although some species of salamanders and all caecilians are limbless. Amphibians have a moist, permeable skin used for cutaneous respiration. Amphibia can be divided into three clades: salamanders (Urodela), frogs (Anura), and caecilians (Apoda). The life cycle of amphibians consists of two distinct stages: the larval stage and metamorphosis to an adult stage.

The amniotes are distinguished from amphibians by the presence of a terrestrially adapted egg protected by amniotic membranes. The amniotes include reptiles, birds, and mammals. A key adaptation that permitted reptiles to live on land was the development of scaly skin. Reptilia includes four living clades: Crocodilia (crocodiles and alligators), Sphenodontia (tuataras), Squamata (lizards and snakes), and Testudines (turtles).

Birds are endothermic amniotes. Feathers act as insulation and allow for flight. Birds have pneumatic bones that are hollow rather than tissue-filled. Airflow through bird lungs travels in one direction. Birds evolved from dinosaurs.

Mammals have hair and mammary glands. Mammalian skin includes various secretory glands. Mammals are endothermic, like birds. There are three groups of mammals living today: monotremes, marsupials, and eutherians. Monotremes are unique among mammals as they lay eggs, rather than giving birth to live young. Eutherian mammals have a complex placenta.

There are 16 extant (living) orders of eutherian mammals. Humans are most closely related to Primates, all of which have adaptations for climbing trees, although not all species are arboreal. Other characteristics of primates are brains that are larger than those of other mammals, claws that have been modified into flattened nails, and typically one young per pregnancy, stereoscopic vision, and a trend toward holding the body upright. Primates are divided into two groups: prosimians and anthropoids.

ART CONNECTION QUESTIONS

1. Figure 15.3 Which of the following statements is false?

    a. Eumetazoa have specialized tissues and Parazoa do not.
    b. Both acoelomates and pseudocoelomates have a body cavity.
    c. Chordates are more closely related to echinoderms than to rotifers according to the figure.
    d. Some animals have radial symmetry, and some animals have bilateral symmetry.

2. Figure 15.24 Which of the following statements about the anatomy of a mollusk is false?
    a. Mollusks have a radula for scraping food.
    b. Mollusks have ventral nerve cords.
    c. The tissue beneath the shell is called the mantle.
    d. The mantle cavity contains hemolymph.

3. Figure 15.33 Which of the following statements about common features of chordates is true?
    a. The dorsal hollow nerve cord is part of the chordate central nervous system.
    b. In vertebrate fishes, the pharyngeal slits become the gills.
    c. Humans are not chordates because humans do not have a tail.
    d. Vertebrates do not have a notochord at any point in their development; instead, they have a vertebral column.

REVIEW QUESTIONS

4. Which of the following is not a feature common to *most* animals?
    a. development into a fixed body plan
    b. asexual reproduction
    c. specialized tissues
    d. heterotrophic nutrient sourcing

5. Which of the following does not occur?
    a. radially symmetrical diploblast
    b. diploblastic eucoelomate
    c. protostomic coelomate
    d. bilaterally symmetrical deuterostome

6. The large central opening in the poriferan body is called the _____.
   a. emmule
   b. picule
   c. stia
   d. osculum

7. Cnidocytes are found in _____.
   a. phylum Porifera
   b. phylum Nemertea
   c. phylum Nematoda
   d. phylum Cnidaria

8. Cubozoans are _____.
   a. polyps
   b. medusoids
   c. polymorphs
   d. sponges

9. Which group of flatworms are primarily external parasites of fish?
   a. monogeneans
   b. trematodes
   c. cestodes
   d. turbellarians

10. Crustaceans are _____.
   a. ecdysozoans
   b. nematodes
   c. arachnids
   d. parazoans

11. A mantle and mantle cavity are present in _____.

   a. class Oligochaeta
   b. class Bivalvia
   c. class Polychaeta
   d. class Hirudinea

12. Annelids have a _____.
   a. pseudocoelom
   b. a true coelom
   c. no coelom
   d. none of the above

13. Echinoderms in their larval state have _____.
   a. triangular symmetry
   b. radial symmetry
   c. hexagonal symmetry
   d. bilateral symmetry

14. The circulatory fluid in echinoderms is _____.

   a. blood
   b. mesohyl
   c. water
   d. saline

15. Which of the following is *not* a member of the phylum Chordata?
   a. Cephalochordata
   b. Echinodermata
   c. Urochordata
   d. Vertebrata

16. Members of Chondrichthyes differ from members of Osteichthyes by having a _____.
   a. jaw
   b. bony skeleton
   c. cartilaginous skeleton
   d. two sets of paired fins

17. Squamata includes _____.
   a. crocodiles and alligators
   b. turtles
   c. tuataras
   d. lizards and snakes

18. Sudoriferous glands produce _____.
   a. sweat
   b. lipids
   c. sebum
   d. milk

19. Which of the following is a Monotreme?
   a. kangaroo
   b. koala
   c. bandicoot
   d. platypus

CRITICAL THINKING QUESTIONS

20. How are specialized tissues important for animal function and complexity?

21. Using the following terms, explain what classifications and groups humans fall into, from the most general to the most specific: symmetry, germ layers, coelom, embryological development.

22. Describe the feeding mechanism of sponges and identify how it is different from other animals.

23. Compare the structural differences between Porifera and Cnidaria.

24. Speculate as to what advantage(s) a complete digestive system has over an incomplete digestive system?

25. Describe a potential advantage and disadvantage of the cuticle of ecdysozoans.

26. Describe the morphology and anatomy of mollusks.

27. Sessile adult tunicates lose the notochord; what does this suggest about one function of this structure?

28. During embryonic development, what features do we share with tunicates or lancelets?

29. What can be inferred about the evolution of the cranium and the vertebral column from examining hagfishes and lampreys?

30. Explain why frogs are restricted to a moist environment.

31. Describe three adaptations that allow for flight in birds.

16 | THE BODY'S SYSTEMS

Figure 16.1 An arctic fox is a complex animal, well adapted to its environment. (credit: Keith Morehouse, USFWS)

Chapter Outline
16.1: Homeostasis and Osmoregulation
16.2: Digestive System
16.3: Circulatory and Respiratory Systems
16.4: Endocrine System
16.5: Musculoskeletal System
16.6: Nervous System

Introduction

The arctic fox, a complex animal that has adapted to its environment, illustrates the relationships between an animal's form and function. The multicellular bodies of animals consist of tissues that make up more complex organs and organ systems. The organ systems of an animal maintain homeostasis within the multicellular body. These systems are adapted to obtain the necessary nutrients and other resources needed by the cells of the body, to remove the wastes those cells produce, to coordinate the activities of the cells, tissues, and organs throughout the body, and to coordinate the many responses of the individual organism to its environment.

16.1 | Homeostasis and Osmoregulation

By the end of this section, you will be able to:

- Explain the concept of homeostasis
- Describe thermoregulation of endothermic and ectothermic animals
- Explain how the kidneys serve as the main osmoregulatory organs in the human body

Homeostasis refers to the relatively stable state inside the body of an animal. Animal organs and organ systems constantly adjust to internal and external changes in order to maintain this steady state. Examples of internal conditions maintained homeostatically are the level of blood glucose, body temperature, blood calcium level. These conditions remain stable because of physiologic processes that result in negative feedback relationships. If the blood glucose or calcium rises, this sends a signal to organs responsible for lowering blood glucose or calcium. The signals that restore the normal levels are examples of negative feedback. When homeostatic mechanisms fail, the results can be unfavorable for the animal. Homeostatic mechanisms keep the body in dynamic equilibrium by constantly adjusting to the changes that the body's systems encounter. Even an animal that is apparently inactive is maintaining this homeostatic equilibrium. Two examples of factors that are regulated homeostatically are temperature and water content. The processes that maintain homeostasis of these two factors are called thermoregulation and osmoregulation.

Homeostasis

The goal of homeostasis is the maintenance of equilibrium around a specific value of some aspect of the body or its cells called a **set point**. While there are normal fluctuations from the set point, the body's systems will usually attempt to go back to this point. A change in the internal or external environment is called a stimulus and is detected by a receptor; the response of the system is to adjust the activities of the system so the value moves back toward the set point. For instance, if the body becomes too warm, adjustments are made to cool the animal. If glucose levels in the blood rise after a meal, adjustments are made to lower them and to get the nutrient into tissues that need it or to store it for later use.

When a change occurs in an animal's environment, an adjustment must be made so that the internal environment of the body and cells remains stable. The receptor that senses the change in the environment is part of a feedback mechanism. The stimulus—temperature, glucose, or calcium levels—is detected by the receptor. The receptor sends information to a control center, often the brain, which relays appropriate signals to an effector organ that is able to cause an appropriate change, either up or down, depending on the information the sensor was sending.

Thermoregulation

Animals can be divided into two groups: those that maintain a constant body temperature in the face of differing environmental temperatures, and those that have a body temperature that is the same as their environment and thus varies with the environmental temperature. Animals that do not have internal control of their body temperature are called **ectotherms**. The body temperature of these organisms is generally similar to the temperature of the environment, although the individual organisms may do things that keep their bodies slightly below or above the environmental temperature. This can include burrowing underground on a hot day or resting in the sunlight on a cold day. The ectotherms have been called cold-blooded, a term that may not apply to an animal in the desert with a very warm body temperature.

An animal that maintains a constant body temperature in the face of environmental changes is called an **endotherm**. These animals are able to maintain a level of activity that an ectothermic animal cannot because they generate internal heat that keeps their cellular processes operating optimally even when the environment is cold.

Watch this Discovery Channel video (http://openstaxcollege.org/l/thermoregulate2) on thermoregulation to see illustrations of the process in a variety of animals.

Animals conserve or dissipate heat in a variety of ways. Endothermic animals have some form of insulation. They have fur, fat, or feathers. Animals with thick fur or feathers create an insulating layer of air between their skin and internal organs. Polar bears and seals live and swim in a subfreezing environment and yet maintain a constant, warm, body temperature. The arctic fox, for example, uses its fluffy tail as extra insulation when it curls up to sleep in cold weather. Mammals can increase body heat production by shivering, which is an involuntary increase in muscle activity. In addition, arrector pili muscles can contract causing individual hairs to stand up when the individual is cold. This increases the insulating effect of the hair. Humans retain this reaction, which does not have the intended effect on our relatively hairless bodies; it causes "goose bumps" instead. Mammals use layers of fat as insulation also. Loss of significant amounts of body fat will compromise an individual's ability to conserve heat.

Ectotherms and endotherms use their circulatory systems to help maintain body temperature. Vasodilation, the opening up of arteries to the skin by relaxation of their smooth muscles, brings more blood and heat to the body surface, facilitating radiation and evaporative heat loss, cooling the body. Vasoconstriction, the narrowing of blood vessels to the skin by contraction of their smooth muscles, reduces blood flow in peripheral blood vessels, forcing blood toward the core and vital organs, conserving heat. Some animals have adaptions to their circulatory system that enable them to transfer heat from arteries to veins that are flowing next to each other, warming blood returning to the heart. This is called a countercurrent heat exchange; it prevents the cold venous blood from cooling the heart and other internal organs. The countercurrent adaptation is found in dolphins, sharks, bony fish, bees, and hummingbirds.

Some ectothermic animals use changes in their behavior to help regulate body temperature. They simply seek cooler areas during the hottest part of the day in the desert to keep from getting too warm. The same animals may climb onto rocks in the evening to capture heat on a cold desert night before entering their burrows.

Thermoregulation is coordinated by the nervous system (Figure 16.2). The processes of temperature control are centered in the hypothalamus of the advanced animal brain. The hypothalamus maintains the set point for body temperature through reflexes that cause vasodilation or vasoconstriction and shivering or sweating. The sympathetic nervous system under control of the hypothalamus directs the responses that effect the changes in temperature loss or gain that return the body to the set point. The set point may be adjusted in some instances. During an infection, compounds called pyrogens are produced and circulate to the hypothalamus resetting the thermostat to a higher value. This allows the body's temperature to increase to a new homeostatic equilibrium point in what is commonly called a fever. The increase in body heat makes the body less optimal for bacterial growth and increases the activities of cells so they are better able to fight the infection.

art CONNECTION

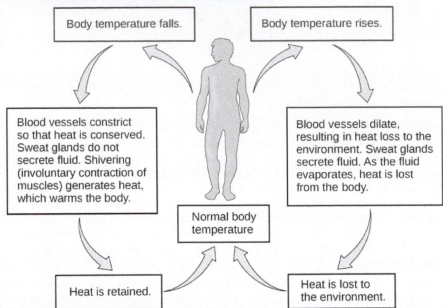

Figure 16.2 The body is able to regulate temperature in response to signals from the nervous system.

When bacteria are destroyed by leukocytes, pyrogens are released into the blood. Pyrogens reset the body's thermostat to a higher temperature, resulting in fever. How might pyrogens cause the body temperature to rise?

Osmoregulation

Osmoregulation is the process of maintaining salt and water balance (**osmotic balance**) across membranes within the body. The fluids inside and surrounding cells are composed of water, electrolytes, and nonelectrolytes. An electrolyte is a compound that dissociates into ions when dissolved in water. A nonelectrolyte, in contrast, does not dissociate into ions in water. The body's fluids include blood plasma, fluid that exists within cells, and the **interstitial fluid** that exists in the spaces between cells and tissues of the body. The membranes of the body (both the membranes around cells and the "membranes" made of cells lining body cavities) are semipermeable membranes. Semipermeable membranes are permeable to certain types of solutes and to water, but typically cell membranes are impermeable to solutes.

The body does not exist in isolation. There is a constant input of water and electrolytes into the system. Excess water, electrolytes, and wastes are transported to the kidneys and excreted, helping to maintain osmotic balance. Insufficient fluid intake results in fluid conservation by the kidneys. Biological systems constantly interact and exchange water and nutrients with the environment by way of consumption of food and water and through excretion in the form of sweat, urine, and feces. Without a mechanism to regulate osmotic pressure, or when a disease damages this mechanism, there is a tendency to accumulate toxic waste and water, which can have dire consequences.

Mammalian systems have evolved to regulate not only the overall osmotic pressure across membranes, but also specific concentrations of important electrolytes in the three major fluid compartments: blood plasma, interstitial fluid, and intracellular fluid. Since osmotic pressure is regulated by the movement of water across membranes, the volume of the fluid compartments can also change temporarily. Since blood plasma is one of the fluid components, osmotic pressures have a direct bearing on blood pressure.

Excretory System

The human excretory system functions to remove waste from the body through the skin as sweat, the lungs in the form of exhaled carbon dioxide, and through the urinary system in the form of urine. All three of these systems participate in osmoregulation and waste removal. Here we focus on the urinary system, which is comprised of the paired kidneys, the ureter, urinary bladder and urethra (Figure 16.3). The **kidneys** are a pair of bean-shaped structures that are located just below the liver in the body cavity. Each of the kidneys contains more than a million tiny units called nephrons that filter blood containing the metabolic wastes from cells. All the blood in the human body is filtered about 60 times a day by the kidneys. The nephrons remove wastes, concentrate them, and form urine that is collected in the bladder.

Internally, the kidney has three regions—an outer cortex, a medulla in the middle, and the renal pelvis, which is the expanded end of the ureter. The renal cortex contains the **nephrons**—the functional unit of the kidney. The renal pelvis collects the urine and leads to the **ureter** on the outside of the kidney. The ureters are urine-bearing tubes that exit the kidney and empty into the **urinary bladder**.

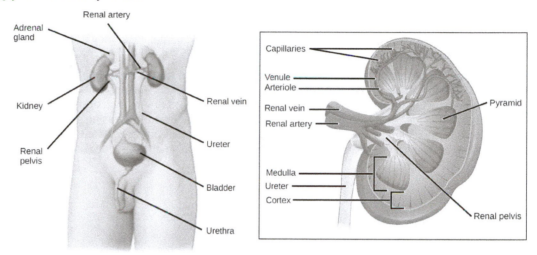

Figure 16.3 The human excretory system is made up of the kidneys, ureter, urinary bladder, and urethra. The kidneys filter blood and form urine, which is stored in the bladder until it is eliminated through the urethra. On the right, the internal structure of the kidney is shown. (credit: modification of work by NCI, NIH)

Blood enters each kidney from the aorta, the main artery supplying the body below the heart, through a **renal artery**. It is distributed in smaller vessels until it reaches each nephron in capillaries. Within the nephron the blood comes in intimate contact with the waste-collecting tubules in a structure called the glomerulus. Water and many solutes present in the blood, including ions of sodium, calcium, magnesium, and others; as well as wastes and valuable substances such as amino acids, glucose and vitamins, leave the blood and enter the tubule system of the nephron. As materials pass through the tubule much of the water, required ions, and useful compounds are reabsorbed back into the capillaries that surround the tubules leaving the wastes behind. Some of this reabsorption requires active transport and consumes ATP. Some wastes, including ions and some drugs remaining in the blood, diffuse out of the capillaries into the interstitial fluid and are taken up by the tubule cells. These wastes are then actively secreted into the tubules. The blood then collects in larger and larger vessels and leaves the kidney in the renal vein. The **renal vein** joins the inferior vena cava, the main vein that returns blood to the heart from the lower body. The amounts of water and ions reabsorbed into the circulatory system are carefully regulated and this is an important way the body regulates its water content and ion levels. The waste is collected in larger tubules and then leaves the kidney in the ureter, which leads to the bladder where urine, the combination of waste materials and water, is stored.

The bladder contains sensory nerves, stretch receptors that signal when it needs to be emptied. These signals create the urge to urinate, which can be voluntarily suppressed up to a limit. The conscious decision to urinate sets in play signals that open the sphincters, rings of smooth muscle that close off the opening, to the **urethra** that allows urine to flow out of the bladder and the body.

Dialysis Technician

Dialysis is a medical process of removing wastes and excess water from the blood by diffusion and ultrafiltration. When kidney function fails, dialysis must be done to artificially rid the body of wastes and fluids. This is a vital process to keep patients alive. In some cases, the patients undergo artificial dialysis until they are eligible for a kidney transplant. In others who are not candidates for kidney transplants, dialysis is a lifelong necessity.

Dialysis technicians typically work in hospitals and clinics. While some roles in this field include equipment development and maintenance, most dialysis technicians work in direct patient care. Their on-the-job duties, which typically occur under the direct supervision of a registered nurse, focus on providing dialysis treatments. This can include reviewing patient history and current condition, assessing and responding to patient needs before and during treatment, and monitoring the dialysis process. Treatment may include taking and reporting a patient's vital signs, preparing solutions and equipment to ensure accurate and sterile procedures.

16.2 | Digestive System

By the end of this section, you will be able to:

- Explain the processes of digestion and absorption

- Explain the specialized functions of the organs involved in processing food in the body

- Describe the ways in which organs work together to digest food and absorb nutrients

- Describe the essential nutrients required for cellular function that cannot be synthesized by the animal body

- Describe how excess carbohydrates and energy are stored in the body

All living organisms need nutrients to survive. While plants can obtain nutrients from their roots and the energy molecules required for cellular function through the process of photosynthesis, animals obtain their nutrients by the consumption of other organisms. At the cellular level, the biological molecules necessary for animal function are amino acids, lipid molecules, nucleotides, and simple sugars. However, the food consumed consists of protein, fat, and complex carbohydrates. Animals must convert these macromolecules into the simple molecules required for maintaining cellular function. The conversion of the food consumed to the nutrients required is a multistep process involving digestion and absorption. During digestion, food particles are broken down to smaller components, which are later absorbed by the body. This happens by both physical means, such as chewing, and by chemical means.

One of the challenges in human nutrition is maintaining a balance between food intake, storage, and energy expenditure. Taking in more food energy than is used in activity leads to storage of the excess in the form of fat deposits. The rise in obesity and the resulting diseases like type 2 diabetes makes understanding the role of diet and nutrition in maintaining good health all the more important.

The Human Digestive System

The process of digestion begins in the mouth with the intake of food (Figure 16.4). The teeth play an important role in masticating (chewing) or physically breaking food into smaller particles. The enzymes present in saliva also begin to chemically break down food. The food is then swallowed and enters the **esophagus**—a long tube that connects the mouth to the stomach. Using **peristalsis**, or wave-like smooth-muscle contractions, the muscles of the esophagus push the food toward the stomach. The stomach contents are extremely acidic, with a pH between 1.5 and 2.5. This acidity kills microorganisms, breaks down food tissues, and activates digestive enzymes. Further breakdown of food takes place in the small intestine where bile produced by the liver, and enzymes produced by the small intestine and the pancreas, continue the process of digestion. The smaller molecules are absorbed into the blood stream through the epithelial cells lining the walls of the small intestine. The waste material travels on to the large intestine where water is absorbed and the drier waste material is compacted into feces; it is stored until it is excreted through the anus.

Download for free at https://openstax.org/details/books/concepts-biology

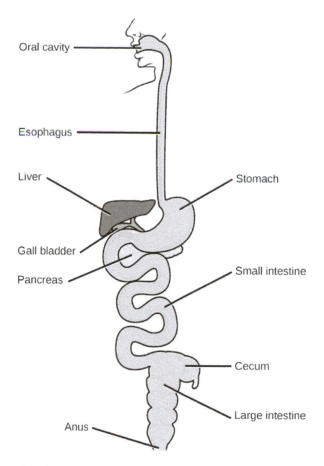

Figure 16.4 The components of the human digestive system are shown.

Oral Cavity

Both physical and chemical digestion begin in the mouth or **oral cavity**, which is the point of entry of food into the digestive system. The food is broken into smaller particles by mastication, the chewing action of the teeth. All mammals have teeth and can chew their food to begin the process of physically breaking it down into smaller particles.

The chemical process of digestion begins during chewing as food mixes with saliva, produced by the **salivary glands** (Figure 16.5). Saliva contains mucus that moistens food and buffers the pH of the food. Saliva also contains lysozyme, which has antibacterial action. It also contains an enzyme called salivary **amylase** that begins the process of converting starches in the food into a disaccharide called maltose. Another enzyme called lipase is produced by cells in the tongue to break down fats. The chewing and wetting action provided by the teeth and saliva prepare the food into a mass called the **bolus** for swallowing. The tongue helps in swallowing—moving the bolus from the mouth into the pharynx. The pharynx opens to two passageways: the esophagus and the trachea. The esophagus leads to the stomach and the trachea leads to the lungs. The epiglottis is a flap of tissue that covers the tracheal opening during swallowing to prevent food from entering the lungs.

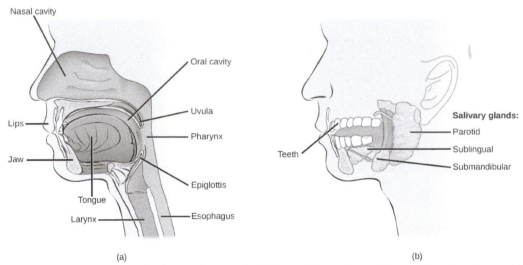

Figure 16.5 (a) Digestion of food begins in the mouth. (b) Food is masticated by teeth and moistened by saliva secreted from the salivary glands. Enzymes in the saliva begin to digest starches and fats. With the help of the tongue, the resulting bolus is moved into the esophagus by swallowing. (credit: modification of work by Mariana Ruiz Villareal)

Esophagus

The esophagus is a tubular organ that connects the mouth to the stomach. The chewed and softened food passes through the esophagus after being swallowed. The smooth muscles of the esophagus undergo peristalsis that pushes the food toward the stomach. The peristaltic wave is unidirectional—it moves food from the mouth the stomach, and reverse movement is not possible, except in the case of the vomit reflex. The peristaltic movement of the esophagus is an involuntary reflex; it takes place in response to the act of swallowing.

Ring-like muscles called sphincters form valves in the digestive system. The gastro-esophageal sphincter (or cardiac sphincter) is located at the stomach end of the esophagus. In response to swallowing and the pressure exerted by the bolus of food, this sphincter opens, and the bolus enters the stomach. When there is no swallowing action, this sphincter is shut and prevents the contents of the stomach from traveling up the esophagus. Acid reflux or "heartburn" occurs when the acidic digestive juices escape into the esophagus.

Stomach

A large part of protein digestion occurs in the stomach (Figure 16.7). The **stomach** is a saclike organ that secretes gastric digestive juices.

Protein digestion is carried out by an enzyme called **pepsin** in the stomach chamber. The highly acidic environment kills many microorganisms in the food and, combined with the action of the enzyme pepsin, results in the catabolism of protein in the food. Chemical digestion is facilitated by the churning action of the stomach caused by contraction and relaxation of smooth muscles. The partially digested food and gastric juice mixture is called **chyme**. Gastric emptying occurs within two to six hours after a meal. Only a small amount of chyme is released into the small intestine at a time. The movement of chyme from the stomach into the small intestine is regulated by hormones, stomach distension and muscular reflexes that influence the pyloric sphincter.

The stomach lining is unaffected by pepsin and the acidity because pepsin is released in an inactive form and the stomach has a thick mucus lining that protects the underlying tissue.

Small Intestine

Chyme moves from the stomach to the small intestine. The **small intestine** is the organ where the digestion of protein, fats, and carbohydrates is completed. The small intestine is a long tube-like organ with a highly folded surface containing finger-like projections called the villi. The top surface of each villus has many microscopic projections called microvilli. The epithelial cells of these structures absorb nutrients from the digested food and release them to the bloodstream on the other side. The villi and microvilli, with their many folds, increase the surface area of the small intestine and increase absorption efficiency of the nutrients.

The human small intestine is over 6 m (19.6 ft) long and is divided into three parts: the duodenum, the jejunum and the ileum. The duodenum is separated from the stomach by the pyloric sphincter. The chyme is mixed with pancreatic juices, an alkaline solution rich in bicarbonate that neutralizes the acidity of chyme from the stomach. Pancreatic juices contain several digestive enzymes that break down starches, disaccharides, proteins, and fats. **Bile** is produced in the liver and stored and concentrated in the gallbladder; it enters the duodenum through the bile duct. Bile contains bile salts, which make lipids accessible to the water-soluble enzymes. The monosaccharides, amino acids, bile salts, vitamins, and other nutrients are absorbed by the cells of the intestinal lining.

The undigested food is sent to the colon from the ileum via peristaltic movements. The ileum ends and the large intestine begins at the ileocecal valve. The vermiform, "worm-like," appendix is located at the ileocecal valve. The appendix of humans has a minor role in immunity.

Large Intestine

The **large intestine** reabsorbs the water from indigestible food material and processes the waste material (Figure 16.6). The human large intestine is much smaller in length compared to the small intestine but larger in diameter. It has three parts: the cecum, the colon, and the rectum. The cecum joins the ileum to the colon and is the receiving pouch for the waste matter. The colon is home to many bacteria or "intestinal flora" that aid in the digestive processes. The **colon** has four regions, the ascending colon, the transverse colon, the descending colon and the sigmoid colon. The main functions of the colon are to extract the water and mineral salts from undigested food, and to store waste material.

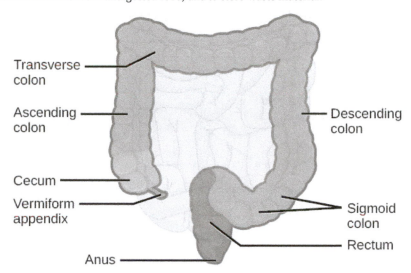

Figure 16.6 The large intestine reabsorbs water from undigested food and stores waste until it is eliminated. (credit: modification of work by Mariana Ruiz Villareal)

The **rectum** (Figure 16.6) stores feces until defecation. The feces are propelled using peristaltic movements during elimination. The **anus** is an opening at the far-end of the digestive tract and is the exit point for the waste material. Two sphincters regulate the exit of feces, the inner sphincter is involuntary and the outer sphincter is voluntary.

Accessory Organs

The organs discussed above are the organs of the digestive tract through which food passes. Accessory organs add secretions and enzymes that break down food into nutrients. Accessory organs include the salivary glands, the liver, the pancreas, and the gall bladder. The secretions of the liver, pancreas, and gallbladder are regulated by hormones in response to food consumption.

The **liver** is the largest internal organ in humans and it plays an important role in digestion of fats and detoxifying blood. The liver produces bile, a digestive juice that is required for the breakdown of fats in the duodenum. The liver also processes the absorbed vitamins and fatty acids and synthesizes many plasma proteins. The **gallbladder** is a small organ that aids the liver by storing bile and concentrating bile salts.

The **pancreas** secretes bicarbonate that neutralizes the acidic chyme and a variety of enzymes for the digestion of protein and carbohydrates.

a r t CONNECTION

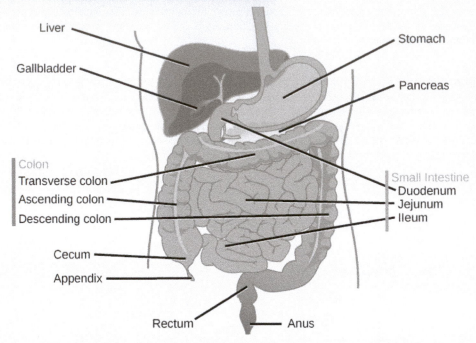

Figure 16.7 The stomach has an extremely acidic environment where most of the protein gets digested. (credit: modification of work by Mariana Ruiz Villareal)

Which of the following statements about the digestive system is false?

a. Chyme is a mixture of food and digestive juices that is produced in the stomach.

b. Food enters the large intestine before the small intestine.

c. In the small intestine, chyme mixes with bile, which emulsifies fats.

d. The stomach is separated from the small intestine by the pyloric sphincter.

Nutrition

The human diet should be well balanced to provide nutrients required for bodily function and the minerals and vitamins required for maintaining structure and regulation necessary for good health and reproductive capability (Figure 16.8).

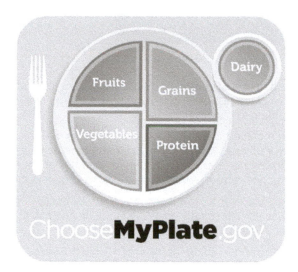

Figure 16.8 For humans, a balanced diet includes fruits, vegetables, grains, protein, and dairy. (credit: USDA)

Explore this interactive United States Department of Agriculture website (http://openstaxcollege.org/l/food_groups2) to learn more about each food group and the recommended daily amounts.

The organic molecules required for building cellular material and tissues must come from food. During digestion, digestible carbohydrates are ultimately broken down into glucose and used to provide energy within the cells of the body. Complex carbohydrates, including polysaccharides, can be broken down into glucose through biochemical modification; however, humans do not produce the enzyme necessary to digest cellulose (fiber). The intestinal flora in the human gut are able to extract some nutrition from these plant fibers. These plant fibers are known as dietary fiber and are an important component of the diet. The excess sugars in the body are converted into glycogen and stored for later use in the liver and muscle tissue. Glycogen stores are used to fuel prolonged exertions, such as long-distance running, and to provide energy during food shortage. Fats are stored under the skin of mammals for insulation and energy reserves.

Proteins in food are broken down during digestion and the resulting amino acids are absorbed. All of the proteins in the body must be formed from these amino-acid constituents; no proteins are obtained directly from food.

Fats add flavor to food and promote a sense of satiety or fullness. Fatty foods are also significant sources of energy, and fatty acids are required for the construction of lipid membranes. Fats are also required in the diet to aid the absorption of fat-soluble vitamins and the production of fat-soluble hormones.

While the animal body can synthesize many of the molecules required for function from precursors, there are some nutrients that must be obtained from food. These nutrients are termed **essential nutrients**, meaning they must be eaten, because the body cannot produce them.

The fatty acids omega-3 alpha-linolenic acid and omega-6 linoleic acid are essential fatty acids needed to make some membrane phospholipids. **Vitamins** are another class of essential organic molecules that are required in small quantities. Many of these assist enzymes in their function and, for this reason, are called coenzymes. Absence or low levels of vitamins can have a dramatic effect on health. **Minerals** are another set of inorganic essential nutrients that must be obtained from food. Minerals perform many functions, from muscle and nerve function, to acting as enzyme cofactors. Certain amino acids also must be procured from food and cannot be synthesized by the body. These amino acids are the "essential" amino acids. The human body can synthesize only 11 of the 20 required amino acids; the rest must be obtained from food.

bi⊕logy IN ACTION

Obesity

With obesity at high rates in the United States, there is a public health focus on reducing obesity and associated health risks, which include diabetes, colon and breast cancer, and cardiovascular disease. How does the food consumed contribute to obesity?

Fatty foods are calorie-dense, meaning that they have more calories per unit mass than carbohydrates or proteins. One gram of carbohydrates has four calories, one gram of protein has four calories, and one gram of fat has nine calories. Animals tend to seek lipid-rich food for their higher energy content. Greater amounts of food energy taken in than the body's requirements will result in storage of the excess in fat deposits.

Excess carbohydrate is used by the liver to synthesize glycogen. When glycogen stores are full, additional glucose is converted into fatty acids. These fatty acids are stored in adipose tissue cells—the fat cells in the mammalian body whose primary role is to store fat for later use.

The rate of obesity among children is rapidly rising in the United States. To combat childhood obesity and ensure that children get a healthy start in life, in 2010 First Lady Michelle Obama launched the Let's Move! campaign. The goal of this campaign is to educate parents and caregivers on providing healthy nutrition and encouraging active lifestyles in future generations. This program aims to involve the entire community, including parents, teachers, and healthcare providers to ensure that children have access to healthy foods—more fruits, vegetables, and whole grains—and consume fewer calories from processed foods. Another goal is to ensure that children get physical activity. With the increase in television viewing and stationary pursuits such as video games, sedentary lifestyles have become the norm. Visit www.letsmove.gov to learn more.

16.3 | Circulatory and Respiratory Systems

> By the end of this section, you will be able to:
> - Describe the passage of air from the outside environment to the lungs
> - Describe the function of the circulatory system
> - Describe the cardiac cycle
> - Explain how blood flows through the body

Animals are complex multicellular organisms that require a mechanism for transporting nutrients throughout their bodies and removing wastes. The human circulatory system has a complex network of blood vessels that reach all parts of the body. This extensive network supplies the cells, tissues, and organs with oxygen and nutrients, and removes carbon dioxide and waste compounds.

The medium for transport of gases and other molecules is the blood, which continually circulates through the system. Pressure differences within the system cause the movement of the blood and are created by the pumping of the heart.

Gas exchange between tissues and the blood is an essential function of the circulatory system. In humans, other mammals, and birds, blood absorbs oxygen and releases carbon dioxide in the lungs. Thus the circulatory and respiratory system, whose function is to obtain oxygen and discharge carbon dioxide, work in tandem.

The Respiratory System

Take a breath in and hold it. Wait several seconds and then let it out. Humans, when they are not exerting themselves, breathe approximately 15 times per minute on average. This equates to about 900 breaths an hour or 21,600 breaths per day. With every inhalation, air fills the lungs, and with every exhalation, it rushes back out. That air is doing more than just inflating and deflating the lungs in the chest cavity. The air contains oxygen that crosses the lung tissue, enters the bloodstream, and travels to organs and tissues. There, oxygen is exchanged for carbon dioxide, which is a cellular waste material. Carbon dioxide exits the cells, enters the bloodstream, travels back to the lungs, and is expired out of the body during exhalation.

Download for free at https://openstax.org/details/books/concepts-biology

Breathing is both a voluntary and an involuntary event. How often a breath is taken and how much air is inhaled or exhaled is regulated by the respiratory center in the brain in response to signals it receives about the carbon dioxide content of the blood. However, it is possible to override this automatic regulation for activities such as speaking, singing and swimming under water.

During inhalation the **diaphragm** descends creating a negative pressure around the lungs and they begin to inflate, drawing in air from outside the body. The air enters the body through the **nasal cavity** located just inside the nose (Figure 16.9). As the air passes through the nasal cavity, the air is warmed to body temperature and humidified by moisture from mucous membranes. These processes help equilibrate the air to the body conditions, reducing any damage that cold, dry air can cause. Particulate matter that is floating in the air is removed in the nasal passages by hairs, mucus, and cilia. Air is also chemically sampled by the sense of smell.

From the nasal cavity, air passes through the **pharynx** (throat) and the **larynx** (voice box) as it makes its way to the **trachea** (Figure 16.9). The main function of the trachea is to funnel the inhaled air to the lungs and the exhaled air back out of the body. The human trachea is a cylinder, about 25 to 30 cm (9.8–11.8 in) long, which sits in front of the esophagus and extends from the pharynx into the chest cavity to the lungs. It is made of incomplete rings of cartilage and smooth muscle. The cartilage provides strength and support to the trachea to keep the passage open. The trachea is lined with cells that have cilia and secrete mucus. The mucus catches particles that have been inhaled, and the cilia move the particles toward the pharynx.

The end of the trachea divides into two bronchi that enter the right and left lung. Air enters the lungs through the **primary bronchi**. The primary bronchus divides, creating smaller and smaller diameter **bronchi** until the passages are under 1 mm (.03 in) in diameter when they are called **bronchioles** as they split and spread through the lung. Like the trachea, the bronchus and bronchioles are made of cartilage and smooth muscle. Bronchi are innervated by nerves of both the parasympathetic and sympathetic nervous systems that control muscle contraction (parasympathetic) or relaxation (sympathetic) in the bronchi and bronchioles, depending on the nervous system's cues. The final bronchioles are the respiratory bronchioles. Alveolar ducts are attached to the end of each respiratory bronchiole. At the end of each duct are alveolar sacs, each containing 20 to 30 **alveoli**. Gas exchange occurs only in the alveoli. The alveoli are thin-walled and look like tiny bubbles within the sacs. The alveoli are in direct contact with capillaries of the circulatory system. Such intimate contact ensures that oxygen will diffuse from the alveoli into the blood. In addition, carbon dioxide will diffuse from the blood into the alveoli to be exhaled. The anatomical arrangement of capillaries and alveoli emphasizes the structural and functional relationship of the respiratory and circulatory systems. Estimates for the surface area of alveoli in the lungs vary around 100 m^2. This large area is about the area of half a tennis court. This large surface area, combined with the thin-walled nature of the alveolar cells, allows gases to easily diffuse across the cells.

a r t CONNECTION

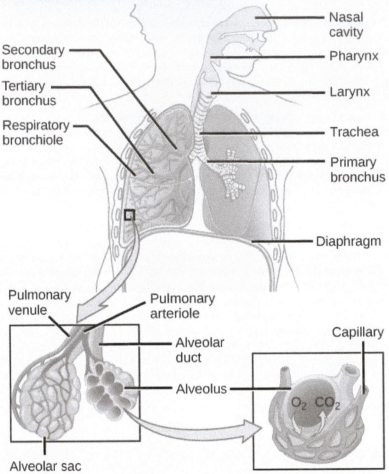

Figure 16.9 Air enters the respiratory system through the nasal cavity, and then passes through the pharynx and the trachea into the lungs. (credit: modification of work by NCI)

Which of the following statements about the human respiratory system is false?

a. When we breathe in, air travels from the pharynx to the trachea.

b. The bronchioles branch into bronchi.

c. Alveolar ducts connect to alveolar sacs.

d. Gas exchange between the lungs and blood takes place in the alveolus.

Watch this video (http://openstaxcollege.org/l/lungs_pulmonar2) for a review of the respiratory system.

The Circulatory System

The circulatory system is a network of vessels—the arteries, veins, and capillaries—and a pump, the heart. In all vertebrate organisms this is a closed-loop system, in which the blood is largely separated from the body's other extracellular fluid compartment, the interstitial fluid, which is the fluid bathing the cells. Blood circulates inside blood vessels and circulates unidirectionally from the heart around one of two circulatory routes, then returns to the heart again; this is a **closed circulatory system**. **Open circulatory systems** are found in invertebrate animals in which the circulatory fluid bathes the internal organs directly even though it may be moved about with a pumping heart.

The Heart

The heart is a complex muscle that consists of two pumps: one that pumps blood through **pulmonary circulation** to the lungs, and the other that pumps blood through **systemic circulation** to the rest of the body's tissues (and the heart itself).

The heart is asymmetrical, with the left side being larger than the right side, correlating with the different sizes of the pulmonary and systemic circuits (Figure 16.10). In humans, the heart is about the size of a clenched fist; it is divided into four chambers: two atria and two ventricles. There is one **atrium** and one **ventricle** on the right side and one atrium and one ventricle on the left side. The right atrium receives deoxygenated blood from the systemic circulation through the major veins: the **superior vena cava**, which drains blood from the head and from the veins that come from the arms, as well as the **inferior vena cava**, which drains blood from the veins that come from the lower organs and the legs. This deoxygenated blood then passes to the right ventricle through the **tricuspid valve**, which prevents the backflow of blood. After it is filled, the right ventricle contracts, pumping the blood to the lungs for reoxygenation. The left atrium receives the oxygen-rich blood from the lungs. This blood passes through the **bicuspid valve** to the left ventricle where the blood is pumped into the **aorta**. The aorta is the major artery of the body, taking oxygenated blood to the organs and muscles of the body. This pattern of pumping is referred to as double circulation and is found in all mammals. (Figure 16.10).

art CONNECTION

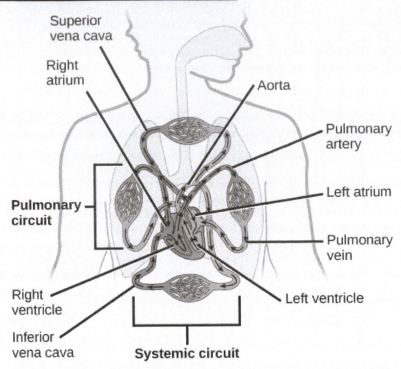

Figure 16.10 The heart is divided into four chambers, two atria, and two ventricles. Each chamber is separated by one-way valves. The right side of the heart receives deoxygenated blood from the body and pumps it to the lungs. The left side of the heart pumps blood to the rest of the body.

Which of the following statements about the circulatory system is false?

a. Blood in the pulmonary vein is deoxygenated.

b. Blood in the inferior vena cava is deoxygenated.

c. Blood in the pulmonary artery is deoxygenated.

d. Blood in the aorta is oxygenated.

The Cardiac Cycle

The main purpose of the heart is to pump blood through the body; it does so in a repeating sequence called the cardiac cycle. The **cardiac cycle** is the flow of blood through the heart coordinated by electrochemical signals that cause the heart muscle to contract and relax. In each cardiac cycle, a sequence of contractions pushes out the blood, pumping it through the body; this is followed by a relaxation phase, where the heart fills with blood. These two phases are called the **systole** (contraction) and **diastole** (relaxation), respectively (Figure 16.11). The signal for contraction begins at a location on the outside of the right atrium. The electrochemical signal moves from there across the atria causing them to contract. The contraction of the atria forces blood through the valves into the ventricles. Closing of these valves caused by the contraction of the ventricles produces a "lub" sound. The signal has, by this time, passed down the walls of the heart, through a point between the right atrium and right ventricle. The signal then causes the ventricles to contract. The ventricles contract together forcing blood into the aorta and the pulmonary arteries. Closing of the valves to these arteries caused by blood being drawn back toward the heart during ventricular relaxation produces a monosyllabic "dub" sound.

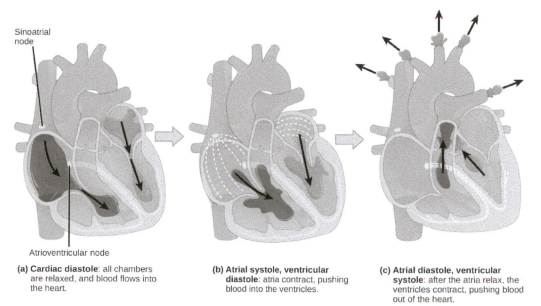

Sinoatrial node

Atrioventricular node

(a) Cardiac diastole: all chambers are relaxed, and blood flows into the heart.

(b) Atrial systole, ventricular diastole: atria contract, pushing blood into the ventricles.

(c) Atrial diastole, ventricular systole: after the atria relax, the ventricles contract, pushing blood out of the heart.

Figure 16.11 In each cardiac cycle, a series of contractions (systoles) and relaxations (diastoles) pumps blood through the heart and through the body. (a) During cardiac diastole, blood flows into the heart while all chambers are relaxed. (b) Then the ventricles remain relaxed while atrial systole pushes blood into the ventricles. (c) Once the atria relax again, ventricle systole pushes blood out of the heart.

The pumping of the heart is a function of the cardiac muscle cells, or cardiomyocytes, that make up the heart muscle. Cardiomyocytes are distinctive muscle cells that are striated like skeletal muscle but pump rhythmically and involuntarily like smooth muscle; adjacent cells are connected by intercalated disks found only in cardiac muscle. These connections allow the electrical signal to travel directly to neighboring muscle cells.

The electrical impulses in the heart produce electrical currents that flow through the body and can be measured on the skin using electrodes. This information can be observed as an **electrocardiogram (ECG)** a recording of the electrical impulses of the cardiac muscle.

Visit the following website (http://openstaxcollege.org/l/electric_heart2) to see the heart's pacemaker, or electrocardiogram system, in action.

Blood Vessels

The blood from the heart is carried through the body by a complex network of blood vessels (Figure 16.12). **Arteries** take blood away from the heart. The main artery of the systemic circulation is the aorta; it branches into major arteries that take blood to different limbs and organs. The aorta and arteries near the heart have heavy but elastic walls that respond to and smooth out the pressure differences caused by the beating heart. Arteries farther away from the heart have more muscle tissue in their walls that can constrict to affect flow rates of blood. The major arteries diverge into minor arteries, and then smaller vessels called arterioles, to reach more deeply into the muscles and organs of the body.

Arterioles diverge into capillary beds. Capillary beds contain a large number, 10's to 100's of **capillaries** that branch among the cells of the body. Capillaries are narrow-diameter tubes that can fit single red blood cells and are the sites for the exchange of nutrients, waste, and oxygen with tissues at the cellular level. Fluid also leaks from the blood into the interstitial space from the capillaries. The capillaries converge again into venules that connect to minor veins that finally connect to major veins. **Veins** are blood vessels that bring blood high in carbon dioxide back to the heart. Veins are not as thick-walled as arteries, since pressure is lower, and they have valves along their length that prevent backflow of blood away from the heart. The major veins drain blood from the same organs and limbs that the major arteries supply.

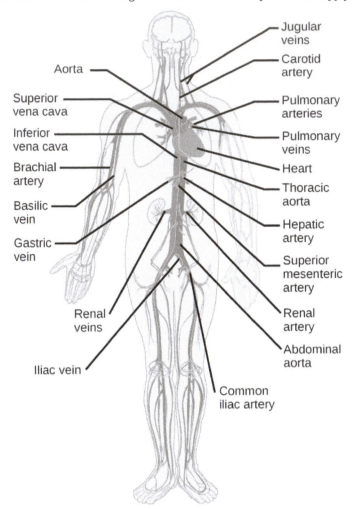

Figure 16.12 The arteries of the body, indicated in red, start at the aortic arch and branch to supply the organs and muscles of the body with oxygenated blood. The veins of the body, indicated in blue, return blood to the heart. The pulmonary arteries are blue to reflect the fact that they are deoxygenated, and the pulmonary veins are red to reflect that they are oxygenated. (credit: modification of work by Mariana Ruiz Villareal)

16.4 | Endocrine System

By the end of this section, you will be able to:

- List the different types of hormones and explain their roles in maintaining homeostasis
- Explain how hormones work
- Explain how hormone production is regulated
- Describe the role of different glands in the endocrine system
- Explain how the different glands work together to maintain homeostasis

The endocrine system produces hormones that function to control and regulate many different body processes. The endocrine system coordinates with the nervous system to control the functions of the other organ systems. Cells of the endocrine system produce molecular signals called hormones. These cells may compose endocrine glands, may be tissues or may be located in organs or tissues that have functions in addition to hormone production. Hormones circulate throughout the body and stimulate a response in cells that have receptors able to bind with them. The changes brought about in the receiving cells affect the functioning of the organ system to which they belong. Many of the hormones are secreted in response to signals from the nervous system, thus the two systems act in concert to effect changes in the body.

Hormones

Maintaining homeostasis within the body requires the coordination of many different systems and organs. One mechanism of communication between neighboring cells, and between cells and tissues in distant parts of the body, occurs through the release of chemicals called hormones. **Hormones** are released into body fluids, usually blood, which carries them to their target cells where they elicit a response. The cells that secrete hormones are often located in specific organs, called **endocrine glands**, and the cells, tissues, and organs that secrete hormones make up the endocrine system. Examples of endocrine organs include the pancreas, which produces the hormones insulin and glucagon to regulate blood-glucose levels, the adrenal glands, which produce hormones such as epinephrine and norepinephrine that regulate responses to stress, and the thyroid gland, which produces thyroid hormones that regulate metabolic rates.

The endocrine glands differ from the exocrine glands. **Exocrine glands** secrete chemicals through ducts that lead outside the gland (not to the blood). For example, sweat produced by sweat glands is released into ducts that carry sweat to the surface of the skin. The pancreas has both endocrine and exocrine functions because besides releasing hormones into the blood. It also produces digestive juices, which are carried by ducts into the small intestine.

careers IN ACTION

Endocrinologist

An endocrinologist is a medical doctor who specializes in treating endocrine disorders. An endocrine surgeon specializes in the surgical treatment of endocrine diseases and glands. Some of the diseases that are managed by endocrinologists include disorders of the pancreas (diabetes mellitus), disorders of the pituitary (gigantism, acromegaly, and pituitary dwarfism), disorders of the thyroid gland (goiter and Graves' disease), and disorders of the adrenal glands (Cushing's disease and Addison's disease).

Endocrinologists are required to assess patients and diagnose endocrine disorders through extensive use of laboratory tests. Many endocrine diseases are diagnosed using tests that stimulate or suppress endocrine organ functioning. Blood samples are then drawn to determine the effect of stimulating or suppressing an endocrine organ on the production of hormones. For example, to diagnose diabetes mellitus, patients are required to fast for 12 to 24 hours. They are then given a sugary drink, which stimulates the pancreas to produce insulin to decrease blood-glucose levels. A blood sample is taken one to two hours after the sugar drink is consumed. If the pancreas is functioning properly, the blood-glucose level will be within a normal range. Another example is the A1C test, which can be performed during blood screening. The A1C test measures average blood-glucose levels over the past two to three months. The A1C test is an indicator of how well blood glucose is being managed over a long time.

Once a disease such as diabetes has been diagnosed, endocrinologists can prescribe lifestyle changes and medications to treat the disease. Some cases of diabetes mellitus can be managed by exercise, weight loss, and a healthy diet; in other cases, medications may be required to enhance insulin's production or effect. If the disease cannot be controlled by these means, the endocrinologist may prescribe insulin injections.

In addition to clinical practice, endocrinologists may also be involved in primary research and development activities. For example, ongoing islet transplant research is investigating how healthy pancreas islet cells may be transplanted into diabetic patients. Successful islet transplants may allow patients to stop taking insulin injections.

How Hormones Work

Hormones cause changes in target cells by binding to specific cell-surface or **intracellular hormone receptors**, molecules embedded in the cell membrane or floating in the cytoplasm with a binding site that matches a binding site on the hormone molecule. In this way, even though hormones circulate throughout the body and come into contact with many different cell types, they only affect cells that possess the necessary receptors. Receptors for a specific hormone may be found on or in many different cells or may be limited to a small number of specialized cells. For example, thyroid hormones act on many different tissue types, stimulating metabolic activity throughout the body. Cells can have many receptors for the same hormone but often also possess receptors for different types of hormones. The number of receptors that respond to a hormone determines the cell's sensitivity to that hormone, and the resulting cellular response. Additionally, the number of receptors available to respond to a hormone can change over time, resulting in increased or decreased cell sensitivity. In **up-regulation**, the number of receptors increases in response to rising hormone levels, making the cell more sensitive to the hormone and allowing for more cellular activity. When the number of receptors decreases in response to rising hormone levels, called **down-regulation**, cellular activity is reduced.

Endocrine Glands

The endocrine glands secrete hormones into the surrounding interstitial fluid; those hormones then diffuse into blood and are carried to various organs and tissues within the body. The endocrine glands include the pituitary, thyroid, parathyroid, adrenal glands, gonads, pineal, and pancreas.

The **pituitary gland**, sometimes called the hypophysis, is located at the base of the brain (Figure 16.13a). It is attached to the hypothalamus. The posterior lobe stores and releases oxytocin and antidiuretic hormone produced by the hypothalamus. The anterior lobe responds to hormones produced by the hypothalamus by producing its own hormones, most of which regulate other hormone-producing glands.

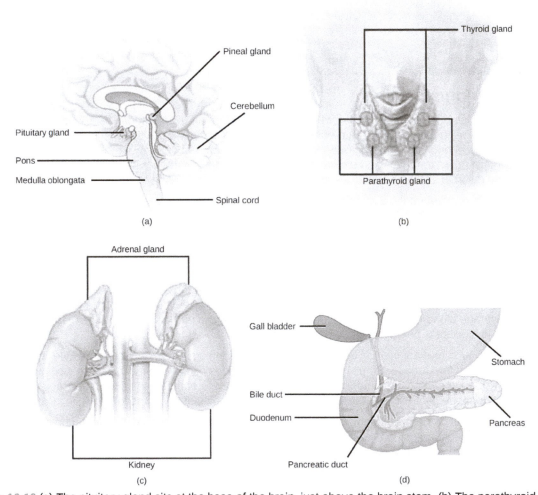

Figure 16.13 (a) The pituitary gland sits at the base of the brain, just above the brain stem. (b) The parathyroid glands are located on the posterior of the thyroid gland. (c) The adrenal glands are on top of the kidneys. d) The pancreas is found between the stomach and the small intestine. (credit: modification of work by NCI, NIH)

The anterior pituitary produces six hormones: growth hormone, prolactin, thyroid-stimulating hormone, adrenocorticotropic hormone, follicle-stimulating hormone, and luteinizing hormone. Growth hormone stimulates cellular activities like protein synthesis that promote growth. Prolactin stimulates the production of milk by the mammary glands. The other hormones produced by the anterior pituitary regulate the production of hormones by other endocrine tissues (Table 16.1). The posterior pituitary is significantly different in structure from the anterior pituitary. It is a part of the brain, extending down from the hypothalamus, and contains mostly nerve fibers that extend from the hypothalamus to the posterior pituitary.

The **thyroid gland** is located in the neck, just below the larynx and in front of the trachea (Figure 16.13b). It is a butterfly-shaped gland with two lobes that are connected. The thyroid follicle cells synthesize the hormone thyroxine, which is also known as T_4 because it contains four atoms of iodine, and triiodothyronine, also known as T_3 because it contains three atoms of iodine. T_3 and T_4 are released by the thyroid in response to thyroid-stimulating hormone produced by the anterior pituitary, and both T_3 and T_4 have the effect of stimulating metabolic activity in the body and increasing energy use. A third hormone, calcitonin, is also produced by the thyroid. Calcitonin is released in response to rising calcium ion concentrations in the blood and has the effect of reducing those levels.

Most people have four **parathyroid glands**; however, the number can vary from two to six. These glands are located on the posterior surface of the thyroid gland (Figure 16.13b).

The parathyroid glands produce parathyroid hormone. Parathyroid hormone increases blood calcium concentrations when calcium ion levels fall below normal.

The **adrenal glands** are located on top of each kidney (Figure 16.13c). The adrenal glands consist of an outer adrenal cortex and an inner adrenal medulla. These regions secrete different hormones.

The adrenal cortex produces mineralocorticoids, glucocorticoids, and androgens. The main mineralocorticoid is aldosterone, which regulates the concentration of ions in urine, sweat, and saliva. Aldosterone release from the adrenal cortex is stimulated by a decrease in blood concentrations of sodium ions, blood volume, or blood pressure, or by an increase in blood potassium levels. The glucocorticoids maintain proper blood-glucose levels between meals. They also control a response to stress by increasing glucose synthesis from fats and proteins and interact with epinephrine to cause vasoconstriction. Androgens are sex hormones that are produced in small amounts by the adrenal cortex. They do not normally affect sexual characteristics and may supplement sex hormones released from the gonads. The adrenal medulla contains two types of secretory cells: one that produces epinephrine (adrenaline) and another that produces norepinephrine (noradrenaline). Epinephrine and norepinephrine cause immediate, short-term changes in response to stressors, inducing the so-called fight-or-flight response. The responses include increased heart rate, breathing rate, cardiac muscle contractions, and blood-glucose levels. They also accelerate the breakdown of glucose in skeletal muscles and stored fats in adipose tissue, and redirect blood flow toward skeletal muscles and away from skin and viscera. The release of epinephrine and norepinephrine is stimulated by neural impulses from the sympathetic nervous system that originate from the hypothalamus.

The **pancreas** is an elongate organ located between the stomach and the proximal portion of the small intestine (Figure 16.13d). It contains both exocrine cells that excrete digestive enzymes and endocrine cells that release hormones.

The endocrine cells of the pancreas form clusters called pancreatic islets or the islets of Langerhans. Among the cell types in each pancreatic islet are the alpha cells, which produce the hormone glucagon, and the beta cells, which produce the hormone insulin. These hormones regulate blood-glucose levels. Alpha cells release glucagon as blood-glucose levels decline. When blood-glucose levels rise, beta cells release insulin. Glucagon causes the release of glucose to the blood from the liver, and insulin facilitates the uptake of glucose by the body's cells.

The gonads—the male testes and female ovaries—produce steroid hormones. The testes produce androgens, testosterone being the most prominent, which allow for the development of secondary sex characteristics and the production of sperm cells. The ovaries produce estrogen and progesterone, which cause secondary sex characteristics, regulate production of eggs, control pregnancy, and prepare the body for childbirth.

There are several organs whose primary functions are non-endocrine but that also possess endocrine functions. These include the heart, kidneys, intestines, thymus, and adipose tissue. The heart has endocrine cells in the walls of the atria that release a hormone in response to increased blood volume. It causes a reduction in blood volume and blood pressure, and reduces the concentration of Na^+ in the blood.

The gastrointestinal tract produces several hormones that aid in digestion. The endocrine cells are located in the mucosa of the GI tract throughout the stomach and small intestine. They trigger the release of gastric juices, which help to break down and digest food in the GI tract.

The kidneys also possess endocrine function. Two of these hormones regulate ion concentrations and blood volume or pressure. Erythropoietin (EPO) is released by kidneys in response to low oxygen levels. EPO triggers the formation of red blood cells in the bone marrow. EPO has been used by athletes to improve performance. But EPO doping has its risks, since it thickens the blood and increases strain on the heart; it also increases the risk of blood clots and therefore heart attacks and stroke.

The **thymus** is found behind the sternum. The thymus produces hormones referred to as thymosins, which contribute to the development of the immune response in infants. Adipose tissue, or fat tissue, produces the hormone leptin in response to food intake. Leptin produces a feeling of satiety after eating, reducing the urge for further eating.

Endocrine Glands and Their Associated Hormones

Endocrine Gland	Associated Hormones	Effect
Pituitary (anterior)	growth hormone	promotes growth of body tissues
	prolactin	promotes milk production
	thyroid-stimulating hormone	stimulates thyroid hormone release
	adrenocorticotropic hormone	stimulates hormone release by adrenal cortex
	follicle-stimulating hormone	stimulates gamete production
	luteinizing hormone	stimulates androgen production by gonads in males; stimulates ovulation and production of estrogen and progesterone in females
Pituitary (posterior)	antidiuretic hormone	stimulates water reabsorption by kidneys
	oxytocin	stimulates uterine contractions during childbirth
Thyroid	thyroxine, triiodothyronine	stimulate metabolism
	calcitonin	reduces blood Ca^{2+} levels
Parathyroid	parathyroid hormone	increases blood Ca^{2+} levels
Adrenal (cortex)	aldosterone	increases blood Na^{+} levels
	cortisol, corticosterone, cortisone	increase blood-glucose levels
Adrenal (medulla)	epinephrine, norepinephrine	stimulate fight-or-flight response
Pancreas	insulin	reduces blood-glucose levels
	glucagon	increases blood-glucose levels

Table 16.1

Regulation of Hormone Production

Hormone production and release are primarily controlled by negative feedback, as described in the discussion on homeostasis. In this way, the concentration of hormones in blood is maintained within a narrow range. For example, the anterior pituitary signals the thyroid to release thyroid hormones. Increasing levels of these hormones in the blood then give feedback to the hypothalamus and anterior pituitary to inhibit further signaling to the thyroid gland (Figure 16.14).

Thyroid System

Hypothalamus

Anterior pituitary gland

Thyrotropin-releasing hormone (TRH)

Thyroid-stimulating hormone (TSH)

Negative feedback

Thyroid gland

Thyroid hormones (T3 and T4)

Increased metabolism

Growth and development

Figure 16.14 The anterior pituitary stimulates the thyroid gland to release thyroid hormones T_3 and T_4. Increasing levels of these hormones in the blood result in feedback to the hypothalamus and anterior pituitary to inhibit further signaling to the thyroid gland. (credit: modification of work by Mikael Häggström)

Goiter, a disease caused by iodine deficiency, results in the inability of the thyroid gland to form T_3 and T_4. The body typically attempts to compensate by producing greater amounts of TSH. Which of the following symptoms would you expect goiter to cause?

a. Hypothyroidism, resulting in weight gain, cold sensitivity, and reduced mental activity.

b. Hyperthyroidism, resulting in weight loss, profuse sweating, and increased heart rate.

c. Hyperthyroidism, resulting in weight gain, cold sensitivity, and reduced mental activity.

d. Hypothyroidism, resulting in weight loss, profuse sweating, and increased heart rate.

16.5 | Musculoskeletal System

By the end of this section, you will be able to:

- Discuss the axial and appendicular parts of the skeletal system
- Explain the role of joints in skeletal movement
- Explain the role of muscles in locomotion

The muscular and skeletal systems provide support to the body and allow for movement. The bones of the skeleton protect the body's internal organs and support the weight of the body. The muscles of the muscular system contract and pull on the bones, allowing for movements as diverse as standing, walking, running, and grasping items.

Injury or disease affecting the musculoskeletal system can be very debilitating. The most common musculoskeletal diseases worldwide are caused by malnutrition, which can negatively affect development and maintenance of bones and muscles. Other diseases affect the joints, such as arthritis, which can make movement difficult and, in advanced cases, completely impair mobility.

Progress in the science of prosthesis design has resulted in the development of artificial joints, with joint replacement surgery in the hips and knees being the most common. Replacement joints for shoulders, elbows, and fingers are also available.

Skeletal System

The human skeleton is an endoskeleton that consists of 206 bones in the adult. An endoskeleton develops within the body rather than outside like the exoskeleton of insects. The skeleton has five main functions: providing support to the body, storing minerals and lipids, producing blood cells, protecting internal organs, and allowing for movement. The skeletal system in vertebrates is divided into the axial skeleton (which consists of the skull, vertebral column, and rib cage), and the appendicular skeleton (which consists of limb bones, the pectoral or shoulder girdle, and the pelvic girdle).

Explore the human skeleton by viewing the following video (http://openstaxcollege.org/l/human_skeleton) with digital 3D sculpturing.

The **axial skeleton** forms the central axis of the body and includes the bones of the skull, ossicles of the middle ear, hyoid bone of the throat, vertebral column, and the thoracic cage (rib cage) (Figure 16.15).

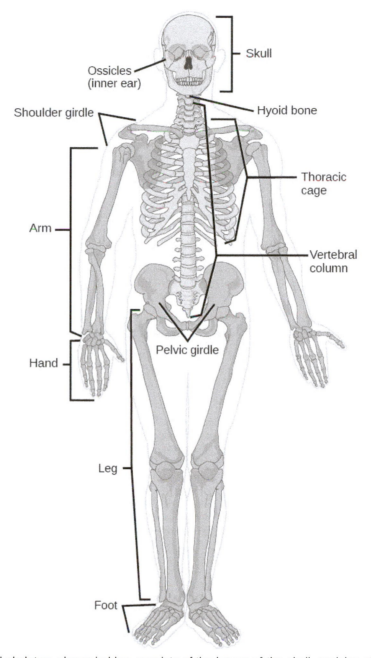

Figure 16.15 The axial skeleton, shown in blue, consists of the bones of the skull, ossicles of the middle ear, hyoid bone, vertebral column, and thoracic cage. The appendicular skeleton, shown in red, consists of the bones of the pectoral limbs, pectoral girdle, pelvic limb, and pelvic girdle. (credit: modification of work by Mariana Ruiz Villareal)

The bones of the **skull** support the structures of the face and protect the brain. The skull consists of cranial bones and facial bones. The cranial bones form the cranial cavity, which encloses the brain and serves as an attachment site for muscles of the head and neck. In the adult they are tightly jointed with connective tissue and adjoining bones do not move.

The **auditory ossicles** of the middle ear transmit sounds from the air as vibrations to the fluid-filled cochlea. The auditory ossicles consist of two malleus (hammer) bones, two incus (anvil) bones, and two stapes (stirrups), one on each side. Facial bones provide cavities for the sense organs (eyes, mouth, and nose), and serve as attachment points for facial muscles.

The **hyoid bone** lies below the mandible in the front of the neck. It acts as a movable base for the tongue and is connected to muscles of the jaw, larynx, and tongue. The mandible forms a joint with the base of the skull. The mandible controls the opening to the mouth and hence, the airway and gut.

The **vertebral column**, or spinal column, surrounds and protects the spinal cord, supports the head, and acts as an attachment point for ribs and muscles of the back and neck. It consists of 26 bones: the 24 vertebrae, the sacrum, and the coccyx. Each vertebral body has a large hole in the center through which the spinal cord passes down to the level of the first lumbar vertebra. Below this level, the hole contains spinal nerves which exit between the vertebrae. There is a notch on each side of the hole through which the spinal nerves, can exit from the spinal cord to serve different regions of the body. The vertebral column is approximately 70 cm (28 in) in adults and is curved, which can be seen from a side view.

Intervertebral discs composed of fibrous cartilage lie between adjacent vertebrae from the second cervical vertebra to the sacrum. Each disc helps form a slightly moveable joint and acts as a cushion to absorb shocks from movements such as walking and running.

The **thoracic cage**, also known as the rib cage consists of the ribs, sternum, thoracic vertebrae, and costal cartilages. The thoracic cage encloses and protects the organs of the thoracic cavity including the heart and lungs. It also provides support for the shoulder girdles and upper limbs and serves as the attachment point for the diaphragm, muscles of the back, chest, neck, and shoulders. Changes in the volume of the thorax enable breathing. The sternum, or breastbone, is a long flat bone located at the anterior of the chest. Like the skull, it is formed from many bones in the embryo, which fuse in the adult. The ribs are 12 pairs of long curved bones that attach to the thoracic vertebrae and curve toward the front of the body, forming the ribcage. Costal cartilages connect the anterior ends of most ribs to the sternum.

The **appendicular skeleton** is composed of the bones of the upper and lower limbs. It also includes the pectoral, or shoulder girdle, which attaches the upper limbs to the body, and the pelvic girdle, which attaches the lower limbs to the body (Figure 16.15).

The **pectoral girdle** bones transfer force generated by muscles acting on the upper limb to the thorax. It consists of the clavicles (or collarbones) in the anterior, and the scapulae (or shoulder blades) in the posterior.

The upper limb contains bones of the arm (shoulder to elbow), the forearm, and the hand. The humerus is the largest and longest bone of the upper limb. It forms a joint with the shoulder and with the forearm at the elbow. The forearm extends from the elbow to the wrist and consists of two bones. The hand includes the bones of the wrist, the palm, and the bones of the fingers.

The **pelvic girdle** attaches to the lower limbs of the axial skeleton. Since it is responsible for bearing the weight of the body and for locomotion, the pelvic girdle is securely attached to the axial skeleton by strong ligaments. It also has deep sockets with robust ligaments that securely attach to the femur. The pelvic girdle is mainly composed of two large hip bones. The hip bones join together in the anterior of the body at a joint called the pubic symphysis and with the bones of the sacrum at the posterior of the body.

The lower limb consists of the thigh, the leg, and the foot. The bones of the lower limbs are thicker and stronger than the bones of the upper limbs to support the entire weight of the body and the forces from locomotion. The femur, or thighbone, is the longest, heaviest, and strongest bone in the body. The femur and pelvis form the hip joint. At its other end, the femur, along with the shinbone and kneecap, form the knee joint.

Joints and Skeletal Movement

The point at which two or more bones meet is called a **joint**, or articulation. Joints are responsible for movement, such as the movement of limbs, and stability, such as the stability found in the bones of the skull.

There are two ways to classify joints: based on their structure or based on their function. The structural classification divides joints into fibrous, cartilaginous, and synovial joints depending on the material composing the joint and the presence or absence of a cavity in the joint. The bones of **fibrous joints** are held together by fibrous connective tissue. There is no cavity, or space, present between the bones, so most fibrous joints do not move at all, or are only capable of minor movements. The joints between the bones in the skull and between the teeth and the bone of their sockets are examples of fibrous joints (Figure 16.16**a**).

Cartilaginous joints are joints in which the bones are connected by cartilage (Figure 16.16**b**). An example is found at the joints between vertebrae, the so-called "disks" of the backbone. Cartilaginous joints allow for very little movement.

Synovial joints are the only joints that have a space between the adjoining bones (Figure 16.16**c**). This space is referred to as the joint cavity and is filled with fluid. The fluid lubricates the joint, reducing friction between the bones and allowing for greater movement. The ends of the bones are covered with cartilage and the entire joint is surrounded by a capsule. Synovial joints are capable of the greatest movement of the joint types. Knees, elbows, and shoulders are examples of synovial joints.

Download for free at https://openstax.org/details/books/concepts-biology

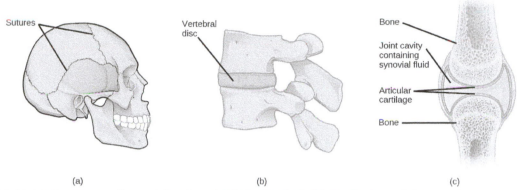

Figure 16.16 (a) Sutures are fibrous joints found only in the skull. (b) Cartilaginous joints are bones connected by cartilage, such as between vertebrae. (c) Synovial joints are the only joints that have a space or "synovial cavity" in the joint.

The wide range of movement allowed by synovial joints produces different types of movements. Angular movements are produced when the angle between the bones of a joint changes. Flexion, or bending, occurs when the angle between the bones decreases. Moving the forearm upward at the elbow is an example of flexion. Extension is the opposite of flexion in that the angle between the bones of a joint increases. Rotational movement is the movement of a bone as it rotates around its own longitudinal axis. Movement of the head as in saying "no" is an example of rotation.

careers IN ACTION

Rheumatologist

Rheumatologists are medical doctors who specialize in the diagnosis and treatment of disorders of the joints, muscles, and bones. They diagnose and treat diseases such as arthritis, musculoskeletal disorders, osteoporosis, plus autoimmune diseases like ankylosing spondylitis, a chronic spinal inflammatory disease and rheumatoid arthritis.

Rheumatoid arthritis (RA) is an inflammatory disorder that primarily affects synovial joints of the hands, feet, and cervical spine. Affected joints become swollen, stiff, and painful. Although it is known that RA is an autoimmune disease in which the body's immune system mistakenly attacks healthy tissue, the exact cause of RA remains unknown. Immune cells from the blood enter joints and the joint capsule causing cartilage breakdown and swelling of the joint lining. Breakdown of cartilage causes bones to rub against each other causing pain. RA is more common in women than men and the age of onset is usually between 40 to 50 years.

Rheumatologists can diagnose RA based on symptoms such as joint inflammation and pain, x-ray and MRI imaging, and blood tests. Arthrography is a type of medical imaging of joints that uses a contrast agent, such as a dye that is opaque to x-rays. This allows the soft tissue structures of joints—such as cartilage, tendons, and ligaments—to be visualized. An arthrogram differs from a regular x-ray by showing the surface of soft tissues lining the joint in addition to joint bones. An arthrogram allows early degenerative changes in joint cartilage to be detected before bones become affected.

There is currently no cure for RA; however, rheumatologists have a number of treatment options available. Treatments are divided into those that reduce the symptoms of the disease and those that reduce the damage to bone and cartilage caused by the disease. Early stages can be treated with rest of the affected joints through the use of a cane, or with joint splints that minimize inflammation. When inflammation has decreased, exercise can be used to strengthen muscles that surround the joint and to maintain joint flexibility. If joint damage is more extensive, medications can be used to relieve pain and decrease inflammation. Anti-inflammatory drugs that may be used include aspirin, topical pain relievers, and corticosteroid injections. Surgery may be required in cases where joint damage is severe. Physicians are now using drugs that reduce the damage to bones and cartilage caused by the disease to slow its development. These drugs are diverse in their mechanisms but they all act to reduce the impact of the autoimmune response, for example by inhibiting the inflammatory response or reducing the number of T lymphocytes, a cell of the immune system.

Muscles

Muscles allow for movement such as walking, and they also facilitate bodily processes such as respiration and digestion. The body contains three types of muscle tissue: skeletal muscle, cardiac muscle, and smooth muscle (Figure 16.17).

 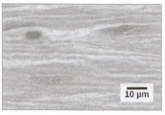 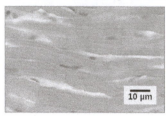

| Skeletal muscle | Smooth muscle | Cardiac muscle |

Figure 16.17 The body contains three types of muscle tissue: skeletal muscle, smooth muscle, and cardiac muscle. Notice that skeletal muscle cells are long and cylindrical, they have multiple nuclei, and the small, dark nuclei are pushed to the periphery of the cell. Smooth muscle cells are short, tapered at each end, and have only one nucleus each. Cardiac muscle cells are also cylindrical, but short. The cytoplasm may branch, and they have one or two nuclei in the center of the cell. (credit: modification of work by NCI, NIH; scale-bar data from Matt Russell)

Skeletal muscle tissue forms skeletal muscles, which attach to bones and sometimes the skin and control locomotion and any other movement that can be consciously controlled. Because it can be controlled intentionally, skeletal muscle is also called voluntary muscle. When viewed under a microscope, skeletal muscle tissue has a striped or striated appearance. This appearance results from the arrangement of the proteins inside the cell that are responsible for contraction. The cells of skeletal muscle are long and tapered and have multiple nuclei on the periphery of each cell.

Smooth muscle tissue occurs in the walls of hollow organs such as the intestines, stomach, and urinary bladder, and around passages such as in the respiratory tract and blood vessels. Smooth muscle has no striations, is not under voluntary control, and is called involuntary muscle. Smooth muscle cells have a single nucleus.

Cardiac muscle tissue is only found in the heart. The contractions of cardiac muscle tissue pump blood throughout the body and maintain blood pressure. Like skeletal muscle, cardiac muscle is striated, but unlike skeletal muscle, cardiac muscle cannot be consciously controlled and is called involuntary muscle. The cells of cardiac muscle tissue are connected to each other through intercalated disks and usually have just one nucleus per cell.

Skeletal Muscle Fiber Structure and Function

Each skeletal muscle fiber is a skeletal muscle cell. Within each muscle fiber are **myofibrils**, long cylindrical structures that lie parallel to the muscle fiber. Myofibrils run the entire length of the muscle fiber. They attach to the plasma membrane, called the **sarcolemma**, at their ends, so that as myofibrils shorten, the entire muscle cell contracts (Figure 16.18).

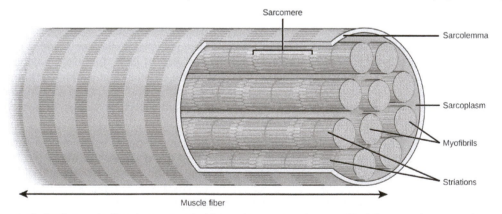

Figure 16.18 A skeletal muscle fiber is surrounded by a plasma membrane called the sarcolemma, with a cytoplasm called the sarcoplasm. A muscle fiber is composed of many fibrils packaged into orderly units. The orderly arrangement of the proteins in each unit, shown as red and blue lines, gives the cell its striated appearance.

The striated appearance of skeletal muscle tissue is a result of repeating bands of the proteins actin and myosin that occur along the length of myofibrils.

Myofibrils are composed of smaller structures called **myofilaments**. There are two main types of myofilaments: thick filaments and thin filaments. Thick filaments are composed of the protein myosin. The primary component of thin filaments is the protein actin.

The thick and thin filaments alternate with each other in a structure called a **sarcomere**. The sarcomere is the unit of contraction in a muscle cell. Contraction is stimulated by an electrochemical signal from a nerve cell associated with the muscle fiber. For a muscle cell to contract, the sarcomere must shorten. However, thick and thin filaments do not shorten. Instead, they slide by one another, causing the sarcomere to shorten while the filaments remain the same length. The sliding is accomplished when a molecular extension of myosin, called the myosin head, temporarily binds to an actin filament next to it and through a change in conformation, bends, dragging the two filaments in opposite directions. The myosin head then releases its actin filament, relaxes, and then repeats the process, dragging the two filaments further along each other. The combined activity of many binding sites and repeated movements within the sarcomere causes it to contract. The coordinated contractions of many sarcomeres in a myofibril leads to contraction of the entire muscle cell and ultimately the muscle itself. The movement of the myosin head requires ATP, which provides the energy for the contraction.

View this animation (http://openstaxcollege.org/l/skeletal_muscl2) to see how muscle fibers are organized.

16.6 | Nervous System

By the end of this section, you will be able to:

- Describe the form and function of a neuron
- Describe the basic parts and functions of the central nervous system
- Describe the basic parts and functions of the peripheral nervous system

As you read this, your nervous system is performing several functions simultaneously. The visual system is processing what is seen on the page; the motor system controls your eye movements and the turn of the pages (or click of the mouse); the prefrontal cortex maintains attention. Even fundamental functions, like breathing and regulation of body temperature, are controlled by the nervous system. The nervous system is one of two systems that exert control over all the organ systems of the body; the other is the endocrine system. The nervous system's control is much more specific and rapid than the hormonal system. It communicates signals through cells and the tiny gaps between them rather than through the circulatory system as in the endocrine system. It uses a combination of chemical and electrochemical signals, rather than purely chemical signals used by the endocrine system to cover long distances quickly. The nervous system acquires information from sensory organs, processes it and then may initiate a response either through motor function, leading to movement, or in a change in the organism's physiological state.

Nervous systems throughout the animal kingdom vary in structure and complexity. Some organisms, like sea sponges, lack a true nervous system. Others, like jellyfish, lack a true brain and instead have a system of separate but connected nerve cells (neurons) called a "nerve net." Flatworms have both a central nervous system (CNS), made up of a ganglion (clusters of connected neurons) and two nerve cords, and a peripheral nervous system (PNS) containing a system of nerves that extend throughout the body. The insect nervous system is more complex but also fairly decentralized. It contains a brain, ventral nerve cord, and ganglia. These ganglia can control movements and behaviors without input from the brain.

Compared to invertebrates, vertebrate nervous systems are more complex, centralized, and specialized. While there is great diversity among different vertebrate nervous systems, they all share a basic structure: a CNS that contains a brain and spinal cord and a PNS made up of peripheral sensory and motor nerves. One interesting difference between the nervous systems of invertebrates and vertebrates is that the nerve cords of many invertebrates are located ventrally (toward the stomach) whereas the vertebrate spinal cords are located dorsally (toward the back). There is debate among evolutionary biologists as to whether these different nervous system plans evolved separately or whether the invertebrate body plan arrangement somehow "flipped" during the evolution of vertebrates.

The nervous system is made up of **neurons**, specialized cells that can receive and transmit chemical or electrical signals, and **glia**, cells that provide support functions for the neurons. There is great diversity in the types of neurons and glia that are present in different parts of the nervous system.

Neurons and Glial Cells

The nervous system of the common laboratory fly, *Drosophila melanogaster,* contains around 100,000 neurons, the same number as a lobster. This number compares to 75 million in the mouse and 300 million in the octopus. A human brain contains around 86 billion neurons. Despite these very different numbers, the nervous systems of these animals control many of the same behaviors—from basic reflexes to more complicated behaviors like finding food and courting mates. The ability of neurons to communicate with each other as well as with other types of cells underlies all of these behaviors.

Most neurons share the same cellular components. But neurons are also highly specialized—different types of neurons have different sizes and shapes that relate to their functional roles.

Like other cells, each neuron has a cell body (or soma) that contains a nucleus, smooth and rough endoplasmic reticulum, Golgi apparatus, mitochondria, and other cellular components. Neurons also contain unique structures for receiving and sending the electrical signals that make communication between neurons possible (Figure 16.19). **Dendrites** are tree-like structures that extend away from the cell body to receive messages from other neurons at specialized junctions called **synapses**. Although some neurons do not have any dendrites, most have one or many dendrites.

The bilayer lipid membrane that surrounds a neuron is impermeable to ions. To enter or exit the neuron, ions must pass through ion channels that span the membrane. Some ion channels need to be activated to open and allow ions to pass into or out of the cell. These ion channels are sensitive to the environment and can change their shape accordingly. Ion channels that change their structure in response to voltage changes are called voltage-gated ion channels. The difference in total charge between the inside and outside of the cell is called the membrane potential.

A neuron at rest is negatively charged: the inside of a cell is approximately 70 millivolts more negative than the outside (–70 mV). This voltage is called the resting membrane potential; it is caused by differences in the concentrations of ions inside and outside the cell and the selective permeability created by ion channels. Sodium-potassium pumps in the membrane produce the different ion concentrations inside and outside of the cell by bringing in two K^+ ions and removing three Na^+ ions. The actions of this pump are costly: one molecule of ATP is used up for each turn. Up to 50 percent of a neuron's ATP is used in maintaining its membrane resting potential. Potassium ions (K^+), which are higher inside the cell, move fairly freely out of the neuron through potassium channels; this loss of positive charge produces a net negative charge inside the cell. Sodium ions (Na^+), which are low inside, have a driving force to enter but move less freely. Their channels are voltage dependent and will open when a slight change in the membrane potential triggers them.

A neuron can receive input from other neurons and, if this input is strong enough, send the signal to downstream neurons. Transmission of a signal between neurons is generally carried by a chemical, called a neurotransmitter, which diffuses from the axon of one neuron to the dendrite of a second neuron. When neurotransmitter molecules bind to receptors located on a neuron's dendrites, the neurotransmitter opens ion channels in the dendrite's plasma membrane. This opening allows sodium ions to enter the neuron and results in **depolarization** of the membrane—a decrease in the voltage across the neuron membrane. Once a signal is received by the dendrite, it then travels passively to the cell body. A large enough signal from neurotransmitters will reach the axon. If it is strong enough (that is, if the **threshold of excitation**, a depolarization to around –60mV is reached), then depolarization creates a positive feedback loop: as more Na^+ ions enter the cell, the axon becomes further depolarized, opening even more sodium channels at further distances from the cell body. This will cause voltage dependent Na^+ channels further down the axon to open and more positive ions to enter the cell. In the axon, this "signal" will become a self-propagating brief reversal of the resting membrane potential called an **action potential**.

An action potential is an all-or-nothing event; it either happens or it does not. The threshold of excitation must be reached for the neuron to "fire" an action potential. As sodium ions rush into the cell, depolarization actually reverses the charge across the membrane form -70mv to +30mV. This change in the membrane potential causes voltage-gated K^+ channels to open, and K^+ begins to leave the cell, repolarizing it. At the same time, Na^+ channels inactivate so no more Na^+ enters the cell. K^+ ions continue to leave the cell and the membrane potential returns to the resting potential. At the resting potential, the K^+ channels close and Na^+ channels reset. The depolarization of the membrane proceeds in a wave down the length of the axon. It travels in only one direction because the sodium channels have been inactivated and unavailable until the membrane potential is near the resting potential again; at this point they are reset to closed and can be opened again.

An **axon** is a tube-like structure that propagates the signal from the cell body to specialized endings called axon terminals. These terminals in turn then synapse with other neurons, muscle, or target organs. When the action potential reaches the axon terminal, this causes the release of neurotransmitter onto the dendrite of another neuron. Neurotransmitters released at axon terminals allow signals to be communicated to these other cells, and the process begins again. Neurons usually have one or two axons, but some neurons do not contain any axons.

Some axons are covered with a special structure called a **myelin sheath**, which acts as an insulator to keep the electrical signal from dissipating as it travels down the axon. This insulation is important, as the axon from a human motor neuron can be as long as a meter (3.2 ft)—from the base of the spine to the toes. The myelin sheath is produced by glial cells. Along the axon there are periodic gaps in the myelin sheath. These gaps are called nodes of Ranvier and are sites where the signal is "recharged" as it travels along the axon.

It is important to note that a single neuron does not act alone—neuronal communication depends on the connections that neurons make with one another (as well as with other cells, like muscle cells). Dendrites from a single neuron may receive synaptic contact from many other neurons. For example, dendrites from a Purkinje cell in the cerebellum are thought to receive contact from as many as 200,000 other neurons.

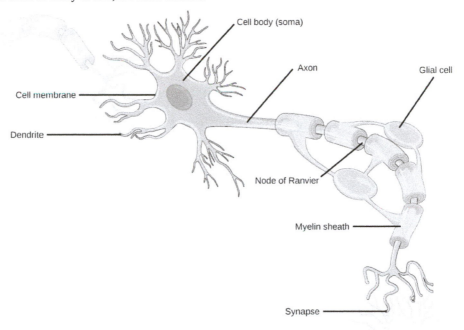

Figure 16.19 Neurons contain organelles common to other cells, such as a nucleus and mitochondria. They also have more specialized structures, including dendrites and axons.

biology IN ACTION

Neurogenesis

At one time, scientists believed that people were born with all the neurons they would ever have. Research performed during the last few decades indicates that neurogenesis, the birth of new neurons, continues into adulthood. Neurogenesis was first discovered in songbirds that produce new neurons while learning songs. For mammals, new neurons also play an important role in learning: about 1,000 new neurons develop in the hippocampus (a brain structure involved in learning and memory) each day. While most of the new neurons will die, researchers found that an increase in the number of surviving new neurons in the hippocampus correlated with how well rats learned a new task. Interestingly, both exercise and some antidepressant medications also promote neurogenesis in the hippocampus. Stress has the opposite effect. While neurogenesis is quite limited compared to regeneration in other tissues, research in this area may lead to new treatments for disorders such as Alzheimer's, stroke, and epilepsy.

How do scientists identify new neurons? A researcher can inject a compound called bromodeoxyuridine (BrdU) into the brain of an animal. While all cells will be exposed to BrdU, BrdU will only be incorporated into the DNA of newly generated cells that are in S phase. A technique called immunohistochemistry can be used to attach a fluorescent label to the incorporated BrdU, and a researcher can use fluorescent microscopy to visualize the presence of BrdU, and thus new neurons, in brain tissue (Figure 16.20).

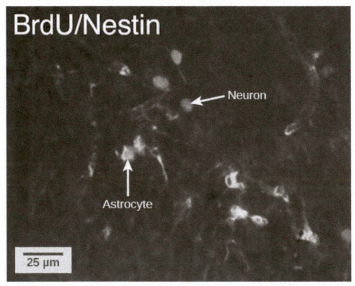

Figure 16.20 This image shows new neurons in a rat hippocampus. New neurons tagged with BrdU glow red in this micrograph. (credit: modification of work by Dr. Maryam Faiz, University of Barcelona)

Visit this link interactive lab (http://openstaxcollege.org/l/neurogenesis2) to see more information about neurogenesis, including an interactive laboratory simulation and a video that explains how BrdU labels new cells.

While glial cells are often thought of as the supporting cast of the nervous system, the number of glial cells in the brain actually outnumbers the number of neurons by a factor of 10. Neurons would be unable to function without the vital roles that are fulfilled by these glial cells. Glia guide developing neurons to their destinations, buffer ions and chemicals that

Download for free at https://openstax.org/details/books/concepts-biology

would otherwise harm neurons, and provide myelin sheaths around axons. When glia do not function properly, the result can be disastrous—most brain tumors are caused by mutations in glia.

How Neurons Communicate

All functions performed by the nervous system—from a simple motor reflex to more advanced functions like making a memory or a decision—require neurons to communicate with one another. Neurons communicate between the axon of one neuron and the dendrites, and sometimes the cell body, of another neuron across the gap between them, known as the **synaptic cleft**. When an action potential reaches the end of an axon it stimulates the release of neurotransmitter molecules into the synaptic cleft between the synaptic knob of the axon and the post-synaptic membrane of the dendrite or soma of the next cell. The neurotransmitter is released through exocytosis of vesicles containing the neurotransmitter molecules. The neurotransmitter diffuses across the synaptic cleft and binds to receptors in the post-synaptic membrane. These receptor molecules are chemically regulated ion channels and will open, allowing sodium to enter the cell. If sufficient neurotransmitter has been released an action potential may be initiated in the next cell, but this is not guaranteed. If insufficient neurotransmitter is released the nerve signal will die at this point. There are a number of different neurotransmitters that are specific to neuron types that have specific functions.

The Central Nervous System

The **central nervous system (CNS)** is made up of the brain and spinal cord and is covered with three layers of protective coverings called **meninges** ("meninges" is derived from the Greek and means "membranes") (Figure 16.21). The outermost layer is the dura mater, the middle layer is the web-like arachnoid mater, and the inner layer is the pia mater, which directly contacts and covers the brain and spinal cord. The space between the arachnoid and pia maters is filled with **cerebrospinal fluid (CSF)**. The brain floats in CSF, which acts as a cushion and shock absorber.

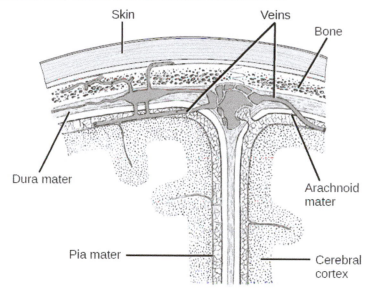

Figure 16.21 The cerebral cortex is covered by three layers of meninges: the dura, arachnoid, and pia maters. (credit: modification of work by Gray's Anatomy)

The Brain

The brain is the part of the central nervous system that is contained in the cranial cavity of the skull. It includes the cerebral cortex, limbic system, basal ganglia, thalamus, hypothalamus, cerebellum, brainstem, and retinas. The outermost part of the brain is a thick piece of nervous system tissue called the **cerebral cortex**. The cerebral cortex, limbic system, and basal ganglia make up the two cerebral hemispheres. A thick fiber bundle called the **corpus callosum** (corpus = "body"; callosum = "tough") connects the two hemispheres. Although there are some brain functions that are localized more to one hemisphere than the other, the functions of the two hemispheres are largely redundant. In fact, sometimes (very rarely) an entire hemisphere is removed to treat severe epilepsy. While patients do suffer some deficits following the surgery, they can have surprisingly few problems, especially when the surgery is performed on children who have very immature nervous systems.

In other surgeries to treat severe epilepsy, the corpus callosum is cut instead of removing an entire hemisphere. This causes a condition called split-brain, which gives insights into unique functions of the two hemispheres. For example, when an object is presented to patients' left visual field, they may be unable to verbally name the object (and may claim to not have seen an object at all). This is because the visual input from the left visual field crosses and enters the right hemisphere and cannot then signal to the speech center, which generally is found in the left side of the brain. Remarkably, if a split-brain patient is asked to pick up a specific object out of a group of objects with the left hand, the patient will be able to do so but will still be unable to verbally identify it.

Visit the following website (http://openstaxcollege.org/l/split-brain2) to learn more about split-brain patients and to play a game where you can model split-brain experiments yourself.

Each hemisphere contains regions called lobes that are involved in different functions. Each hemisphere of the mammalian cerebral cortex can be broken down into four functionally and spatially defined lobes: frontal, parietal, temporal, and occipital (Figure 16.22).

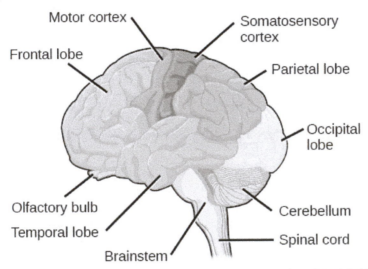

Figure 16.22 The human cerebral cortex includes the frontal, parietal, temporal, and occipital lobes.

The **frontal lobe** is located at the front of the brain, over the eyes. This lobe contains the olfactory bulb, which processes smells. The frontal lobe also contains the motor cortex, which is important for planning and implementing movement. Areas within the motor cortex map to different muscle groups. Neurons in the frontal lobe also control cognitive functions like maintaining attention, speech, and decision-making. Studies of humans who have damaged their frontal lobes show that parts of this area are involved in personality, socialization, and assessing risk. The **parietal lobe** is located at the top of the brain. Neurons in the parietal lobe are involved in speech and also reading. Two of the parietal lobe's main functions are processing somatosensation—touch sensations like pressure, pain, heat, cold—and processing proprioception—the sense of how parts of the body are oriented in space. The parietal lobe contains a somatosensory map of the body similar to the motor cortex. The **occipital lobe** is located at the back of the brain. It is primarily involved in vision—seeing, recognizing, and identifying the visual world. The **temporal lobe** is located at the base of the brain and is primarily involved in processing and interpreting sounds. It also contains the **hippocampus** (named from the Greek for "seahorse," which it resembles in shape) a structure that processes memory formation. The role of the hippocampus in memory was partially determined by studying one famous epileptic patient, HM, who had both sides of his hippocampus removed in an attempt to cure his epilepsy. His seizures went away, but he could no longer form new memories (although he could remember some facts from before his surgery and could learn new motor tasks).

Interconnected brain areas called the **basal ganglia** play important roles in movement control and posture. The basal ganglia also regulate motivation.

The **thalamus** acts as a gateway to and from the cortex. It receives sensory and motor inputs from the body and also receives feedback from the cortex. This feedback mechanism can modulate conscious awareness of sensory and motor inputs depending on the attention and arousal state of the animal. The thalamus helps regulate consciousness, arousal, and sleep states.

Below the thalamus is the **hypothalamus**. The hypothalamus controls the endocrine system by sending signals to the pituitary gland. Among other functions, the hypothalamus is the body's thermostat—it makes sure the body temperature is kept at appropriate levels. Neurons within the hypothalamus also regulate circadian rhythms, sometimes called sleep cycles.

The **limbic system** is a connected set of structures that regulates emotion, as well as behaviors related to fear and motivation. It plays a role in memory formation and includes parts of the thalamus and hypothalamus as well as the hippocampus. One important structure within the limbic system is a temporal lobe structure called the **amygdala**. The two amygdala (one on each side) are important both for the sensation of fear and for recognizing fearful faces.

The **cerebellum** (cerebellum = "little brain") sits at the base of the brain on top of the brainstem. The cerebellum controls balance and aids in coordinating movement and learning new motor tasks. The cerebellum of birds is large compared to other vertebrates because of the coordination required by flight.

The **brainstem** connects the rest of the brain with the spinal cord and regulates some of the most important and basic functions of the nervous system including breathing, swallowing, digestion, sleeping, walking, and sensory and motor information integration.

Spinal cord

Connecting to the brainstem and extending down the body through the spinal column is the spinal cord. The spinal cord is a thick bundle of nerve tissue that carries information about the body to the brain and from the brain to the body. The spinal cord is contained within the meninges and the bones of the vertebral column but is able to communicate signals to and from the body through its connections with spinal nerves (part of the peripheral nervous system). A cross-section of the spinal cord looks like a white oval containing a gray butterfly-shape (Figure 16.23). Axons make up the "white matter" and neuron and glia cell bodies (and interneurons) make up the "gray matter." Axons and cell bodies in the dorsa spinal cord convey mostly sensory information from the body to the brain. Axons and cell bodies in the spinal cord primarily transmit signals controlling movement from the brain to the body.

The spinal cord also controls motor reflexes. These reflexes are quick, unconscious movements—like automatically removing a hand from a hot object. Reflexes are so fast because they involve local synaptic connections. For example, the knee reflex that a doctor tests during a routine physical is controlled by a single synapse between a sensory neuron and a motor neuron. While a reflex may only require the involvement of one or two synapses, synapses with interneurons in the spinal column transmit information to the brain to convey what happened (the knee jerked, or the hand was hot).

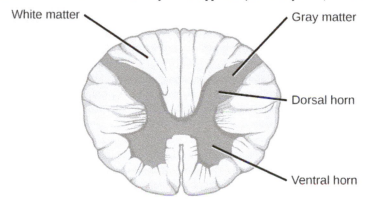

Figure 16.23 A cross-section of the spinal cord shows gray matter (containing cell bodies and interneurons) and white matter (containing myelinated axons).

The Peripheral Nervous System

The **peripheral nervous system (PNS)** is the connection between the central nervous system and the rest of the body. The PNS can be broken down into the **autonomic nervous system**, which controls bodily functions without conscious control, and the **sensory-somatic nervous system**, which transmits sensory information from the skin, muscles, and sensory organs to the CNS and sends motor commands from the CNS to the muscles.

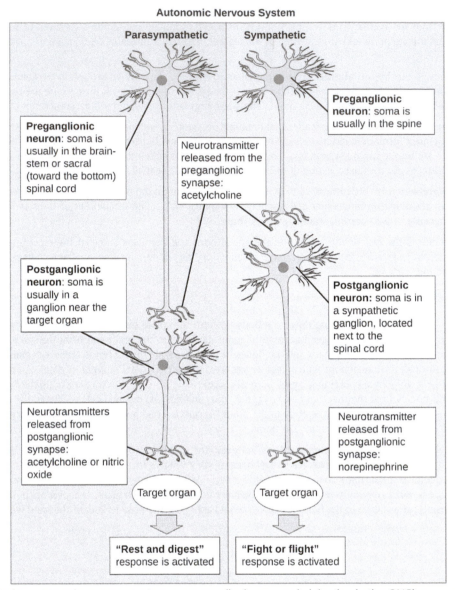

Figure 16.24 In the autonomic nervous system, a preganglionic neuron (originating in the CNS) synapses to a neuron in a ganglion that, in turn, synapses on a target organ. Activation of the sympathetic nervous system causes release of norepinephrine on the target organ. Activation of the parasympathetic nervous system causes release of acetylcholine on the target organ.

The autonomic nervous system serves as the relay between the CNS and the internal organs. It controls the lungs, the heart, smooth muscle, and exocrine and endocrine glands. The autonomic nervous system controls these organs largely without conscious control; it can continuously monitor the conditions of these different systems and implement changes as needed. Signaling to the target tissue usually involves two synapses: a preganglionic neuron (originating in the CNS) synapses to a neuron in a ganglion that, in turn, synapses on the target organ (Figure 16.24). There are two divisions of the autonomic nervous system that often have opposing effects: the sympathetic nervous system and the parasympathetic nervous system.

The **sympathetic nervous system** is responsible for the immediate responses an animal makes when it encounters a dangerous situation. One way to remember this is to think of the "fight-or-flight" response a person feels when encountering a snake ("snake" and "sympathetic" both begin with "s"). Examples of functions controlled by the sympathetic nervous system include an accelerated heart rate and inhibited digestion. These functions help prepare an organism's body for the physical strain required to escape a potentially dangerous situation or to fend off a predator.

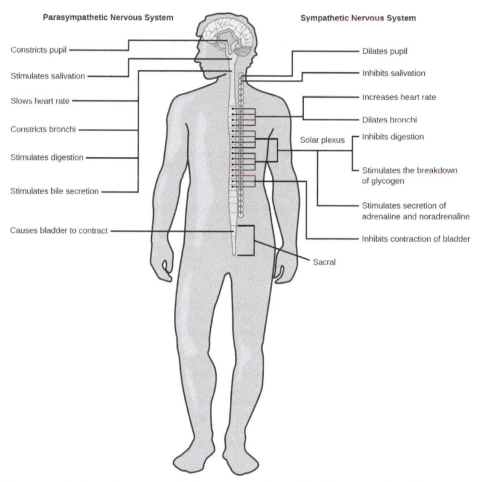

Parasympathetic Nervous System

Constricts pupil

Stimulates salivation

Slows heart rate

Constricts bronchi

Stimulates digestion

Stimulates bile secretion

Causes bladder to contract

Sympathetic Nervous System

Dilates pupil

Inhibits salivation

Increases heart rate

Dilates bronchi

Solar plexus

Inhibits digestion

Stimulates the breakdown of glycogen

Stimulates secretion of adrenaline and noradrenaline

Inhibits contraction of bladder

Sacral

Figure 16.25 The sympathetic and parasympathetic nervous systems often have opposing effects on target organs.

While the sympathetic nervous system is activated in stressful situations, the **parasympathetic nervous system** allows an animal to "rest and digest." One way to remember this is to think that during a restful situation like a picnic, the parasympathetic nervous system is in control ("picnic" and "parasympathetic" both start with "p"). Parasympathetic preganglionic neurons have cell bodies located in the brainstem and in the sacral (toward the bottom) spinal cord (Figure 16.25). The parasympathetic nervous system resets organ function after the sympathetic nervous system is activated including slowing of heart rate, lowered blood pressure, and stimulation of digestion.

The sensory-somatic nervous system is made up of cranial and spinal nerves and contains both sensory and motor neurons. Sensory neurons transmit sensory information from the skin, skeletal muscle, and sensory organs to the CNS. Motor neurons transmit messages about desired movement from the CNS to the muscles to make them contract. Without its sensory-somatic nervous system, an animal would be unable to process any information about its environment (what it sees, feels, hears, and so on) and could not control motor movements. Unlike the autonomic nervous system, which usually has two synapses between the CNS and the target organ, sensory and motor neurons usually have only one synapse—one ending of the neuron is at the organ and the other directly contacts a CNS neuron.

KEY TERMS

action potential a momentary change in the electrical potential of a neuron (or muscle) membrane

adrenal gland the endocrine gland associated with the kidneys

alveolus (plural: alveoli) (also, air sacs) the terminal structure of the lung passage where gas exchange occurs

amygdala a structure within the limbic system that processes fear

amylase an enzyme found in saliva and secreted by the pancreas that converts carbohydrates to maltose

anus the exit point of the digestive system for waste material

aorta the major artery that takes blood away from the heart to the systemic circulatory system

appendicular skeleton the skeleton composed of the bones of the upper limbs, which function to grasp and manipulate objects, and the lower limbs, which permit locomotion

artery a blood vessel that takes blood away from the heart

atrium (plural: atria) a chamber of the heart that receives blood from the veins

auditory ossicles (also, middle ear bones) the bones that transduce sounds from the air into vibrations in the fluid-filled cochlea

autonomic nervous system the part of the peripheral nervous system that controls bodily functions

axial skeleton skeleton that forms the central axis of the body and includes the bones of the skull, the ossicles of the middle ear, the hyoid bone of the throat, the vertebral column, and the thoracic cage (ribcage)

axon a tube-like structure that propagates a signal from a neuron's cell body to axon terminals

basal ganglia an interconnected collections of cells in the brain that are involved in movement and motivation

bicuspid valve a one-way opening between the atrium and the ventricle in the left side of the heart

bile a digestive juice produced by the liver; important for digestion of lipids

bolus a mass of food resulting from chewing action and wetting by saliva

brainstem a portion of brain that connects with the spinal cord; controls basic nervous system functions like breathing and swallowing

bronchi (singular: bronchus) smaller branches of cartilaginous tissue that stem off of the trachea; air is funneled through the bronchi to the region where gas exchange occurs in the alveoli

bronchiole an airway that extends from the main bronchus to the alveolar sac

capillary the smallest blood vessel that allows the passage of individual blood cells and the site of diffusion of oxygen and nutrient exchange

cardiac cycle the filling and emptying the heart of blood caused by electrical signals that cause the heart muscles to contract and relax

cardiac muscle tissue the muscle tissue found only in the heart; cardiac contractions pump blood throughout the body and maintain blood pressure

cartilaginous joint a joint in which the bones are connected by cartilage

central nervous system (CNS) the nervous system made up of the brain and spinal cord; covered with three layers of protective meninges

Download for free at https://openstax.org/details/books/concepts-biology

cerebellum the brain structure involved in posture, motor coordination, and learning new motor actions

cerebral cortex the outermost sheet of brain tissue; involved in many higher-order functions

cerebrospinal fluid (CSF) a clear liquid that surrounds the brain and fills its ventricles and acts as a shock absorber

chyme a mixture of partially digested food and stomach juices

closed circulatory system a system that has the blood separated from the bodily interstitial fluid and contained in blood vessels

colon the largest portion of the large intestine consisting of the ascending colon, transverse colon, and descending colon

corpus callosum a thick nerve bundle that connects the cerebral hemispheres

dendrite a structure that extends away from the cell body to receive messages from other neurons

depolarization a change in the membrane potential to a less negative value

diaphragm a skeletal muscle located under lungs that encloses the lungs in the thorax

diastole the relaxation phase of the cardiac cycle when the heart is relaxed and the ventricles are filling with blood

down-regulation a decrease in the number of hormone receptors in response to increased hormone levels

ectotherm an organism that relies primarily on environmental heat sources to maintain its body temperature

electrocardiogram (ECG) a recording of the electrical impulses of the cardiac muscle

endocrine gland the gland that secretes hormones into the surrounding interstitial fluid, which then diffuse into blood and are carried to various organs and tissues within the body

endotherm an organism that relies primarily on internal heat sources to maintain its body temperature

esophagus a tubular organ that connects the mouth to the stomach

essential nutrient a nutrient that cannot be synthesized by the body; it must be obtained from food

exocrine gland the gland that secretes chemicals through ducts that lead to skin surfaces, body cavities, and organ cavities.

fibrous joint a joint held together by fibrous connective tissue

frontal lobe the part of the cerebral cortex that contains the motor cortex and areas involved in planning, attention, and language

gallbladder the organ that stores and concentrates bile

glia (also, glial cells) the cells that provide support functions for neurons

hippocampus the brain structure in the temporal lobe involved in processing memories

hormone a chemical released by cells in one area of the body that affects cells in other parts of the body

hyoid bone the bone that lies below the mandible in the front of the neck

hypothalamus the brain structure that controls hormone release and body homeostasis

inferior vena cava the major vein of the body returning blood from the lower parts of the body to the right atrium

interstitial fluid the fluid found between cells in the body, similar in constitution to the fluid component of blood, but without the high concentrations of proteins

intracellular hormone receptor a hormone receptor in the cytoplasm or nucleus of a cell

joint the point at which two or more bones meet

kidney the organ that performs excretory and osmoregulatory functions

large intestine a digestive system organ that reabsorbs water from undigested material and processes waste matter

larynx the voice box, located within the throat

limbic system a connected brain area that processes emotion and motivation

liver an organ that produces bile for digestion and processes vitamins and lipids

membrane potential a difference in electrical potential between the inside and outside of a cell

meninges (singular: meninx) the membranes that cover and protect the central nervous system

mineral an inorganic, elemental molecule that carries out important roles in the body

myelin sheath a cellular extension containing a fatty substance produced by glia that surrounds and insulates axons

myofibril the long cylindrical structures that lie parallel to the muscle fiber

myofilament the small structures that make up myofibrils

nasal cavity an opening of the respiratory system to the outside environment

nephron the functional unit of the kidney

neuron a specialized cell that can receive and transmit electrical and chemical signals

occipital lobe the part of the cerebral cortex that contains visual cortex and processes visual stimuli

open circulatory system a circulatory system that has the blood mixed with interstitial fluid in the body cavity and directly bathes the organs

oral cavity the point of entry of food into the digestive system

osmoregulation the mechanism by which water and solute concentrations are maintained at desired levels

osmotic balance the appropriate values of water and solute concentrations for a healthy organism

pancreas a gland that secretes digestive juices

pancreas the organ located between the stomach and the small intestine that contains exocrine and endocrine cells

parasympathetic nervous system the division of autonomic nervous system that regulates visceral functions during relaxation

parathyroid gland the gland located on the surface of the thyroid that produces parathyroid hormone

parietal lobe the part of the cerebral cortex involved in processing touch and the sense of the body in space

pectoral girdle the bones that transmit the force generated by the upper limbs to the axial skeleton

pelvic girdle the bones that transmit the force generated by the lower limbs to the axial skeleton

pepsin an enzyme found in the stomach whose main role is protein digestion

peripheral nervous system (PNS) the nervous system that serves as the connection between the central nervous system and the rest of the body; consists of the autonomic nervous system and the sensory-somatic nervous system

peristalsis wave-like movements of muscle tissue

pharynx the throat

Download for free at https://openstax.org/details/books/concepts-biology

pituitary gland the endocrine gland located at the base of the brain composed of an anterior and posterior region; also called hypophysis

primary bronchus (also, main bronchus) a region of the airway within the lung that attaches to the trachea and bifurcates to form the bronchioles

pulmonary circulation the flow of blood away from the heart through the lungs where oxygenation occurs and then back to the heart

rectum the area of the body where feces is stored until elimination

renal artery the artery that delivers blood to the kidney

renal vein the vein that drains blood from the kidney

salivary gland one of three pairs of exocrine glands in the mammalian mouth that secretes saliva, a mix of watery mucus and enzymes

sarcolemma the plasma membrane of a skeletal muscle fiber

sarcomere the functional unit of skeletal muscle

sensory-somatic nervous system the system of sensory and motor nerves

set point the target value of a physiological state in homeostasis

skeletal muscle tissue forms skeletal muscles, which attach to bones and control locomotion and any movement that can be consciously controlled

skull the bone that supports the structures of the face and protects the brain

small intestine the organ where digestion of protein, fats, and carbohydrates is completed

smooth muscle tissue the muscle that occurs in the walls of hollow organs such as the intestines, stomach, and urinary bladder, and around passages such as the respiratory tract and blood vessels

spinal cord a thick fiber bundle that connects the brain with peripheral nerves; transmits sensory and motor information; contains neurons that control motor reflexes

stomach a saclike organ containing acidic digestive juices

superior vena cava the major vein of the body returning blood from the upper part of the body to the right atrium

sympathetic nervous system the division of autonomic nervous system activated during stressful "fight-or-flight" situations

synapse a junction between two neurons where neuronal signals are communicated

synaptic cleft a space between the presynaptic and postsynaptic membranes

synovial joints the only joints that have a space between the adjoining bones

systemic circulation the flow of blood away from the heart to the brain, liver, kidneys, stomach, and other organs, the limbs, and the muscles of the body, and then back to the heart

systole the contraction phase of cardiac cycle when the ventricles are pumping blood into the arteries

temporal lobe the part of the cerebral cortex that processes auditory input; parts of the temporal lobe are involved in speech, memory, and emotion processing

thalamus the brain area that relays sensory information to the cortex

thoracic cage (also, ribcage) the skeleton of the chest, which consists of the ribs, thoracic vertebrae, sternum, and costal cartilages

threshold of excitation the level of depolarization needed for an action potential to fire

thymus the gland located behind the sternum that produces thymosin hormones that contribute to the development of the immune system

thyroid gland an endocrine gland located in the neck that produces thyroid hormones thyroxine and triiodothyronine

trachea the cartilaginous tube that transports air from the throat to the lungs

tricuspid valve a one-way opening between the atrium and the ventricle in the right side of the heart

up-regulation an increase in the number of hormone receptors in response to increased hormone levels

ureter the urine-bearing tubes coming out of the kidney

urethra the tube that conducts urine from the urinary bladder to the external environment

urinary bladder the structure that the ureters empty the urine into

vein a blood vessel that brings blood back to the heart

ventricle (of the heart) a large chamber of the heart that pumps blood into arteries

vertebral column (also, spine) the column that surrounds and protects the spinal cord, supports the head, and acts as an attachment point for ribs and muscles of the back and neck

vitamin an organic substance necessary in small amounts to sustain life

CHAPTER SUMMARY

16.1 Homeostasis and Osmoregulation

Homeostasis is a dynamic equilibrium that is maintained in body tissues and organs. It is dynamic because it is constantly adjusting to the changes that the systems encounter. It is an equilibrium because body functions are kept within a normal range, with some fluctuations around a set point. The kidneys are the main osmoregulatory organs in mammalian systems; they function to filter blood and maintain the dissolved ion concentrations of body fluids. They are made up internally of three distinct regions—the cortex, medulla, and pelvis. The blood vessels that transport blood into and out of the kidneys arise from and merge with the aorta and inferior vena cava, respectively. The nephron is the functional unit of the kidney, which actively filters blood and generates urine. The urine leaves the kidney through the ureter and is stored in the urinary bladder. Urine is voided from the body through the urethra.

16.2 Digestive System

There are many organs that work together to digest food and absorb nutrients. The mouth is the point of ingestion and the location where both mechanical and chemical breakdown of food begins. Saliva contains an enzyme called amylase that breaks down carbohydrates. The food bolus travels through the esophagus by peristaltic movements to the stomach. The stomach has an extremely acidic environment. The enzyme pepsin digests protein in the stomach. Further digestion and absorption take place in the small intestine. The large intestine reabsorbs water from the undigested food and stores waste until elimination.

Carbohydrates, proteins, and fats are the primary components of food. Some essential nutrients are required for cellular function but cannot be produced by the animal body. These include vitamins, minerals, some fatty acids, and some amino acids. Food intake in more than necessary amounts is stored as glycogen in the liver and muscle cells, and in adipose tissue. Excess adipose storage can lead to obesity and serious health problems.

16.3 Circulatory and Respiratory Systems

Animal respiratory systems are designed to facilitate gas exchange. In mammals, air is warmed and humidified in the nasal cavity. Air then travels down the pharynx and larynx, through the trachea, and into the lungs. In the lungs, air passes through the branching bronchi, reaching the respiratory bronchioles. The respiratory bronchioles open up into the alveolar ducts, alveolar sacs, and alveoli. Because there are so many alveoli and alveolar sacs in the lung, the surface area for gas exchange is very large.

The mammalian circulatory system is a closed system with double circulation passing through the lungs and the body. It consists of a network of vessels containing blood that circulates because of pressure differences generated by the heart.

The heart contains two pumps that move blood through the pulmonary and systemic circulations. There is one atrium and one ventricle on the right side and one atrium and one ventricle on the left side. The pumping of the heart is a function of cardiomyocytes, distinctive muscle cells that are striated like skeletal muscle but pump rhythmically and involuntarily like smooth muscle. The signal for contraction begins in the wall of the right atrium. The electrochemical signal causes the two atria to contract in unison; then the signal causes the ventricles to contract. The blood from the heart is carried through the body by a complex network of blood vessels; arteries take blood away from the heart, and veins bring blood back to the heart.

16.4 Endocrine System

Hormones cause cellular changes by binding to receptors on or in target cells. The number of receptors on a target cell can increase or decrease in response to hormone activity.

Hormone levels are primarily controlled through negative feedback, in which rising levels of a hormone inhibit its further release.

The pituitary gland is located at the base of the brain. The anterior pituitary receives signals from the hypothalamus and produces six hormones. The posterior pituitary is an extension of the brain and releases hormones (antidiuretic hormone and oxytocin) produced by the hypothalamus. The thyroid gland is located in the neck and is composed of two lobes. The thyroid produces the hormones thyroxine and triiodothyronine. The thyroid also produces calcitonin. The parathyroid glands lie on the posterior surface of the thyroid gland and produce parathyroid hormone.

The adrenal glands are located on top of the kidneys and consist of the adrenal cortex and adrenal medulla. The adrenal cortex produces the corticosteroids, glucocorticoids and mineralocorticoids. The adrenal medulla is the inner part of the adrenal gland and produces epinephrine and norepinephrine.

The pancreas lies in the abdomen between the stomach and the small intestine. Clusters of endocrine cells in the pancreas form the islets of Langerhans, which contain alpha cells that release glucagon and beta cells that release insulin. Some organs possess endocrine activity as a secondary function but have another primary function. The heart produces the hormone atrial natriuretic peptide, which functions to reduce blood volume, pressure, and Na^+ concentration. The gastrointestinal tract produces various hormones that aid in digestion. The kidneys produce erythropoietin. The thymus produces hormones that aid in the development of the immune system. The gonads produce steroid hormones, including testosterone in males and estrogen and progesterone in females. Adipose tissue produces leptin, which promotes satiety signals in the brain.

16.5 Musculoskeletal System

The human skeleton is an endoskeleton that is composed of the axial and appendicular skeleton. The axial skeleton is composed of the bones of the skull, ossicles of the ear, hyoid bone, vertebral column, and ribcage. The skull consists of eight cranial bones and 14 facial bones. Six bones make up the ossicles of the middle ear, while the hyoid bone is located in the neck under the mandible. The vertebral column contains 26 bones and surrounds and protects the spinal cord. The thoracic cage consists of the sternum, ribs, thoracic vertebrae, and costal cartilages. The appendicular skeleton is made up of the upper and lower limbs. The pectoral girdle is composed of the clavicles and the scapulae. The upper limb contains 30 bones in the arm, the forearm, and the hand. The pelvic girdle attaches the lower limbs to the axial skeleton. The lower limb includes the bones of the thigh, the leg, and the foot.

The structural classification of joints divides them into fibrous, cartilaginous, and synovial joints. The bones of fibrous joints are held together by fibrous connective tissue. Cartilaginous joints are joints in which the bones are connected by cartilage. Synovial joints are joints that have a space between the adjoining bones. The movement of synovial joints includes angular and rotational. Angular movements are produced when the angle between the bones of a joint changes. Rotational movement is the movement of a bone as it rotates around its own longitudinal axis.

The body contains three types of muscle tissue: skeletal muscle, cardiac muscle, and smooth muscle. Muscles are composed of individual cells called muscle fibers. Muscle fibers consist of myofilaments composed of the proteins actin and myosin arranged in units called sarcomeres. Contraction of the muscle occurs by the combined action of myosin and actin fibers sliding past each other when the myosin heads bind to the actin fiber, bend, disengage, and then repeat the process.

16.6 Nervous System

The nervous system is made up of neurons and glia. Neurons are specialized cells that are capable of sending electrical as well as chemical signals. Most neurons contain dendrites, which receive these signals, and axons that send signals to other neurons or tissues. Glia are non-neuronal cells in the nervous system that support neuronal development and signaling. There are several types of glia that serve different functions.

Neurons have a resting potential across their membranes and when they are stimulated by a strong enough signal from another neuron an action potential may carry an electrochemical signal along the neuron to a synapse with another neuron. Neurotransmitters carry signals across synapses to initiate a response in another neuron.

The vertebrate central nervous system contains the brain and the spinal cord, which are covered and protected by three meninges. The brain contains structurally and functionally defined regions. In mammals, these include the cortex (which can be broken down into four primary functional lobes: frontal, temporal, occipital, and parietal), basal ganglia, thalamus, hypothalamus, limbic system, cerebellum, and brainstem—although structures in some of these designations overlap. While functions may be primarily localized to one structure in the brain, most complex functions, like language and sleep, involve neurons in multiple brain regions. The spinal cord is the information superhighway that connects the brain with the rest of the body through its connections with peripheral nerves. It transmits sensory and motor input and also controls motor reflexes.

The peripheral nervous system contains both the autonomic and sensory-somatic nervous systems. The autonomic nervous system provides unconscious control over visceral functions and has two divisions: the sympathetic and parasympathetic nervous systems. The sympathetic nervous system is activated in stressful situations to prepare the animal for a "fight-or-flight" response. The parasympathetic nervous system is active during restful periods. The sensory-somatic nervous system is made of cranial and spinal nerves that transmit sensory information from skin and muscle to the CNS and motor commands from the CNS to the muscles.

ART CONNECTION QUESTIONS

1. Figure 16.2 When bacteria are destroyed by leukocytes, pyrogens are released into the blood. Pyrogens reset the body's thermostat to a higher temperature, resulting in fever. How might pyrogens cause the body temperature to rise?

2. Figure 16.7 Which of the following statements about the digestive system is false?

    a. Chyme is a mixture of food and digestive juices that is produced in the stomach.

    b. Food enters the large intestine before the small intestine.

    c. In the small intestine, chyme mixes with bile, which emulsifies fats.

    d. The stomach is separated from the small intestine by the pyloric sphincter.

3. Figure 16.9 Which of the following statements about the human respiratory system is false?

    a. When we breathe in, air travels from the pharynx to the trachea.

    b. The bronchioles branch into bronchi.

    c. Alveolar ducts connect to alveolar sacs.

    d. Gas exchange between the lungs and blood takes place in the alveolus.

4. Figure 16.10 Which of the following statements about the circulatory system is false?

    a. Blood in the pulmonary vein is deoxygenated.

    b. Blood in the inferior vena cava is deoxygenated.

    c. Blood in the pulmonary artery is deoxygenated.

    d. Blood in the aorta is oxygenated.

5. Figure 16.14 Goiter, a disease caused by iodine deficiency, results in the inability of the thyroid gland to form T_3 and T_4. The body typically attempts to compensate by producing greater amounts of TSH. Which of the following symptoms would you expect goiter to cause?

    a. Hypothyroidism, resulting in weight gain, cold sensitivity, and reduced mental activity.

    b. Hyperthyroidism, resulting in weight loss, profuse sweating and increased heart rate.

    c. Hyperthyroidism, resulting in weight gain, cold sensitivity, and reduced mental activity.

    d. Hypothyroidism, resulting in weight loss, profuse sweating and increased heart rate.

REVIEW QUESTIONS

6. When faced with a sudden drop in environmental temperature, an endothermic animal will _____.

    a. experience a drop in its body temperature

    b. wait to see if it goes lower

    c. increase muscle activity to generate heat

    d. add fur or fat to increase insulation

7. How are wastes carried to the kidney for removal?

    a. in cells

b.   in the urine
c.   in blood
d.   in interstitial fluid

8. What is the cause of a fever of 38.3 °C (101 °F)?

a.   too much heat produced by the body
b.   upward adjustment of the body temperature set point
c.   inadequate cooling mechanisms in the body
d.   the heat caused by a viral or bacterial infection

9. Where does the majority of fat digestion take place?

a.   mouth
b.   stomach
c.   small intestine
d.   large intestine

10. The bile from the liver is delivered to the _____.

a.   stomach
b.   liver
c.   small intestine
d.   colon

11. Which of the following statements is not true?

a.   Essential nutrients can be synthesized by the body.
b.   Vitamins are required in small quantities for bodily function.
c.   Some amino acids can be synthesized by the body, while others need to be obtained from diet.
d.   Vitamins come in two categories: fat-soluble and water-soluble.

12. The respiratory system _____.
a.   provides body tissues with oxygen
b.   provides body tissues with oxygen and carbon dioxide
c.   establishes how many breaths are taken per minute
d.   provides the body with carbon dioxide

13. Which is the order of airflow during inhalation?

a.   nasal cavity, trachea, larynx, bronchi, bronchioles, alveoli
b.   nasal cavity, larynx, trachea, bronchi, bronchioles, alveoli
c.   nasal cavity, larynx, trachea, bronchioles, bronchi, alveoli
d.   nasal cavity, trachea, larynx, bronchi, bronchioles, alveoli

14. Where does the right ventricle send blood?
a.   the head
b.   the upper body
c.   the lungs
d.   the lower body

15. During the systolic phase of the cardiac cycle, the heart is _____.
a.   contracting
b.   relaxing
c.   contracting and relaxing
d.   filling with blood

16. How do arteries differ from veins?
a.   Arteries have thicker wall layers to accommodate the changes in pressure from the heart.
b.   Arteries carry blood.
c.   Arteries have thinner wall layers and valves and move blood by the action of skeletal muscle.
d.   Arteries are thin walled and are used for gas exchange.

17. Most of the hormones produced by the anterior pituitary perform what function?
a.   regulate growth
b.   regulate the sleep cycle
c.   regulate production of other hormones
d.   regulate blood volume and blood pressure

18. What is the function of the hormone erythropoietin?

a.   stimulates production of red blood cells
b.   stimulates muscle growth
c.   causes the fight-or-flight response
d.   causes testosterone production

19. Which endocrine glands are associated with the kidneys?
a.   thyroid glands
b.   pituitary glands
c.   adrenal glands
d.   gonads

20. Among other bones, the axial skeleton includes the _____.
a.   thoracic cage and vertebral column
b.   thoracic cage and pectoral girdle
c.   skull and pelvic girdle
d.   pectoral and pelvic girdles

21. The pectoral girdle supports the _____.
a.   arms
b.   legs
c.   skull
d.   thoracic cage

22. Which component is responsible for initially stimulating a muscle contraction?
a.   proteins
b.   electrochemical signals
c.   plasma membranes
d.   striations

23. What kind of muscle tissue is found surrounding the urinary bladder?
a.   cardiac
b.   skeletal
c.   striated

d.   smooth

24. Neurons contain _____, which can receive signals from other neurons.
   a.   axons
   b.   mitochondria
   c.   dendrites
   d.   Golgi bodies

25. The part of the brain that is responsible for coordination during movement is the _____.
   a.   limbic system

b.   thalamus
c.   cerebellum
d.   parietal lobe

26. Which part of the nervous system directly controls the digestive system?
   a.   parasympathetic nervous system
   b.   central nervous system
   c.   spinal cord
   d.   sensory-somatic nervous system

CRITICAL THINKING QUESTIONS

27. Describe how the body's mechanisms maintain homeostasis?

28. Why is excretion important in order to achieve osmotic balance?

29. What is the role of the accessory organs in digestion?

30. What is the role of minerals in maintaining good health?

31. Discuss why obesity is a growing epidemic.

32. Describe the function of these terms and describe where they are located: main bronchus, trachea, alveoli.

33. How does the structure of alveoli maximize gas exchange?

34. Describe the cardiac cycle.

35. What is a similarity and a difference between an exocrine gland and an endocrine gland?

36. Describe how hormone receptors can play a role in affecting the size of the responses of tissues to hormones.

37. Many hormone systems regulate body functions through opposing hormone actions. Describe how opposing hormone actions regulate blood-glucose levels?

38. What movements occur at the hip joint and knees as you bend down to pick something up?

39. How are neurons similar to other cells? How are they unique?

40. What are the main functions of the spinal cord?

41. What are the main differences between the sympathetic and parasympathetic branches of the autonomic nervous system?

42. What are the main functions of the sensory-somatic nervous system?

17 | THE IMMUNE SYSTEM AND DISEASE

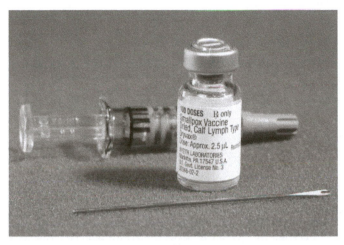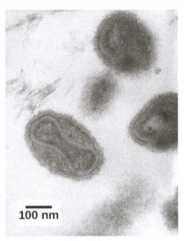

(a)   (b)

Figure 17.1 (a) This smallpox (variola) vaccine is derived from calves exposed to cowpox virus. Vaccines provoke a reaction in the immune system that prepares it for a subsequent infection by smallpox. (b) Viewed under a transmission electron microscope, you can see the variola's dumbbell-shaped structure that contains the viral DNA. (credit a: modification of work by James Gathany, CDC; credit b: modification of work by Dr. Fred Murphy; Sylvia Whitfield, CDC; scale-bar data from Matt Russell)

Chapter Outline

17.1: Viruses

17.2: Innate Immunity

17.3: Adaptive Immunity

17.4: Disruptions in the Immune System

Introduction

Organisms have a wide array of adaptations for preventing attacks of parasites and diseases. The vertebrate defense systems, including those of humans, are complex and multilayered, with defenses unique to vertebrates. These unique vertebrate defenses interact with other defense systems inherited from ancestral lineages, and include complex and specific pathogen recognition and memory mechanisms. Research continues to unravel the complexities and vulnerabilities of the immune system.

Despite a poor understanding of the workings of the body in the early 18th century in Europe, the practice of inoculation as a method to prevent the often-deadly effects of smallpox was introduced from the courts of the Ottoman Empire. The method involved causing limited infection with the smallpox virus by introducing the pus of an affected individual to a scratch in an uninfected person. The resulting infection was milder than if it had been caught naturally and mortality rates were shown to be about two percent rather than 30 percent from natural infections. Moreover, the inoculation gave the individual immunity to the disease. It was from these early experiences with inoculation that the methods of vaccination were developed, in which a weakened or relatively harmless (killed) derivative of a pathogen is introduced into the individual. The vaccination induces immunity to the disease with few of the risks of being infected. A modern understanding of the causes of the infectious disease and the mechanisms of the immune system began in the late 19th century and continues to grow today.

17.1 | Viruses

By the end of this section, you will be able to:

- Describe how viruses were first discovered and how they are detected
- Explain the detailed steps of viral replication
- Describe how vaccines are used in prevention and treatment of viral diseases

(a)   (b)

Figure 17.2 (a) The tobacco mosaic virus, seen by transmission electron microscopy, was the first virus to be discovered. (b) The leaves of an infected plant are shown. (credit a: scale-bar data from Matt Russell; credit b: modification of work by USDA, Department of Plant Pathology Archive, North Carolina State University)

No one knows exactly when viruses emerged or from where they came, since viruses do not leave historical footprints such as fossils. Modern viruses are thought to be a mosaic of bits and pieces of nucleic acids picked up from various sources along their respective evolutionary paths. Viruses are **acellular**, parasitic entities that are not classified within any domain because they are not considered alive. They have no plasma membrane, internal organelles, or metabolic processes, and they do not divide. Instead, they infect a host cell and use the host's replication processes to produce progeny virus particles. Viruses infect all forms of organisms including bacteria, archaea, fungi, plants, and animals. Living things grow, metabolize, and reproduce. Viruses replicate, but to do so, they are entirely dependent on their host cells. They do not metabolize or grow, but are assembled in their mature form.

Viruses are diverse. They vary in their structure, their replication methods, and in their target hosts or even host cells. While most biological diversity can be understood through evolutionary history, such as how species have adapted to conditions and environments, much about virus origins and evolution remains unknown.

How Viruses Replicate

Viruses were first discovered after the development of a porcelain filter, called the Chamberland-Pasteur filter, which could remove all bacteria visible under the microscope from any liquid sample. In 1886, Adolph Meyer demonstrated that a disease of tobacco plants, tobacco mosaic disease, could be transferred from a diseased plant to a healthy one through liquid plant extracts. In 1892, Dmitri Ivanowski showed that this disease could be transmitted in this way even after the Chamberland-Pasteur filter had removed all viable bacteria from the extract. Still, it was many years before it was proven that these "filterable" infectious agents were not simply very small bacteria but were a new type of tiny, disease-causing particle.

Virions, single virus particles, are very small, about 20–250 nanometers (1 nanometer = 1/1,000,000 mm). These individual virus particles are the infectious form of a virus outside the host cell. Unlike bacteria (which are about 100 times larger), we cannot see viruses with a light microscope, with the exception of some large virions of the poxvirus family (Figure 17.3).

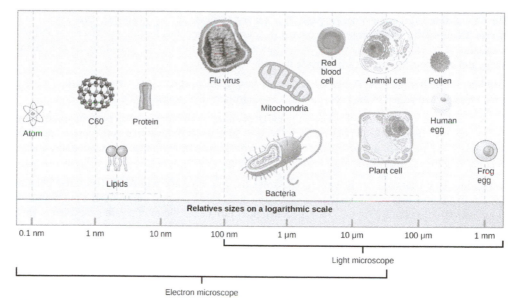

Figure 17.3 The size of a virus is very small relative to the size of cells and organelles.

It was not until the development of the electron microscope in the 1940s that scientists got their first good view of the structure of the tobacco mosaic virus (Figure 17.2) and others. The surface structure of virions can be observed by both scanning and transmission electron microscopy, whereas the internal structures of the virus can only be observed in images from a transmission electron microscope (Figure 17.4).

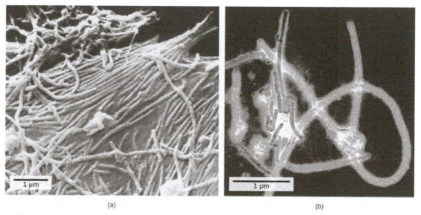

Figure 17.4 The ebola virus is shown here as visualized through (a) a scanning electron micrograph and (b) a transmission electron micrograph. (credit a: modification of work by Cynthia Goldsmith, CDC; credit b: modification of work by Thomas W. Geisbert, Boston University School of Medicine; scale-bar data from Matt Russell)

The use of this technology has allowed for the discovery of many viruses of all types of living organisms. They were initially grouped by shared morphology, meaning their size, shape, and distinguishing structures. Later, groups of viruses were classified by the type of nucleic acid they contained, DNA or RNA, and whether their nucleic acid was single- or double-stranded. More recently, molecular analysis of viral replication cycles has further refined their classification.

A **virion** consists of a nucleic-acid core, an outer protein coating, and sometimes an outer envelope made of protein and phospholipid membranes derived from the host cell. The most visible difference between members of viral families is their morphology, which is quite diverse. An interesting feature of viral complexity is that the complexity of the host does not correlate to the complexity of the virion. Some of the most complex virion structures are observed in bacteriophages, viruses that infect the simplest living organisms, bacteria.

Viruses come in many shapes and sizes, but these are consistent and distinct for each viral family (Figure 17.5). All virions have a nucleic-acid genome covered by a protective layer of protein, called a **capsid**. The capsid is made of protein subunits called capsomeres. Some viral capsids are simple polyhedral "spheres," whereas others are quite complex in structure. The outer structure surrounding the capsid of some viruses is called the **viral envelope**. All viruses use some sort of **glycoprotein** to attach to their host cells at molecules on the cell called viral receptors. The virus exploits these cell-surface molecules, which the cell uses for some other purpose, as a way to recognize and infect specific cell types. For example, the measles virus uses a cell-surface glycoprotein in humans that normally functions in immune reactions and possibly in the sperm-egg interaction at fertilization. Attachment is a requirement for viruses to later penetrate the cell membrane, inject the viral genome, and complete their replication inside the cell.

The T4 bacteriophage, which infects the *E. coli* bacterium, is among the most complex virion known; T4 has a protein tail structure that the virus uses to attach to the host cell and a head structure that houses its DNA.

Adenovirus, a nonenveloped animal virus that causes respiratory illnesses in humans, uses protein spikes protruding from its capsomeres to attach to the host cell. Nonenveloped viruses also include those that cause polio (poliovirus), plantar warts (papillomavirus), and hepatitis A (hepatitis A virus). Nonenveloped viruses tend to be more robust and more likely to survive under harsh conditions, such as the gut.

Enveloped virions like HIV (human immunodeficiency virus), the causative agent in AIDS (acquired immune deficiency syndrome), consist of nucleic acid (RNA in the case of HIV) and capsid proteins surrounded by a phospholipid bilayer envelope and its associated proteins (Figure 17.5). Chicken pox, influenza, and mumps are examples of diseases caused by viruses with envelopes. Because of the fragility of the envelope, nonenveloped viruses are more resistant to changes in temperature, pH, and some disinfectants than enveloped viruses.

Overall, the shape of the virion and the presence or absence of an envelope tells us little about what diseases the viruses may cause or what species they might infect, but is still a useful means to begin viral classification.

art CONNECTION

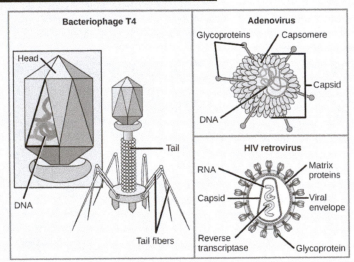

Figure 17.5 Viruses can be complex in shape or relatively simple. This figure shows three relatively complex virions: the bacteriophage T4, with its DNA-containing head group and tail fibers that attach to host cells; adenovirus, which uses spikes from its capsid to bind to the host cells; and HIV, which uses glycoproteins embedded in its envelope to do so. Notice that HIV has proteins called matrix proteins, internal to the envelope, which help stabilize virion shape. HIV is a retrovirus, which means it reverse transcribes its RNA genome into DNA, which is then spliced into the host's DNA. (credit "bacteriophage, adenovirus": modification of work by NCBI, NIH; credit "HIV retrovirus": modification of work by NIAID, NIH)

Which of the following statements about virus structure is true?

a. All viruses are encased in a viral membrane.

b. The capsomere is made up of small protein subunits called capsids.

c. DNA is the genetic material in all viruses.

d. Glycoproteins help the virus attach to the host cell.

Unlike all living organisms that use DNA as their genetic material, viruses may use either DNA or RNA as theirs. The virus core contains the genome or total genetic content of the virus. Viral genomes tend to be small compared to bacteria or eukaryotes, containing only those genes that code for proteins the virus cannot get from the host cell. This genetic material may be single-stranded or double-stranded. It may also be linear or circular. While most viruses contain a single segment of nucleic acid, others have genomes that consist of several segments.

DNA viruses have a DNA core. The viral DNA directs the host cell's replication proteins to synthesize new copies of the viral genome and to transcribe and translate that genome into viral proteins. DNA viruses cause human diseases such as chickenpox, hepatitis B, and some venereal diseases like herpes and genital warts.

RNA viruses contain only RNA in their cores. To replicate their genomes in the host cell, the genomes of RNA viruses encode enzymes not found in host cells. RNA polymerase enzymes are not as stable as DNA polymerases and often make mistakes during transcription. For this reason, mutations, changes in the nucleotide sequence, in RNA viruses occur more frequently than in DNA viruses. This leads to more rapid evolution and change in RNA viruses. For example, the fact that influenza is an RNA virus is one reason a new flu vaccine is needed every year. Human diseases caused by RNA viruses include hepatitis C, measles, and rabies.

Viruses can be seen as obligate intracellular parasites. The virus must attach to a living cell, be taken inside, manufacture its proteins and copy its genome, and find a way to escape the cell so the virus can infect other cells and ultimately other individuals. Viruses can infect only certain species of hosts and only certain cells within that host. The molecular basis for this specificity is that a particular surface molecule, known as the viral receptor, must be found on the host cell surface for the virus to attach. Also, metabolic differences seen in different cell types based on differential gene expression are a likely factor in which cells a virus may use to replicate. The cell must be making the substances the virus needs, such as enzymes the virus genome itself does not have genes for, or the virus will not be able to replicate using that cell.

Steps of Virus Infections

A virus must "take over" a cell to replicate. The viral replication cycle can produce dramatic biochemical and structural changes in the host cell, which may cause cell damage. These changes, called **cytopathic** effects, can change cell functions or even destroy the cell. Some infected cells, such as those infected by the common cold virus (rhinovirus), die through lysis (bursting) or **apoptosis** (programmed cell death or "cell suicide"), releasing all the progeny virions at once. The symptoms of viral diseases result from the immune response to the virus, which attempts to control and eliminate the virus from the body, and from cell damage caused by the virus. Many animal viruses, such as HIV (human immunodeficiency virus), leave the infected cells of the immune system by a process known as budding, where virions leave the cell individually. During the budding process, the cell does not undergo lysis and is not immediately killed. However, the damage to the cells that HIV infects may make it impossible for the cells to function as mediators of immunity, even though the cells remain alive for a period of time. Most productive viral infections follow similar steps in the virus replication cycle: attachment, penetration, uncoating, replication, assembly, and release.

A virus attaches to a specific receptor site on the host-cell membrane through attachment proteins in the capsid or proteins embedded in its envelope. The attachment is specific, and typically a virus will only attach to cells of one or a few species and only certain cell types within those species with the appropriate receptors.

View this video (http://openstaxcollege.org/l/influenza2) for a visual explanation of how influenza attacks the body.

Unlike animal viruses, the nucleic acid of bacteriophages is injected into the host cell naked, leaving the capsid outside the cell. Plant and animal viruses can enter their cells through endocytosis, in which the cell membrane surrounds and engulfs the entire virus. Some enveloped viruses enter the cell when the viral envelope fuses directly with the cell membrane. Once inside the cell, the viral capsid is degraded and the viral nucleic acid is released, which then becomes available for replication and transcription.

The replication mechanism depends on the viral genome. DNA viruses usually use host cell proteins and enzymes to make additional DNA that is used to copy the genome or be transcribed to messenger RNA (mRNA), which is then used in protein synthesis. RNA viruses, such as the influenza virus, usually use the RNA core as a template for synthesis of viral genomic RNA and mRNA. The viral mRNA is translated into viral enzymes and capsid proteins to assemble new virions (Figure 17.6). Of course, there are exceptions to this pattern. If a host cell does not provide the enzymes necessary for viral replication, viral genes supply the information to direct synthesis of the missing proteins. Retroviruses, such as HIV, have an RNA genome that must be reverse transcribed to make DNA, which then is inserted into the host's DNA. To convert RNA into DNA, retroviruses contain genes that encode the virus-specific enzyme reverse transcriptase that transcribes an RNA template to DNA. The fact that HIV produces some of its own enzymes, which are not found in the host, has allowed researchers to develop drugs that inhibit these enzymes. These drugs, including the reverse transcriptase inhibitor AZT, inhibit HIV replication by reducing the activity of the enzyme without affecting the host's metabolism.

The last stage of viral replication is the release of the new virions into the host organism, where they are able to infect adjacent cells and repeat the replication cycle. Some viruses are released when the host cell dies and other viruses can leave infected cells by budding through the membrane without directly killing the cell.

art CONNECTION

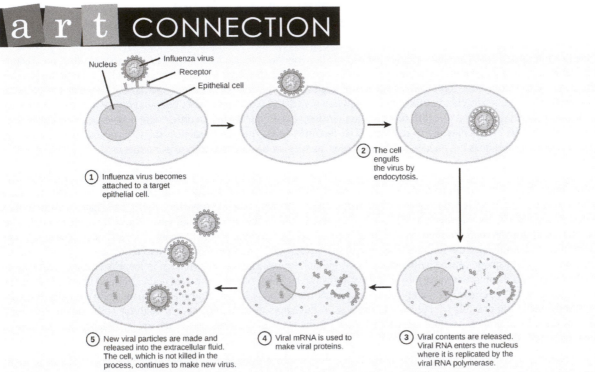

Figure 17.6 In influenza virus infection, glycoproteins attach to a host epithelial cell. As a result, the virus is engulfed. RNA and proteins are made and assembled into new virions.

Influenza virus is packaged in a viral envelope, which fuses with the plasma membrane. This way, the virus can exit the host cell without killing it. What advantage does the virus gain by keeping the host cell alive?

Click through this tutorial (http://openstaxcollege.org/l/viruses2) on viruses to identify structures, modes of transmission, replication, and more.

Viruses and Disease

Viruses cause a variety of diseases in animals, including humans, ranging from the common cold to potentially fatal illnesses like meningitis (Figure 17.7). These diseases can be treated by antiviral drugs or by vaccines, but some viruses, such as HIV, are capable of avoiding the immune response and mutating so as to become resistant to antiviral drugs.

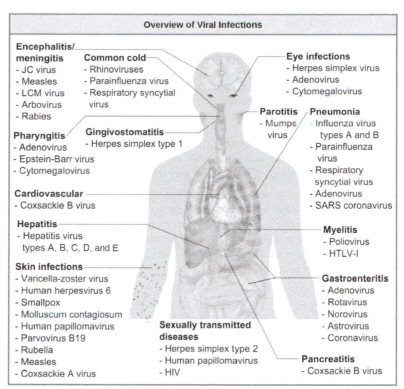

Figure 17.7 Viruses are the cause of dozens of ailments in humans, ranging from mild illnesses to serious diseases. (credit: modification of work by Mikael Häggström)

Vaccines for Prevention

While we do have limited numbers of effective antiviral drugs, such as those used to treat HIV and influenza, the primary method of controlling viral disease is by vaccination, which is intended to prevent outbreaks by building immunity to a virus or virus family. A **vaccine** may be prepared using weakened live viruses, killed viruses, or molecular subunits of the virus. In general, live viruses lead to better immunity, but have the possibility of causing disease at some low frequency. Killed viral vaccine and the subunit viruses are both incapable of causing disease, but in general lead to less effective or long-lasting immunity.

Weakened live viral vaccines are designed in the laboratory to cause few symptoms in recipients while giving them immunity against future infections. Polio was one disease that represented a milestone in the use of vaccines. Mass immunization campaigns in the U.S. in the 1950s (killed vaccine) and 1960s (live vaccine) essentially eradicated the disease, which caused muscle paralysis in children and generated fear in the general population when regional epidemics occurred. The success of the polio vaccine paved the way for the routine dispensation of childhood vaccines against measles, mumps, rubella, chickenpox, and other diseases.

Live vaccines are usually made by **attenuation** (weakening) of the "wild-type" (disease-causing) virus by growing it in the laboratory in tissues or at temperatures different from what the virus is accustomed to in the host. For example, the virus may be grown in cells in a test tube, in bird embryos, or in live animals. The adaptation to these new cells or temperature induces mutations in the virus' genomes, allowing them to grow better in the laboratory while inhibiting their ability to cause disease when reintroduced into the conditions found in the host. These attenuated viruses thus still cause an infection, but they do not grow very well, allowing the immune response to develop in time to prevent major disease. The danger of using live vaccines, which are usually more effective than killed vaccines, is the low but significant risk that these viruses will revert back to their disease-causing form by back mutations. Back mutations occur when the vaccine undergoes mutations in the host such that it readapts to the host and can again cause disease, which can then be spread to other humans in an epidemic. This happened as recently as 2007 in Nigeria where mutations in a polio vaccine led to an epidemic of polio in that country.

Some vaccines are in continuous development because certain viruses, such as influenza and HIV, have a high mutation rate compared to other viruses or host cells. With influenza, mutation in genes for the surface molecules helps the virus evade the protective immunity that may have been obtained in a previous influenza season, making it necessary for individuals to get vaccinated every year. Other viruses, such as those that cause the childhood diseases measles, mumps, and rubella, mutate so little that the same vaccine is used year after year.

Vaccines and Antiviral Drugs for Treatment

In some cases, vaccines can be used to treat an active viral infection. In the case of rabies, a fatal neurological disease transmitted in the saliva of rabies virus-infected animals, the progression of the disease from the time of the animal bite to the time it enters the central nervous system may be two weeks or longer. This is enough time to vaccinate an individual who suspects being bitten by a rabid animal, and the boosted immune response from the vaccination is enough to prevent the virus from entering nervous tissue. Thus, the fatal neurological consequences of the disease are averted and the individual only has to recover from the infected bite. This approach is also being used for the treatment of Ebola, one of the fastest and most deadly viruses affecting humans, though usually infecting limited populations. Ebola is also a leading cause of death in gorillas. Transmitted by bats and great apes, this virus can cause death in 70–90 percent of the infected within two weeks. Using newly developed vaccines that boost the immune response, there is hope that immune systems of affected individuals will be better able to control the virus, potentially reducing mortality rates.

Another way of treating viral infections is the use of antiviral drugs. These drugs often have limited ability to cure viral disease but have been used to control and reduce symptoms for a wide variety of viral diseases. For most viruses, these drugs inhibit the virus by blocking the actions of one or more of its proteins. It is important that the targeted proteins be encoded for by viral genes and that these molecules are not present in a healthy host cell. In this way, viral growth is inhibited without damaging the host. There are large numbers of antiviral drugs available to treat infections, some specific for a particular virus and others that can affect multiple viruses.

Antivirals have been developed to treat genital herpes (herpes simplex II) and influenza. For genital herpes, drugs such as acyclovir can reduce the number and duration of the episodes of active viral disease during which patients develop viral lesions in their skins cells. As the virus remains latent in nervous tissue of the body for life, this drug is not a cure but can make the symptoms of the disease more manageable. For influenza, drugs like Tamiflu can reduce the duration of "flu" symptoms by one or two days, but the drug does not prevent symptoms entirely. Other antiviral drugs, such as Ribavirin, have been used to treat a variety of viral infections.

By far the most successful use of antivirals has been in the treatment of the retrovirus HIV, which causes a disease that, if untreated, is usually fatal within 10–12 years after being infected. Anti-HIV drugs have been able to control viral replication to the point that individuals receiving these drugs survive for a significantly longer time than the untreated.

Anti-HIV drugs inhibit viral replication at many different phases of the HIV replicative cycle. Drugs have been developed that inhibit the fusion of the HIV viral envelope with the plasma membrane of the host cell (fusion inhibitors), the conversion of its RNA genome to double-stranded DNA (reverse transcriptase inhibitors), the integration of the viral DNA into the host genome (integrase inhibitors), and the processing of viral proteins (protease inhibitors).

When any of these drugs are used individually, the virus' high mutation rate allows the virus to rapidly evolve resistance to the drug. The breakthrough in the treatment of HIV was the development of highly active anti-retroviral therapy (HAART), which involves a mixture of different drugs, sometimes called a drug "cocktail." By attacking the virus at different stages of its replication cycle, it is difficult for the virus to develop resistance to multiple drugs at the same time. Still, even with the use of combination HAART therapy, there is concern that, over time, the virus will evolve resistance to this therapy. Thus, new anti-HIV drugs are constantly being developed with the hope of continuing the battle against this highly fatal virus.

17.2 | Innate Immunity

By the end of this section, you will be able to:

- Describe the body's innate physical and chemical defenses
- Explain the inflammatory response
- Describe the complement system

The vertebrate, including human, immune system is a complex multilayered system for defending against external and internal threats to the integrity of the body. The system can be divided into two types of defense systems: the innate immune system, which is nonspecific toward a particular kind of pathogen, and the adaptive immune system, which is specific (Figure 17.8). **Innate immunity** is not caused by an infection or vaccination and depends initially on physical and chemical barriers that work on all pathogens, sometimes called the first line of defense. The second line of defense of the innate system includes chemical signals that produce inflammation and fever responses as well as mobilizing protective cells and other chemical defenses. The adaptive immune system mounts a highly specific response to substances and organisms that do not belong in the body. The adaptive system takes longer to respond and has a memory system that allows it to respond with greater intensity should the body reencounter a pathogen even years later.

Vertebrate Immunity		
Innate Immune System		Adaptive Immune System
Physical Barriers	Internal Defenses	
• Skin, hair, cilia • Mucus membranes • Mucus and chemical secretions • Digestive enzymes in mouth • Stomach acid	• Inflammatory response • Complement proteins • Phagocytic cells • Natural killer (NK) cells	• Antibodies and the humoral immune response • Cell-mediated immune response • Memory response

Figure 17.8 There are two main parts to the vertebrate immune system. The innate immune system, which is made up of physical barriers and internal defenses, responds to all pathogens. The adaptive immune system is highly specific.

External and Chemical Barriers

The body has significant physical barriers to potential pathogens. The skin contains the protein keratin, which resists physical entry into cells. Other body surfaces, particularly those associated with body openings, are protected by the mucous membranes. The sticky mucus provides a physical trap for pathogens, preventing their movement deeper into the body. The openings of the body, such as the nose and ears, are protected by hairs that catch pathogens, and the mucous membranes of the upper respiratory tract have cilia that constantly move pathogens trapped in the mucus coat up to the mouth.

The skin and mucous membranes also create a chemical environment that is hostile to many microorganisms. The surface of the skin is acidic, which prevents bacterial growth. Saliva, mucus, and the tears of the eye contain an enzyme that breaks down bacterial cell walls. The stomach secretions create a highly acidic environment, which kills many pathogens entering the digestive system.

Finally, the surface of the body and the lower digestive system have a community of microorganisms such as bacteria, archaea, and fungi that coexist without harming the body. There is evidence that these organisms are highly beneficial to their host, combating disease-causing organisms and outcompeting them for nutritional resources provided by the host body. Despite these defenses, pathogens may enter the body through skin abrasions or punctures, or by collecting on mucosal surfaces in large numbers that overcome the protections of mucus or cilia.

Internal Defenses

When pathogens enter the body, the innate immune system responds with a variety of internal defenses. These include the inflammatory response, phagocytosis, natural killer cells, and the complement system. White blood cells in the blood and lymph recognize pathogens as foreign to the body. A **white blood cell** is larger than a red blood cell, is nucleated, and is typically able to move using amoeboid locomotion. Because they can move on their own, white blood cells can leave the blood to go to infected tissues. For example, a **monocyte** is a type of white blood cell that circulates in the blood and lymph and develops into a macrophage after it moves into infected tissue. A **macrophage** is a large cell that engulfs foreign particles and pathogens. **Mast cells** are produced in the same way as white blood cells, but unlike circulating white blood cells, mast cells take up residence in connective tissues and especially mucosal tissues. They are responsible for releasing chemicals in response to physical injury. They also play a role in the allergic response, which will be discussed later in the chapter.

When a pathogen is recognized as foreign, chemicals called cytokines are released. A **cytokine** is a chemical messenger that regulates cell differentiation (form and function), proliferation (production), and gene expression to produce a variety of immune responses. Approximately 40 types of cytokines exist in humans. In addition to being released from white blood cells after pathogen recognition, cytokines are also released by the infected cells and bind to nearby uninfected cells, inducing those cells to release cytokines. This positive feedback loop results in a burst of cytokine production.

One class of early-acting cytokines is the interferons, which are released by infected cells as a warning to nearby uninfected cells. An **interferon** is a small protein that signals a viral infection to other cells. The interferons stimulate uninfected cells to produce compounds that interfere with viral replication. Interferons also activate macrophages and other cells.

The Inflammatory Response and Phagocytosis

The first cytokines to be produced encourage **inflammation**, a localized redness, swelling, heat, and pain. Inflammation is a response to physical trauma, such as a cut or a blow, chemical irritation, and infection by pathogens (viruses, bacteria, or fungi). The chemical signals that trigger an inflammatory response enter the extracellular fluid and cause capillaries to dilate (expand) and capillary walls to become more permeable, or leaky. The serum and other compounds leaking from capillaries cause swelling of the area, which in turn causes pain. Various kinds of white blood cells are attracted to the area of inflammation. The types of white blood cells that arrive at an inflamed site depend on the nature of the injury or infecting pathogen. For example, a **neutrophil** is an early arriving white blood cell that engulfs and digests pathogens. Neutrophils are the most abundant white blood cells of the immune system (Figure 17.9). Macrophages follow neutrophils and take over the phagocytosis function and are involved in the resolution of an inflamed site, cleaning up cell debris and pathogens.

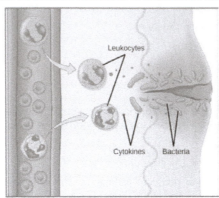

Figure 17.9 White blood cells (leukocytes) release chemicals to stimulate the inflammatory response following a cut in the skin.

Cytokines also send feedback to cells of the nervous system to bring about the overall symptoms of feeling sick, which include lethargy, muscle pain, and nausea. Cytokines also increase the core body temperature, causing a fever. The elevated temperatures of a fever inhibit the growth of pathogens and speed up cellular repair processes. For these reasons, suppression of fevers should be limited to those that are dangerously high.

Check out this 23-second, stop-motion video (http://openstaxcollege.org/l/neutrophil) showing a neutrophil that searches and engulfs fungus spores during an elapsed time of 79 minutes.

Natural Killer Cells

A **lymphocyte** is a white blood cell that contains a large nucleus (Figure 17.10). Most lymphocytes are associated with the adaptive immune response, but infected cells are identified and destroyed by natural killer cells, the only lymphocytes of the innate immune system. A **natural killer (NK) cell** is a lymphocyte that can kill cells infected with viruses (or cancerous cells). NK cells identify intracellular infections, especially from viruses, by the altered expression of **major histocompatibility class (MHC) I molecules** on the surface of infected cells. MHC class I molecules are proteins on the surfaces of all nucleated cells that provide a sample of the cell's internal environment at any given time. Unhealthy cells, whether infected or cancerous, display an altered MHC class I complement on their cell surfaces.

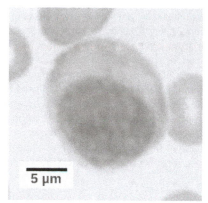

Figure 17.10 Lymphocytes, such as NK cells, are characterized by their large nuclei that actively absorb Wright stain and therefore appear dark colored under a microscope. (credit: scale-bar data from Matt Russell)

After the NK cell detects an infected or tumor cell, it induces programmed cell death, or apoptosis. Phagocytic cells then come along and digest the cell debris left behind. NK cells are constantly patrolling the body and are an effective mechanism for controlling potential infections and preventing cancer progression. The various types of immune cells are shown in Figure 17.11.

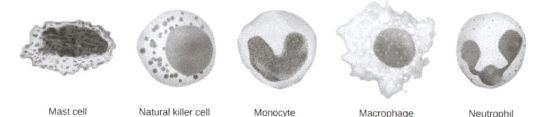

| Mast cell | Natural killer cell | Monocyte | Macrophage | Neutrophil |

Figure 17.11 Cells involved in the innate immune response include mast cells, natural killer cells, and white blood cells, such as monocytes, macrophages and neutrophils.

Complement

An array of approximately 20 types of proteins, called a **complement system**, is also activated by infection or the activity of the cells of the adaptive immune system and functions to destroy extracellular pathogens. Liver cells and macrophages synthesize inactive forms of complement proteins continuously; these proteins are abundant in the blood serum and are capable of responding immediately to infecting microorganisms. The complement system is so named because it is complementary to the innate and adaptive immune system. Complement proteins bind to the surfaces of microorganisms and are particularly attracted to pathogens that are already tagged by the adaptive immune system. This "tagging" involves the attachment of specific proteins called antibodies (discussed in detail later) to the pathogen. When they attach, the antibodies change shape providing a binding site for one of the complement proteins. After the first few complement proteins bind, a cascade of binding in a specific sequence of proteins follows in which the pathogen rapidly becomes coated in complement proteins.

Complement proteins perform several functions, one of which is to serve as a marker to indicate the presence of a pathogen to phagocytic cells and enhance engulfment. Certain complement proteins can combine to open pores in microbial cell membranes and cause lysis of the cells.

17.3 | Adaptive Immunity

By the end of this section, you will be able to:

- Explain adaptive immunity

- Describe cell-mediated immune response and humoral immune response

- Describe immune tolerance

The adaptive, or acquired, immune response takes days or even weeks to become established—much longer than the innate response; however, adaptive immunity is more specific to an invading pathogen. **Adaptive immunity** is an immunity that occurs after exposure to an antigen either from a pathogen or a vaccination. An **antigen** is a molecule that stimulates a response in the immune system. This part of the immune system is activated when the innate immune response is insufficient to control an infection. In fact, without information from the innate immune system, the adaptive response could not be mobilized. There are two types of adaptive responses: the **cell-mediated immune response**, which is controlled by activated **T cells**, and the **humoral immune response**, which is controlled by activated **B cells** and antibodies. Activated T and B cells whose surface binding sites are specific to the molecules on the pathogen greatly increase in numbers and attack the invading pathogen. Their attack can kill pathogens directly or they can secrete antibodies that enhance the phagocytosis of pathogens and disrupt the infection. Adaptive immunity also involves a memory to give the host long-term protection from reinfection with the same type of pathogen; on reexposure, this host memory will facilitate a rapid and powerful response.

B and T Cells

Lymphocytes, which are white blood cells, are formed with other blood cells in the red bone marrow found in many flat bones, such as the shoulder or pelvic bones. The two types of lymphocytes of the adaptive immune response are B and T cells (Figure 17.12). Whether an immature lymphocyte becomes a B cell or T cell depends on where in the body it matures. The B cells remain in the bone marrow to mature (hence the name "B" for "bone marrow"), while T cells migrate to the thymus, where they mature (hence the name "T" for "thymus").

Maturation of a B or T cell involves becoming immunocompetent, meaning that it can recognize, by binding, a specific molecule or antigen (discussed below). During the maturation process, B and T cells that bind too strongly to the body's own cells are eliminated in order to minimize an immune response against the body's own tissues. Those cells that react weakly to the body's own cells, but have highly specific receptors on their cell surfaces that allow them to recognize a foreign molecule, or antigen, remain. This process occurs during fetal development and continues throughout life. The specificity of this receptor is determined by the genetics of the individual and is present before a foreign molecule is introduced to the body or encountered. Thus, it is genetics and not experience that initially provides a vast array of cells, each capable of binding to a different specific foreign molecule. Once they are immunocompetent, the T and B cells will migrate to the spleen and lymph nodes where they will remain until they are called on during an infection. B cells are involved in the humoral immune response, which targets pathogens loose in blood and lymph, and T cells are involved in the cell-mediated immune response, which targets infected cells.

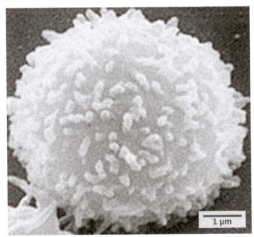

Figure 17.12 This scanning electron micrograph shows a T lymphocyte. T and B cells are indistinguishable by light microscopy but can be differentiated experimentally by probing their surface receptors. (credit: modification of work by NCI; scale-bar data from Matt Russell)

Humoral Immune Response

As mentioned, an antigen is a molecule that stimulates a response in the immune system. Not every molecule is antigenic. B cells participate in a chemical response to antigens present in the body by producing specific antibodies that circulate throughout the body and bind with the antigen whenever it is encountered. This is known as the humoral immune response. As discussed, during maturation of B cells, a set of highly specific B cells are produced that have many antigen receptor molecules in their membrane (Figure 17.13).

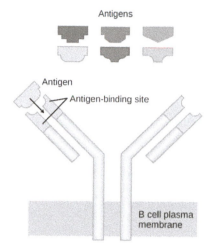

Figure 17.13 B cell receptors are embedded in the membranes of B cells and bind a variety of antigens through their variable regions.

Each B cell has only one kind of antigen receptor, which makes every B cell different. Once the B cells mature in the bone marrow, they migrate to lymph nodes or other lymphatic organs. When a B cell encounters the antigen that binds to its receptor, the antigen molecule is brought into the cell by endocytosis and reappears on the surface of the cell bound to an **MHC class II molecule**. When this process is complete, the B cell is sensitized. In most cases, the sensitized B cell must then encounter a specific kind of T cell, called a helper T cell, before it is activated. The helper T cell must already have been activated through an encounter with the antigen (discussed below).

The helper T cell binds to the antigen-MHC class II complex and is induced to release cytokines that induce the B cell to divide rapidly, which makes thousands of identical (clonal) cells. These daughter cells become either plasma cells or memory B cells. The memory B cells remain inactive at this point, until another later encounter with the antigen, caused by a reinfection by the same bacteria or virus, results in them dividing into a new population of plasma cells. The plasma cells, on the other hand, produce and secrete large quantities, up to 100 million molecules per hour, of antibody molecules. An **antibody**, also known as an immunoglobulin (Ig), is a protein that is produced by plasma cells after stimulation by an antigen. Antibodies are the agents of humoral immunity. Antibodies occur in the blood, in gastric and mucus secretions, and in breast milk. Antibodies in these bodily fluids can bind pathogens and mark them for destruction by phagocytes before they can infect cells.

These antibodies circulate in the blood stream and lymphatic system and bind with the antigen whenever it is encountered. The binding can fight infection in several ways. Antibodies can bind to viruses or bacteria and interfere with the chemical interactions required for them to infect or bind to other cells. The antibodies may create bridges between different particles containing antigenic sites clumping them all together and preventing their proper functioning. The antigen-antibody complex stimulates the complement system described previously, destroying the cell bearing the antigen. Phagocytic cells, such as those already described, are attracted by the antigen-antibody complexes, and phagocytosis is enhanced when the complexes are present. Finally, antibodies stimulate inflammation, and their presence in mucus and on the skin prevents pathogen attack.

Antibodies coat extracellular pathogens and neutralize them by blocking key sites on the pathogen that enhance their infectivity (such as receptors that "dock" pathogens on host cells) (Figure 17.14). Antibody neutralization can prevent pathogens from entering and infecting host cells. The neutralized antibody-coated pathogens can then be filtered by the spleen and eliminated in urine or feces.

Antibodies also mark pathogens for destruction by phagocytic cells, such as macrophages or neutrophils, in a process called opsonization. In a process called complement fixation, some antibodies provide a place for complement proteins to bind. The combination of antibodies and complement promotes rapid clearing of pathogens.

The production of antibodies by plasma cells in response to an antigen is called **active immunity** and describes the host's active response of the immune system to an infection or to a vaccination. There is also a **passive immune** response where antibodies come from an outside source, instead of the individual's own plasma cells, and are introduced into the host. For example, antibodies circulating in a pregnant woman's body move across the placenta into the developing fetus. The child benefits from the presence of these antibodies for up to several months after birth. In addition, a passive immune response is possible by injecting antibodies into an individual in the form of an antivenom to a snake-bite toxin or antibodies in blood serum to help fight a hepatitis infection. This gives immediate protection since the body does not need the time required to mount its own response.

(a) Neutralization Antibodies prevent a virus or toxic protein from binding their target.

Antibody

Virus

Diptheria toxin

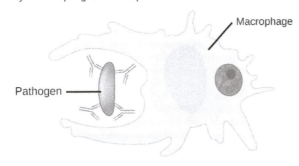

(b) Opsonization A pathogen tagged by antibodies is consumed by a macrophage or neutrophil.

Macrophage

Pathogen

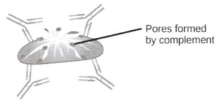

(c) Complement activation Antibodies attached to the surface of a pathogen cell activate the complement system.

Pores formed by complement

Figure 17.14 Antibodies may inhibit infection by (a) preventing the antigen from binding its target, (b) tagging a pathogen for destruction by macrophages or neutrophils, or (c) activating the complement cascade.

Cell-Mediated Immunity

Unlike B cells, T lymphocytes are unable to recognize pathogens without assistance. Instead, dendritic cells and macrophages first engulf and digest pathogens into hundreds or thousands of antigens. Then, an **antigen-presenting cell (APC)** detects, engulfs, and informs the adaptive immune response about an infection. When a pathogen is detected, these APCs will engulf and break it down through phagocytosis. Antigen fragments will then be transported to the surface of the APC, where they will serve as an indicator to other immune cells. A **dendritic cell** is an immune cell that mops up antigenic materials in its surroundings and presents them on its surface. Dendritic cells are located in the skin, the linings of the nose, lungs, stomach, and intestines. These positions are ideal locations to encounter invading pathogens. Once they are activated by pathogens and mature to become APCs they migrate to the spleen or a lymph node. Macrophages also function as APCs. After phagocytosis by a macrophage, the phagocytic vesicle fuses with an intracellular lysosome. Within the resulting phagolysosome, the components are broken down into fragments; the fragments are then loaded onto MHC class II molecules and are transported to the cell surface for antigen presentation (Figure 17.15). Helper T cells cannot properly respond to an antigen unless it is processed and embedded in an MHC class II molecule. The APCs express MHC class II on their surfaces, and when combined with a foreign antigen, these complexes signal an invader.

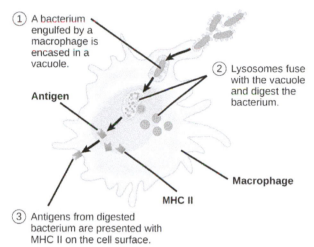

① A bacterium engulfed by a macrophage is encased in a vacuole.

② Lysosomes fuse with the vacuole and digest the bacterium.

Antigen

Macrophage

MHC II

③ Antigens from digested bacterium are presented with MHC II on the cell surface.

Figure 17.15 An antigen-presenting cell (APC), such as a macrophage, engulfs a foreign antigen, partially digests it in a lysosome, and then embeds it in an MHC class II molecule for presentation at the cell surface. Lymphocytes of the adaptive immune response must interact with antigen-embedded MHC class II molecules to mature into functional immune cells.

View this animation from Rockefeller University (http://openstaxcollege.org/l/immune_system2) to see how dendritic cells act as sentinels in the body's immune system.

T cells have many functions. Some respond to APCs of the innate immune system and indirectly induce immune responses by releasing cytokines. Others stimulate B cells to start the humoral response as described previously. Another type of T cell detects APC signals and directly kills the infected cells, while some are involved in suppressing inappropriate immune reactions to harmless or "self" antigens.

There are two main types of T cells: helper T lymphocytes (T_H) and the cytotoxic T lymphocytes (T_C). The T_H lymphocytes function indirectly to tell other immune cells about potential pathogens. T_H lymphocytes recognize specific antigens presented by the MHC class II complexes of APCs. There are two populations of T_H cells: T_H1 and T_H2. T_H1 cells secrete cytokines to enhance the activities of macrophages and other T cells. T_H2 cells stimulate naïve B cells to secrete antibodies. Whether a T_H1 or a T_H2 immune response develops depends on the specific types of cytokines secreted by cells of the innate immune system, which in turn depends on the nature of the invading pathogen.

Cytotoxic T cells (T_C) are the key component of the cell-mediated part of the adaptive immune system and attack and destroy infected cells. T_C cells are particularly important in protecting against viral infections; this is because viruses replicate within cells where they are shielded from extracellular contact with circulating antibodies. Once activated, the T_C creates a large clone of cells with one specific set of cell-surface receptors, as in the case with proliferation of activated B cells. As with B cells, the clone includes active T_C cells and inactive memory T_C cells. The resulting active T_C cells then identify infected host cells. Because of the time required to generate a population of clonal T and B cells, there is a delay in the adaptive immune response compared to the innate immune response.

T_C cells attempt to identify and destroy infected cells before the pathogen can replicate and escape, thereby halting the progression of intracellular infections. T_C cells also support NK lymphocytes to destroy early cancers. Cytokines secreted by the T_H1 response that stimulates macrophages also stimulate T_C cells and enhance their ability to identify and destroy infected cells and tumors. A summary of how the humoral and cell-mediated immune responses are activated appears in Figure 17.16.

B plasma cells and T$_C$ cells are collectively called **effector cells** because they are involved in "effecting" (bringing about) the immune response of killing pathogens and infected host cells.

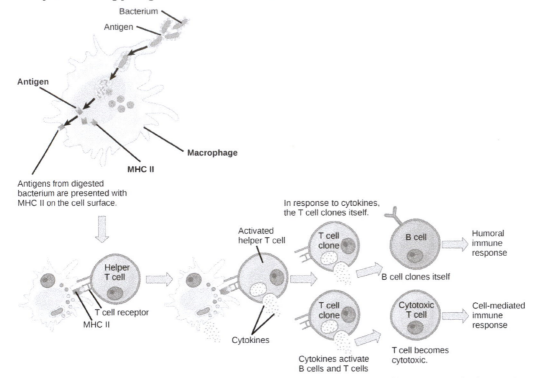

Figure 17.16 A helper T cell becomes activated by binding to an antigen presented by an APC via the MHCII receptor, causing it to release cytokines. Depending on the cytokines released, this activates either the humoral or the cell-mediated immune response.

Immunological Memory

The adaptive immune system has a memory component that allows for a rapid and large response upon reinvasion of the same pathogen. During the adaptive immune response to a pathogen that has not been encountered before, known as the **primary immune response**, plasma cells secreting antibodies and differentiated T cells increase, then plateau over time. As B and T cells mature into effector cells, a subset of the naïve populations differentiates into B and T memory cells with the same antigen specificities (Figure 17.17). A **memory cell** is an antigen-specific B or T lymphocyte that does not differentiate into an effector cell during the primary immune response, but that can immediately become an effector cell on reexposure to the same pathogen. As the infection is cleared and pathogenic stimuli subside, the effectors are no longer needed and they undergo apoptosis. In contrast, the memory cells persist in the circulation.

a r t CONNECTION

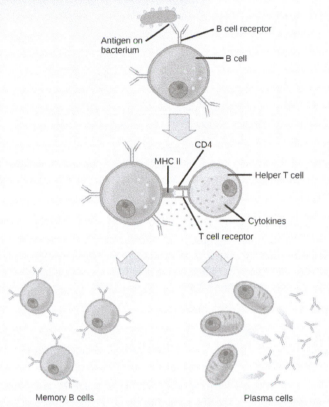

Figure 17.17 After initially binding an antigen to the B cell receptor, a B cell internalizes the antigen and presents it on MHC class II. A helper T cell recognizes the MHC class II- antigen complex and activates the B cell. As a result, memory B cells and plasma cells are made.

The Rh antigen is found on Rh-positive red blood cells. An Rh-negative female can usually carry an Rh-positive fetus to term without difficulty. However, if she has a second Rh-positive fetus, her body may launch an immune attack that causes hemolytic disease of the newborn. Why do you think hemolytic disease is only a problem during the second or subsequent pregnancies?

If the pathogen is never encountered again during the individual's lifetime, B and T memory cells will circulate for a few years or even several decades and will gradually die off, having never functioned as effector cells. However, if the host is re-exposed to the same pathogen type, circulating memory cells will immediately differentiate into plasma cells and T$_C$ cells without input from APCs or T$_H$ cells. This is known as the **secondary immune response**. One reason why the adaptive immune response is delayed is because it takes time for naïve B and T cells with the appropriate antigen specificities to be identified, activated, and proliferate. On reinfection, this step is skipped, and the result is a more rapid production of immune defenses. Memory B cells that differentiate into plasma cells output tens to hundreds-fold greater antibody amounts than were secreted during the primary response (Figure 17.18). This rapid and dramatic antibody response may stop the infection before it can even become established, and the individual may not realize they had been exposed.

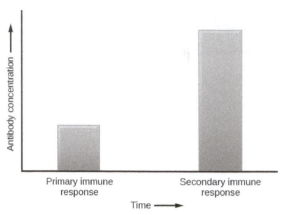

Figure 17.18 In the primary response to infection, antibodies are secreted first from plasma cells. Upon re-exposure to the same pathogen, memory cells differentiate into antibody-secreting plasma cells that output a greater amount of antibody for a longer period of time.

Vaccination is based on the knowledge that exposure to noninfectious antigens, derived from known pathogens, generates a mild primary immune response. The immune response to vaccination may not be perceived by the host as illness but still confers immune memory. When exposed to the corresponding pathogen to which an individual was vaccinated, the reaction is similar to a secondary exposure. Because each reinfection generates more memory cells and increased resistance to the pathogen, some vaccine courses involve one or more booster vaccinations to mimic repeat exposures.

The Lymphatic System

Lymph is the watery fluid that bathes tissues and organs and contains protective white blood cells but does not contain erythrocytes. Lymph moves about the body through the lymphatic system, which is made up of vessels, lymph ducts, lymph glands, and organs, such as tonsils, adenoids, thymus, and spleen.

Although the immune system is characterized by circulating cells throughout the body, the regulation, maturation, and intercommunication of immune factors occur at specific sites. The blood circulates immune cells, proteins, and other factors through the body. Approximately 0.1 percent of all cells in the blood are leukocytes, which include monocytes (the precursor of macrophages) and lymphocytes. Most cells in the blood are red blood cells. Cells of the immune system can travel between the distinct lymphatic and blood circulatory systems, which are separated by interstitial space, by a process called extravasation (passing through to surrounding tissue).

Recall that cells of the immune system originate from stem cells in the bone marrow. B cell maturation occurs in the bone marrow, whereas progenitor cells migrate from the bone marrow and develop and mature into naïve T cells in the organ called the thymus.

On maturation, T and B lymphocytes circulate to various destinations. Lymph nodes scattered throughout the body house large populations of T and B cells, dendritic cells, and macrophages (Figure 17.19). Lymph gathers antigens as it drains from tissues. These antigens then are filtered through lymph nodes before the lymph is returned to circulation. APCs in the lymph nodes capture and process antigens and inform nearby lymphocytes about potential pathogens.

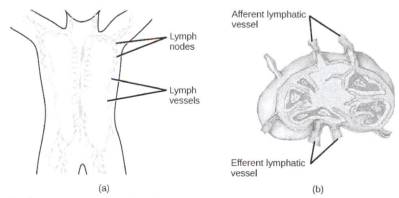

Figure 17.19 (a) Lymphatic vessels carry a clear fluid called lymph throughout the body. The liquid passes through (b) lymph nodes that filter the lymph that enters the node through afferent vessels and leaves through efferent vessels; lymph nodes are filled with lymphocytes that purge infecting cells. (credit a: modification of work by NIH; credit b: modification of work by NCI, NIH)

The spleen houses B and T cells, macrophages, dendritic cells, and NK cells (Figure 17.20). The spleen is the site where APCs that have trapped foreign particles in the blood can communicate with lymphocytes. Antibodies are synthesized and secreted by activated plasma cells in the spleen, and the spleen filters foreign substances and antibody-complexed pathogens from the blood. Functionally, the spleen is to the blood as lymph nodes are to the lymph.

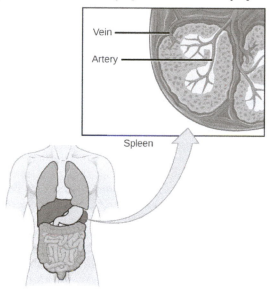

Figure 17.20 The spleen functions to immunologically filter the blood and allow for communication between cells corresponding to the innate and adaptive immune responses. (credit: modification of work by NCI, NIH)

Mucosal Immune System

The innate and adaptive immune responses compose the systemic immune system (affecting the whole body), which is distinct from the mucosal immune system. Mucosa associated lymphoid tissue (MALT) is a crucial component of a functional immune system because mucosal surfaces, such as the nasal passages, are the first tissues onto which inhaled or ingested pathogens are deposited. The mucosal tissue includes the mouth, pharynx, and esophagus, and the gastrointestinal, respiratory, and urogenital tracts.

Mucosal immunity is formed by MALT, which functions independently of the systemic immune system, and which has its own innate and adaptive components. MALT is a collection of lymphatic tissue that combines with epithelial tissue lining the mucosa throughout the body. This tissue functions as the immune barrier and response in areas of the body with direct contact to the external environment. The systemic and mucosal immune systems use many of the same cell types. Foreign particles that make their way to MALT are taken up by absorptive epithelial cells and delivered to APCs located directly below the mucosal tissue. APCs of the mucosal immune system are primarily dendritic cells, with B cells and macrophages having minor roles. Processed antigens displayed on APCs are detected by T cells in the MALT and at the tonsils, adenoids, appendix, or the mesenteric lymph nodes of the intestine. Activated T cells then migrate through the lymphatic system and into the circulatory system to mucosal sites of infection.

Immune Tolerance

The immune system has to be regulated to prevent wasteful, unnecessary responses to harmless substances, and more importantly, so that it does not attack "self." The acquired ability to prevent an unnecessary or harmful immune response to a detected foreign substance known not to cause disease, or self-antigens, is described as **immune tolerance**. The primary mechanism for developing immune tolerance to self-antigens occurs during the selection for weakly self-binding cells during T and B lymphocyte maturation. There are populations of T cells that suppress the immune response to self-antigens and that suppress the immune response after the infection has cleared to minimize host cell damage induced by inflammation and cell lysis. Immune tolerance is especially well developed in the mucosa of the upper digestive system because of the tremendous number of foreign substances (such as food proteins) that APCs of the oral cavity, pharynx, and gastrointestinal mucosa encounter. Immune tolerance is brought about by specialized APCs in the liver, lymph nodes, small intestine, and lung that present harmless antigens to a diverse population of regulatory T (T_{reg}) cells, specialized lymphocytes that suppress local inflammation and inhibit the secretion of stimulatory immune factors. The combined result of T_{reg} cells is to prevent immunologic activation and inflammation in undesired tissue compartments and to allow the immune system to focus on pathogens instead.

17.4 | Disruptions in the Immune System

By the end of this section, you will be able to:

- Describe hypersensitivity
- Define autoimmunity

A functioning immune system is essential for survival, but even the sophisticated cellular and molecular defenses of the mammalian immune response can be defeated by pathogens at virtually every step. In the competition between immune protection and pathogen evasion, pathogens have the advantage of more rapid evolution because of their shorter generation time, large population sizes and often higher mutation rates. Thus pathogens have evolved a diverse array of immune escape mechanisms. For instance, *Streptococcus pneumoniae* (the bacterium that causes pneumonia and meningitis) surrounds itself with a capsule that inhibits phagocytes from engulfing it and displaying antigens to the adaptive immune system. *Staphylococcus aureus* (the bacterium that can cause skin infections, abscesses, and meningitis) synthesizes a toxin called leukocidin that kills phagocytes after they engulf the bacterium. Other pathogens can also hinder the adaptive immune system. HIV infects T_H cells using their CD4 surface molecules, gradually depleting the number of T_H cells in the body (Figure 17.21); this inhibits the adaptive immune system's capacity to generate sufficient responses to infection or tumors. As a result, HIV-infected individuals often suffer from infections that would not cause illness in people with healthy immune systems but which can cause devastating illness to immune-compromised individuals.

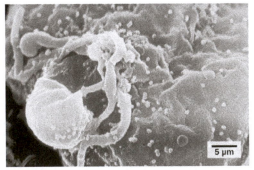

Figure 17.21 HIV (green) is shown budding from a lymphocyte cell (red) in culture. (credit: modification of work by C. Goldsmith, CDC; scale-bar data from Matt Russell)

Inappropriate responses of immune cells and molecules themselves can also disrupt the proper functioning of the entire system, leading to host-cell damage that can become fatal.

Immunodeficiency

Immunodeficiency is a failure, insufficiency, or delay in the response of the immune system, which may be acquired or inherited. Immunodeficiency can allow pathogens or tumor cells to gain a foothold and replicate or proliferate to high enough levels so that the immune system becomes overwhelmed. Immunodeficiency can be acquired as a result of infection with certain pathogens that attack the cells of the immune system itself (such as HIV), chemical exposure (including certain medical treatments such as chemotherapy), malnutrition, or extreme stress. For instance, radiation exposure can destroy populations of lymphocytes and elevate an individual's susceptibility to infections and cancer. Rarely, primary immunodeficiencies that are present from birth may also occur. For example, severe combined immunodeficiency disease (SCID) is a condition in which children are born without functioning B or T cells.

Hypersensitivities

A maladaptive immune response toward harmless foreign substances or self-antigens that occur after tissue sensitization is termed a **hypersensitivity**. Types of hypersensitivities include immediate, delayed, and autoimmune. A large proportion of the human population is affected by one or more types of hypersensitivity.

Allergies

The immune reaction that results from immediate hypersensitivities in which an antibody-mediated immune response occurs within minutes of exposure to a usually harmless antigen is called an **allergy**. In the United States, 20 percent of the population exhibits symptoms of allergy or asthma, whereas 55 percent test positive against one or more allergens. On initial exposure to a potential allergen, an allergic individual synthesizes antibodies through the typical process of APCs presenting processed antigen to T_H cells that stimulate B cells to produce the antibodies. The antibody molecules interact with mast cells embedded in connective tissues. This process primes, or sensitizes, the tissue. On subsequent exposure to the same allergen, antibody molecules on mast cells bind the antigen and stimulate the mast cell to release histamine and other inflammatory chemicals; these chemical mediators then recruit eosinophils (a type of white blood cell), which also appear to be adapted to responding to parasitic worms (Figure 17.22). Eosinophils release factors that enhance the inflammatory response and the secretions of mast cells. The effects of an allergic reaction range from mild symptoms like sneezing and itchy, watery eyes to more severe or even life-threatening reactions involving intensely itchy welts or hives, airway constriction with severe respiratory distress, and plummeting blood pressure caused by dilating blood vessels and fluid loss from the circulatory system. This extreme reaction, typically in response to an allergen introduced to the circulatory system, is known as anaphylactic shock. Antihistamines are an insufficient counter to anaphylactic shock and if not treated with epinephrine to counter the blood pressure and breathing effects, this condition can be fatal.

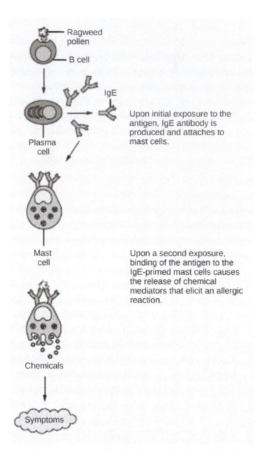

Figure 17.22 On first exposure to an allergen, an antibody is synthesized by plasma cells in response to a harmless antigen. The antibodies bind to mast cells, and on secondary exposure, the mast cells release histamines and other modulators that cause the symptoms of allergy. (credit: modification of work by NIH)

Delayed hypersensitivity is a cell-mediated immune response that takes approximately one to two days after secondary exposure for a maximal reaction. This type of hypersensitivity involves the T_H1 cytokine-mediated inflammatory response and may cause local tissue lesions or contact dermatitis (rash or skin irritation). Delayed hypersensitivity occurs in some individuals in response to contact with certain types of jewelry or cosmetics. Delayed hypersensitivity facilitates the immune response to poison ivy and is also the reason why the skin test for tuberculosis results in a small region of inflammation on individuals who were previously exposed to *Mycobacterium tuberculosis*, the organism that causes tuberculosis.

Try your hand at diagnosing an allergic reaction by selecting one of the interactive case studies (http://openstaxcollege.org/l/world_allergy) at the World Allergy Organization website.

Autoimmunity

Autoimmunity is a type of hypersensitivity to self-antigens that affects approximately five percent of the population. Most types of autoimmunity involve the humoral immune response. An antibody that inappropriately marks self-components as foreign is termed an **autoantibody**. In patients with myasthenia gravis, an autoimmune disease, muscle-cell receptors that induce contraction in response to acetylcholine are targeted by antibodies. The result is muscle weakness that may include marked difficultly with fine or gross motor functions. In systemic lupus erythematosus, a diffuse autoantibody response to the individual's own DNA and proteins results in various systemic diseases (Figure 17.23). Systemic lupus erythematosus may affect the heart, joints, lungs, skin, kidneys, central nervous system, or other tissues, causing tissue damage through antibody binding, complement recruitment, lysis, and inflammation.

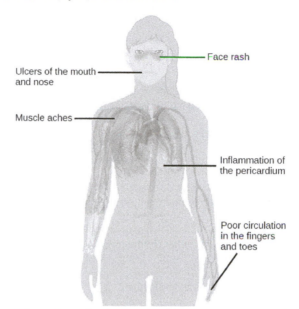

Figure 17.23 Systemic lupus erythematosus is characterized by autoimmunity to the individual's own DNA and/or proteins, which leads to varied dysfunction of the organs. (credit: modification of work by Mikael Häggström)

Autoimmunity can develop with time and its causes may be rooted in molecular mimicry, a situation in which one molecule is similar enough in shape to another molecule that it binds the same immune receptors. Antibodies and T-cell receptors may bind self-antigens that are structurally similar to pathogen antigens. As an example, infection with *Streptococcus pyogenes* (the bacterium that causes strep throat) may generate antibodies or T cells that react with heart muscle, which has a similar structure to the surface of *S. pyogenes*. These antibodies can damage heart muscle with autoimmune attacks, leading to rheumatic fever. Insulin-dependent (Type 1) diabetes mellitus arises from a destructive inflammatory T_H1 response against insulin-producing cells of the pancreas. Patients with this autoimmunity must be treated with regular insulin injections.

KEY TERMS

acellular lacking cells

active immunity an immunity that occurs as a result of the activity of the body's own cells rather than from antibodies acquired from an external source

adaptive immunity a specific immune response that occurs after exposure to an antigen either from a pathogen or a vaccination

allergy an immune reaction that results from immediate hypersensitivities in which an antibody-mediated immune response occurs within minutes of exposure to a harmless antigen

antibody a protein that is produced by plasma cells after stimulation by an antigen; also known as an immunoglobulin

antigen a macromolecule that reacts with cells of the immune system and which may or may not have a stimulatory effect

antigen-presenting cell (APC) an immune cell that detects, engulfs, and informs the adaptive immune response about an infection by presenting the processed antigen on its cell surface

apoptosis the cell death caused by induction of a cell's own internal mechanisms either as a natural step in the development of a multicellular organism or by other environmental factors such as signals from cells of the immune system

attenuation the weakening of a virus during vaccine development

autoantibody an antibody that incorrectly marks "self" components as foreign and stimulates the immune response

autoimmunity a type of hypersensitivity to self-antigens

B cell a lymphocyte that matures in the bone marrow

capsid the protein coating of the viral core

cell-mediated immune response an adaptive immune response that is controlled by T cells

complement system an array of approximately 20 soluble proteins of the innate immune system that enhance phagocytosis, bore holes in pathogens, and recruit lymphocytes

cytokine a chemical messenger that regulates cell differentiation, proliferation, and gene expression to effect immune responses

cytopathic causing cell damage

cytotoxic T lymphocyte (T$_C$) an adaptive immune cell that directly kills infected cells via enzymes, and that releases cytokines to enhance the immune response

dendritic cell an immune cell that processes antigen material and presents it on the surface of its cell in MHC class II molecules and induces an immune response in other cells

effector cell a lymphocyte that has differentiated, such as a B cell, plasma cell, or cytotoxic T cell

glycoprotein a protein molecule with attached carbohydrate molecules

helper T lymphocyte (T$_H$) a cell of the adaptive immune system that binds APCs via MHC class II molecules and stimulates B cells or secretes cytokines to initiate the immune response

humoral immune response the adaptive immune response that is controlled by activated B cells and antibodies

hypersensitivity a spectrum of inappropriate immune responses toward harmless foreign particles or self-antigens; occurs after tissue sensitization and includes immediate-type (allergy), delayed-type, and autoimmunity

immune tolerance an acquired ability to prevent an unnecessary or harmful immune response to a detected foreign body known not to cause disease

immunodeficiency a failure, insufficiency, or delay at any level of the immune system, which may be acquired or inherited

inflammation the localized redness, swelling, heat, and pain that results from the movement of leukocytes through opened capillaries to a site of infection

innate immunity an immunity that occurs naturally because of genetic factors or physiology, and is not caused by infection or vaccination

interferon a cytokine that inhibits viral replication

lymph the watery fluid present in the lymphatic circulatory system that bathes tissues and organs with protective white blood cells and does not contain erythrocytes

lymphocyte a type of white blood cell that includes natural killer cells of the innate immune system and B and T cells of the adaptive immune system

macrophage a large phagocytic cell that engulfs foreign particles and pathogens

major histocompatibility class (MHC) I a group of proteins found on the surface of all nucleated cells that signals to immune cells whether the cell is normal or is infected or cancerous; it also provides the appropriate sites into which antigens can be loaded for recognition by lymphocytes

major histocompatibility class (MHC) II molecule a protein found on the surface of antigen-presenting cells that signals to immune cells whether the cell is normal or is infected or cancerous; it provides the appropriate template into which antigens can be loaded for recognition by lymphocytes

mast cell a leukocyte that produces inflammatory molecules, such as histamine, in response to large pathogens

memory cell an antigen-specific B or T lymphocyte that does not differentiate into an effector cell during the primary immune response but that can immediately become an effector cell on reexposure to the same pathogen

monocyte a type of white blood cell that circulates in the blood and lymph and differentiates into a macrophage after it moves into infected tissue

natural killer (NK) cell a lymphocyte that can kill cells infected with viruses or tumor cells

neutrophil a phagocytic leukocyte that engulfs and digests pathogens

passive immunity an immunity that does not result from the activity of the body's own immune cells but by transfer of antibodies from one individual to another

primary immune response the response of the adaptive immune system to the first exposure to an antigen

secondary immune response the response of the adaptive immune system to a second or later exposure to an antigen mediated by memory cells

T cell a lymphocyte that matures in the thymus gland

vaccine a weakened solution of virus components, viruses, or other agents that produce an immune response

viral envelope a lipid bilayer that envelops some viruses

virion an individual virus particle outside a host cell

white blood cell a nucleated cell found in the blood that is a part of the immune system; also called leukocytes

CHAPTER SUMMARY

17.1 Viruses

Viruses are acellular entities that can usually only be seen with an electron microscope. Their genomes contain either DNA or RNA, and they replicate using the replication proteins of a host cell. Viruses are diverse, infecting archaea, bacteria, fungi, plants, and animals. Viruses consist of a nucleic-acid core surrounded by a protein capsid with or without an outer lipid envelope.

Viral replication within a living cell always produces changes in the cell, sometimes resulting in cell death and sometimes slowly killing the infected cells. There are six basic stages in the virus replication cycle: attachment, penetration, uncoating, replication, assembly, and release. A viral infection may be productive, resulting in new virions, or nonproductive, meaning the virus remains inside the cell without producing new virions.

Viruses cause a variety of diseases in humans. Many of these diseases can be prevented by the use of viral vaccines, which stimulate protective immunity against the virus without causing major disease. Viral vaccines may also be used in active viral infections, boosting the ability of the immune system to control or destroy the virus. Antiviral drugs that target enzymes and other protein products of viral genes have been developed and used with mixed success. Combinations of anti-HIV drugs have been used to effectively control the virus, extending the lifespan of infected individuals.

17.2 Innate Immunity

The innate immune system consists first of physical and chemical barriers to infection including the skin and mucous membranes and their secretions, ciliated surfaces, and body hairs. The second line of defense is an internal defense system designed to counter pathogenic threats that bypass the physical and chemical barriers of the body. Using a combination of cellular and molecular responses, the innate immune system identifies the nature of a pathogen and responds with inflammation, phagocytosis, cytokine release, destruction by NK cells, or the complement system.

17.3 Adaptive Immunity

The adaptive immune response is a slower-acting, longer-lasting, and more specific response than the innate response. However, the adaptive response requires information from the innate immune system to function. APCs display antigens on MHC molecules to naïve T cells. T cells with cell-surface receptors that bind a specific antigen will bind to that APC. In response, the T cells differentiate and proliferate, becoming T_H cells or T_C cells. T_H cells stimulate B cells that have engulfed and presented pathogen-derived antigens. B cells differentiate into plasma cells that secrete antibodies, whereas T_C cells destroy infected or cancerous cells. Memory cells are produced by activated and proliferating B and T cells and persist after a primary exposure to a pathogen. If re-exposure occurs, memory cells differentiate into effector cells without input from the innate immune system. The mucosal immune system is largely independent of the systemic immune system but functions in parallel to protect the extensive mucosal surfaces of the body. Immune tolerance is brought about by T_{reg} cells to limit reactions to harmless antigens and the body's own molecules.

17.4 Disruptions in the Immune System

Immune disruptions may involve insufficient immune responses or inappropriate immune responses. Immunodeficiency increases an individual's susceptibility to infections and cancers. Hypersensitivities are misdirected responses either to harmless foreign particles, as in the case of allergies, or to the individual's own tissues, as in the case of autoimmunity. Reactions to self-components may be the result of molecular mimicry.

ART CONNECTION QUESTIONS

1. Figure 17.5 Which of the following statements about virus structure is true?
   a. All viruses are encased in a viral membrane.
   b. The capsomere is made up of small protein subunits called capsids.
   c. DNA is the genetic material in all viruses.
   d. Glycoproteins help the virus attach to the host cell.

2. Figure 17.6 Influenza virus is packaged in a viral envelope, which fuses with the plasma membrane. This way, the virus can exit the host cell without killing it. What advantage does the virus gain by keeping the host cell alive?

3. Figure 17.17 The Rh antigen is found on Rh-positive red blood cells. An Rh-negative female can usually carry an Rh-positive fetus to term without difficulty. However, if she has a second Rh-positive fetus, her body may launch an immune attack that causes hemolytic disease of the newborn. Why do you think hemolytic disease is only a problem during the second or subsequent pregnancies?

REVIEW QUESTIONS

4. Which statement is true?
  a. A virion contains DNA and RNA.
  b. Viruses are acellular.
  c. Viruses replicate outside of the cell.
  d. Most viruses are easily visualized with a light microscope.

5. The viral _____ plays a role in attaching a virion to the host cell.
  a. core
  b. capsid
  c. envelope
  d. both b and c

6. Which statement is true of viral replication?
  a. In the process of apoptosis, the cell survives.
  b. During attachment, the virus attaches at specific sites on the cell surface.
  c. The viral capsid helps the host cell produce more copies of the viral genome.
  d. mRNA works outside of the host cell to produce enzymes and proteins.

7. Which of the following is a barrier against pathogens provided by the skin?
  a. low pH
  b. mucus
  c. tears
  d. cilia

8. Although interferons have several effects, they are particularly useful against infections with which type of pathogen?
  a. bacteria
  b. viruses
  c. fungi
  d. helminths

9. Which innate immune system component uses MHC class I molecules directly in its defense strategy?

  a. macrophages
  b. neutrophils
  c. NK cells

  d. interferon

10. The humoral immune response depends on which cells?
  a. T_C cells
  b. B cells
  c. B and T_H cells
  d. T_C and T_H cells

11. The fact that the body does not normally mount an immune response to the molecules in food is an example of _____.
  a. secondary immune response
  b. immunological memory
  c. immune tolerance
  d. passive immunity

12. Foreign particles circulating in the blood are filtered by the _____.
  a. spleen
  b. lymph nodes
  c. MALT
  d. lymph

13. Allergy to pollen is classified as _____.
  a. an autoimmune reaction
  b. immunodeficiency
  c. delayed hypersensitivity
  d. immediate hypersensitivity

14. A potential cause of acquired autoimmunity is _____.
  a. tissue hypersensitivity
  b. molecular mimicry
  c. histamine release
  d. radiation exposure

15. Autoantibodies are probably involved in _____.

  a. reactions to poison ivy
  b. pollen allergies
  c. systemic lupus erythematosus
  d. HIV/AIDS

CRITICAL THINKING QUESTIONS

16. Why can't dogs catch the measles?

17. Why is immunization after being bitten by a rabid animal so effective?

18. Different MHC class I molecules between donor and recipient cells can lead to rejection of a transplanted organ or tissue. Suggest a reason for this.

19. If a series of genetic mutations prevented some, but not all, of the complement proteins from binding antibodies or pathogens, would the entire complement system be compromised?

20. How do B and T cells differ with respect to antigens that they bind?

21. Why is the immune response after reinfection much faster than the adaptive immune response after the initial infection?

22. Some photographers develop a sensitivity to certain film developing chemicals leading to severe rashes on their hands such that they are unable to work with them. Explain what is probably happening.

18 | ANIMAL REPRODUCTION AND DEVELOPMENT

Figure 18.1 Female seahorses produce eggs that are then fertilized by the male. Unlike with almost all other animals, the young then develop in a pouch of the male seahorse until birth. (credit: "cliff1066"/Flickr)

Chapter Outline
18.1: How Animals Reproduce
18.2: Development and Organogenesis
18.3: Human Reproduction

Introduction

In the animal kingdom, each species has its unique adaptations for reproduction. Asexual reproduction produces genetically identical offspring (clones), whereas in sexual reproduction, the genetic material of two individuals combines to produce offspring that are genetically different from their parents. During sexual reproduction the male gamete (sperm) may be placed inside the female's body for internal fertilization, the sperm may be left in the environment for the female to pick up and place in her body, or both sperm and eggs may be released into the environment for external fertilization. Seahorses provide an example of the latter, but with a twist (Figure 18.1). Following a mating dance, the female releases eggs into the male seahorse's abdominal brood pouch and the male releases sperm into the water, which then find their way into the brood pouch to fertilize the eggs. The fertilized eggs develop in the pouch for several weeks.

18.1 | How Animals Reproduce

By the end of this section, you will be able to:

- Describe advantages and disadvantages of asexual and sexual reproduction
- Discuss asexual reproduction methods
- Discuss sexual reproduction methods
- Discuss internal and external methods of fertilization

Some animals produce offspring through asexual reproduction while other animals produce offspring through sexual reproduction. Both methods have advantages and disadvantages. **Asexual reproduction** produces offspring that are genetically identical to the parent because the offspring are all clones of the original parent. A single individual can produce offspring asexually and large numbers of offspring can be produced quickly; these are two advantages that asexually reproducing organisms have over sexually reproducing organisms. In a stable or predictable environment, asexual reproduction is an effective means of reproduction because all the offspring will be adapted to that environment. In an unstable or unpredictable environment, species that reproduce asexually may be at a disadvantage because all the offspring are genetically identical and may not be adapted to different conditions.

During **sexual reproduction**, the genetic material of two individuals is combined to produce genetically diverse offspring that differ from their parents. The genetic diversity of sexually produced offspring is thought to give sexually reproducing individuals greater fitness because more of their offspring may survive and reproduce in an unpredictable or changing environment. Species that reproduce sexually (and have separate sexes) must maintain two different types of individuals, males and females. Only half the population (females) can produce the offspring, so fewer offspring will be produced when compared to asexual reproduction. This is a disadvantage of sexual reproduction compared to asexual reproduction.

Asexual Reproduction

Asexual reproduction occurs in prokaryotic microorganisms (bacteria and archaea) and in many eukaryotic, single-celled and multi-celled organisms. There are several ways that animals reproduce asexually, the details of which vary among individual species.

Fission

Fission, also called binary fission, occurs in some invertebrate, multi-celled organisms. It is in some ways analogous to the process of binary fission of single-celled prokaryotic organisms. The term fission is applied to instances in which an organism appears to split itself into two parts and, if necessary, regenerate the missing parts of each new organism. For example, species of turbellarian flatworms commonly called the planarians, such as *Dugesia dorotocephala*, are able to separate their bodies into head and tail regions and then regenerate the missing half in each of the two new organisms. Sea anemones (Cnidaria), such as species of the genus *Anthopleura* (Figure 18.2), will divide along the oral-aboral axis, and sea cucumbers (Echinodermata) of the genus *Holothuria*, will divide into two halves across the oral-aboral axis and regenerate the other half in each of the resulting individuals.

Figure 18.2 The *Anthopleura artemisia* sea anemone can reproduce through fission.

Budding

Budding is a form of asexual reproduction that results from the outgrowth of a part of the body leading to a separation of the "bud" from the original organism and the formation of two individuals, one smaller than the other. Budding occurs commonly in some invertebrate animals such as hydras and corals. In hydras, a bud forms that develops into an adult and breaks away from the main body (Figure 18.3).

(a)   (b)

Figure 18.3 (a) Hydra reproduce asexually through budding: a bud forms on the tubular body of an adult hydra, develops a mouth and tentacles, and then detaches from its parent. The new hydra is fully developed and will find its own location for attachment. (b) Some coral, such as the *Lophelia pertusa* shown here, can reproduce through budding. (credit b: modification of work by Ed Bowlby, NOAA/Olympic Coast NMS; NOAA/OAR/Office of Ocean Exploration)

View this video (http://openstaxcollege.org/l/budding_hydra2) to see a hydra budding.

Fragmentation

Fragmentation is the breaking of an individual into parts followed by regeneration. If the animal is capable of fragmentation, and the parts are big enough, a separate individual will regrow from each part. Fragmentation may occur through accidental damage, damage from predators, or as a natural form of reproduction. Reproduction through fragmentation is observed in sponges, some cnidarians, turbellarians, echinoderms, and annelids. In some sea stars, a new individual can be regenerated from a broken arm and a piece of the central disc. This sea star (Figure 18.4) is in the process of growing a complete sea star from an arm that has been cut off. Fisheries workers have been known to try to kill the sea stars eating their clam or oyster beds by cutting them in half and throwing them back into the ocean. Unfortunately for the workers, the two parts can each regenerate a new half, resulting in twice as many sea stars to prey upon the oysters and clams.

(a)   (b)

Figure 18.4 (a) *Linckia multifora* is a species of sea star that can reproduce asexually via fragmentation. In this process, (b) an arm that has been shed grows into a new sea star. (credit a: modification of work by Dwayne Meadows, NOAA/NMFS/OPR)

Parthenogenesis

Parthenogenesis is a form of asexual reproduction in which an egg develops into an individual without being fertilized. The resulting offspring can be either haploid or diploid, depending on the process in the species. Parthenogenesis occurs in invertebrates such as water fleas, rotifers, aphids, stick insects, and ants, wasps, and bees. Ants, bees, and wasps use parthenogenesis to produce haploid males (drones). The diploid females (workers and queens) are the result of a fertilized egg.

Some vertebrate animals—such as certain reptiles, amphibians, and fish—also reproduce through parthenogenesis. Parthenogenesis has been observed in species in which the sexes were separated in terrestrial or marine zoos. Two female Komodo dragons, a hammerhead shark, and a blacktop shark have produced parthenogenic young when the females have been isolated from males. It is possible that the asexual reproduction observed occurred in response to unusual circumstances and would normally not occur.

Sexual Reproduction

Sexual reproduction is the combination of reproductive cells from two individuals to form genetically unique offspring. The nature of the individuals that produce the two kinds of gametes can vary, having for example separate sexes or both sexes in each individual. Sex determination, the mechanism that determines which sex an individual develops into, also can vary.

Hermaphroditism

Hermaphroditism occurs in animals in which one individual has both male and female reproductive systems. Invertebrates such as earthworms, slugs, tapeworms, and snails (Figure 18.5) are often hermaphroditic. Hermaphrodites may self-fertilize, but typically they will mate with another of their species, fertilizing each other and both producing offspring. Self-fertilization is more common in animals that have limited mobility or are not motile, such as barnacles and clams. Many species have specific mechanisms in place to prevent self-fertilization, because it is an extreme form of inbreeding and usually produces less fit offspring.

Figure 18.5 Many (a) snails are hermaphrodites. When two individuals (b) mate, they can produce up to 100 eggs each. (credit a: modification of work by Assaf Shtilman; credit b: modification of work by "Schristia"/Flickr)

Sex Determination

Mammalian sex is determined genetically by the combination of X and Y chromosomes. Individuals homozygous for X (XX) are female and heterozygous individuals (XY) are male. In mammals, the presence of a Y chromosome causes the development of male characteristics and its absence results in female characteristics. The XY system is also found in some insects and plants.

Bird **sex determination** is dependent on the combination of Z and W chromosomes. Homozygous for Z (ZZ) results in a male and heterozygous (ZW) results in a female. Notice that this system is the opposite of the mammalian system because in birds the female is the sex with the different sex chromosomes. The W appears to be essential in determining the sex of the individual, similar to the Y chromosome in mammals. Some fish, crustaceans, insects (such as butterflies and moths), and reptiles use the ZW system.

More complicated chromosomal sex determining systems also exist. For example, some swordtail fish have three sex chromosomes in a population.

The sex of some other species is not determined by chromosomes, but by some aspect of the environment. Sex determination in alligators, some turtles, and tuataras, for example, is dependent on the temperature during the middle third of egg development. This is referred to as environmental sex determination, or more specifically, as temperature-dependent sex determination. In many turtles, cooler temperatures during egg incubation produce males and warm temperatures produce females, while in many other species of turtles, the reverse is true. In some crocodiles and some turtles, moderate temperatures produce males and both warm and cool temperatures produce females.

Individuals of some species change their sex during their lives, switching from one to the other. If the individual is female first, it is termed protogyny or "first female," if it is male first, it is termed protandry or "first male." Oysters are born male, grow in size, and become female and lay eggs. The wrasses, a family of reef fishes, are all sequential hermaphrodites. Some of these species live in closely coordinated schools with a dominant male and a large number of smaller females. If the male dies, a female increases in size, changes sex, and becomes the new dominant male.

Fertilization

The fusion of a sperm and an egg is a process called fertilization. This can occur either inside (**internal fertilization**) or outside (**external fertilization**) the body of the female. Humans provide an example of the former, whereas frog reproduction is an example of the latter.

External Fertilization

External fertilization usually occurs in aquatic environments where both eggs and sperm are released into the water. After the sperm reaches the egg, fertilization takes place. Most external fertilization happens during the process of spawning where one or several females release their eggs and the male(s) release sperm in the same area, at the same time. The spawning may be triggered by environmental signals, such as water temperature or the length of daylight. Nearly all fish spawn, as do crustaceans (such as crabs and shrimp), mollusks (such as oysters), squid, and echinoderms (such as sea urchins and sea cucumbers). Frogs, corals, molluscs, and sea cucumbers also spawn (Figure 18.6).

Figure 18.6 During sexual reproduction in toads, the male grasps the female from behind and externally fertilizes the eggs as they are deposited. (credit: Bernie Kohl)

Internal Fertilization

Internal fertilization occurs most often in terrestrial animals, although some aquatic animals also use this method. Internal fertilization may occur by the male directly depositing sperm in the female during mating. It may also occur by the male depositing sperm in the environment, usually in a protective structure, which a female picks up to deposit the sperm in her reproductive tract. There are three ways that offspring are produced following internal fertilization. In **oviparity**, fertilized eggs are laid outside the female's body and develop there, receiving nourishment from the yolk that is a part of the egg (Figure 18.7**a**). This occurs in some bony fish, some reptiles, a few cartilaginous fish, some amphibians, a few mammals, and all birds. Most non-avian reptiles and insects produce leathery eggs, while birds and some turtles produce eggs with high concentrations of calcium carbonate in the shell, making them hard. Chicken eggs are an example of a hard shell. The eggs of the egg-laying mammals such as the platypus and echidna are leathery.

In **ovoviparity**, fertilized eggs are retained in the female, and the embryo obtains its nourishment from the egg's yolk. The eggs are retained in the female's body until they hatch inside of her, or she lays the eggs right before they hatch. This process helps protect the eggs until hatching. This occurs in some bony fish (like the platyfish *Xiphophorus maculatus*, Figure 18.7**b**), some sharks, lizards, some snakes (garter snake *Thamnophis sirtalis*), some vipers, and some invertebrate animals (Madagascar hissing cockroach *Gromphadorhina portentosa*).

In **viviparity** the young are born alive. They obtain their nourishment from the female and are born in varying states of maturity. This occurs in most mammals (Figure 18.7**c**), some cartilaginous fish, and a few reptiles.

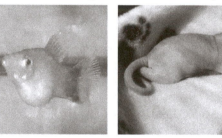

(a)   (b)   (c)

Figure 18.7 In (a) oviparity, young develop in eggs outside the female body, as with these *Harmonia axydridis* beetles hatching. Some aquatic animals, like this (b) pregnant *Xiphophorus maculatus* are ovoviparous, with the egg developing inside the female and nutrition supplied primarily from the yolk. In mammals, nutrition is supported by the placenta, as was the case with this (c) newborn squirrel. (credit b: modification of work by Gourami Watcher; credit c: modification of work by "audreyjm529"/Flickr)

18.2 | Development and Organogenesis

By the end of this section, you will be able to:

- Explain how the embryo forms from the zygote
- Discuss the role of cleavage and gastrulation in animal development
- Describe organogenesis

The process by which an organism develops from a single-celled zygote to a multi-cellular organism is complex and well regulated. The regulation occurs through signaling between cells and tissues and responses in the form of differential gene expression.

Early Embryonic Development

Fertilization is the process in which gametes (an egg and sperm) fuse to form a zygote (Figure 18.8). To ensure that the offspring has only one complete diploid set of chromosomes, only one sperm must fuse with one egg. In mammals, a layer called the **zona pellucida** protects the egg. At the tip of the head of a sperm cell is a structure like a lysosome called the acrosome, which contains enzymes. When a sperm binds to the zona pellucida, a series of events, called the acrosomal reactions, take place. These reactions, involving enzymes from the acrosome, allow the sperm plasma membrane to fuse with the egg plasma membrane and permit the sperm nucleus to transfer into the ovum. The nuclear membranes of the egg and sperm break down and the two haploid nuclei fuse to form a diploid nucleus or genome.

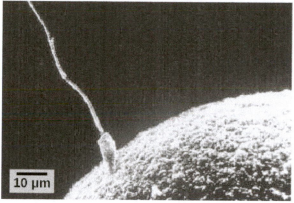

Figure 18.8 Fertilization is the process in which sperm and egg fuse to form a zygote. (credit: scale-bar data from Matt Russell)

To ensure that no more than one sperm fertilizes the egg, once the acrosomal reactions take place at one location of the egg membrane, the egg releases proteins in other locations to prevent other sperm from fusing with the egg.

The development of multi-cellular organisms begins from this single-celled zygote, which undergoes rapid cell division, called cleavage (Figure 18.9**a**), to form a hollow ball of cells called a blastula (Figure 18.9**b**).

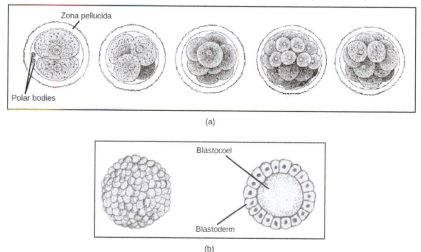

(a)

(b)

Figure 18.9 (a) During cleavage, the zygote rapidly divides into multiple cells. (b) The cells rearrange themselves to form a hollow ball called the blastula. (credit a: modification of work by Gray's Anatomy; credit b: modification of work by Pearson Scott Foresman; donated to the Wikimedia Foundation)

In mammals, the blastula forms the **blastocyst** in the next stage of development. Here the cells in the blastula arrange themselves in two layers: the **inner cell mass**, and an outer layer called the **trophoblast**. The inner cell mass will go on to form the embryo. The trophoblast secretes enzymes that allow implantation of the blastocyst into the endometrium of the uterus. The trophoblast will contribute to the placenta and nourish the embryo.

Visit the Virtual Human Embryo project (http://openstaxcollege.org/l/human_embryo2) at the Endowment for Human Development site to click through an interactive of the stages of embryo development, including micrographs and rotating 3-D images.

The cells in the blastula then rearrange themselves spatially to form three layers of cells. This process is called **gastrulation**. During gastrulation, the blastula folds in on itself and cells migrate to form the three layers of cells (Figure 18.10) in a structure, the gastrula, with a hollow space that will become the digestive tract. Each of the layers of cells is called a germ layer and will differentiate into different organ systems.

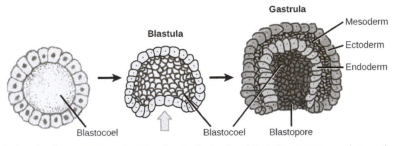

Figure 18.10 Gastrulation is the process wherein the cells in the blastula rearrange themselves to form the germ layers. (credit: modification of work by Abigail Pyne)

The three germ layers are the endoderm, the ectoderm, and the mesoderm. Cells in each germ layer differentiate into tissues and embryonic organs. The ectoderm gives rise to the nervous system and the epidermis, among other tissues. The mesoderm gives rise to the muscle cells and connective tissue in the body. The endoderm gives rise to the gut and many internal organs.

Organogenesis

Gastrulation leads to the formation of the three germ layers that give rise during further development to the different organs in the animal body. This process is called **organogenesis**.

Organs develop from the germ layers through the process of differentiation. During differentiation, the embryonic stem cells express specific sets of genes that will determine their ultimate cell type. For example, some cells in the ectoderm will express the genes specific to skin cells. As a result, these cells will take on the shape and characteristics of epidermal cells. The process of differentiation is regulated by location-specific chemical signals from the cell's embryonic environment that sets in play a cascade of events that regulates gene expression.

18.3 | Human Reproduction

By the end of this section, you will be able to:

- Describe human male and female reproductive anatomies
- Describe spermatogenesis and oogenesis and discuss their differences and similarities
- Describe the role of hormones in human reproduction
- Describe the roles of male and female reproductive hormones

As in all animals, the adaptations for reproduction in humans are complex. They involve specialized and different anatomies in the two sexes, a hormone regulation system, and specialized behaviors regulated by the brain and endocrine system.

Download for free at https://openstax.org/details/books/concepts-biology

Human Reproductive Anatomy

The reproductive tissues of male and female humans develop similarly *in utero* until about the seventh week of gestation when a low level of the hormone testosterone is released from the gonads of the developing male. Testosterone causes the primitive gonads to differentiate into male sexual organs. When testosterone is absent, the primitive gonads develop into ovaries. Tissues that produce a penis in males produce a clitoris in females. The tissue that will become the scrotum in a male becomes the labia in a female. Thus the male and female anatomies arise from a divergence in the development of what were once common embryonic structures.

Male Reproductive Anatomy

Sperm are immobile at body temperature; therefore, the testes are external to the body so that a correct temperature is maintained for motility. In land mammals, including humans, the pair of testes must be suspended outside the body so the environment of the sperm is about 2 °C lower than body temperature to produce viable sperm. If the testes do not descend through the abdominal cavity during fetal development, the individual has reduced fertility.

The **scrotum** houses the testicles or **testes** (singular: testis), and provides passage for blood vessels, nerves, and muscles related to testicular function. The testes are a pair of male gonads that produce sperm and reproductive hormones. Each testis is approximately 2.5 by 3.8 cm (1.5 by 1 inch) in size and divided into wedge-shaped lobes by septa. Coiled in each wedge are seminiferous tubules that produce sperm.

The **penis** drains urine from the urinary bladder and is a copulatory organ during intercourse (Figure 18.12; Table 18.1). The penis contains three tubes of erectile tissue that become engorged with blood, making the penis erect, in preparation for intercourse. The organ is inserted into the vagina culminating with an ejaculation. During orgasm, the accessory organs and glands connected to the testes contract and empty the semen (containing sperm) into the urethra and the fluid is expelled from the body by muscular contractions causing ejaculation. After intercourse, the blood drains from the erectile tissue and the penis becomes flaccid.

Semen is a mixture of sperm (about five percent of the total) and fluids from accessory glands that contribute most of the semen's volume. Sperm are haploid cells, consisting of a flagellum for motility, a neck that contains the cell's energy-producing mitochondria, and a head that contains the genetic material (Figure 18.11). An acrosome (acrosomal vesicle) is found at the top of the head of the sperm. This structure contains enzymes that can digest the protective coverings that surround the egg and allow the sperm to fuse with the egg. An ejaculate will contain from two to five milliliters of fluid and from 50–120 million sperm per milliliter.

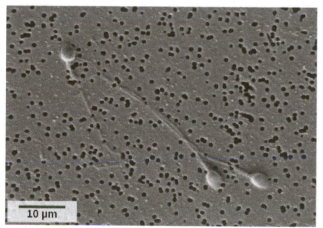

Figure 18.11 As seen in this scanning electron micrograph, human sperm has a flagellum, neck, and head. (credit: scale-bar data from Matt Russell)

Sperm form in the walls of **seminiferous tubules** that are coiled inside the testes (Figure 18.12; Table 18.1). The walls of the seminiferous tubules are made up of the developing sperm cells, with the least developed sperm at the periphery of the tubule and the fully developed sperm next to the lumen. The sperm cells are associated with **Sertoli cells** that nourish and promote the development of the sperm. Other cells present between the walls of the tubules are the **interstitial cells of Leydig**, which produce testosterone once the male reaches adolescence.

When the sperm have developed flagella they leave the seminiferous tubules and enter the epididymis (Figure 18.12; Table 18.1). This structure lies along the top and posterior of the testes and is the site of sperm maturation. The sperm leave the epididymis and enter the vas deferens, which carries the sperm behind the bladder, and forms the ejaculatory duct with the duct from the seminal vesicles. During a vasectomy, a section of the vas deferens is removed, preventing sperm (but not the secretions of the accessory glands) from being passed out of the body during ejaculation and preventing fertilization.

The bulk of the semen comes from the accessory glands associated with the male reproductive system. These are the **seminal vesicles**, the **prostate gland**, and the **bulbourethral gland** (Figure 18.12; Table 18.1). The secretions from the accessory glands provide important compounds for the sperm including nutrients, electrolytes, and pH buffering. There are also coagulation factors that affect sperm delivery and motility.

art CONNECTION

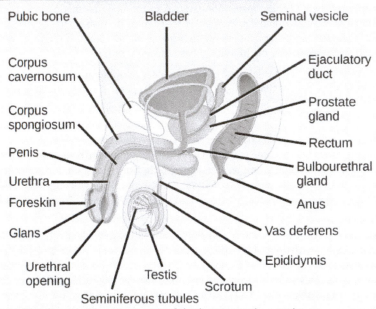

Figure 18.12 The reproductive structures of the human male are shown.

Which of the following statements about the male reproductive system is false?

a. The vas deferens carries sperm from the testes to the seminal vesicles.

b. The ejaculatory duct joins the urethra.

c. Both the prostate and the bulbourethral glands produce components of the semen.

d. The prostate gland is located in the testes.

Male Reproductive Anatomy

Organ	Location	Function
Scrotum	External	Supports testes and regulates their temperature
Penis	External	Delivers urine, copulating organ
Testes	Internal	Produce sperm and male hormones
Seminal Vesicles	Internal	Contribute to semen production
Prostate Gland	Internal	Contributes to semen production
Bulbourethral Glands	Internal	Neutralize urine in urethra

Table 18.1

Female Reproductive Anatomy

A number of female reproductive structures are exterior to the body. These include the breasts and the vulva, which consists of the mons pubis, **clitoris**, **labia majora**, **labia minora**, and the vestibular glands (Figure 18.13; Table 18.2).

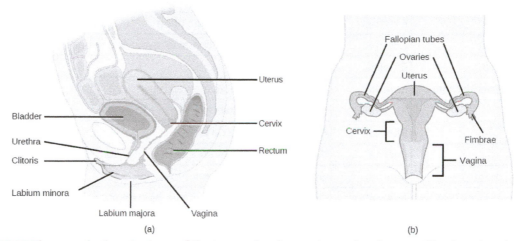

Figure 18.13 The reproductive structures of the human female are shown. (credit a: modification of work by Gray's Anatomy; credit b: modification of work by CDC)

The breasts consist of mammary glands and fat. Each gland consists of 15 to 25 lobes that have ducts that empty at the nipple and that supply the nursing child with nutrient- and antibody-rich milk to aid development and protect the child.

Internal female reproductive structures include ovaries, oviducts, the uterus, and the vagina (Figure 18.13; Table 18.2). The pair of ovaries is held in place in the abdominal cavity by a system of ligaments. The outermost layer of the ovary is made up of follicles, each consisting of one or more follicular cells that surround, nourish, and protect a single egg. During the menstrual period, a batch of follicular cells develops and prepares their eggs for release. At ovulation, one follicle ruptures and one egg is released. Following ovulation, the follicular tissue that surrounded the ovulated egg stays within the ovary and grows to form a solid mass called the **corpus luteum**. The corpus luteum secretes additional estrogen and the hormone progesterone that helps maintain the uterine lining during pregnancy. The ovaries also produce hormones, such as estrogen.

The **oviducts**, or fallopian tubes, extend from the uterus in the lower abdominal cavity to the ovaries, but they are not in contact with the ovaries. The lateral ends of the oviducts flare out into a trumpet-like structure and have a fringe of finger-like projections called fimbrae. When an egg is released at ovulation, the fimbrae help the nonmotile egg enter into the tube. The walls of the oviducts have a ciliated epithelium over smooth muscle. The cilia beat, and the smooth muscle contracts, moving the egg toward the uterus. Fertilization usually takes place within the oviduct and the developing embryo is moved toward the uterus. It usually takes the egg or embryo a week to travel through the oviduct.

Sterilization in women is called a tubal ligation; it is analogous to a vasectomy in males in that the oviducts are severed and sealed, preventing sperm from reaching the egg.

The **uterus** is a structure about the size of a woman's fist. The uterus has a thick muscular wall and is lined with an endometrium rich in blood vessels and mucus glands that develop and thicken during the female cycle. Thickening of the endometrium prepares the uterus to receive the fertilized egg or zygote, which will then implant itself in the endometrium. The uterus supports the developing embryo and fetus during gestation. Contractions of the smooth muscle in the uterus aid in forcing the baby through the vagina during labor. If fertilization does not occur, a portion of the lining of the uterus sloughs off during each menstrual period. The endometrium builds up again in preparation for implantation. Part of the uterus, called the cervix, protrudes into the top of the vagina.

The **vagina** is a muscular tube that serves several purposes. It allows menstrual flow to leave the body. It is the receptacle for the penis during intercourse and the pathway for the delivery of offspring.

Female Reproductive Anatomy

Organ	Location	Function
Clitoris	External	Sensory organ
Mons pubis	External	Fatty area overlying pubic bone
Labia majora	External	Covers labia minora; contains sweat and sebaceous glands
Labia minora	External	Covers vestibule
Greater vestibular glands	External	Secrete mucus; lubricate vagina
Breast	External	Produces and delivers milk
Ovaries	Internal	Produce and develop eggs
Oviducts	Internal	Transport egg to uterus; site of fertilization
Uterus	Internal	Supports developing embryo
Vagina	Internal	Common tube for intercourse, birth canal, passing menstrual flow

Table 18.2

Gametogenesis (Spermatogenesis and Oogenesis)

Gametogenesis, the production of sperm and eggs, involves the process of meiosis. During meiosis, two nuclear divisions separate the paired chromosomes in the nucleus and then separate the chromatids that were made during an earlier stage of the cell's life cycle. Meiosis and its associated cell divisions produces haploid cells with half of each pair of chromosomes normally found in diploid cells. The production of sperm is called **spermatogenesis** and the production of eggs is called **oogenesis**.

Spermatogenesis

Spermatogenesis occurs in the wall of the seminiferous tubules, with the most primitive cells at the periphery of the tube and the most mature sperm at the lumen of the tube (Figure 18.14). Immediately under the capsule of the tubule are diploid, undifferentiated cells. These stem cells, each called a spermatogonium (pl. spermatogonia), go through mitosis to produce one cell that remains as a stem cell and a second cell called a primary spermatocyte that will undergo meiosis to produce sperm.

The diploid primary spermatocyte goes through meiosis I to produce two haploid cells called secondary spermatocytes. Each secondary spermatocyte divides after meiosis II to produce two cells called spermatids. The spermatids eventually reach the lumen of the tubule and grow a flagellum, becoming sperm cells. Four sperm result from each primary spermatocyte that goes through meiosis.

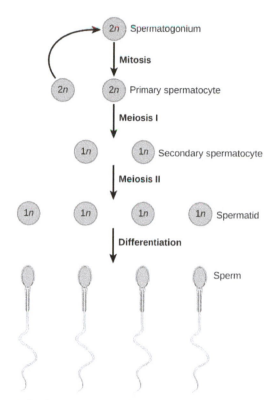

Figure 18.14 During spermatogenesis, four sperm result from each primary spermatocyte. The process also maps onto the physical structure of the wall of the seminiferous tubule, with the spermatogonia on the outer side of the tubule, and the sperm with their developing tails extended into the lumen of the tubule.

Visit this site (http://openstaxcollege.org/l/spermatogenes2) to see the process of spermatogenesis.

Oogenesis

Oogenesis occurs in the outermost layers of the ovaries. As with sperm production, oogenesis starts with a germ cell. In oogenesis, this germ cell is called an oogonium and forms during the embryological development of the individual. The oogonium undergoes mitosis to produce about one to two million oocytes by the time of birth.

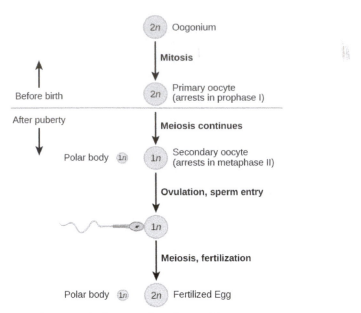

Figure 18.15 The process of oogenesis occurs in the ovary's outermost layer.

The primary oocytes begin meiosis before birth (Figure 18.15). However, the meiotic division is arrested in its progress in the first prophase stage. At the time of birth, all future eggs are in prophase I. This situation is in contrast with the male reproductive system in which sperm are produced continuously throughout the life of the individual. Starting at adolescence, anterior pituitary hormones cause the development of a few follicles in an ovary each month. This results in a primary oocyte finishing the first meiotic division. The cell divides unequally, with most of the cytoplasm and organelles going to one cell, called a secondary oocyte, and only one set of chromosomes and a small amount of cytoplasm going to the other cell. This second cell is called a polar body and usually dies. Cell division is again arrested, this time at metaphase II. At ovulation, this secondary oocyte is released and travels toward the uterus through the oviduct. If the secondary oocyte is fertilized, the cell continues through meiosis II, producing a second polar body and haploid egg, which fuses with the haploid sperm to form a fertilized egg (zygote) containing all 46 chromosomes.

Hormonal Control of Reproduction

The human male and female reproductive cycles are controlled by the interaction of hormones from the hypothalamus and anterior pituitary with hormones from reproductive tissues and organs. In both sexes, the hypothalamus monitors and causes the release of hormones from the anterior pituitary gland. When the reproductive hormone is required, the hypothalamus sends a **gonadotropin-releasing hormone (GnRH)** to the anterior pituitary. This causes the release of **follicle stimulating hormone (FSH)** and **luteinizing hormone (LH)** from the anterior pituitary into the blood. Although these hormones are named after their functions in female reproduction, they are produced in both sexes and play important roles in controlling reproduction. Other hormones have specific functions in the male and female reproductive systems.

Male Hormones

At the onset of puberty, the hypothalamus causes the release of FSH and LH into the male system for the first time. FSH enters the testes and stimulates the Sertoli cells located in the walls of the seminiferous tubules to begin promoting spermatogenesis (Figure 18.16). LH also enters the testes and stimulates the interstitial cells of Leydig, located in between the walls of the seminiferous tubules, to make and release testosterone into the testes and the blood.

Testosterone stimulates spermatogenesis. This hormone is also responsible for the secondary sexual characteristics that develop in the male during adolescence. The secondary sex characteristics in males include a deepening of the voice, the growth of facial, axillary, and pubic hair, an increase in muscle bulk, and the beginnings of the sex drive.

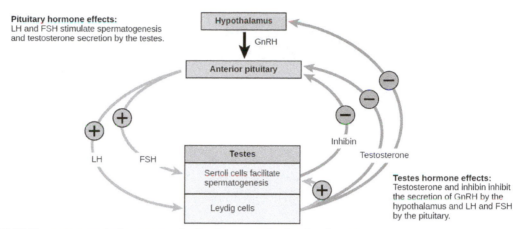

Figure 18.16 Hormones control sperm production in a negative feedback system.

A negative feedback system occurs in the male with rising levels of testosterone acting on the hypothalamus and anterior pituitary to inhibit the release of GnRH, FSH, and LH. In addition, the Sertoli cells produce the hormone **inhibin**, which is released into the blood when the sperm count is too high. This inhibits the release of GnRH and FSH, which will cause spermatogenesis to slow down. If the sperm count reaches a low of 20 million/mL, the Sertoli cells cease the release of inhibin, and the sperm count increases.

Female Hormones

The control of reproduction in females is more complex. The female reproductive cycle is divided into the ovarian cycle and the menstrual cycle. The **ovarian cycle** governs the preparation of endocrine tissues and release of eggs, while the **menstrual cycle** governs the preparation and maintenance of the uterine lining (Figure 18.17). These cycles are coordinated over a 22–32 day cycle, with an average length of 28 days.

As with the male, the GnRH from the hypothalamus causes the release of the hormones FSH and LH from the anterior pituitary. In addition, **estrogen** and **progesterone** are released from the developing follicles. As with testosterone in males, estrogen is responsible for the secondary sexual characteristics of females. These include breast development, flaring of the hips, and a shorter period for bone growth.

The Ovarian Cycle and the Menstrual Cycle

The ovarian and menstrual cycles are regulated by hormones of the hypothalamus, pituitary, and ovaries (Figure 18.17). The ebb and flow of the hormones causes the ovarian and menstrual cycles to advance. The ovarian and menstrual cycles occur concurrently. The first half of the ovarian cycle is the follicular phase. Slowly rising levels of FSH cause the growth of follicles on the surface of the ovary. This process prepares the egg for ovulation. As the follicles grow, they begin releasing estrogen. The first few days of this cycle coincide with menstruation or the sloughing off of the functional layer of the endometrium in the uterus. After about five days, estrogen levels rise and the menstrual cycle enters the proliferative phase. The endometrium begins to regrow, replacing the blood vessels and glands that deteriorated during the end of the last cycle.

art CONNECTION

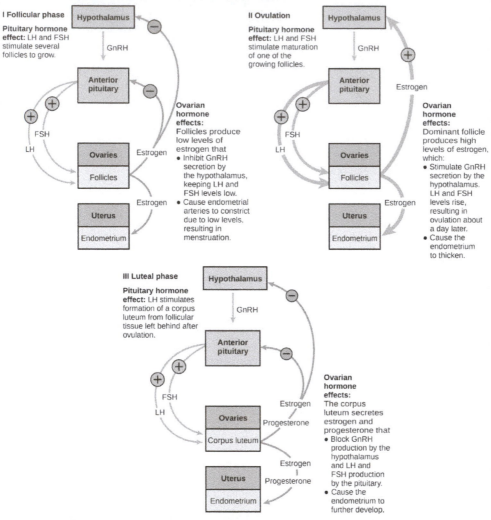

Figure 18.17 The ovarian and menstrual cycles of female reproduction are regulated by hormones produced by the hypothalamus, pituitary, and ovaries.

Which of the following statements about hormone regulation of the female reproductive cycle is false?

a. LH and FSH are produced in the pituitary, and estrogen and progesterone are produced in the ovaries.

b. Estradiol and progesterone secreted from the corpus luteum cause the endometrium to thicken.

c. Both progesterone and estrogen are produced by the follicles.

d. Secretion of GnRH by the hypothalamus is inhibited by low levels of estrogen but stimulated by high levels of estrogen.

Just prior to the middle of the cycle (approximately day 14), the high level of estrogen causes FSH and especially LH to rise rapidly then fall. The spike in LH causes the most mature follicle to rupture and release its egg. This is **ovulation**. The follicles that did not rupture degenerate and their eggs are lost. The level of estrogen decreases when the extra follicles degenerate.

Following ovulation, the ovarian cycle enters its luteal phase and the menstrual cycle enters its secretory phase, both of which run from about day 15 to 28. The luteal and secretory phases refer to changes in the ruptured follicle. The cells in the follicle undergo physical changes and produce a structure called a corpus luteum. The corpus luteum produces estrogen and progesterone. The progesterone facilitates the regrowth of the uterine lining and inhibits the release of further FSH and LH. The uterus is being prepared to accept a fertilized egg, should it occur during this cycle. The inhibition of FSH and LH prevents any further eggs and follicles from developing, while the progesterone is elevated. The level of estrogen produced by the corpus luteum increases to a steady level for the next few days.

If no fertilized egg is implanted into the uterus, the corpus luteum degenerates and the levels of estrogen and progesterone decrease. The endometrium begins to degenerate as the progesterone levels drop, initiating the next menstrual cycle. The decrease in progesterone also allows the hypothalamus to send GnRH to the anterior pituitary, releasing FSH and LH and starting the cycles again.

Reproductive Endocrinologist

A reproductive endocrinologist is a physician who treats a variety of hormonal disorders related to reproduction and infertility in both men and women. The disorders include menstrual problems, infertility, pregnancy loss, sexual dysfunction, and menopause. Doctors may use fertility drugs, surgery, or assisted reproductive techniques (ART) in their therapy. ART involves the use of procedures to manipulate the egg or sperm to facilitate reproduction, such as *in vitro* fertilization.

Reproductive endocrinologists undergo extensive medical training, first in a four-year residency in obstetrics and gynecology, then in a three-year fellowship in reproductive endocrinology. To be board certified in this area, the physician must pass written and oral exams in both areas.

Gestation

Pregnancy begins with the fertilization of an egg and continues through to the birth of the individual. The length of time of **gestation**, or the **gestation period**, in humans is 266 days and is similar in other great apes.

Within 24 hours of fertilization, the egg nucleus has finished meiosis and the egg and sperm nuclei fuse. With fusion, the cell is known as a zygote. The zygote initiates cleavage and the developing embryo travels through the oviduct to the uterus. The developing embryo must implant into the wall of the uterus within seven days, or it will deteriorate and die. The outer layers of the developing embryo or blastocyst grow into the endometrium by digesting the endometrial cells, and healing of the endometrium closes up the blastocyst into the tissue. Another layer of the blastocyst, the chorion, begins releasing a hormone called **human beta chorionic gonadotropin (β-HCG)**, which makes its way to the corpus luteum and keeps that structure active. This ensures adequate levels of progesterone that will maintain the endometrium of the uterus for the support of the developing embryo. Pregnancy tests determine the level of β-HCG in urine or serum. If the hormone is present, the test is positive.

The gestation period is divided into three equal periods or trimesters. During the first two-to-four weeks of the first trimester, nutrition and waste are handled by the endometrial lining through diffusion. As the trimester progresses, the outer layer of the embryo begins to merge with the endometrium, and the placenta forms. The **placenta** takes over the nutrient and waste requirements of the embryo and fetus, with the mother's blood passing nutrients to the placenta and removing waste from it. Chemicals from the fetus, such as bilirubin, are processed by the mother's liver for elimination. Some of the mother's immunoglobulins will pass through the placenta, providing passive immunity against some potential infections.

Internal organs and body structures begin to develop during the first trimester. By five weeks, limb buds, eyes, the heart, and liver have been basically formed. By eight weeks, the term fetus applies, and the body is essentially formed (Figure 18.18a). The individual is about five centimeters (two inches) in length and many of the organs, such as the lungs and liver, are not yet functioning. Exposure to any toxins is especially dangerous during the first trimester, as all of the body's organs and structures are going through initial development. Anything that interferes with chemical signaling during that development can have a severe effect on the fetus' survival.

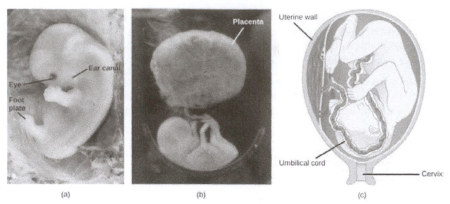

Figure 18.18 (a) Fetal development is shown at nine weeks gestation. (b) This fetus is just entering the second trimester, when the placenta takes over more of the functions performed as the baby develops. (c) There is rapid fetal growth during the third trimester. (credit a: modification of work by Ed Uthman; credit b: modification of work by National Museum of Health and Medicine; credit c: modification of work by Gray's Anatomy)

During the second trimester, the fetus grows to about 30 cm (about 12 inches) (Figure 18.18**b**). It becomes active and the mother usually feels the first movements. All organs and structures continue to develop. The placenta has taken over the functions of nutrition and waste elimination and the production of estrogen and progesterone from the corpus luteum, which has degenerated. The placenta will continue functioning up through the delivery of the baby. During the third trimester, the fetus grows to 3 to 4 kg (6.5–8.5 lbs.) and about 50 cm (19–20 inches) long (Figure 18.18**c**). This is the period of the most rapid growth during the pregnancy as all organ systems continue to grow and develop.

Visit this website (http://openstaxcollege.org/l/embryo_fetus2) to see the stages of human fetal development.

Labor is the muscular contractions to expel the fetus and placenta from the uterus. Toward the end of the third trimester, estrogen causes receptors on the uterine wall to develop and bind the hormone oxytocin. At this time, the baby reorients, facing forward and down with the back or crown of the head engaging the cervix (uterine opening). This causes the cervix to stretch and nerve impulses are sent to the hypothalamus, which signals the release of oxytocin from the posterior pituitary. Oxytocin causes smooth muscle in the uterine wall to contract. At the same time, the placenta releases prostaglandins into the uterus, increasing the contractions. A positive feedback relay occurs between the uterus, hypothalamus, and the posterior pituitary to assure an adequate supply of oxytocin. As more smooth muscle cells are recruited, the contractions increase in intensity and force.

There are three stages to labor. During stage one, the cervix thins and dilates. This is necessary for the baby and placenta to be expelled during birth. The cervix will eventually dilate to about 10 cm. During stage two, the baby is expelled from the uterus. The uterus contracts and the mother pushes as she compresses her abdominal muscles to aid the delivery. The last stage is the passage of the placenta after the baby has been born and the organ has completely disengaged from the uterine wall. If labor should stop before stage two is reached, synthetic oxytocin, known as Pitocin, can be administered to restart and maintain labor.

Download for free at https://openstax.org/details/books/concepts-biology

KEY TERMS

asexual reproduction a mechanism that produces offspring that are genetically identical to the parent

blastocyst the structure formed when cells in the mammalian blastula separate into an inner and outer layer

budding a form of asexual reproduction that results from the outgrowth of a part of an organism leading to a separation from the original animal into two individuals

bulbourethral gland the paired glands in the human male that produce a secretion that cleanses the urethra prior to ejaculation

clitoris a sensory and erectile structure in female mammals, homologous to the male penis, stimulated during sexual arousal

corpus luteum the endocrine tissue that develops from an ovarian follicle after ovulation; secretes progesterone and estrogen during pregnancy

estrogen a reproductive hormone in females that assists in endometrial regrowth, ovulation, and calcium absorption

external fertilization the fertilization of eggs by sperm outside an animal's body, often during spawning

fission (also, binary fission) a form of asexual reproduction in which an organism splits into two separate organisms or two parts that regenerate the missing portions of the body

follicle stimulating hormone (FSH) a reproductive hormone that causes sperm production in men and follicle development in women

fragmentation the breaking of an organism into parts and the growth of a separate individual from each part

gastrulation the process in which the blastula folds over itself to form the three germ layers

gestation the development before birth of a viviparous animal

gestation period the length of time of development, from conception to birth, of the young of a viviparous animal

gonadotropin-releasing hormone (GnRH) a hormone from the hypothalamus that causes the release of FSH and LH from the anterior pituitary

hermaphroditism the state of having both male and female reproductive structures within the same individual

human beta chorionic gonadotropin (β-HCG) a hormone produced by the chorion of the zygote that helps to maintain the corpus luteum and elevated levels of progesterone

inhibin a hormone made by Sertoli cells, provides negative feedback to hypothalamus in control of FSH and GnRH release

inner cell mass the inner layer of cells in the blastocyst, which becomes the embryo

internal fertilization the fertilization of eggs by sperm inside the body of the female

interstitial cell of Leydig a cell type found next to the seminiferous tubules that makes testosterone

labia majora the large folds of tissue covering inguinal area

labia minora the smaller folds of tissue within labia majora

luteinizing hormone (LH) a reproductive hormone in both men and women, causes testosterone production in men and ovulation and lactation in women

menstrual cycle the cycle of the degradation and re-growth of the endometrium

oogenesis the process of producing haploid eggs

organogenesis the process of organ formation during development

ovarian cycle the cycle of preparation of egg for ovulation and the conversion of the follicle to the corpus luteum

oviduct (also, fallopian tube) the muscular tube connecting uterus with ovary area

oviparity a process by which fertilized eggs are laid outside the female's body and develop there, receiving nourishment from the yolk that is a part of the egg

ovoviparity a process by which fertilized eggs are retained within the female; the embryo obtains its nourishment from the egg's yolk, and the young are fully developed when they are hatched

ovulation the release of an oocyte from a mature follicle in the ovary of a vertebrate

parthenogenesis a form of asexual reproduction in which an egg develops into a complete individual without being fertilized

penis the male reproductive structure for urine elimination and copulation

placenta the organ that supports the transport of nutrients and waste between the mothers and fetus' blood in eutherian mammals

progesterone a reproductive hormone in women; assists in endometrial regrowth and inhibition of FSH and LH release

prostate gland a structure that is a mixture of smooth muscle and glandular material and that contributes to semen

scrotum a sac containing testes, exterior to body

semen a fluid mixture of sperm and supporting materials

seminal vesicle a secretory accessory gland in male; contributes to semen

seminiferous tubule the structures within which sperm production occurs in the testes

Sertoli cell a cell in the walls of the seminiferous tubules that assists developing sperm and secretes inhibin

sex determination the mechanism by which the sex of individuals in sexually reproducing organisms is initially established

sexual reproduction a form of reproduction in which cells containing genetic material from two individuals combines to produce genetically unique offspring

spermatogenesis the process of producing haploid sperm

testes a pair of male reproductive organs

testosterone a reproductive hormone in men that assists in sperm production and promoting secondary sexual characteristics

trophoblast the outer layer of cells in the blastocyst, which gives rise to the embryo's contribution to the placenta

uterus a female reproductive structure in which an embryo develops

vagina a muscular tube for the passage of menstrual flow, copulation, and birth of offspring

viviparity a process in which the young develop within the female and are born in a nonembryonic state

zona pellucida the protective layer around the mammalian egg

CHAPTER SUMMARY

18.1 How Animals Reproduce

Reproduction may be asexual when one individual produces genetically identical offspring, or sexual when the genetic material from two individuals is combined to produce genetically diverse offspring. Asexual reproduction in animals occurs through fission, budding, fragmentation, and parthenogenesis. Sexual reproduction may involve fertilization inside the body or in the external environment. A species may have separate sexes or combined sexes; when the sexes are combined they may be expressed at different times in the life cycle. The sex of an individual may be determined by various chromosomal systems or environmental factors such as temperature.

Sexual reproduction starts with the combination of a sperm and an egg in a process called fertilization. This can occur either outside the bodies or inside the female. The method of fertilization varies among animals. Some species release the egg and sperm into the environment, some species retain the egg and receive the sperm into the female body and then expel the developing embryo covered with shell, while still other species retain the developing offspring throughout the gestation period.

18.2 Development and Organogenesis

The early stages of embryonic development begin with fertilization. The process of fertilization is tightly controlled to ensure that only one sperm fuses with one egg. After fertilization, the zygote undergoes cleavage to form the blastula. The blastula, which in some species is a hollow ball of cells, undergoes a process called gastrulation, during which the three germ layers form. The ectoderm gives rise to the nervous system and the epidermal skin cells, the mesoderm gives rise to the muscle cells and connective tissue in the body, and the endoderm gives rise to the digestive system and other internal organs. Organogenesis is the formation of organs from the germ layers. Each germ layer gives rise to specific tissue types.

18.3 Human Reproduction

The reproductive structures that evolved in land animals allow males and females to mate, fertilize internally, and support the growth and development of offspring. Gametogenesis, the production of sperm (spermatogenesis) and eggs (oogenesis), takes place through the process of meiosis.

The male and female reproductive cycles are controlled by hormones released from the hypothalamus and anterior pituitary and hormones from reproductive tissues and organs. The hypothalamus monitors the need for FSH and LH production and release from the anterior pituitary. FSH and LH affect reproductive structures to cause the formation of sperm and the preparation of eggs for release and possible fertilization. In the male, FSH and LH stimulate Sertoli cells and interstitial cells of Leydig in the testes to facilitate sperm production. The Leydig cells produce testosterone, which also is responsible for the secondary sexual characteristics of males. In females, FSH and LH cause estrogen and progesterone to be produced. They regulate the female reproductive cycle, which is divided into the ovarian cycle and the menstrual cycle.

Human pregnancy begins with fertilization of an egg and proceeds through the three trimesters of gestation. The first trimester lays down the basic structures of the body, including the limb buds, heart, eyes, and the liver. The second trimester continues the development of all of the organs and systems. The third trimester exhibits the greatest growth of the fetus and culminates in labor and delivery. The labor process has three stages (contractions, delivery of the fetus, and expulsion of the placenta), each propelled by hormones.

ART CONNECTION QUESTIONS

1. Figure 18.12 Which of the following statements about the male reproductive system is false?
   a. The vas deferens carries sperm from the testes to the seminal vesicles.
   b. The ejaculatory duct joins the urethra.
   c. Both the prostate and the bulbourethral glands produce components of the semen.
   d. The prostate gland is located in the testes.

2. Figure 18.17 Which of the following statements about hormone regulation of the female reproductive cycle is false?

   a. LH and FSH are produced in the pituitary, and estrogen and progesterone are produced in the ovaries.
   b. Estradiol and progesterone secreted from the corpus luteum cause the endometrium to thicken.
   c. Both progesterone and estrogen are produced by the follicles.
   d. Secretion of GnRH by the hypothalamus is inhibited by low levels of estrogen but stimulated by high levels of estrogen.

REVIEW QUESTIONS

3. In which group is parthenogenesis a normal event?

    a. chickens
    b. bees
    c. rabbits
    d. sea stars

4. Genetically unique individuals are produced through _____.

    a. sexual reproduction
    b. parthenogenesis
    c. budding
    d. fragmentation

5. External fertilization occurs in which type of environment?

    a. aquatic
    b. forested
    c. savanna
    d. steppe

6. The process of gastrulation forms the _____.

    a. blastula
    b. zygote
    c. organs
    d. germ layers

7. Which of the following gives rise to the skin cells?

    a. ectoderm
    b. endoderm
    c. mesoderm
    d. none of the above

8. Sperm are produced in the _____.
    a. scrotum
    b. seminal vesicles
    c. seminiferous tubules
    d. prostate gland

9. Which female organ has an endometrial lining that will support a developing baby?
    a. labia minora
    b. breast
    c. ovaries
    d. uterus

10. Which hormone causes FSH and LH to be released?

    a. testosterone
    b. estrogen
    c. GnRH
    d. progesterone

11. Nutrient and waste requirements for the developing fetus are handled during the first few weeks by _____.

    a. the placenta
    b. diffusion through the endometrium
    c. the chorion
    d. the blastocyst

12. Which hormone is primarily responsible for the contractions during labor?
    a. oxytocin
    b. estrogen
    c. β-HCG
    d. progesterone

CRITICAL THINKING QUESTIONS

13. What might be a disadvantage to temperature-dependent sex determination?

14. Compared to separate sexes and assuming self-fertilizing is not possible, what might be one advantage and one disadvantage to hermaphroditism?

15. What do you think would happen if multiple sperm fused with one egg?

16. Compare spermatogenesis and oogenesis as to timing of the processes, and the number and type of cells finally produced.

17. Describe the events in the ovarian cycle leading up to ovulation.

18. Describe the stages of labor.

19 | POPULATION AND COMMUNITY ECOLOGY

Figure 19.1 Asian carp jump out of the water in response to electrofishing. The Asian carp in the inset photograph were harvested from the Little Calumet River in Illinois in May, 2010, using rotenone, a toxin often used as an insecticide, in an effort to learn more about the population of the species. (credit main image: modification of work by USGS; credit inset: modification of work by Lt. David French, USCG)

Chapter Outline

19.1: Population Demographics and Dynamics

19.2: Population Growth and Regulation

19.3: The Human Population

19.4: Community Ecology

Introduction

Imagine sailing down a river in a small motorboat on a weekend afternoon; the water is smooth, and you are enjoying the sunshine and cool breeze when suddenly you are hit in the head by a 20-pound silver carp. This is a risk now on many rivers and canal systems in Illinois and Missouri because of the presence of Asian carp.

This fish—actually a group of species including the silver, black, grass, and big head carp—has been farmed and eaten in China for over 1,000 years. It is one of the most important aquaculture food resources worldwide. In the United States, however, Asian carp is considered a dangerous invasive species that disrupts ecological community structure to the point of threatening native species.

The effects of invasive species (such as the Asian carp, kudzu vine, predatory snakehead fish, and zebra mussel) are just one aspect of what ecologists study to understand how populations interact within ecological communities, and what impact natural and human-induced disturbances have on the characteristics of communities.

19.1 | Population Demographics and Dynamics

By the end of this section, you will be able to:

- Describe how ecologists measure population size and density
- Describe three different patterns of population distribution
- Use life tables to calculate mortality rates
- Describe the three types of survivorship curves and relate them to specific populations

Populations are dynamic entities. Their size and composition fluctuate in response to numerous factors, including seasonal and yearly changes in the environment, natural disasters such as forest fires and volcanic eruptions, and competition for resources between and within species. The statistical study of populations is called **demography**: a set of mathematical tools designed to describe populations and investigate how they change. Many of these tools were actually designed to study human populations. For example, **life tables**, which detail the life expectancy of individuals within a population, were initially developed by life insurance companies to set insurance rates. In fact, while the term "demographics" is sometimes assumed to mean a study of human populations, all living populations can be studied using this approach.

Population Size and Density

Populations are characterized by their **population size** (total number of individuals) and their **population density** (number of individuals per unit area). A population may have a large number of individuals that are distributed densely, or sparsely. There are also populations with small numbers of individuals that may be dense or very sparsely distributed in a local area. Population size can affect potential for adaptation because it affects the amount of genetic variation present in the population. Density can have effects on interactions within a population such as competition for food and the ability of individuals to find a mate. Smaller organisms tend to be more densely distributed than larger organisms (Figure 19.2).

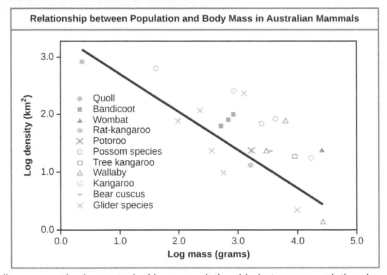

Figure 19.2 Australian mammals show a typical inverse relationship between population density and body size.

As this graph shows, population density typically decreases with increasing body size. Why do you think this is the case?

Estimating Population Size

The most accurate way to determine population size is to count all of the individuals within the area. However, this method is usually not logistically or economically feasible, especially when studying large areas. Thus, scientists usually study populations by sampling a representative portion of each habitat and use this sample to make inferences about the population as a whole. The methods used to sample populations to determine their size and density are typically tailored to the characteristics of the organism being studied. For immobile organisms such as plants, or for very small and slow-moving organisms, a quadrat may be used. A **quadrat** is a wood, plastic, or metal square that is randomly located on the ground and used to count the number of individuals that lie within its boundaries. To obtain an accurate count using this method, the square must be placed at random locations within the habitat enough times to produce an accurate estimate. This counting method will provide an estimate of both population size and density. The number and size of quadrat samples depends on the type of organisms and the nature of their distribution.

For smaller mobile organisms, such as mammals, a technique called **mark and recapture** is often used. This method involves marking a sample of captured animals in some way and releasing them back into the environment to mix with the rest of the population; then, a new sample is captured and scientists determine how many of the marked animals are in the new sample. This method assumes that the larger the population, the lower the percentage of marked organisms that will be recaptured since they will have mixed with more unmarked individuals. For example, if 80 field mice are captured, marked, and released into the forest, then a second trapping 100 field mice are captured and 20 of them are marked, the population size (N) can be determined using the following equation:

$$\frac{\text{number marked first catch} \times \text{total number second catch}}{\text{number marked second catch}} = N$$

Using our example, the population size would be 400.

$$\frac{80 \times 100}{20} = 400$$

These results give us an estimate of 400 total individuals in the original population. The true number usually will be a bit different from this because of chance errors and possible bias caused by the sampling methods.

Species Distribution

In addition to measuring density, further information about a population can be obtained by looking at the distribution of the individuals throughout their range. A **species distribution pattern** is the distribution of individuals within a habitat at a particular point in time—broad categories of patterns are used to describe them.

Individuals within a population can be distributed at random, in groups, or equally spaced apart (more or less). These are known as random, clumped, and uniform distribution patterns, respectively (Figure 19.3). Different distributions reflect important aspects of the biology of the species; they also affect the mathematical methods required to estimate population sizes. An example of random distribution occurs with dandelion and other plants that have wind-dispersed seeds that germinate wherever they happen to fall in favorable environments. A clumped distribution, may be seen in plants that drop their seeds straight to the ground, such as oak trees; it can also be seen in animals that live in social groups (schools of fish or herds of elephants). Uniform distribution is observed in plants that secrete substances inhibiting the growth of nearby individuals (such as the release of toxic chemicals by sage plants). It is also seen in territorial animal species, such as penguins that maintain a defined territory for nesting. The territorial defensive behaviors of each individual create a regular pattern of distribution of similar-sized territories and individuals within those territories. Thus, the distribution of the individuals within a population provides more information about how they interact with each other than does a simple density measurement. Just as lower density species might have more difficulty finding a mate, solitary species with a random distribution might have a similar difficulty when compared to social species clumped together in groups.

Figure 19.3 Species may have a random, clumped, or uniform distribution. Plants such as (a) dandelions with wind-dispersed seeds tend to be randomly distributed. Animals such as (b) elephants that travel in groups exhibit a clumped distribution. Territorial birds such as (c) penguins tend to have a uniform distribution. (credit a: modification of work by Rosendahl; credit b: modification of work by Rebecca Wood; credit c: modification of work by Ben Tubby)

Demography

While population size and density describe a population at one particular point in time, scientists must use demography to study the dynamics of a population. Demography is the statistical study of population changes over time: birth rates, death rates, and life expectancies. These population characteristics are often displayed in a life table.

Life Tables

Life tables provide important information about the life history of an organism and the life expectancy of individuals at each age. They are modeled after actuarial tables used by the insurance industry for estimating human life expectancy. Life tables may include the probability of each age group dying before their next birthday, the percentage of surviving individuals dying at a particular age interval (their **mortality rate**, and their life expectancy at each interval. An example of a life table is shown in Table 19.1 from a study of Dall mountain sheep, a species native to northwestern North America. Notice that the population is divided into age intervals (column A). The mortality rate (per 1000) shown in column D is based on the number of individuals dying during the age interval (column B), divided by the number of individuals surviving at the beginning of the interval (Column C) multiplied by 1000.

$$\text{mortality rate} \ = \ \frac{\text{number of individuals dying}}{\text{number of individuals surviving}} \times 1000$$

For example, between ages three and four, 12 individuals die out of the 776 that were remaining from the original 1000 sheep. This number is then multiplied by 1000 to give the mortality rate per thousand.

$$\text{mortality rate} \ = \ \frac{12}{776} \times 1000 \ \approx \ 15.5$$

As can be seen from the mortality rate data (column D), a high death rate occurred when the sheep were between six months and a year old, and then increased even more from 8 to 12 years old, after which there were few survivors. The data indicate that if a sheep in this population were to survive to age one, it could be expected to live another 7.7 years on average, as shown by the life-expectancy numbers in column E.

Life Table of Dall Mountain Sheep[1]

A	B	C	D	E
Age interval (years)	Number dying in age interval out of 1000 born	Number surviving at beginning of age interval out of 1000 born	Mortality rate per 1000 alive at beginning of age interval	Life expectancy or mean lifetime remaining to those attaining age interval
0–0.5	54	1000	54.0	7.06
0.5–1	145	946	153.3	—
1–2	12	801	15.0	7.7
2–3	13	789	16.5	6.8
3–4	12	776	15.5	5.9
4–5	30	764	39.3	5.0
5–6	46	734	62.7	4.2
6–7	48	688	69.8	3.4
7–8	69	640	107.8	2.6
8–9	132	571	231.2	1.9
9–10	187	439	426.0	1.3
10–11	156	252	619.0	0.9
11–12	90	96	937.5	0.6
12–13	3	6	500.0	1.2
13–14	3	3	1000	0.7

Table 19.1 This life table of *Ovis dalli* shows the number of deaths, number of survivors, mortality rate, and life expectancy at each age interval for Dall mountain sheep.

Survivorship Curves

Another tool used by population ecologists is a **survivorship curve**, which is a graph of the number of individuals surviving at each age interval versus time. These curves allow us to compare the life histories of different populations (Figure 19.4). There are three types of survivorship curves. In a type I curve, mortality is low in the early and middle years and occurs mostly in older individuals. Organisms exhibiting a type I survivorship typically produce few offspring and provide good care to the offspring increasing the likelihood of their survival. Humans and most mammals exhibit a type I survivorship curve. In type II curves, mortality is relatively constant throughout the entire life span, and mortality is equally likely to occur at any point in the life span. Many bird populations provide examples of an intermediate or type II survivorship curve. In type III survivorship curves, early ages experience the highest mortality with much lower mortality rates for organisms that make it to advanced years. Type III organisms typically produce large numbers of offspring, but provide very little or no care for them. Trees and marine invertebrates exhibit a type III survivorship curve because very few of these organisms survive their younger years, but those that do make it to an old age are more likely to survive for a relatively long period of time.

---

1. Data Adapted from Edward S. Deevey, Jr., "Life Tables for Natural Populations of Animals," *The Quarterly Review of Biology* 22, no. 4 (December 1947): 283-314.

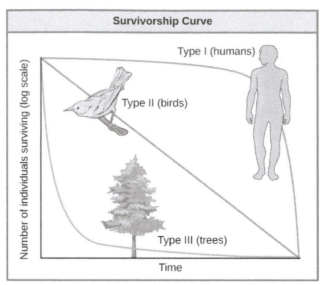

Figure 19.4 Survivorship curves show the distribution of individuals in a population according to age. Humans and most mammals have a Type I survivorship curve, because death primarily occurs in the older years. Birds have a Type II survivorship curve, as death at any age is equally probable. Trees have a Type III survivorship curve because very few survive the younger years, but after a certain age, individuals are much more likely to survive.

19.2 | Population Growth and Regulation

By the end of this section, you will be able to:

- Explain the characteristics of and differences between exponential and logistic growth patterns
- Give examples of exponential and logistic growth in natural populations
- Give examples of how the carrying capacity of a habitat may change
- Compare and contrast density-dependent growth regulation and density-independent growth regulation giving examples

Population ecologists make use of a variety of methods to model population dynamics. An accurate model should be able to describe the changes occurring in a population and predict future changes.

Population Growth

The two simplest models of population growth use deterministic equations (equations that do not account for random events) to describe the rate of change in the size of a population over time. The first of these models, exponential growth, describes theoretical populations that increase in numbers without any limits to their growth. The second model, logistic growth, introduces limits to reproductive growth that become more intense as the population size increases. Neither model adequately describes natural populations, but they provide points of comparison.

Exponential Growth

Charles Darwin, in developing his theory of natural selection, was influenced by the English clergyman Thomas Malthus. Malthus published his book in 1798 stating that populations with abundant natural resources grow very rapidly; however, they limit further growth by depleting their resources. The early pattern of accelerating population size is called **exponential growth**.

The best example of exponential growth in organisms is seen in bacteria. Bacteria are prokaryotes that reproduce largely by binary fission. This division takes about an hour for many bacterial species. If 1000 bacteria are placed in a large flask with an abundant supply of nutrients (so the nutrients will not become quickly depleted), the number of bacteria will have doubled from 1000 to 2000 after just an hour. In another hour, each of the 2000 bacteria will divide, producing 4000 bacteria. After the third hour, there should be 8000 bacteria in the flask. The important concept of exponential growth is that the growth rate—the number of organisms added in each reproductive generation—is itself increasing; that is, the population size is increasing at a greater and greater rate. After 24 of these cycles, the population would have increased from 1000 to more than 16 billion bacteria. When the population size, N, is plotted over time, a **J-shaped growth curve** is produced (Figure 19.5**a**).

The bacteria-in-a-flask example is not truly representative of the real world where resources are usually limited. However, when a species is introduced into a new habitat that it finds suitable, it may show exponential growth for a while. In the case of the bacteria in the flask, some bacteria will die during the experiment and thus not reproduce; therefore, the growth rate is lowered from a maximal rate in which there is no mortality. The growth rate of a population is largely determined by subtracting the **death rate**, D, (number organisms that die during an interval) from the **birth rate**, B, (number organisms that are born during an interval). The growth rate can be expressed in a simple equation that combines the birth and death rates into a single factor: r. This is shown in the following formula:

$$\text{Population growth } = \text{ } rN$$

The value of r can be positive, meaning the population is increasing in size (the rate of change is positive); or negative, meaning the population is decreasing in size; or zero, in which case the population size is unchanging, a condition known as **zero population growth**.

Logistic Growth

Extended exponential growth is possible only when infinite natural resources are available; this is not the case in the real world. Charles Darwin recognized this fact in his description of the "struggle for existence," which states that individuals will compete (with members of their own or other species) for limited resources. The successful ones are more likely to survive and pass on the traits that made them successful to the next generation at a greater rate (natural selection). To model the reality of limited resources, population ecologists developed the **logistic growth** model.

Carrying Capacity and the Logistic Model

In the real world, with its limited resources, exponential growth cannot continue indefinitely. Exponential growth may occur in environments where there are few individuals and plentiful resources, but when the number of individuals gets large enough, resources will be depleted and the growth rate will slow down. Eventually, the growth rate will plateau or level off (Figure 19.5**b**). This population size, which is determined by the maximum population size that a particular environment can sustain, is called the **carrying capacity**, or K. In real populations, a growing population often overshoots its carrying capacity, and the death rate increases beyond the birth rate causing the population size to decline back to the carrying capacity or below it. Most populations usually fluctuate around the carrying capacity in an undulating fashion rather than existing right at it.

The formula used to calculate logistic growth adds the carrying capacity as a moderating force in the growth rate. The expression "$K - N$" is equal to the number of individuals that may be added to a population at a given time, and "$K - N$" divided by "K" is the fraction of the carrying capacity available for further growth. Thus, the exponential growth model is restricted by this factor to generate the logistic growth equation:

$$\text{Population growth } = \text{ } rN\left[\frac{K-N}{K}\right]$$

Notice that when N is almost zero the quantity in brackets is almost equal to 1 (or K/K) and growth is close to exponential. When the population size is equal to the carrying capacity, or $N = K$, the quantity in brackets is equal to zero and growth is equal to zero. A graph of this equation (logistic growth) yields the **S-shaped curve** (Figure 19.5**b**). It is a more realistic model of population growth than exponential growth. There are three different sections to an S-shaped curve. Initially, growth is exponential because there are few individuals and ample resources available. Then, as resources begin to become limited, the growth rate decreases. Finally, the growth rate levels off at the carrying capacity of the environment, with little change in population number over time.

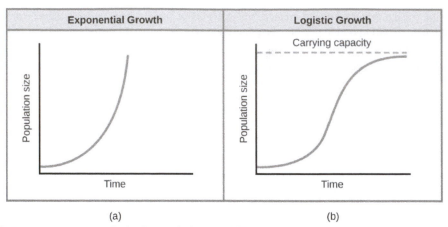

Figure 19.5 When resources are unlimited, populations exhibit (a) exponential growth, shown in a J-shaped curve. When resources are limited, populations exhibit (b) logistic growth. In logistic growth, population expansion decreases as resources become scarce, and it levels off when the carrying capacity of the environment is reached. The logistic growth curve is S-shaped.

Role of Intraspecific Competition

The logistic model assumes that every individual within a population will have equal access to resources and, thus, an equal chance for survival. For plants, the amount of water, sunlight, nutrients, and space to grow are the important resources, whereas in animals, important resources include food, water, shelter, nesting space, and mates.

In the real world, phenotypic variation among individuals within a population means that some individuals will be better adapted to their environment than others. The resulting competition for resources among population members of the same species is termed **intraspecific competition**. Intraspecific competition may not affect populations that are well below their carrying capacity, as resources are plentiful and all individuals can obtain what they need. However, as population size increases, this competition intensifies. In addition, the accumulation of waste products can reduce carrying capacity in an environment.

Examples of Logistic Growth

Yeast, a microscopic fungus used to make bread and alcoholic beverages, exhibits the classical S-shaped curve when grown in a test tube (Figure 19.6a). Its growth levels off as the population depletes the nutrients that are necessary for its growth. In the real world, however, there are variations to this idealized curve. Examples in wild populations include sheep and harbor seals (Figure 19.6b). In both examples, the population size exceeds the carrying capacity for short periods of time and then falls below the carrying capacity afterwards. This fluctuation in population size continues to occur as the population oscillates around its carrying capacity. Still, even with this oscillation, the logistic model is confirmed.

art CONNECTION

Figure 19.6 (a) Yeast grown in ideal conditions in a test tube shows a classical S-shaped logistic growth curve, whereas (b) a natural population of seals shows real-world fluctuation. The yeast is visualized using differential interference contrast light micrography. (credit a: scale-bar data from Matt Russell)

If the major food source of seals declines due to pollution or overfishing, which of the following would likely occur?

a. The carrying capacity of seals would decrease, as would the seal population.

b. The carrying capacity of seals would decrease, but the seal population would remain the same.

c. The number of seal deaths would increase, but the number of births would also increase, so the population size would remain the same.

d. The carrying capacity of seals would remain the same, but the population of seals would decrease.

Population Dynamics and Regulation

The logistic model of population growth, while valid in many natural populations and a useful model, is a simplification of real-world population dynamics. Implicit in the model is that the carrying capacity of the environment does not change, which is not the case. The carrying capacity varies annually. For example, some summers are hot and dry whereas others are cold and wet; in many areas, the carrying capacity during the winter is much lower than it is during the summer. Also, natural events such as earthquakes, volcanoes, and fires can alter an environment and hence its carrying capacity.

Additionally, populations do not usually exist in isolation. They share the environment with other species, competing with them for the same resources (interspecific competition). These factors are also important to understanding how a specific population will grow.

Population growth is regulated in a variety of ways. These are grouped into **density-dependent** factors, in which the density of the population affects growth rate and mortality, and **density-independent** factors, which cause mortality in a population regardless of population density. Wildlife biologists, in particular, want to understand both types because this helps them manage populations and prevent extinction or overpopulation.

Density-dependent Regulation

Most density-dependent factors are biological in nature and include predation, inter- and intraspecific competition, and parasites. Usually, the denser a population is, the greater its mortality rate. For example, during intra- and interspecific competition, the reproductive rates of the species will usually be lower, reducing their populations' rate of growth. In addition, low prey density increases the mortality of its predator because it has more difficulty locating its food source. Also, when the population is denser, diseases spread more rapidly among the members of the population, which affect the mortality rate.

Density dependent regulation was studied in a natural experiment with wild donkey populations on two sites in Australia.[2] On one site the population was reduced by a population control program; the population on the other site received no interference. The high-density plot was twice as dense as the low-density plot. From 1986 to 1987 the high-density plot saw no change in donkey density, while the low-density plot saw an increase in donkey density. The difference in the growth rates of the two populations was caused by mortality, not by a difference in birth rates. The researchers found that numbers of offspring birthed by each mother was unaffected by density. Growth rates in the two populations were different mostly because of juvenile mortality caused by the mother's malnutrition due to scarce high-quality food in the dense population. Figure 19.7 shows the difference in age-specific mortalities in the two populations.

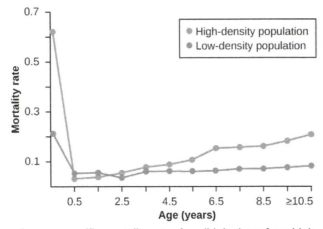

Figure 19.7 This graph shows the age-specific mortality rates for wild donkeys from high- and low-density populations. The juvenile mortality is much higher in the high-density population because of maternal malnutrition caused by a shortage of high-quality food.

Density-independent Regulation and Interaction with Density-dependent Factors

Many factors that are typically physical in nature cause mortality of a population regardless of its density. These factors include weather, natural disasters, and pollution. An individual deer will be killed in a forest fire regardless of how many deer happen to be in that area. Its chances of survival are the same whether the population density is high or low. The same holds true for cold winter weather.

In real-life situations, population regulation is very complicated and density-dependent and independent factors can interact. A dense population that suffers mortality from a density-independent cause will be able to recover differently than a sparse population. For example, a population of deer affected by a harsh winter will recover faster if there are more deer remaining to reproduce.

---

2. David Choquenot, "Density-Dependent Growth, Body Condition, and Demography in Feral Donkeys: Testing the Food Hypothesis," *Ecology* 72, no. 3 (June 1991):805–813.

Why Did the Woolly Mammoth Go Extinct?

Figure 19.8 The three images include: (a) 1916 mural of a mammoth herd from the American Museum of Natural History, (b) the only stuffed mammoth in the world is in the Museum of Zoology located in St. Petersburg, Russia, and (c) a one-month-old baby mammoth, named Lyuba, discovered in Siberia in 2007. (credit a: modification of work by Charles R. Knight; credit b: modification of work by "Tanapon"/Flickr; credit c: modification of work by Matt Howry)

Woolly mammoths began to go extinct about 10,000 years ago, soon after paleontologists believe humans able to hunt them began to colonize North America and northern Eurasia (Figure 19.8). A mammoth population survived on Wrangel Island, in the East Siberian Sea, and was isolated from human contact until as recently as 1700 BC. We know a lot about these animals from carcasses found frozen in the ice of Siberia and other northern regions.

It is commonly thought that climate change and human hunting led to their extinction. A 2008 study estimated that climate change reduced the mammoth's range from 3,000,000 square miles 42,000 years ago to 310,000 square miles 6,000 years ago.[3] Through archaeological evidence of kill sites, it is also well documented that humans hunted these animals. A 2012 study concluded that no single factor was exclusively responsible for the extinction of these magnificent creatures.[4] In addition to climate change and reduction of habitat, scientists demonstrated another important factor in the mammoth's extinction was the migration of human hunters across the Bering Strait to North America during the last ice age 20,000 years ago.

The maintenance of stable populations was and is very complex, with many interacting factors determining the outcome. It is important to remember that humans are also part of nature. Once we contributed to a species' decline using primitive hunting technology only.

---

3. David Nogués-Bravo et al., "Climate Change, Humans, and the Extinction of the Woolly Mammoth." *PLoS Biol* 6 (April 2008): e79, doi:10.1371/journal.pbio.0060079.

4. G.M. MacDonald et al., "Pattern of Extinction of the Woolly Mammoth in Beringia." *Nature Communications* 3, no. 893 (June 2012), doi:10.1038/ncomms1881.

Demographic-Based Population Models

Population ecologists have hypothesized that suites of characteristics may evolve in species that lead to particular adaptations to their environments. These adaptations impact the kind of population growth their species experience. Life history characteristics such as birth rates, age at first reproduction, the numbers of offspring, and even death rates evolve just like anatomy or behavior, leading to adaptations that affect population growth. Population ecologists have described a continuum of life-history "strategies" with K-selected species on one end and r-selected species on the other. **K-selected species** are adapted to stable, predictable environments. Populations of K-selected species tend to exist close to their carrying capacity. These species tend to have larger, but fewer, offspring and contribute large amounts of resources to each offspring. Elephants would be an example of a K-selected species. **r-selected species** are adapted to unstable and unpredictable environments. They have large numbers of small offspring. Animals that are r-selected do not provide a lot of resources or parental care to offspring, and the offspring are relatively self-sufficient at birth. Examples of r-selected species are marine invertebrates such as jellyfish and plants such as the dandelion. The two extreme strategies are at two ends of a continuum on which real species life histories will exist. In addition, life history strategies do not need to evolve as suites, but can evolve independently of each other, so each species may have some characteristics that trend toward one extreme or the other.

19.3 | The Human Population

By the end of this section, you will be able to:

- Discuss how human population growth can be exponential
- Explain how humans have expanded the carrying capacity of their habitat
- Relate population growth and age structure to the level of economic development in different countries
- Discuss the long-term implications of unchecked human population growth

Concepts of animal population dynamics can be applied to human population growth. Humans are not unique in their ability to alter their environment. For example, beaver dams alter the stream environment where they are built. Humans, however, have the ability to alter their environment to increase its carrying capacity, sometimes to the detriment of other species. Earth's human population and their use of resources are growing rapidly, to the extent that some worry about the ability of Earth's environment to sustain its human population. Long-term exponential growth carries with it the potential risks of famine, disease, and large-scale death, as well as social consequences of crowding such as increased crime.

Human technology and particularly our harnessing of the energy contained in fossil fuels have caused unprecedented changes to Earth's environment, altering ecosystems to the point where some may be in danger of collapse. Changes on a global scale including depletion of the ozone layer, desertification and topsoil loss, and global climate change are caused by human activities.

The world's human population is presently growing exponentially (Figure 19.9).

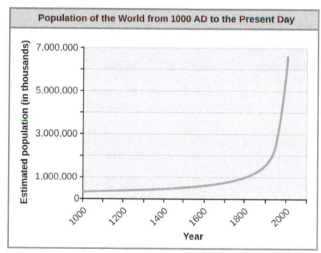

Figure 19.9 Human population growth since 1000 AD is exponential.

A consequence of exponential growth rate is that the time that it takes to add a particular number of humans to the population is becoming shorter. Figure 19.10 shows that 123 years were necessary to add 1 billion humans between 1804 and 1930, but it only took 24 years to add the two billion people between 1975 and 1999. This acceleration in growth rate will likely begin to decrease in the coming decades. Despite this, the population will continue to increase and the threat of overpopulation remains, particularly because the damage caused to ecosystems and biodiversity is lowering the human carrying capacity of the planet.

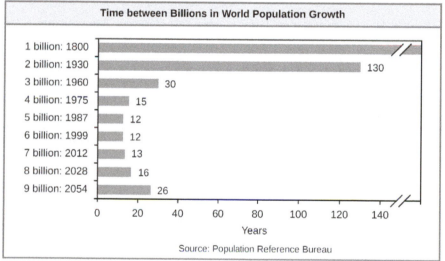

Figure 19.10 The time between the addition of each billion human beings to Earth decreases over time. (credit: modification of work by Ryan T. Cragun)

Click through this interactive view (http://openstaxcollege.org/l/human_growth2) of how human populations have changed over time.

Overcoming Density-Dependent Regulation

Humans are unique in their ability to alter their environment in myriad ways. This ability is responsible for human population growth because it resets the carrying capacity and overcomes density-dependent growth regulation. Much of this ability is related to human intelligence, society, and communication. Humans construct shelters to protect themselves from the elements and have developed agriculture and domesticated animals to increase their food supplies. In addition, humans use language to communicate this technology to new generations, allowing them to improve upon previous accomplishments.

Other factors in human population growth are migration and public health. Humans originated in Africa, but we have since migrated to nearly all inhabitable land on Earth, thus, increasing the area that we have colonized. Public health, sanitation, and the use of antibiotics and vaccines have decreased the ability of infectious disease to limit human population growth in developed countries. In the past, diseases such as the bubonic plaque of the fourteenth century killed between 30 and 60 percent of Europe's population and reduced the overall world population by as many as one hundred million people. Infectious disease continues to have an impact on human population growth. For example, life expectancy in sub-Saharan Africa, which was increasing from 1950 to 1990, began to decline after 1985 largely as a result of HIV/AIDS mortality. The reduction in life expectancy caused by HIV/AIDS was estimated to be 7 years for 2005.[5]

Declining life expectancy is an indicator of higher mortality rates and leads to lower birth rates.

The fundamental cause of the acceleration of growth rate for humans in the past 200 years has been the reduced death rate due to a development of the technological advances of the industrial age, urbanization that supported those technologies, and especially the exploitation of the energy in fossil fuels. Fossil fuels are responsible for dramatically increasing the resources available for human population growth through agriculture (mechanization, pesticides, and fertilizers) and harvesting wild populations.

Age Structure, Population Growth, and Economic Development

The age structure of a population is an important factor in population dynamics. **Age structure** is the proportion of a population in different age classes. Models that incorporate age structure allow better prediction of population growth, plus the ability to associate this growth with the level of economic development in a region. Countries with rapid growth have a pyramidal shape in their age structure diagrams, showing a preponderance of younger individuals, many of whom are of reproductive age (Figure 19.11). This pattern is most often observed in underdeveloped countries where individuals do not live to old age because of less-than-optimal living conditions, and there is a high birth rate. Age structures of areas with slow growth, including developed countries such as the United States, still have a pyramidal structure, but with many fewer young and reproductive-aged individuals and a greater proportion of older individuals. Other developed countries, such as Italy, have zero population growth. The age structure of these populations is more conical, with an even greater percentage of middle-aged and older individuals. The actual growth rates in different countries are shown in Figure 19.12, with the highest rates tending to be in the less economically developed countries of Africa and Asia.

5.　Danny Dorling, Mary Shaw, and George Davey Smith, "Global Inequality of Life Expectancy due to AIDS," *BMJ* 332, no. 7542 (March 2006): 662-664, doi: 10.1136/bmj.332.7542.662.

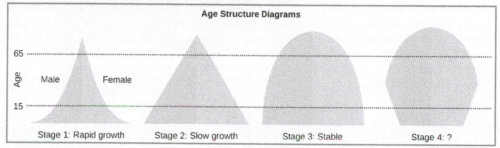

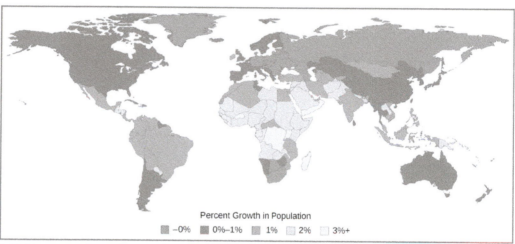

Figure 19.11 Typical age structure diagrams are shown. The rapid growth diagram narrows to a point, indicating that the number of individuals decreases rapidly with age. In the slow growth model, the number of individuals decreases steadily with age. Stable population diagrams are rounded on the top, showing that the number of individuals per age group decreases gradually, and then increases for the older part of the population.

Age structure diagrams for rapidly growing, slow growing, and stable populations are shown in stages 1 through 3. What type of population change do you think stage 4 represents?

Figure 19.12 The percent growth rate of population in different countries is shown. Notice that the highest growth is occurring in less economically developed countries in Africa and Asia.

Long-Term Consequences of Exponential Human Population Growth

Many dire predictions have been made about the world's population leading to a major crisis called the "population explosion." In the 1968 book *The Population Bomb*, biologist Dr. Paul R. Ehrlich wrote, "The battle to feed all of humanity is over. In the 1970s hundreds of millions of people will starve to death in spite of any crash programs embarked upon now. At this late date nothing can prevent a substantial increase in the world death rate."[6] While many critics view this statement as an exaggeration, the laws of exponential population growth are still in effect, and unchecked human population growth cannot continue indefinitely.

Efforts to moderate population control led to the **one-child policy** in China, which imposes fines on urban couples who have more than one child. Due to the fact that some couples wish to have a male heir, many Chinese couples continue to have more than one child. The effectiveness of the policy in limiting overall population growth is controversial, as is the policy itself. Moreover, there are stories of female infanticide having occurred in some of the more rural areas of the country. Family planning education programs in other countries have had highly positive effects on limiting population growth rates and increasing standards of living. In spite of population control policies, the human population continues to grow. Because of the subsequent need to produce more and more food to feed our population, inequalities in access to food and other resources will continue to widen. The United Nations estimates the future world population size could vary from 6 billion (a

6. Paul R. Erlich, prologue to *The Population Bomb*, (1968; repr., New York: Ballantine, 1970).

decrease) to 16 billion people by the year 2100. There is no way to know whether human population growth will moderate to the point where the crisis described by Dr. Ehrlich will be averted.

Another consequence of population growth is the change and degradation of the natural environment. Many countries have attempted to reduce the human impact on climate change by limiting their emission of greenhouse gases. However, a global climate change treaty remains elusive, and many underdeveloped countries trying to improve their economic condition may be less likely to agree with such provisions without compensation if it means slowing their economic development. Furthermore, the role of human activity in causing climate change has become a hotly debated socio-political issue in some developed countries, including the United States. Thus, we enter the future with considerable uncertainty about our ability to curb human population growth and protect our environment to maintain the carrying capacity for the human species.

Visit this website (http://openstaxcollege.org/l/populations2) and select "Launch the movie" for an animation discussing the global impacts of human population growth.

19.4 | Community Ecology

By the end of this section, you will be able to:

- Discuss the predator-prey cycle
- Give examples of defenses against predation and herbivory
- Describe the competitive exclusion principle
- Give examples of symbiotic relationships between species
- Describe community structure and succession

In general, populations of one species never live in isolation from populations of other species. The interacting populations occupying a given habitat form an ecological community. The number of species occupying the same habitat and their relative abundance is known as the diversity of the community. Areas with low species diversity, such as the glaciers of Antarctica, still contain a wide variety of living organisms, whereas the diversity of tropical rainforests is so great that it cannot be accurately assessed. Scientists study ecology at the community level to understand how species interact with each other and compete for the same resources.

Predation and Herbivory

Perhaps the classical example of species interaction is the predator-prey relationship. The narrowest definition of the predator-prey interaction describes individuals of one population that kill and then consume the individuals of another population. Population sizes of predators and prey in a community are not constant over time, and they may vary in cycles that appear to be related. The most often cited example of predator-prey population dynamics is seen in the cycling of the lynx (predator) and the snowshoe hare (prey), using 100 years of trapping data from North America (Figure 19.13). This cycling of predator and prey population sizes has a period of approximately ten years, with the predator population lagging one to two years behind the prey population. An apparent explanation for this pattern is that as the hare numbers increase, there is more food available for the lynx, allowing the lynx population to increase as well. When the lynx population grows to a threshold level, however, they kill so many hares that hare numbers begin to decline, followed by a decline in the lynx population because of scarcity of food. When the lynx population is low, the hare population size begins to increase due, in part, to low predation pressure, starting the cycle anew.

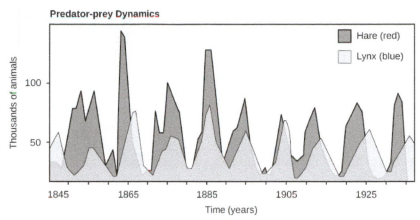

Figure 19.13 The cycling of snowshoe hare and lynx populations in Northern Ontario is an example of predator-prey dynamics.

Defense Mechanisms against Predation and Herbivory

Predation and predator avoidance are strong selective agents. Any heritable character that allows an individual of a prey population to better evade its predators will be represented in greater numbers in later generations. Likewise, traits that allow a predator to more efficiently locate and capture its prey will lead to a greater number of offspring and an increase in the commonness of the trait within the population. Such ecological relationships between specific populations lead to adaptations that are driven by reciprocal evolutionary responses in those populations. Species have evolved numerous mechanisms to escape predation and herbivory (the consumption of plants for food). Defenses may be mechanical, chemical, physical, or behavioral.

Mechanical defenses, such as the presence of armor in animals or thorns in plants, discourage predation and herbivory by discouraging physical contact (Figure 19.14a). Many animals produce or obtain chemical defenses from plants and store them to prevent predation. Many plant species produce secondary plant compounds that serve no function for the plant except that they are toxic to animals and discourage consumption. For example, the foxglove produces several compounds, including digitalis, that are extremely toxic when eaten (Figure 19.14b). (Biomedical scientists have purposed the chemical produced by foxglove as a heart medication, which has saved lives for many decades.)

(a)　　　(b)

Figure 19.14 The (a) honey locust tree uses thorns, a mechanical defense, against herbivores, while the (b) foxglove uses a chemical defense: toxins produces by the plant can cause nausea, vomiting, hallucinations, convulsions, or death when consumed. (credit a: modification of work by Huw Williams; credit b: modification of work by Philip Jägenstedt)

Many species use their body shape and coloration to avoid being detected by predators. The tropical walking stick is an insect with the coloration and body shape of a twig, which makes it very hard to see when it is stationary against a background of real twigs (Figure 19.15a). In another example, the chameleon can change its color to match its surroundings (Figure 19.15b).

(a)   (b)

Figure 19.15 (a) The tropical walking stick and (b) the chameleon use their body shape and/or coloration to prevent detection by predators. (credit a: modification of work by Linda Tanner; credit b: modification of work by Frank Vassen)

Some species use coloration as a way of warning predators that they are distasteful or poisonous. For example, the monarch butterfly caterpillar sequesters poisons from its food (plants and milkweeds) to make itself poisonous or distasteful to potential predators. The caterpillar is bright yellow and black to advertise its toxicity. The caterpillar is also able to pass the sequestered toxins on to the adult monarch, which is also dramatically colored black and red as a warning to potential predators. Fire-bellied toads produce toxins that make them distasteful to their potential predators. They have bright red or orange coloration on their bellies, which they display to a potential predator to advertise their poisonous nature and discourage an attack. These are only two examples of warning coloration, which is a relatively common adaptation. Warning coloration only works if a predator uses eyesight to locate prey and can learn—a naïve predator must experience the negative consequences of eating one before it will avoid other similarly colored individuals (Figure 19.16).

Figure 19.16 The fire-bellied toad has bright coloration on its belly that serves to warn potential predators that it is toxic. (credit: modification of work by Roberto Verzo)

While some predators learn to avoid eating certain potential prey because of their coloration, other species have evolved mechanisms to mimic this coloration to avoid being eaten, even though they themselves may not be unpleasant to eat or contain toxic chemicals. In some cases of **mimicry**, a harmless species imitates the warning coloration of a harmful species. Assuming they share the same predators, this coloration then protects the harmless ones. Many insect species mimic the coloration of wasps, which are stinging, venomous insects, thereby discouraging predation (Figure 19.17).

Download for free at https://openstax.org/details/books/concepts-biology

(a)   (b)

Figure 19.17 One form of mimicry is when a harmless species mimics the coloration of a harmful species, as is seen with the (a) wasp (*Polistes* sp.) and the (b) hoverfly (*Syrphus* sp.). (credit: modification of work by Tom Ings)

In other cases of mimicry, multiple species share the same warning coloration, but all of them actually have defenses. The commonness of the signal improves the compliance of all the potential predators. Figure 19.18 shows a variety of foul-tasting butterflies with similar coloration.

Figure 19.18 Several unpleasant-tasting *Heliconius* butterfly species share a similar color pattern with better-tasting varieties, an example of mimicry. (credit: Joron M, Papa R, Beltrán M, Chamberlain N, Mavárez J, et al.)

Go to this website (http://openstaxcollege.org/l/find_the_mimic2) to view stunning examples of mimicry.

Competitive Exclusion Principle

Resources are often limited within a habitat and multiple species may compete to obtain them. Ecologists have come to understand that all species have an ecological niche. A niche is the unique set of resources used by a species, which includes its interactions with other species. The **competitive exclusion principle** states that two species cannot occupy the same niche in a habitat: in other words, different species cannot coexist in a community if they are competing for all the same resources. This principle works because if there is an overlap in resource use and therefore competition between two species, then traits that lessen reliance on the shared resource will be selected for leading to evolution that reduces the overlap. If either species is unable to evolve to reduce competition, then the species that most efficiently exploits the resource will drive the other species to extinction. An experimental example of this principle is shown in Figure 19.19 with two protozoan species: *Paramecium aurelia* and *Paramecium caudatum*. When grown individually in the laboratory, they both thrive. But when they are placed together in the same test tube (habitat), *P. aurelia* outcompetes *P. caudatum* for food, leading to the latter's eventual extinction.

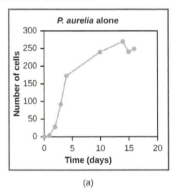
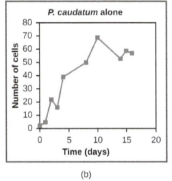
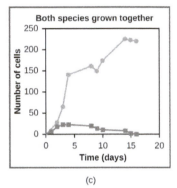

(a) (b) (c)

Figure 19.19 *Paramecium aurelia* and *Paramecium caudatum* grow well individually, but when they compete for the same resources, the *P. aurelia* outcompetes the *P. caudatum*.

Symbiosis

Symbiotic relationships are close, long-term interactions between individuals of different species. Symbioses may be commensal, in which one species benefits while the other is neither harmed nor benefited; mutualistic, in which both species benefit; or parasitic, in which the interaction harms one species and benefits the other.

Commensalism

A commensal relationship occurs when one species benefits from a close prolonged interaction, while the other neither benefits nor is harmed. Birds nesting in trees provide an example of a commensal relationship (Figure 19.20). The tree is not harmed by the presence of the nest among its branches. The nests are light and produce little strain on the structural integrity of the branch, and most of the leaves, which the tree uses to get energy by photosynthesis, are above the nest so they are unaffected. The bird, on the other hand, benefits greatly. If the bird had to nest in the open, its eggs and young would be vulnerable to predators. Many potential commensal relationships are difficult to identify because it is difficult to prove that one partner does not derive some benefit from the presence of the other.

Figure 19.20 The southern masked-weaver is starting to make a nest in a tree in Zambezi Valley, Zambia. This is an example of a commensal relationship, in which one species (the bird) benefits, while the other (the tree) neither benefits nor is harmed. (credit: "Hanay"/Wikimedia Commons)

Mutualism

A second type of symbiotic relationship is called **mutualism**, in which two species benefit from their interaction. For example, termites have a mutualistic relationship with protists that live in the insect's gut (Figure 19.21**a**). The termite benefits from the ability of the protists to digest cellulose. However, the protists are able to digest cellulose only because of the presence of symbiotic bacteria within their cells that produce the cellulase enzyme. The termite itself cannot do this: without the protozoa, it would not be able to obtain energy from its food (cellulose from the wood it chews and eats). The protozoa benefit by having a protective environment and a constant supply of food from the wood chewing actions of the termite. In turn, the protists benefit from the enzymes provided by their bacterial endosymbionts, while the bacteria benefit from a doubly protective environment and a constant source of nutrients from two hosts. Lichen are a mutualistic relationship between a fungus and photosynthetic algae or cyanobacteria (Figure 19.21**b**). The glucose produced by the algae provides nourishment for both organisms, whereas the physical structure of the lichen protects the algae from the elements and makes certain nutrients in the atmosphere more available to the algae. The algae of lichens can live independently given the right environment, but many of the fungal partners are unable to live on their own.

(a)   (b)

Figure 19.21 (a) Termites form a mutualistic relationship with symbiotic protozoa in their guts, which allow both organisms to obtain energy from the cellulose the termite consumes. (b) Lichen is a fungus that has symbiotic photosynthetic algae living in close association. (credit a: modification of work by Scott Bauer, USDA; credit b: modification of work by Cory Zanker)

Parasitism

A **parasite** is an organism that feeds off another without immediately killing the organism it is feeding on. In this relationship, the parasite benefits, but the organism being fed upon, the **host**, is harmed. The host is usually weakened by the parasite as it siphons resources the host would normally use to maintain itself. Parasites may kill their hosts, but there is usually selection to slow down this process to allow the parasite time to complete its reproductive cycle before it or its offspring are able to spread to another host.

The reproductive cycles of parasites are often very complex, sometimes requiring more than one host species. A tapeworm causes disease in humans when contaminated, undercooked meat such as pork, fish, or beef is consumed (Figure 19.22). The tapeworm can live inside the intestine of the host for several years, benefiting from the host's food, and it may grow to be over 50 feet long by adding segments. The parasite moves from one host species to a second host species in order to complete its life cycle. *Plasmodium falciparum* is another parasite: the protists that cause malaria, a significant disease in many parts of the world. Living inside human liver and red blood cells, the organism reproduces asexually in the human host and then sexually in the gut of blood-feeding mosquitoes to complete its life cycle. Thus malaria is spread from human to mosquito and back to human, one of many arthropod-borne infectious diseases of humans.

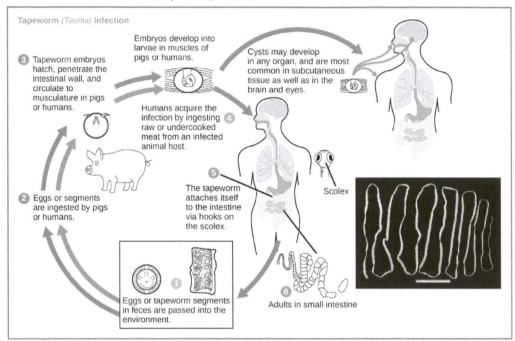

Figure 19.22 This diagram shows the life cycle of the tapeworm, a human worm parasite. (credit: modification of work by CDC)

To learn more about "Symbiosis in the Sea," watch this webisode (http://openstaxcollege.org/l/symbiosis) of Jonathan Bird's Blue World.

Characteristics of Communities

Communities are complex systems that can be characterized by their structure (the number and size of populations and their interactions) and dynamics (how the members and their interactions change over time). Understanding community structure and dynamics allows us to minimize impacts on ecosystems and manage ecological communities we benefit from.

Biodiversity

Ecologists have extensively studied one of the fundamental characteristics of communities: biodiversity. One measure of biodiversity used by ecologists is the number of different species in a particular area and their relative abundance. The area in question could be a habitat, a biome, or the entire biosphere. **Species richness** is the term used to describe the number of species living in a habitat or other unit. Species richness varies across the globe (Figure 19.23). Ecologists have struggled to understand the determinants of biodiversity. Species richness is related to latitude: the greatest species richness occurs near the equator and the lowest richness occurs near the poles. Other factors influence species richness as well. **Island biogeography** attempts to explain the great species richness found in isolated islands, and has found relationships between species richness, island size, and distance from the mainland.

Relative species abundance is the number individuals in a species relative to the total number of individuals in all species within a system. Foundation species, described below, often have the highest relative abundance of species.

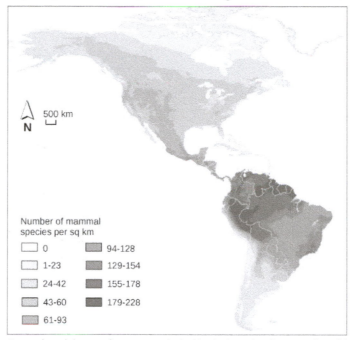

Figure 19.23 The greatest species richness for mammals in North America is associated in the equatorial latitudes. (credit: modification of work by NASA, CIESIN, Columbia University)

Foundation Species

Foundation species are considered the "base" or "bedrock" of a community, having the greatest influence on its overall structure. They are often primary producers, and they are typically an abundant organism. For example, kelp, a species of brown algae, is a foundation species that forms the basis of the kelp forests off the coast of California.

Foundation species may physically modify the environment to produce and maintain habitats that benefit the other organisms that use them. Examples include the kelp described above or tree species found in a forest. The photosynthetic corals of the coral reef also provide structure by physically modifying the environment (Figure 19.24). The exoskeletons of living and dead coral make up most of the reef structure, which protects many other species from waves and ocean currents.

Figure 19.24 Coral is the foundation species of coral reef ecosystems. (credit: Jim E. Maragos, USFWS)

Keystone Species

A **keystone species** is one whose presence has inordinate influence in maintaining the prevalence of various species in an ecosystem, the ecological community's structure, and sometimes its biodiversity. *Pisaster ochraceus*, the intertidal sea star, is a keystone species in the northwestern portion of the United States (Figure 19.25). Studies have shown that when this organism is removed from communities, mussel populations (their natural prey) increase, which completely alters the species composition and reduces biodiversity. Another keystone species is the banded tetra, a fish in tropical streams, which supplies nearly all of the phosphorus, a necessary inorganic nutrient, to the rest of the community. The banded tetra feeds largely on insects from the terrestrial ecosystem and then excretes phosphorus into the aquatic ecosystem. The relationships between populations in the community, and possibly the biodiversity, would change dramatically if these fish were to become extinct.

Figure 19.25 The *Pisaster ochraceus* sea star is a keystone species. (credit: Jerry Kirkhart)

biology IN ACTION

Invasive Species

Invasive species are non-native organisms that, when introduced to an area out of its native range, alter the community they invade. In the United States, invasive species like the purple loosestrife (*Lythrum salicaria*) and the zebra mussel (*Dreissena polymorpha*) have altered aquatic ecosystems, and some forests are threatened by the spread of common buckthorn (*Rhamnus cathartica*) and garlic mustard (*Alliaria petiolata*). Some well-known invasive animals include the emerald ash borer (*Agrilus planipennis*) and the European starling (*Sturnus vulgaris*). Whether enjoying a forest hike, taking a summer boat trip, or simply walking down an urban street, you have likely encountered an invasive species.

One of the many recent proliferations of an invasive species concerns the Asian carp in the United States. Asian carp were introduced to the United States in the 1970s by fisheries (commercial catfish ponds) and by sewage treatment facilities that used the fish's excellent filter feeding abilities to clean their ponds of excess plankton. Some of the fish escaped, and by the 1980s they had colonized many waterways of the Mississippi River basin, including the Illinois and Missouri Rivers.

Voracious feeders and rapid reproducers, Asian carp may outcompete native species for food and could lead to their extinction. One species, the grass carp, feeds on phytoplankton and aquatic plants. It competes with native species for these resources and alters nursery habitats for other fish by removing aquatic plants. Another species, the silver carp, competes with native fish that feed on zooplankton. In some parts of the Illinois River, Asian carp constitute 95 percent of the community's biomass. Although edible, the fish is bony and not desired in the United States. Moreover, their presence now threatens the native fish and fisheries of the Great Lakes, which are important to local economies and recreational anglers. Asian carp have even injured humans. The fish, frightened by the sound of approaching motorboats, thrust themselves into the air, often landing in the boat or directly hitting boaters.

The Great Lakes and their prized salmon and lake trout fisheries are being threatened by Asian carp. The carp are not yet present in the Great Lakes, and attempts are being made to prevent its access to the lakes through the Chicago Ship and Sanitary Canal, which is the only connection between the Mississippi River and Great Lakes basins. To prevent the Asian carp from leaving the canal, a series of electric barriers have been used to discourage their migration; however, the threat is significant enough that several states and Canada have sued to have the Chicago channel permanently cut off from Lake Michigan. Local and national politicians have weighed in on how to solve the problem. In general, governments have been ineffective in preventing or slowing the introduction of invasive species.

The issues associated with Asian carp show how population and community ecology, fisheries management, and politics intersect on issues of vital importance to the human food supply and economy. Socio-political issues like the Asian carp make extensive use of the sciences of population ecology, the study of members of a particular species occupying a habitat; and community ecology, the study of the interaction of all species within a habitat.

Community Dynamics

Community dynamics are the changes in community structure and composition over time, often following **environmental disturbances** such as volcanoes, earthquakes, storms, fires, and climate change. Communities with a relatively constant number of species are said to be at equilibrium. The equilibrium is dynamic with species identities and relationships changing over time, but maintaining relatively constant numbers. Following a disturbance, the community may or may not return to the equilibrium state.

Succession describes the sequential appearance and disappearance of species in a community over time after a severe disturbance. In **primary succession**, newly exposed or newly formed rock is colonized by living organisms; in **secondary succession**, a part of an ecosystem is disturbed and remnants of the previous community remain. In both cases, there is a sequential change in species until a more or less permanent community develops.

Primary Succession and Pioneer Species

Primary succession occurs when new land is formed, for example, following the eruption of volcanoes, such as those on the Big Island of Hawaii. As lava flows into the ocean, new land is continually being formed. On the Big Island, approximately 32 acres of land is added to it its size each year. Weathering and other natural forces break down the rock enough for the establishment of hearty species such as lichens and some plants, known as **pioneer species** (Figure 19.26). These species help to further break down the mineral-rich lava into soil where other, less hardy but more competitive species, such as grasses, shrubs, and trees, will grow and eventually replace the pioneer species. Over time the area will reach an equilibrium state, with a set of organisms quite different from the pioneer species.

Figure 19.26 During primary succession in lava on Maui, Hawaii, succulent plants are the pioneer species. (credit: Forest and Kim Starr)

Secondary succession

A classic example of secondary succession occurs in oak and hickory forests cleared by wildfire (Figure 19.27). Wildfires will burn most vegetation, and unless the animals can flee the area, they are killed. Their nutrients, however, are returned to the ground in the form of ash. Thus, although the community has been dramatically altered, there is a soil ecosystem present that provides a foundation for rapid recolonization.

Before the fire, the vegetation was dominated by tall trees with access to the major plant energy resource: sunlight. Their height gave them access to sunlight while also shading the ground and other low-lying species. After the fire, though, these trees are no longer dominant. Thus, the first plants to grow back are usually annual plants followed within a few years by quickly growing and spreading grasses and other pioneer species. Due, at least in part, to changes in the environment brought on by the growth of grasses and forbs, over many years, shrubs emerge along with small pine, oak, and hickory trees. These organisms are called intermediate species. Eventually, over 150 years, the forest will reach its equilibrium point and resemble the community before the fire. This equilibrium state is referred to as the **climax community**, which will remain until the next disturbance. The climax community is typically characteristic of a given climate and geology. Although the community in equilibrium looks the same once it is attained, the equilibrium is a dynamic one with constant changes in abundance and sometimes species identities. The return of a natural ecosystem after agricultural activities is also a well-documented secondary succession process.

Secondary Succession of an Oak and Hickory Forest

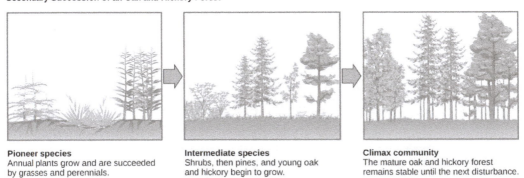

Pioneer species
Annual plants grow and are succeeded by grasses and perennials.

Intermediate species
Shrubs, then pines, and young oak and hickory begin to grow.

Climax community
The mature oak and hickory forest remains stable until the next disturbance.

Figure 19.27 Secondary succession is seen in an oak and hickory forest after a forest fire. A sequence of the community present at three successive times at the same location is depicted.

Download for free at https://openstax.org/details/books/concepts-biology

KEY TERMS

age structure the distribution of the proportion of population members in each age class

birth rate the number of births within a population at a specific point in time

carrying capacity the maximum number of individuals of a population that can be supported by the limited resources of a habitat

climax community the final stage of succession, where a stable community is formed by a characteristic assortment of plant and animal species

competitive exclusion principle no two species within a habitat can coexist indefinitely when they compete for the same resources at the same time and place

death rate the number of deaths within a population at a specific point in time

demography the statistical study of changes in populations over time

density-dependent regulation the regulation of population in which birth and death rates are dependent on population size

density-independent regulation the regulation of population in which the death rate is independent of the population size

environmental disturbance a change in the environment caused by natural disasters or human activities

exponential growth an accelerating growth pattern seen in populations where resources are not limiting

foundation species a species which often forms the major structural portion of the habitat

host an organism a parasite lives on

intraspecific competition the competition among members of the same species

island biogeography the study of life on island chains and how their geography interacts with the diversity of species found there

J-shaped growth curve the shape of an exponential growth curve

K-selected species a species suited to stable environments that produce a few, relatively large offspring and provide parental care

keystone species a species whose presence is key to maintaining biodiversity in an ecosystem and to upholding an ecological community's structure

life table a table showing the life expectancy of a population member based on its age

logistic growth the leveling off of exponential growth due to limiting resources

mark and recapture a method used to determine population size in mobile organisms

mimicry an adaptation in which an organism looks like another organism that is dangerous, toxic, or distasteful to its predators

mortality rate the proportion of population surviving to the beginning of an age interval that dies during that age interval

mutualism a symbiotic relationship between two species where both species benefit

one-child policy a policy in China to limit population growth by limiting urban couples to have only one child or face a penalty of a fine

parasite an organism that uses resources from another species: the host

pioneer species the first species to appear in primary and secondary succession

population density the number of population members divided by the area being measured

population size the number of individuals in a population

primary succession the succession on land that previously has had no life

quadrat a square within which a count of individuals is made that is combined with other such counts to determine population size and density in slow moving or stationary organisms

r-selected species a species suited to changing environments that produce many offspring and provide little or no parental care

relative species abundance the absolute population size of a particular species relative to the population size of other species within the community

S-shaped growth curve the shape of a logistic growth curve

secondary succession the succession in response to environmental disturbances that move a community away from its equilibrium

species distribution pattern the distribution of individuals within a habitat at a given point in time

species richness the number of different species in a community

survivorship curve a graph of the number of surviving population members versus the relative age of the member

zero population growth the steady population size where birth rates and death rates are equal

CHAPTER SUMMARY

19.1 Population Demographics and Dynamics

Populations are individuals of a species that live in a particular habitat. Ecologists measure characteristics of populations: size, density, and distribution pattern. Life tables are useful to calculate life expectancies of individual population members. Survivorship curves show the number of individuals surviving at each age interval plotted versus time.

19.2 Population Growth and Regulation

Populations with unlimited resources grow exponentially—with an accelerating growth rate. When resources become limiting, populations follow a logistic growth curve in which population size will level off at the carrying capacity.

Populations are regulated by a variety of density-dependent and density-independent factors. Life-history characteristics, such as age at first reproduction or numbers of offspring, are characteristics that evolve in populations just as anatomy or behavior can evolve over time. The model of r- and K-selection suggests that characters, and possibly suites of characters, may evolve adaptations to population stability near the carrying capacity (K-selection) or rapid population growth and collapse (r-selection). Species will exhibit adaptations somewhere on a continuum between these two extremes.

19.3 The Human Population

Earth's human population is growing exponentially. Humans have increased their carrying capacity through technology, urbanization, and harnessing the energy of fossil fuels. The age structure of a population allows us to predict population growth. Unchecked human population growth could have dire long-term effects on human welfare and Earth's ecosystems.

19.4 Community Ecology

Communities include all the different species living in a given area. The variety of these species is referred to as biodiversity. Many organisms have developed defenses against predation and herbivory, including mechanical defenses, warning coloration, and mimicry. Two species cannot exist indefinitely in the same habitat competing directly for the same resources. Species may form symbiotic relationships such as commensalism, mutualism, or parasitism. Community structure is described by its foundation and keystone species. Communities respond to environmental disturbances by succession: the predictable appearance of different types of plant species, until a stable community structure is established.

ART CONNECTION QUESTIONS

1. Figure 19.2 As this graph shows, population density typically decreases with increasing body size. Why do you think this is the case?

2. Figure 19.6 If the major food source of seals declines due to pollution or overfishing, which of the following would likely occur?

    a. The carrying capacity of seals would decrease, as would the seal population.

    b. The carrying capacity of seals would decrease, but the seal population would remain the same.

    c. The number of seal deaths would increase, but the number of births would also increase, so the population size would remain the same.

    d. The carrying capacity of seals would remain the same, but the population of seals would decrease.

3. Figure 19.11 Age structure diagrams for rapidly growing, slow growing, and stable populations are shown in stages 1 through 3. What type of population change do you think stage 4 represents?

REVIEW QUESTIONS

4. Which of the following methods will provide information to an ecologist about both the size and density of a population?

    a. mark and recapture

    b. mark and release

    c. quadrat

    d. life table

5. Which of the following is best at showing the life expectancy of an individual within a population?

    a. quadrat

    b. mark and recapture

    c. survivorship curve

    d. life table

6. Human populations have which type of survivorship curve?

    a. Type I

    b. Type II

    c. Type III

    d. Type IV

7. Species with limited resources usually exhibit a(n) _____ growth curve.

    a. logistic

    b. logical

    c. experimental

    d. exponential

8. The maximum growth rate characteristic of a species is called its _____.

    a. limit

    b. carrying capacity

    c. biotic potential

    d. exponential growth pattern

9. The population size of a species capable of being supported by the environment is called its _____.

    a. limit

    b. carrying capacity

    c. biotic potential

    d. logistic growth pattern

10. Species that have many offspring at one time are usually:

    a. *r*-selected

    b. *K*-selected

    c. both *r*- and *K*-selected

    d. not selected

11. A forest fire is an example of _____ regulation.

    a. density-dependent

    b. density-independent

    c. *r*-selected

    d. *K*-selected

12. A country with zero population growth is likely to be _____.

    a. in Africa

    b. in Asia

    c. economically developed

    d. economically underdeveloped

13. Which type of country has the greatest proportion of young individuals?

    a. economically developed

    b. economically underdeveloped

    c. countries with zero population growth

    d. countries in Europe

14. Which of the following is <u>not</u> a way that humans have increased the carrying capacity of the environment?

    a. agriculture

    b. using large amounts of natural resources

    c. domestication of animals

    d. use of language

15. The first species to live on new land, such as that formed from volcanic lava, are called_____.

    a. climax community

    b. keystone species

c. foundation species
d. pioneer species

16. A symbiotic relationship where both of the co-existing species benefit from the interaction is called _____.

a. commensalism
b. parasitism
c. mutualism

d. communism

17. When an invasive species alters the community structure it is introduced to, what can the consequence be?

a. extinction of economically important species
b. reduced predation on some native species
c. increased predation on some native species
d. all of the above

CRITICAL THINKING QUESTIONS

18. Describe how a researcher would determine the size of a penguin population in Antarctica using the mark and release method.

19. Describe the growth at various parts of the S-shaped curve of logistic growth.

20. Give an example of how density-dependent and density-independent factors might interact.

21. Describe the age structures in rapidly growing countries, slowly growing countries, and countries with zero population growth.

22. Describe the competitive exclusion principle and its effects on competing species.

23. Describe the potential effects when a keystone species is removed from a community.

20 | ECOSYSTEMS AND THE BIOSPHERE

(a)   (b)

Figure 20.1 The (a) Karner blue butterfly and (b) wild lupine live in oak-pine barren habitats in North America. (credit a: modification of work by John & Karen Hollingsworth, USFWS)

Chapter Outline
20.1: Energy Flow through Ecosystems
20.2: Biogeochemical Cycles
20.3: Terrestrial Biomes
20.4: Aquatic and Marine Biomes

Introduction

Ecosystem ecology is an extension of organismal, population, and community ecology. The ecosystem comprises all the biotic components (living things) and abiotic components (non-living things) in a particular geographic area. Some of the abiotic components include air, water, soil, and climate. Ecosystem biologists study how nutrients and energy are stored and moved among organisms and the surrounding atmosphere, soil, and water.

Wild lupine and Karner blue butterflies live in an oak-pine barren habitat in portions of Indiana, Michigan, Minnesota, Wisconsin, and New York (Figure 20.1). This habitat is characterized by natural disturbance in the form of fire and nutrient-poor soils that are low in nitrogen—important factors in the distribution of the plants that live in this habitat. Researchers interested in ecosystem ecology study the importance of limited resources in this ecosystem and the movement of resources (such as nutrients) through the biotic and abiotic portions of the ecosystem. Researchers also examine how organisms have adapted to their ecosystem.

20.1 | Energy Flow through Ecosystems

By the end of this section, you will be able to:

* Describe the basic types of ecosystems on Earth
* Differentiate between food chains and food webs and recognize the importance of each
* Describe how organisms acquire energy in a food web and in associated food chains
* Explain how the efficiency of energy transfers between trophic levels effects ecosystem

An **ecosystem** is a community of living organisms and their abiotic (non-living) environment. Ecosystems can be small, such as the tide pools found near the rocky shores of many oceans, or large, such as those found in the tropical rainforest of the Amazon in Brazil (Figure 20.2).

(a)   (b)

Figure 20.2 A (a) tidal pool ecosystem in Matinicus Island, Maine, is a small ecosystem, while the (b) Amazon rainforest in Brazil is a large ecosystem. (credit a: modification of work by Jim Kuhn; credit b: modification of work by Ivan Mlinaric)

There are three broad categories of ecosystems based on their general environment: freshwater, marine, and terrestrial. Within these three categories are individual ecosystem types based on the environmental habitat and organisms present.

Ecology of Ecosystems

Life in an ecosystem often involves competition for limited resources, which occurs both within a single species and between different species. Organisms compete for food, water, sunlight, space, and mineral nutrients. These resources provide the energy for metabolic processes and the matter to make up organisms' physical structures. Other critical factors influencing community dynamics are the components of its physical environment: a habitat's climate (seasons, sunlight, and rainfall), elevation, and geology. These can all be important environmental variables that determine which organisms can exist within a particular area.

Freshwater ecosystems are the least common, occurring on only 1.8 percent of Earth's surface. These systems comprise lakes, rivers, streams, and springs; they are quite diverse, and support a variety of animals, plants, fungi, protists and prokaryotes.

Marine ecosystems are the most common, comprising 75 percent of Earth's surface and consisting of three basic types: shallow ocean, deep ocean water, and deep ocean bottom. Shallow ocean ecosystems include extremely biodiverse coral reef ecosystems, yet the deep ocean water is known for large numbers of plankton and krill (small crustaceans) that support it. These two environments are especially important to aerobic respirators worldwide, as the phytoplankton perform 40 percent of all photosynthesis on Earth. Although not as diverse as the other two, deep ocean bottom ecosystems contain a wide variety of marine organisms. Such ecosystems exist even at depths where light is unable to penetrate through the water.

Terrestrial ecosystems, also known for their diversity, are grouped into large categories called biomes. A **biome** is a large-scale community of organisms, primarily defined on land by the dominant plant types that exist in geographic regions of the planet with similar climatic conditions. Examples of biomes include tropical rainforests, savannas, deserts, grasslands, temperate forests, and tundras. Grouping these ecosystems into just a few biome categories obscures the great diversity of the individual ecosystems within them. For example, the saguaro cacti (*Carnegiea gigantean*) and other plant life in the Sonoran Desert, in the United States, are relatively diverse compared with the desolate rocky desert of Boa Vista, an island off the coast of Western Africa (Figure 20.3).

(a)   (b)

Figure 20.3 Desert ecosystems, like all ecosystems, can vary greatly. The desert in (a) Saguaro National Park, Arizona, has abundant plant life, while the rocky desert of (b) Boa Vista island, Cape Verde, Africa, is devoid of plant life. (credit a: modification of work by Jay Galvin; credit b: modification of work by Ingo Wölbern)

Ecosystems and Disturbance

Ecosystems are complex with many interacting parts. They are routinely exposed to various disturbances: changes in the environment that affect their compositions, such as yearly variations in rainfall and temperature. Many disturbances are a result of natural processes. For example, when lightning causes a forest fire and destroys part of a forest ecosystem, the ground is eventually populated with grasses, followed by bushes and shrubs, and later mature trees: thus, the forest is restored to its former state. This process is so universal that ecologists have given it a name—succession. The impact of environmental disturbances caused by human activities is now as significant as the changes wrought by natural processes. Human agricultural practices, air pollution, acid rain, global deforestation, overfishing, oil spills, and illegal dumping on land and into the ocean all have impacts on ecosystems.

Equilibrium is a dynamic state of an ecosystem in which, despite changes in species numbers and occurrence, biodiversity remains somewhat constant. In ecology, two parameters are used to measure changes in ecosystems: resistance and resilience. The ability of an ecosystem to remain at equilibrium in spite of disturbances is called **resistance**. The speed at which an ecosystem recovers equilibrium after being disturbed is called **resilience**. Ecosystem resistance and resilience are especially important when considering human impact. The nature of an ecosystem may change to such a degree that it can lose its resilience entirely. This process can lead to the complete destruction or irreversible altering of the ecosystem.

Food Chains and Food Webs

A **food chain** is a linear sequence of organisms through which nutrients and energy pass as one organism eats another; the levels in the food chain are producers, primary consumers, higher-level consumers, and finally decomposers. These levels are used to describe ecosystem structure and dynamics. There is a single path through a food chain. Each organism in a food chain occupies a specific **trophic level** (energy level), its position in the food chain or food web.

In many ecosystems, the base, or foundation, of the food chain consists of photosynthetic organisms (plants or phytoplankton), which are called **producers**. The organisms that consume the producers are herbivores: the **primary consumers**. **Secondary consumers** are usually carnivores that eat the primary consumers. **Tertiary consumers** are carnivores that eat other carnivores. Higher-level consumers feed on the next lower trophic levels, and so on, up to the organisms at the top of the food chain: the **apex consumers**. In the Lake Ontario food chain, shown in Figure 20.4, the Chinook salmon is the apex consumer at the top of this food chain.

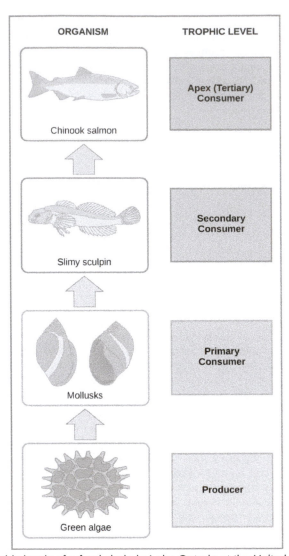

Figure 20.4 These are the trophic levels of a food chain in Lake Ontario at the United States–Canada border. Energy and nutrients flow from photosynthetic green algae at the base to the top of the food chain: the Chinook salmon. (credit: modification of work by National Oceanic and Atmospheric Administration/NOAA)

One major factor that limits the number of steps in a food chain is energy. Energy is lost at each trophic level and between trophic levels as heat and in the transfer to decomposers (Figure 20.5). Thus, after a limited number of trophic energy transfers, the amount of energy remaining in the food chain may not be great enough to support viable populations at yet a higher trophic level.

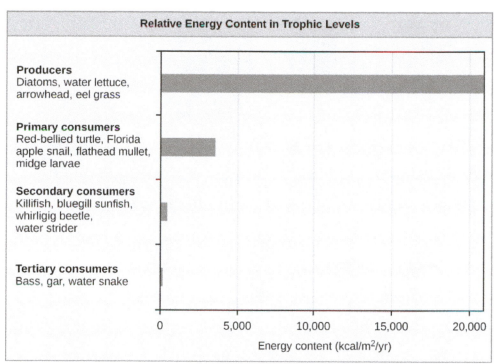

Figure 20.5 The relative energy in trophic levels in a Silver Springs, Florida, ecosystem is shown. Each trophic level has less energy available, and usually, but not always, supports a smaller mass of organisms at the next level.

There is a one problem when using food chains to describe most ecosystems. Even when all organisms are grouped into appropriate trophic levels, some of these organisms can feed on more than one trophic level; likewise, some of these organisms can also be fed on from multiple trophic levels. In addition, species feed on and are eaten by more than one species. In other words, the linear model of ecosystems, the food chain, is a hypothetical, overly simplistic representation of ecosystem structure. A holistic model—which includes all the interactions between different species and their complex interconnected relationships with each other and with the environment—is a more accurate and descriptive model for ecosystems. A **food web** is a concept that accounts for the multiple trophic (feeding) interactions between each species and the many species it may feed on, or that feed on it. In a food web, the several trophic connections between each species and the other species that interact with it may cross multiple trophic levels. The matter and energy movements of virtually all ecosystems are more accurately described by food webs (Figure 20.6).

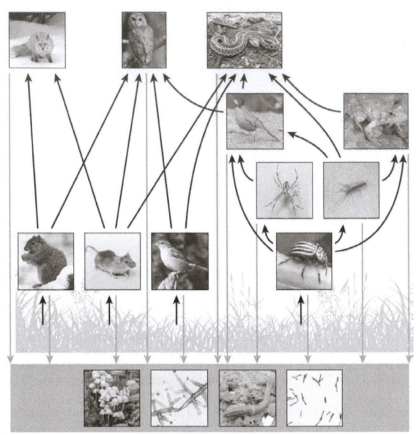

Figure 20.6 This food web shows the interactions between organisms across trophic levels. Arrows point from an organism that is consumed to the organism that consumes it. All the producers and consumers eventually become nourishment for the decomposers (fungi, mold, earthworms, and bacteria in the soil). (credit "fox": modification of work by Kevin Bacher, NPS; credit "owl": modification of work by John and Karen Hollingsworth, USFWS; credit "snake": modification of work by Steve Jurvetson; credit "robin": modification of work by Alan Vernon; credit "frog": modification of work by Alessandro Catenazzi; credit "spider": modification of work by "Sanba38"/Wikimedia Commons; credit "centipede": modification of work by "Bauerph"/Wikimedia Commons; credit "squirrel": modification of work by Dawn Huczek; credit "mouse": modification of work by NIGMS, NIH; credit "sparrow": modification of work by David Friel; credit "beetle": modification of work by Scott Bauer, USDA Agricultural Research Service; credit "mushrooms": modification of work by Chris Wee; credit "mold": modification of work by Dr. Lucille Georg, CDC; credit "earthworm": modification of work by Rob Hille; credit "bacteria": modification of work by Don Stalons, CDC)

C⌖NCEPT in ACTION

Head to this online interactive simulator (http://openstaxcollege.org/l/food_web) to investigate food web function. In the *Interactive Labs* box, under <u>Food Web</u>, click **Step 1**. Read the instructions first, and then click **Step 2** for additional instructions. When you are ready to create a simulation, in the upper-right corner of the *Interactive Labs* box, click **OPEN SIMULATOR**.

Two general types of food webs are often shown interacting within a single ecosystem. A **grazing food web** has plants or other photosynthetic organisms at its base, followed by herbivores and various carnivores. A **detrital food web** consists of a base of organisms that feed on decaying organic matter (dead organisms), including decomposers (which break down dead and decaying organisms) and detritivores (which consume organic detritus). These organisms are usually bacteria, fungi, and invertebrate animals that recycle organic material back into the biotic part of the ecosystem as they themselves are consumed by other organisms. As ecosystems require a method to recycle material from dead organisms, grazing food webs have an associated detrital food web. For example, in a meadow ecosystem, plants may support a grazing food web of different organisms, primary and other levels of consumers, while at the same time supporting a detrital food web of bacteria and fungi feeding off dead plants and animals. Simultaneously, a detrital food web can contribute energy to a grazing food web, as when a robin eats an earthworm.

How Organisms Acquire Energy in a Food Web

All living things require energy in one form or another. Energy is used by most complex metabolic pathways (usually in the form of ATP), especially those responsible for building large molecules from smaller compounds. Living organisms would not be able to assemble macromolecules (proteins, lipids, nucleic acids, and complex carbohydrates) from their monomers without a constant energy input.

Food-web diagrams illustrate how energy flows directionally through ecosystems. They can also indicate how efficiently organisms acquire energy, use it, and how much remains for use by other organisms of the food web. Energy is acquired by living things in two ways: autotrophs harness light or chemical energy and heterotrophs acquire energy through the consumption and digestion of other living or previously living organisms.

Photosynthetic and chemosynthetic organisms are **autotrophs**, which are organisms capable of synthesizing their own food (more specifically, capable of using inorganic carbon as a carbon source). Photosynthetic autotrophs (**photoautotrophs**) use sunlight as an energy source, and chemosynthetic autotrophs (**chemoautotrophs**) use inorganic molecules as an energy source. Autotrophs are critical for most ecosystems: they are the producer trophic level. Without these organisms, energy would not be available to other living organisms, and life itself would not be possible.

Photoautotrophs, such as plants, algae, and photosynthetic bacteria, are the energy source for a majority of the world's ecosystems. These ecosystems are often described by grazing and detrital food webs. Photoautotrophs harness the Sun's solar energy by converting it to chemical energy in the form of ATP (and NADP). The energy stored in ATP is used to synthesize complex organic molecules, such as glucose. The rate at which photosynthetic producers incorporate energy from the Sun is called **gross primary productivity**. However, not all of the energy incorporated by producers is available to the other organisms in the food web because producers must also grow and reproduce, which consumes energy. **Net primary productivity** is the energy that remains in the producers after accounting for these organisms' respiration and heat loss. The net productivity is then available to the primary consumers at the next trophic level.

Chemoautotrophs are primarily bacteria and archaea that are found in rare ecosystems where sunlight is not available, such as those associated with dark caves or hydrothermal vents at the bottom of the ocean (Figure 20.7). Many chemoautotrophs in hydrothermal vents use hydrogen sulfide (H_2S), which is released from the vents as a source of chemical energy; this allows them to synthesize complex organic molecules, such as glucose, for their own energy and, in turn, supplies energy to the rest of the ecosystem.

Figure 20.7 Swimming shrimp, a few squat lobsters, and hundreds of vent mussels are seen at a hydrothermal vent at the bottom of the ocean. As no sunlight penetrates to this depth, the ecosystem is supported by chemoautotrophic bacteria and organic material that sinks from the ocean's surface. This picture was taken in 2006 at the submerged NW Eifuku volcano off the coast of Japan by the National Oceanic and Atmospheric Administration (NOAA). The summit of this highly active volcano lies 1535 m below the surface.

Consequences of Food Webs: Biological Magnification

One of the most important consequences of ecosystem dynamics in terms of human impact is biomagnification. **Biomagnification** is the increasing concentration of persistent, toxic substances in organisms at each successive trophic level. These are substances that are fat soluble, not water soluble, and are stored in the fat reserves of each organism. Many substances have been shown to biomagnify, including classical studies with the pesticide dichlorodiphenyltrichloroethane (DDT), which were described in the 1960s bestseller, *Silent Spring* by Rachel Carson. DDT was a commonly used pesticide before its dangers to apex consumers, such as the bald eagle, became known. In aquatic ecosystems, organisms from each trophic level consumed many organisms in the lower level, which caused DDT to increase in birds (apex consumers) that ate fish. Thus, the birds accumulated sufficient amounts of DDT to cause fragility in their eggshells. This effect increased egg breakage during nesting and was shown to have devastating effects on these bird populations. The use of DDT was banned in the United States in the 1970s.

Other substances that biomagnify are polychlorinated biphenyls (PCB), which were used as coolant liquids in the United States until their use was banned in 1979, and heavy metals, such as mercury, lead, and cadmium. These substances are best studied in aquatic ecosystems, where predatory fish species accumulate very high concentrations of toxic substances that are at quite low concentrations in the environment and in producers. As illustrated in a study performed by the NOAA in the Saginaw Bay of Lake Huron of the North American Great Lakes (Figure 20.8), PCB concentrations increased from the producers of the ecosystem (phytoplankton) through the different trophic levels of fish species. The apex consumer, the walleye, has more than four times the amount of PCBs compared to phytoplankton. Also, based on results from other studies, birds that eat these fish may have PCB levels at least one order of magnitude higher than those found in the lake fish.

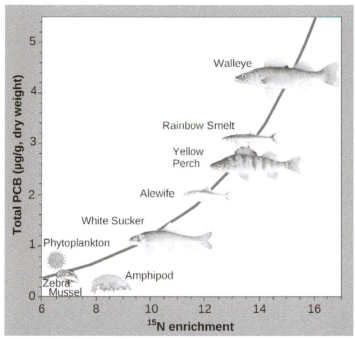

Figure 20.8 This chart shows the PCB concentrations found at the various trophic levels in the Saginaw Bay ecosystem of Lake Huron. Notice that the fish in the higher trophic levels accumulate more PCBs than those in lower trophic levels. (credit: Patricia Van Hoof, NOAA)

Other concerns have been raised by the biomagnification of heavy metals, such as mercury and cadmium, in certain types of seafood. The United States Environmental Protection Agency recommends that pregnant women and young children should not consume any swordfish, shark, king mackerel, or tilefish because of their high mercury content. These individuals are advised to eat fish low in mercury: salmon, shrimp, pollock, and catfish. Biomagnification is a good example of how ecosystem dynamics can affect our everyday lives, even influencing the food we eat.

20.2 | Biogeochemical Cycles

By the end of this section, you will be able to:

- Discuss the biogeochemical cycles of water, carbon, nitrogen, phosphorus, and sulfur
- Explain how human activities have impacted these cycles and the resulting potential consequences for Earth

Energy flows directionally through ecosystems, entering as sunlight (or inorganic molecules for chemoautotrophs) and leaving as heat during the transfers between trophic levels. Rather than flowing through an ecosystem, the matter that makes up living organisms is conserved and recycled. The six most common elements associated with organic molecules—carbon, nitrogen, hydrogen, oxygen, phosphorus, and sulfur—take a variety of chemical forms and may exist for long periods in the atmosphere, on land, in water, or beneath Earth's surface. Geologic processes, such as weathering, erosion, water drainage, and the subduction of the continental plates, all play a role in the cycling of elements on Earth. Because geology and chemistry have major roles in the study of this process, the recycling of inorganic matter between living organisms and their nonliving environment is called a **biogeochemical cycle**.

Water, which contains hydrogen and oxygen, is essential to all living processes. The **hydrosphere** is the area of Earth where water movement and storage occurs: as liquid water on the surface (rivers, lakes, oceans) and beneath the surface (groundwater) or ice, (polar ice caps and glaciers), and as water vapor in the atmosphere. Carbon is found in all organic macromolecules and is an important constituent of fossil fuels. Nitrogen is a major component of our nucleic acids and proteins and is critical to human agriculture. Phosphorus, a major component of nucleic acids, is one of the main ingredients (along with nitrogen) in artificial fertilizers used in agriculture, which has environmental impacts on our surface water. Sulfur, critical to the three-dimensional folding of proteins (as in disulfide binding), is released into the atmosphere by the burning of fossil fuels.

The cycling of these elements is interconnected. For example, the movement of water is critical for the leaching of nitrogen and phosphate into rivers, lakes, and oceans. The ocean is also a major reservoir for carbon. Thus, mineral nutrients are cycled, either rapidly or slowly, through the entire biosphere between the biotic and abiotic world and from one living organism to another.

Head to this website (http://openstaxcollege.org/l/biogeochemical) to learn more about biogeochemical cycles.

The Water Cycle

Water is essential for all living processes. The human body is more than one-half water and human cells are more than 70 percent water. Thus, most land animals need a supply of fresh water to survive. Of the stores of water on Earth, 97.5 percent is salt water (Figure 20.9). Of the remaining water, 99 percent is locked as underground water or ice. Thus, less than one percent of fresh water is present in lakes and rivers. Many living things are dependent on this small amount of surface fresh water supply, a lack of which can have important effects on ecosystem dynamics. Humans, of course, have developed technologies to increase water availability, such as digging wells to harvest groundwater, storing rainwater, and using desalination to obtain drinkable water from the ocean. Although this pursuit of drinkable water has been ongoing throughout human history, the supply of fresh water continues to be a major issue in modern times.

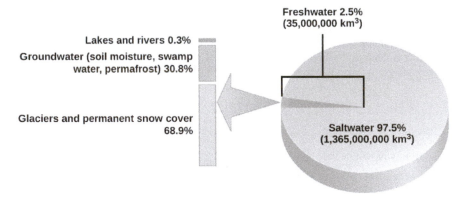

Figure 20.9 Only 2.5 percent of water on Earth is fresh water, and less than 1 percent of fresh water is easily accessible to living things.

The various processes that occur during the cycling of water are illustrated in Figure 20.10. The processes include the following:

- evaporation and sublimation
- condensation and precipitation
- subsurface water flow
- surface runoff and snowmelt
- streamflow

The water cycle is driven by the Sun's energy as it warms the oceans and other surface waters. This leads to evaporation (water to water vapor) of liquid surface water and sublimation (ice to water vapor) of frozen water, thus moving large amounts of water into the atmosphere as water vapor. Over time, this water vapor condenses into clouds as liquid or frozen droplets and eventually leads to precipitation (rain or snow), which returns water to Earth's surface. Rain reaching Earth's surface may evaporate again, flow over the surface, or percolate into the ground. Most easily observed is surface runoff: the flow of fresh water either from rain or melting ice. Runoff can make its way through streams and lakes to the oceans or flow directly to the oceans themselves.

In most natural terrestrial environments rain encounters vegetation before it reaches the soil surface. A significant percentage of water evaporates immediately from the surfaces of plants. What is left reaches the soil and begins to move down. Surface runoff will occur only if the soil becomes saturated with water in a heavy rainfall. Most water in the soil will be taken up by plant roots. The plant will use some of this water for its own metabolism, and some of that will find its way into animals that eat the plants, but much of it will be lost back to the atmosphere through a process known as evapotranspiration. Water enters the vascular system of the plant through the roots and evaporates, or transpires, through the stomata of the leaves. Water in the soil that is not taken up by a plant and that does not evaporate is able to percolate into the subsoil and bedrock. Here it forms groundwater.

Groundwater is a significant reservoir of fresh water. It exists in the pores between particles in sand and gravel, or in the fissures in rocks. Shallow groundwater flows slowly through these pores and fissures and eventually finds its way to a stream or lake where it becomes a part of the surface water again. Streams do not flow because they are replenished from rainwater directly; they flow because there is a constant inflow from groundwater below. Some groundwater is found very deep in the bedrock and can persist there for millennia. Most groundwater reservoirs, or aquifers, are the source of drinking or irrigation water drawn up through wells. In many cases these aquifers are being depleted faster than they are being replenished by water percolating down from above.

Rain and surface runoff are major ways in which minerals, including carbon, nitrogen, phosphorus, and sulfur, are cycled from land to water. The environmental effects of runoff will be discussed later as these cycles are described.

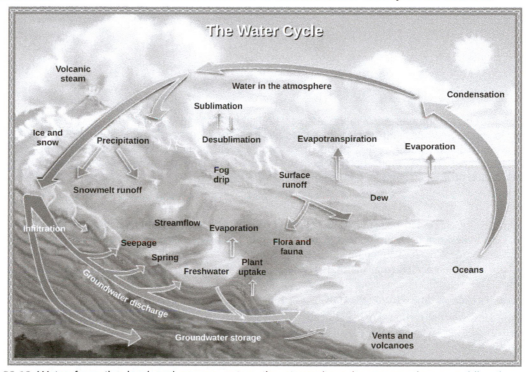

Figure 20.10 Water from the land and oceans enters the atmosphere by evaporation or sublimation, where it condenses into clouds and falls as rain or snow. Precipitated water may enter freshwater bodies or infiltrate the soil. The cycle is complete when surface or groundwater reenters the ocean. (credit: modification of work by John M. Evans and Howard Perlman, USGS)

The Carbon Cycle

Carbon is the fourth most abundant element in living organisms. Carbon is present in all organic molecules, and its role in the structure of macromolecules is of primary importance to living organisms. Carbon compounds contain energy, and many of these compounds from plants and algae have remained stored as fossilized carbon, which humans use as fuel. Since the 1800s, the use of fossil fuels has accelerated. As global demand for Earth's limited fossil fuel supplies has risen since the beginning of the Industrial Revolution, the amount of carbon dioxide in our atmosphere has increased as the fuels are burned. This increase in carbon dioxide has been associated with climate change and is a major environmental concern worldwide.

The carbon cycle is most easily studied as two interconnected subcycles: one dealing with rapid carbon exchange among living organisms and the other dealing with the long-term cycling of carbon through geologic processes. The entire carbon cycle is shown in Figure 20.11.

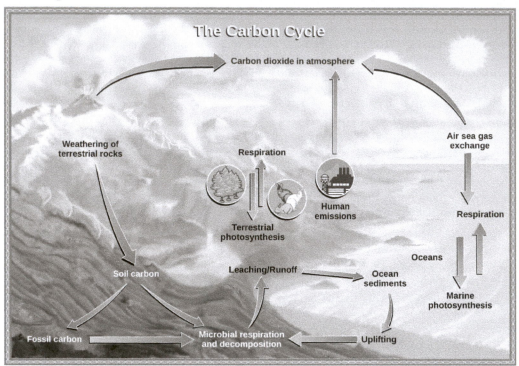

Figure 20.11 Carbon dioxide gas exists in the atmosphere and is dissolved in water. Photosynthesis converts carbon dioxide gas to organic carbon, and respiration cycles the organic carbon back into carbon dioxide gas. Long-term storage of organic carbon occurs when matter from living organisms is buried deep underground and becomes fossilized. Volcanic activity and, more recently, human emissions bring this stored carbon back into the carbon cycle. (credit: modification of work by John M. Evans and Howard Perlman, USGS)

The Biological Carbon Cycle

Living organisms are connected in many ways, even between ecosystems. A good example of this connection is the exchange of carbon between heterotrophs and autotrophs within and between ecosystems by way of atmospheric carbon dioxide. Carbon dioxide is the basic building block that autotrophs use to build multi-carbon, high-energy compounds, such as glucose. The energy harnessed from the Sun is used by these organisms to form the covalent bonds that link carbon atoms together. These chemical bonds store this energy for later use in the process of respiration. Most terrestrial autotrophs obtain their carbon dioxide directly from the atmosphere, while marine autotrophs acquire it in the dissolved form (carbonic acid, HCO_3^-). However the carbon dioxide is acquired, a byproduct of fixing carbon in organic compounds is oxygen. Photosynthetic organisms are responsible for maintaining approximately 21 percent of the oxygen content of the atmosphere that we observe today.

The partners in biological carbon exchange are the heterotrophs (especially the primary consumers, largely herbivores). Heterotrophs acquire the high-energy carbon compounds from the autotrophs by consuming them and breaking them down by respiration to obtain cellular energy, such as ATP. The most efficient type of respiration, aerobic respiration, requires oxygen obtained from the atmosphere or dissolved in water. Thus, there is a constant exchange of oxygen and carbon dioxide between the autotrophs (which need the carbon) and the heterotrophs (which need the oxygen). Autotrophs also respire and consume the organic molecules they form: using oxygen and releasing carbon dioxide. They release more oxygen gas as a waste product of photosynthesis than they use for their own respiration; therefore, there is excess available for the respiration of other aerobic organisms. Gas exchange through the atmosphere and water is one way that the carbon cycle connects all living organisms on Earth.

The Biogeochemical Carbon Cycle

The movement of carbon through land, water, and air is complex, and, in many cases, it occurs much more slowly geologically than the movement between living organisms. Carbon is stored for long periods in what are known as carbon reservoirs, which include the atmosphere, bodies of liquid water (mostly oceans), ocean sediment, soil, rocks (including fossil fuels), and Earth's interior.

As stated, the atmosphere is a major reservoir of carbon in the form of carbon dioxide that is essential to the process of photosynthesis. The level of carbon dioxide in the atmosphere is greatly influenced by the reservoir of carbon in the oceans. The exchange of carbon between the atmosphere and water reservoirs influences how much carbon is found in each, and each one affects the other reciprocally. Carbon dioxide (CO_2) from the atmosphere dissolves in water and, unlike oxygen and nitrogen gas, reacts with water molecules to form ionic compounds. Some of these ions combine with calcium ions in the seawater to form calcium carbonate ($CaCO_3$), a major component of the shells of marine organisms. These organisms eventually form sediments on the ocean floor. Over geologic time, the calcium carbonate forms limestone, which comprises the largest carbon reservoir on Earth.

On land, carbon is stored in soil as organic carbon as a result of the decomposition of living organisms or from weathering of terrestrial rock and minerals. Deeper under the ground, at land and at sea, are fossil fuels, the anaerobically decomposed remains of plants that take millions of years to form. Fossil fuels are considered a non-renewable resource because their use far exceeds their rate of formation. A **non-renewable resource** is either regenerated very slowly or not at all. Another way for carbon to enter the atmosphere is from land (including land beneath the surface of the ocean) by the eruption of volcanoes and other geothermal systems. Carbon sediments from the ocean floor are taken deep within Earth by the process of **subduction**: the movement of one tectonic plate beneath another. Carbon is released as carbon dioxide when a volcano erupts or from volcanic hydrothermal vents.

Carbon dioxide is also added to the atmosphere by the animal husbandry practices of humans. The large number of land animals raised to feed Earth's growing human population results in increased carbon-dioxide levels in the atmosphere caused by their respiration. This is another example of how human activity indirectly affects biogeochemical cycles in a significant way. Although much of the debate about the future effects of increasing atmospheric carbon on climate change focuses on fossils fuels, scientists take natural processes, such as volcanoes, plant growth, soil carbon levels, and respiration, into account as they model and predict the future impact of this increase.

The Nitrogen Cycle

Getting nitrogen into the living world is difficult. Plants and phytoplankton are not equipped to incorporate nitrogen from the atmosphere (which exists as tightly bonded, triple covalent N_2) even though this molecule comprises approximately 78 percent of the atmosphere. Nitrogen enters the living world via free-living and symbiotic bacteria, which incorporate nitrogen into their macromolecules through nitrogen fixation (conversion of N_2). Cyanobacteria live in most aquatic ecosystems where sunlight is present; they play a key role in nitrogen fixation. Cyanobacteria are able to use inorganic sources of nitrogen to "fix" nitrogen. *Rhizobium* bacteria live symbiotically in the root nodules of legumes (such as peas, beans, and peanuts) and provide them with the organic nitrogen they need. Free-living bacteria, such as *Azotobacter*, are also important nitrogen fixers.

Organic nitrogen is especially important to the study of ecosystem dynamics since many ecosystem processes, such as primary production and decomposition, are limited by the available supply of nitrogen. As shown in Figure 20.12, the nitrogen that enters living systems by nitrogen fixation is eventually converted from organic nitrogen back into nitrogen gas by bacteria. This process occurs in three steps in terrestrial systems: ammonification, nitrification, and denitrification. First, the ammonification process converts nitrogenous waste from living animals or from the remains of dead animals into ammonium (NH_4^+) by certain bacteria and fungi. Second, this ammonium is then converted to nitrites (NO_2^-) by nitrifying bacteria, such as *Nitrosomonas*, through nitrification. Subsequently, nitrites are converted to nitrates (NO_3^-) by similar organisms. Lastly, the process of denitrification occurs, whereby bacteria, such as *Pseudomonas* and *Clostridium*, convert the nitrates into nitrogen gas, thus allowing it to re-enter the atmosphere.

art CONNECTION

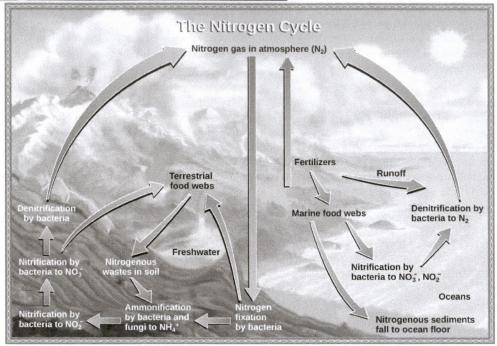

Figure 20.12 Nitrogen enters the living world from the atmosphere through nitrogen-fixing bacteria. This nitrogen and nitrogenous waste from animals is then processed back into gaseous nitrogen by soil bacteria, which also supply terrestrial food webs with the organic nitrogen they need. (credit: modification of work by John M. Evans and Howard Perlman, USGS)

Which of the following statements about the nitrogen cycle is false?

a. Ammonification converts organic nitrogenous matter from living organisms into ammonium (NH_4^+).

b. Denitrification by bacteria converts nitrates (NO_3^-)to nitrogen gas (N_2).

c. Nitrification by bacteria converts nitrates (NO_3^-)to nitrites (NO_2^-)

d. Nitrogen fixing bacteria convert nitrogen gas (N_2) into organic compounds.

Human activity can release nitrogen into the environment by two primary means: the combustion of fossil fuels, which releases different nitrogen oxides, and by the use of artificial fertilizers (which contain nitrogen and phosphorus compounds) in agriculture, which are then washed into lakes, streams, and rivers by surface runoff. Atmospheric nitrogen (other than N_2) is associated with several effects on Earth's ecosystems including the production of acid rain (as nitric acid, HNO_3) and greenhouse gas effects (as nitrous oxide, N_2O), potentially causing climate change. A major effect from fertilizer runoff is saltwater and freshwater **eutrophication**, a process whereby nutrient runoff causes the overgrowth of algae and a number of consequential problems.

A similar process occurs in the marine nitrogen cycle, where the ammonification, nitrification, and denitrification processes are performed by marine bacteria and archaea. Some of this nitrogen falls to the ocean floor as sediment, which can then be moved to land in geologic time by uplift of Earth's surface, and thereby incorporated into terrestrial rock. Although the movement of nitrogen from rock directly into living systems has been traditionally seen as insignificant compared with nitrogen fixed from the atmosphere, a recent study showed that this process may indeed be significant and should be included in any study of the global nitrogen cycle.[1]

---

1. Scott L. Morford, Benjamin Z. Houlton, and Randy A. Dahlgren, "Increased Forest Ecosystem Carbon and Nitrogen Storage from Nitrogen Rich Bedrock," *Nature* 477, no. 7362 (2011): 78–81.

The Phosphorus Cycle

Phosphorus is an essential nutrient for living processes; it is a major component of nucleic acids and phospholipids, and, as calcium phosphate, makes up the supportive components of our bones. Phosphorus is often the limiting nutrient (necessary for growth) in aquatic, particularly freshwater, ecosystems.

Phosphorus occurs in nature as the phosphate ion (PO_4^{3-}). In addition to phosphate runoff as a result of human activity, natural surface runoff occurs when it is leached from phosphate-containing rock by weathering, thus sending phosphates into rivers, lakes, and the ocean. This rock has its origins in the ocean. Phosphate-containing ocean sediments form primarily from the bodies of ocean organisms and from their excretions. However, volcanic ash, aerosols, and mineral dust may also be significant phosphate sources. This sediment then is moved to land over geologic time by the uplifting of Earth's surface. (Figure 20.13)

Phosphorus is also reciprocally exchanged between phosphate dissolved in the ocean and marine organisms. The movement of phosphate from the ocean to the land and through the soil is extremely slow, with the average phosphate ion having an oceanic residence time between 20,000 and 100,000 years.

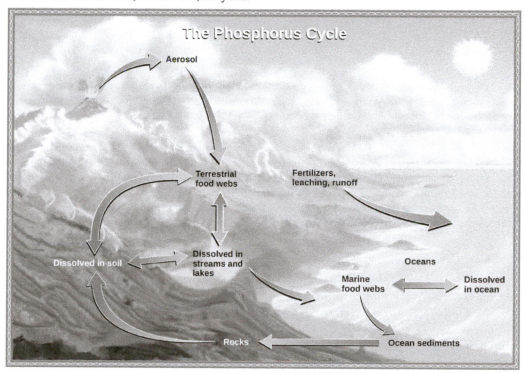

Figure 20.13 In nature, phosphorus exists as the phosphate ion (PO_4^{3-}). Weathering of rocks and volcanic activity releases phosphate into the soil, water, and air, where it becomes available to terrestrial food webs. Phosphate enters the oceans in surface runoff, groundwater flow, and river flow. Phosphate dissolved in ocean water cycles into marine food webs. Some phosphate from the marine food webs falls to the ocean floor, where it forms sediment. (credit: modification of work by John M. Evans and Howard Perlman, USGS)

Excess phosphorus and nitrogen that enter these ecosystems from fertilizer runoff and from sewage cause excessive growth of algae. The subsequent death and decay of these organisms depletes dissolved oxygen, which leads to the death of aquatic organisms, such as shellfish and finfish. This process is responsible for dead zones in lakes and at the mouths of many major rivers and for massive fish kills, which often occur during the summer months (see Figure 20.14).

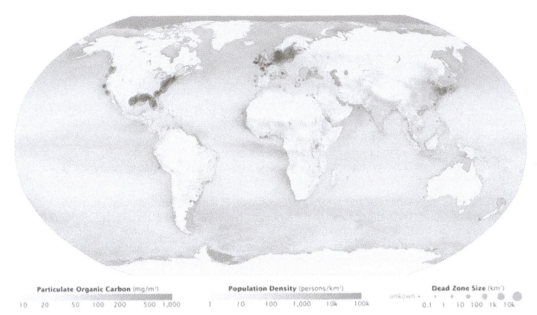

Particulate Organic Carbon (mg/m³)   Population Density (persons/km²)   Dead Zone Size (km²)

10   20   50 100 200 500 1,000   1   10   100 1,000 10k 100k   unkown   0.1   1   10 100 1k 10k

Figure 20.14 Dead zones occur when phosphorus and nitrogen from fertilizers cause excessive growth of microorganisms, which depletes oxygen and kills fauna. Worldwide, large dead zones are found in areas of high population density. (credit: Robert Simmon, Jesse Allen, NASA Earth Observatory)

A **dead zone** is an area in lakes and oceans near the mouths of rivers where large areas are periodically depleted of their normal flora and fauna; these zones can be caused by eutrophication, oil spills, dumping toxic chemicals, and other human activities. The number of dead zones has increased for several years, and more than 400 of these zones were present as of 2008. One of the worst dead zones is off the coast of the United States in the Gulf of Mexico: fertilizer runoff from the Mississippi River basin created a dead zone of over 8,463 square miles. Phosphate and nitrate runoff from fertilizers also negatively affect several lake and bay ecosystems including the Chesapeake Bay in the eastern United States.

ca·eers IN ACTION

Chesapeake Bay

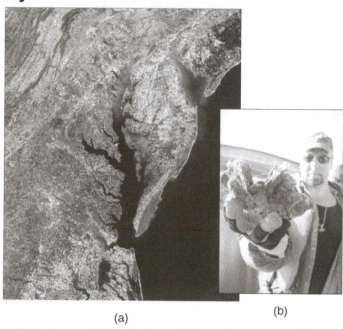

(a)   (b)

Figure 20.15 This (a) satellite image shows the Chesapeake Bay, an ecosystem affected by phosphate and nitrate runoff. A (b) member of the Army Corps of Engineers holds a clump of oysters being used as a part of the oyster restoration effort in the bay. (credit a: modification of work by NASA/MODIS; credit b: modification of work by U.S. Army)

The Chesapeake Bay (Figure 20.15a) is one of the most scenic areas on Earth; it is now in distress and is recognized as a case study of a declining ecosystem. In the 1970s, the Chesapeake Bay was one of the first aquatic ecosystems to have identified dead zones, which continue to kill many fish and bottom-dwelling species such as clams, oysters, and worms. Several species have declined in the Chesapeake Bay because surface water runoff contains excess nutrients from artificial fertilizer use on land. The source of the fertilizers (with high nitrogen and phosphate content) is not limited to agricultural practices. There are many nearby urban areas and more than 150 rivers and streams empty into the bay that are carrying fertilizer runoff from lawns and gardens. Thus, the decline of the Chesapeake Bay is a complex issue and requires the cooperation of industry, agriculture, and individual homeowners.

Of particular interest to conservationists is the oyster population (Figure 20.15b); it is estimated that more than 200,000 acres of oyster reefs existed in the bay in the 1700s, but that number has now declined to only 36,000 acres. Oyster harvesting was once a major industry for Chesapeake Bay, but it declined 88 percent between 1982 and 2007. This decline was caused not only by fertilizer runoff and dead zones, but also because of overharvesting. Oysters require a certain minimum population density because they must be in close proximity to reproduce. Human activity has altered the oyster population and locations, thus greatly disrupting the ecosystem.

The restoration of the oyster population in the Chesapeake Bay has been ongoing for several years with mixed success. Not only do many people find oysters good to eat, but the oysters also clean up the bay. They are filter feeders, and as they eat, they clean the water around them. Filter feeders eat by pumping a continuous stream of water over finely divided appendages (gills in the case of oysters) and capturing prokaryotes, plankton, and fine organic particles in their mucus. In the 1700s, it was estimated that it took only a few days for the oyster population to filter the entire volume of the bay. Today, with the changed water conditions, it is estimated that the present population would take nearly a year to do the same job.

Restoration efforts have been ongoing for several years by non-profit organizations such as the Chesapeake Bay Foundation. The restoration goal is to find a way to increase population density so the oysters can reproduce more efficiently. Many disease-resistant varieties (developed at the Virginia Institute of Marine Science for the College of William and Mary) are now available and have been used in the construction of experimental oyster reefs. Efforts by Virginia and Delaware to clean and restore the bay have been hampered because much of the pollution entering the bay comes from other states, which emphasizes the need for interstate cooperation to gain successful restoration.

The new, hearty oyster strains have also spawned a new and economically viable industry—oyster aquaculture—which not only supplies oysters for food and profit, but also has the added benefit of cleaning the bay.

The Sulfur Cycle

Sulfur is an essential element for the macromolecules of living things. As part of the amino acid cysteine, it is involved in the formation of proteins. As shown in Figure 20.16, sulfur cycles between the oceans, land, and atmosphere. Atmospheric sulfur is found in the form of sulfur dioxide (SO_2), which enters the atmosphere in three ways: first, from the decomposition of organic molecules; second, from volcanic activity and geothermal vents; and, third, from the burning of fossil fuels by humans.

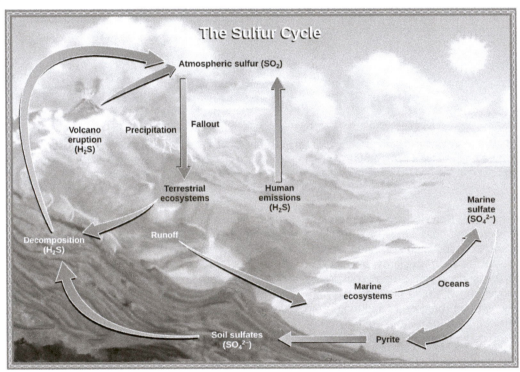

Figure 20.16 Sulfur dioxide from the atmosphere becomes available to terrestrial and marine ecosystems when it is dissolved in precipitation as weak sulfuric acid or when it falls directly to Earth as fallout. Weathering of rocks also makes sulfates available to terrestrial ecosystems. Decomposition of living organisms returns sulfates to the ocean, soil, and atmosphere. (credit: modification of work by John M. Evans and Howard Perlman, USGS)

On land, sulfur is deposited in four major ways: precipitation, direct fallout from the atmosphere, rock weathering, and geothermal vents (Figure 20.17). Atmospheric sulfur is found in the form of sulfur dioxide (SO_2), and as rain falls through the atmosphere, sulfur is dissolved in the form of weak sulfuric acid (H_2SO_4). Sulfur can also fall directly from the atmosphere in a process called **fallout**. Also, as sulfur-containing rocks weather, sulfur is released into the soil. These rocks originate from ocean sediments that are moved to land by the geologic uplifting of ocean sediments. Terrestrial ecosystems can then make use of these soil sulfates (SO_4^{2-}), which enter the food web by being taken up by plant roots. When these plants decompose and die, sulfur is released back into the atmosphere as hydrogen sulfide (H_2S) gas.

Figure 20.17 At this sulfur vent in Lassen Volcanic National Park in northeastern California, the yellowish sulfur deposits are visible near the mouth of the vent. (credit: "Calbear22"/Wikimedia Commons)

Sulfur enters the ocean in runoff from land, from atmospheric fallout, and from underwater geothermal vents. Some ecosystems rely on chemoautotrophs using sulfur as a biological energy source. This sulfur then supports marine ecosystems in the form of sulfates.

Human activities have played a major role in altering the balance of the global sulfur cycle. The burning of large quantities of fossil fuels, especially from coal, releases larger amounts of hydrogen sulfide gas into the atmosphere. As rain falls through this gas, it creates the phenomenon known as acid rain, which damages the natural environment by lowering the pH of lakes, thus killing many of the resident plants and animals. **Acid rain** is corrosive rain caused by rainwater falling to the ground through sulfur dioxide gas, turning it into weak sulfuric acid, which causes damage to aquatic ecosystems. Acid rain also affects the man-made environment through the chemical degradation of buildings. For example, many marble monuments, such as the Lincoln Memorial in Washington, DC, have suffered significant damage from acid rain over the years. These examples show the wide-ranging effects of human activities on our environment and the challenges that remain for our future.

20.3 | Terrestrial Biomes

By the end of this section, you will be able to:

- Identify the two major abiotic factors that determine the type of terrestrial biome in an area
- Recognize distinguishing characteristics of each of the eight major terrestrial biomes

Earth's biomes can be either terrestrial or aquatic. Terrestrial biomes are based on land, while aquatic biomes include both ocean and freshwater biomes. The eight major terrestrial biomes on Earth are each distinguished by characteristic temperatures and amount of precipitation. Annual totals and fluctuations of precipitation affect the kinds of vegetation and animal life that can exist in broad geographical regions. Temperature variation on a daily and seasonal basis is also important for predicting the geographic distribution of a biome. Since a biome is defined by climate, the same biome can occur in geographically distinct areas with similar climates (Figure 20.18). There are also large areas on Antarctica, Greenland, and in mountain ranges that are covered by permanent glaciers and support very little life. Strictly speaking, these are not considered biomes and in addition to extremes of cold, they are also often deserts with very low precipitation.

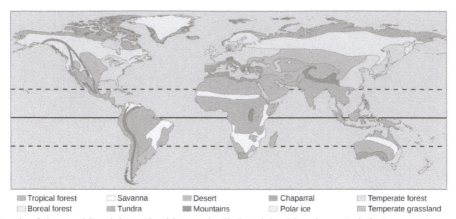

Figure 20.18 Each of the world's eight major biomes is distinguished by characteristic temperatures and amount of precipitation. Polar ice caps and mountains are also shown.

Tropical Forest

Tropical rainforests are also referred to as tropical wet forests. This biome is found in equatorial regions (Figure 20.18). Tropical rainforests are the most diverse terrestrial biome. This biodiversity is still largely unknown to science and is under extraordinary threat primarily through logging and deforestation for agriculture. Tropical rainforests have also been described as nature's pharmacy because of the potential for new drugs that is largely hidden in the chemicals produced by the huge diversity of plants, animals, and other organisms. The vegetation is characterized by plants with spreading roots and broad leaves that fall off throughout the year, unlike the trees of deciduous forests that lose their leaves in one season. These forests are "evergreen," year-round.

The temperature and sunlight profiles of tropical rainforests are stable in comparison to that of other terrestrial biomes, with average temperatures ranging from 20°C to 34°C (68°F to 93°F). Month-to-month temperatures are relatively constant in tropical rainforests, in contrast to forests further from the equator. This lack of temperature seasonality leads to year-round plant growth, rather than the seasonal growth seen in other biomes. In contrast to other ecosystems, a more constant daily amount of sunlight (11–12 hours per day) provides more solar radiation, thereby a longer period of time for plant growth.

The annual rainfall in tropical rainforests ranges from 250 cm to more than 450 cm (8.2–14.8 ft) with considerable seasonal variation. Tropical rainforests have wet months in which there can be more than 30 cm (11–12 in) of precipitation, as well as dry months in which there are fewer than 10 cm (3.5 in) of rainfall. However, the driest month of a tropical rainforest can still exceed the *annual* rainfall of some other biomes, such as deserts.

Tropical rainforests have high net primary productivity because the annual temperatures and precipitation values support rapid plant growth (Figure 20.19). However, the high rainfall quickly leaches nutrients from the soils of these forests, which are typically low in nutrients. Tropical rainforests are characterized by vertical layering of vegetation and the formation of distinct habitats for animals within each layer. On the forest floor is a sparse layer of plants and decaying plant matter. Above that is an understory of short, shrubby foliage. A layer of trees rises above this understory and is topped by a closed upper **canopy**—the uppermost overhead layer of branches and leaves. Some additional trees emerge through this closed upper canopy. These layers provide diverse and complex habitats for the variety of plants, animals, and other organisms within the tropical wet forests. Many species of animals use the variety of plants and the complex structure of the tropical wet forests for food and shelter. Some organisms live several meters above ground rarely ever descending to the forest floor.

Rainforests are not the only forest biome in the tropics; there are also tropical dry forests, which are characterized by a dry season of varying lengths. These forests commonly experience leaf loss during the dry season to one degree or another. The loss of leaves from taller trees during the dry season opens up the canopy and allows sunlight to the forest floor that allows the growth of thick ground-level brush, which is absent in tropical rainforests. Extensive tropical dry forests occur in Africa (including Madagascar), India, southern Mexico, and South America.

Figure 20.19 Species diversity is very high in tropical wet forests, such as these forests of Madre de Dios, Peru, near the Amazon River. (credit: Roosevelt Garcia)

Savannas

Savannas are grasslands with scattered trees, and they are found in Africa, South America, and northern Australia (Figure 20.18). Savannas are hot, tropical areas with temperatures averaging from 24°C –29°C (75°F –84°F) and an annual rainfall of 51–127 cm (20–50 in). Savannas have an extensive dry season and consequent fires. As a result, scattered in the grasses and forbs (herbaceous flowering plants) that dominate the savanna, there are relatively few trees (Figure 20.20). Since fire is an important source of disturbance in this biome, plants have evolved well-developed root systems that allow them to quickly re-sprout after a fire.

Figure 20.20 Although savannas are dominated by grasses, small woodlands, such as this one in Mount Archer National Park in Queensland, Australia, may dot the landscape. (credit: "Ethel Aardvark"/Wikimedia Commons)

Deserts

Subtropical deserts exist between 15° and 30° north and south latitude and are centered on the Tropic of Cancer and the Tropic of Capricorn (Figure 20.18). Deserts are frequently located on the downwind or lee side of mountain ranges, which create a rain shadow after prevailing winds drop their water content on the mountains. This is typical of the North American deserts, such as the Mohave and Sonoran deserts. Deserts in other regions, such as the Sahara Desert in northern Africa or the Namib Desert in southwestern Africa are dry because of the high-pressure, dry air descending at those latitudes. Subtropical deserts are very dry; evaporation typically exceeds precipitation. Subtropical hot deserts can have daytime soil surface temperatures above $60^\circ C$ ($140^\circ F$) and nighttime temperatures approaching $0^\circ C$ ($32^\circ F$). The temperature drops so far because there is little water vapor in the air to prevent radiative cooling of the land surface. Subtropical deserts are characterized by low annual precipitation of fewer than 30 cm (12 in) with little monthly variation and lack of predictability in rainfall. Some years may receive tiny amounts of rainfall, while others receive more. In some cases, the annual rainfall can be as low as 2 cm (0.8 in) in subtropical deserts located in central Australia ("the Outback") and northern Africa.

The low species diversity of this biome is closely related to its low and unpredictable precipitation. Despite the relatively low diversity, desert species exhibit fascinating adaptations to the harshness of their environment. Very dry deserts lack perennial vegetation that lives from one year to the next; instead, many plants are annuals that grow quickly and reproduce when rainfall does occur, then they die. Perennial plants in deserts are characterized by adaptations that conserve water: deep roots, reduced foliage, and water-storing stems (Figure 20.21). Seed plants in the desert produce seeds that can lie dormant for extended periods between rains. Most animal life in subtropical deserts has adapted to a nocturnal life, spending the hot daytime hours beneath the ground. The Namib Desert is the oldest on the planet, and has probably been dry for more than 55 million years. It supports a number of endemic species (species found only there) because of this great age. For example, the unusual gymnosperm *Welwitschia mirabilis* is the only extant species of an entire order of plants. There are also five species of reptiles considered endemic to the Namib.

In addition to subtropical deserts there are cold deserts that experience freezing temperatures during the winter and any precipitation is in the form of snowfall. The largest of these deserts are the Gobi Desert in northern China and southern Mongolia, the Taklimakan Desert in western China, the Turkestan Desert, and the Great Basin Desert of the United States.

Figure 20.21 Many desert plants have tiny leaves or no leaves at all to reduce water loss. The leaves of ocotillo, shown here in the Chihuahuan Desert in Big Bend National Park, Texas, appear only after rainfall and then are shed. (credit "bare ocotillo": "Leaflet"/Wikimedia Commons)

Chaparral

The **chaparral** is also called scrub forest and is found in California, along the Mediterranean Sea, and along the southern coast of Australia (Figure 20.18). The annual rainfall in this biome ranges from 65 cm to 75 cm (25.6–29.5 in) and the majority of the rain falls in the winter. Summers are very dry and many chaparral plants are dormant during the summertime. The chaparral vegetation is dominated by shrubs and is adapted to periodic fires, with some plants producing seeds that germinate only after a hot fire. The ashes left behind after a fire are rich in nutrients like nitrogen that fertilize the soil and promote plant regrowth. Fire is a natural part of the maintenance of this biome and frequently threatens human habitation in this biome in the U.S. (Figure 20.22).

Figure 20.22 The chaparral is dominated by shrubs. (credit: Miguel Vieira)

Temperate Grasslands

Temperate grasslands are found throughout central North America, where they are also known as prairies, and in Eurasia, where they are known as steppes (Figure 20.18). Temperate grasslands have pronounced annual fluctuations in temperature with hot summers and cold winters. The annual temperature variation produces specific growing seasons for plants. Plant growth is possible when temperatures are warm enough to sustain plant growth, which occurs in the spring, summer, and fall.

Annual precipitation ranges from 25.4 cm to 88.9 cm (10–35 in). Temperate grasslands have few trees except for those found growing along rivers or streams. The dominant vegetation tends to consist of grasses. The treeless condition is maintained by low precipitation, frequent fires, and grazing (Figure 20.23). The vegetation is very dense and the soils are fertile because the subsurface of the soil is packed with the roots and rhizomes (underground stems) of these grasses. The roots and rhizomes act to anchor plants into the ground and replenish the organic material (humus) in the soil when they die and decay.

Figure 20.23 The American bison (*Bison bison*), more commonly called the buffalo, is a grazing mammal that once populated American prairies in huge numbers. (credit: Jack Dykinga, USDA ARS)

Fires, which are a natural disturbance in temperate grasslands, can be ignited by lightning strikes. It also appears that the lightning-caused fire regime in North American grasslands was enhanced by intentional burning by humans. When fire is suppressed in temperate grasslands, the vegetation eventually converts to scrub and dense forests. Often, the restoration or management of temperate grasslands requires the use of controlled burns to suppress the growth of trees and maintain the grasses.

Temperate Forests

Temperate forests are the most common biome in eastern North America, Western Europe, Eastern Asia, Chile, and New Zealand (Figure 20.18). This biome is found throughout mid-latitude regions. Temperatures range between -30°C and 30°C (-22°F to 86°F) and drop to below freezing on an annual basis. These temperatures mean that temperate forests have defined growing seasons during the spring, summer, and early fall. Precipitation is relatively constant throughout the year and ranges between 75 cm and 150 cm (29.5–59 in).

Deciduous trees are the dominant plant in this biome with fewer evergreen conifers. Deciduous trees lose their leaves each fall and remain leafless in the winter. Thus, little photosynthesis occurs during the dormant winter period. Each spring, new leaves appear as temperature increases. Because of the dormant period, the net primary productivity of temperate forests is less than that of tropical rainforests. In addition, temperate forests show far less diversity of tree species than tropical rainforest biomes.

The trees of the temperate forests leaf out and shade much of the ground; however, more sunlight reaches the ground in this biome than in tropical rainforests because trees in temperate forests do not grow as tall as the trees in tropical rainforests. The soils of the temperate forests are rich in inorganic and organic nutrients compared to tropical rainforests. This is because of the thick layer of leaf litter on forest floors and reduced leaching of nutrients by rainfall. As this leaf litter decays, nutrients are returned to the soil. The leaf litter also protects soil from erosion, insulates the ground, and provides habitats for invertebrates and their predators (Figure 20.24).

Figure 20.24 Deciduous trees are the dominant plant in the temperate forest. (credit: Oliver Herold)

Boreal Forests

The **boreal forest**, also known as taiga or coniferous forest, is found roughly between 50° and 60° north latitude across most of Canada, Alaska, Russia, and northern Europe (Figure 20.18). Boreal forests are also found above a certain elevation (and below high elevations where trees cannot grow) in mountain ranges throughout the Northern Hemisphere. This biome has cold, dry winters and short, cool, wet summers. The annual precipitation is from 40 cm to 100 cm (15.7–39 in) and usually takes the form of snow; little evaporation occurs because of the cold temperatures.

The long and cold winters in the boreal forest have led to the predominance of cold-tolerant cone-bearing plants. These are evergreen coniferous trees like pines, spruce, and fir, which retain their needle-shaped leaves year-round. Evergreen trees can photosynthesize earlier in the spring than deciduous trees because less energy from the Sun is required to warm a needle-like leaf than a broad leaf. Evergreen trees grow faster than deciduous trees in the boreal forest. In addition, soils in boreal forest regions tend to be acidic with little available nitrogen. Leaves are a nitrogen-rich structure and deciduous trees must produce a new set of these nitrogen-rich structures each year. Therefore, coniferous trees that retain nitrogen-rich needles in a nitrogen limiting environment may have had a competitive advantage over the broad-leafed deciduous trees.

The net primary productivity of boreal forests is lower than that of temperate forests and tropical wet forests. The aboveground biomass of boreal forests is high because these slow-growing tree species are long-lived and accumulate standing biomass over time. Species diversity is less than that seen in temperate forests and tropical rainforests. Boreal forests lack the layered forest structure seen in tropical rainforests or, to a lesser degree, temperate forests. The structure of a boreal forest is often only a tree layer and a ground layer. When conifer needles are dropped, they decompose more slowly than broad leaves; therefore, fewer nutrients are returned to the soil to fuel plant growth (Figure 20.25).

Figure 20.25 The boreal forest (taiga) has low lying plants and conifer trees. (credit: L.B. Brubaker, NOAA)

Arctic Tundra

The **Arctic tundra** lies north of the subarctic boreal forests and is located throughout the Arctic regions of the Northern Hemisphere (Figure 20.18). Tundra also exists at elevations above the tree line on mountains. The average winter temperature is –34°C (–29.2°F) and the average summer temperature is 3°C–12°C (37°F –52°F). Plants in the Arctic tundra have a short growing season of approximately 50–60 days. However, during this time, there are almost 24 hours of daylight and plant growth is rapid. The annual precipitation of the Arctic tundra is low (15–25 cm or 6–10 in) with little annual variation in precipitation. And, as in the boreal forests, there is little evaporation because of the cold temperatures.

Plants in the Arctic tundra are generally low to the ground and include low shrubs, grasses, lichens, and small flowering plants (Figure 20.26). There is little species diversity, low net primary productivity, and low aboveground biomass. The soils of the Arctic tundra may remain in a perennially frozen state referred to as **permafrost**. The permafrost makes it impossible for roots to penetrate far into the soil and slows the decay of organic matter, which inhibits the release of nutrients from organic matter. The melting of the permafrost in the brief summer provides water for a burst of productivity while temperatures and long days permit it. During the growing season, the ground of the Arctic tundra can be completely covered with plants or lichens.

Figure 20.26 Low-growing plants such as shrub willow dominate the tundra landscape during the summer, shown here in the Arctic National Wildlife Refuge. (credit: Arctic National Wildlife Refuge, USFWS)

Watch this Assignment Discovery: Biomes (http://openstaxcollege.org/l/biomes) video for an overview of biomes. To explore further, select one of the biomes on the extended playlist: desert, savanna, temperate forest, temperate grassland, tropic, tundra.

20.4 | Aquatic and Marine Biomes

By the end of this section, you will be able to:

- Describe the effects of abiotic factors on the composition of plant and animal communities in aquatic biomes
- Compare the characteristics of the ocean zones
- Summarize the characteristics of standing water and flowing water in freshwater biomes

Like terrestrial biomes, aquatic biomes are influenced by abiotic factors. In the case of aquatic biomes the abiotic factors include light, temperature, flow regime, and dissolved solids. The aquatic medium—water— has different physical and chemical properties than air. Even if the water in a pond or other body of water is perfectly clear (there are no suspended particles), water, on its own, absorbs light. As one descends deep enough into a body of water, eventually there will be a depth at which the sunlight cannot reach. While there are some abiotic and biotic factors in a terrestrial ecosystem that shade light (like fog, dust, or insect swarms), these are not usually permanent features of the environment. The importance of light in aquatic biomes is central to the communities of organisms found in both freshwater and marine ecosystems because it controls productivity through photosynthesis.

In addition to light, solar radiation warms bodies of water and many exhibit distinct layers of water at differing temperatures. The water temperature affects the organisms' rates of growth and the amount of dissolved oxygen available for respiration.

The movement of water is also important in many aquatic biomes. In rivers, the organisms must obviously be adapted to the constant movement of the water around them, but even in larger bodies of water such as the oceans, regular currents and tides impact availability of nutrients, food resources, and the presence of the water itself.

Finally, all natural water contains dissolved solids, or salts. Fresh water contains low levels of such dissolved substances because the water is rapidly recycled through evaporation and precipitation. The oceans have a relatively constant high salt content. Aquatic habitats at the interface of marine and freshwater ecosystems have complex and variable salt environments that range between freshwater and marine levels. These are known as brackish water environments. Lakes located in closed drainage basins concentrate salt in their waters and can have extremely high salt content that only a few and highly specialized species are able to inhabit.

Marine Biomes

The ocean is a continuous body of salt water that is relatively uniform in chemical composition. It is a weak solution of mineral salts and decayed biological matter. Within the ocean, coral reefs are a second type of marine biome. Estuaries, coastal areas where salt water and fresh water mix, form a third unique marine biome.

The ocean is categorized by several zones (Figure 20.28). All of the ocean's open water is referred to as the **pelagic realm** (or zone). The **benthic realm** (or zone) extends along the ocean bottom from the shoreline to the deepest parts of the ocean floor. From the surface to the bottom or the limit to which photosynthesis occurs is the **photic zone** (approximately 200 m or 650 ft). At depths greater than 200 m, light cannot penetrate; thus, this is referred to as the **aphotic zone**. The majority of the ocean is aphotic and lacks sufficient light for photosynthesis. The deepest part of the ocean, the Challenger Deep (in the Mariana Trench, located in the western Pacific Ocean), is about 11,000 m (about 6.8 mi) deep. To give some perspective on the depth of this trench, the ocean is, on average, 4267 m or 14,000 ft deep.

Ocean

The physical diversity of the ocean has a significant influence on the diversity of organisms that live within it. The ocean is categorized into different zones based on how far light reaches into the water. Each zone has a distinct group of species adapted to the biotic and abiotic conditions particular to that zone.

The **intertidal zone** (Figure 20.28) is the oceanic region that is closest to land. With each tidal cycle, the intertidal zone alternates between being inundated with water and left high and dry. Generally, most people think of this portion of the ocean as a sandy beach. In some cases, the intertidal zone is indeed a sandy beach, but it can also be rocky, muddy, or dense with tangled roots in mangrove forests. The intertidal zone is an extremely variable environment because of tides. Organisms may be exposed to air at low tide and are underwater during high tide. Therefore, living things that thrive in the intertidal zone are often adapted to being dry for long periods of time. The shore of the intertidal zone is also repeatedly struck by waves and the organisms found there are adapted to withstand damage from the pounding action of the waves (Figure 20.27). The exoskeletons of shoreline crustaceans (such as the shore crab, *Carcinus maenas*) are tough and protect them from desiccation (drying out) and wave damage. Another consequence of the pounding waves is that few algae and plants establish themselves in constantly moving sand or mud.

Figure 20.27 Sea stars, sea urchins, and mussel shells are often found in the intertidal zone, shown here in Kachemak Bay, Alaska. (credit: NOAA)

The **neritic zone** (Figure 20.28) extends from the margin of the intertidal zone to depths of about 200 m (or 650 ft) at the edge of the continental shelf. When the water is relatively clear, photosynthesis can occur in the neritic zone. The water contains silt and is well-oxygenated, low in pressure, and stable in temperature. These factors all contribute to the neritic zone having the highest productivity and biodiversity of the ocean. Phytoplankton, including photosynthetic bacteria and larger species of algae, are responsible for the bulk of this primary productivity. Zooplankton, protists, small fishes, and shrimp feed on the producers and are the primary food source for most of the world's fisheries. The majority of these fisheries exist within the neritic zone.

Beyond the neritic zone is the open ocean area known as the **oceanic zone** (Figure 20.28). Within the oceanic zone there is thermal stratification. Abundant phytoplankton and zooplankton support populations of fish and whales. Nutrients are scarce and this is a relatively less productive part of the marine biome. When photosynthetic organisms and the organisms that feed on them die, their bodies fall to the bottom of the ocean where they remain; the open ocean lacks a process for bringing the organic nutrients back up to the surface.

Beneath the pelagic zone is the benthic realm, the deepwater region beyond the continental shelf (Figure 20.28). The bottom of the benthic realm is comprised of sand, silt, and dead organisms. Temperature decreases as water depth increases. This is a nutrient-rich portion of the ocean because of the dead organisms that fall from the upper layers of the ocean. Because of this high level of nutrients, a diversity of fungi, sponges, sea anemones, marine worms, sea stars, fishes, and bacteria exists.

The deepest part of the ocean is the **abyssal zone**, which is at depths of 4000 m or greater. The abyssal zone (Figure 20.28) is very cold and has very high pressure, high oxygen content, and low nutrient content. There are a variety of invertebrates and fishes found in this zone, but the abyssal zone does not have photosynthetic organisms. Chemosynthetic bacteria use the hydrogen sulfide and other minerals emitted from deep hydrothermal vents. These chemosynthetic bacteria use the hydrogen sulfide as an energy source and serve as the base of the food chain found around the vents.

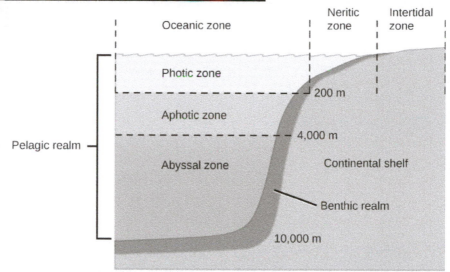

Figure 20.28 The ocean is divided into different zones based on water depth, distance from the shoreline, and light penetration.

In which of the following regions would you expect to find photosynthetic organisms?

a. The aphotic zone, the neritic zone, the oceanic zone, and the benthic realm.
b. The photic zone, the intertidal zone, the neritic zone, and the oceanic zone.
c. The photic zone, the abyssal zone, the neritic zone, and the oceanic zone.
d. The pelagic realm, the aphotic zone, the neritic zone, and the oceanic zone.

Coral Reefs

Coral reefs are ocean ridges formed by marine invertebrates living in warm shallow waters within the photic zone of the ocean. They are found within 30° north and south of the equator. The Great Barrier Reef is a well-known reef system located several miles off the northeastern coast of Australia. Other coral reefs are fringing islands, which are directly adjacent to land, or atolls, which are circular reefs surrounding a former island that is now underwater. The coral-forming colonies of organisms (members of phylum Cnidaria) secrete a calcium carbonate skeleton. These calcium-rich skeletons slowly accumulate, thus forming the underwater reef (Figure 20.29). Corals found in shallower waters (at a depth of approximately 60 m or about 200 ft) have a mutualistic relationship with photosynthetic unicellular protists. The relationship provides corals with the majority of the nutrition and the energy they require. The waters in which these corals live are nutritionally poor and, without this mutualism, it would not be possible for large corals to grow because there are few planktonic organisms for them to feed on. Some corals living in deeper and colder water do not have a mutualistic relationship with protists; these corals must obtain their energy exclusively by feeding on plankton using stinging cells on their tentacles.

In this National Oceanic and Atmospheric Administration (NOAA) video (http://openstaxcollege.org/l/coral_organisms) , marine ecologist Dr. Peter Etnoyer discusses his research on coral organisms.

Coral reefs are one of the most diverse biomes. It is estimated that more than 4000 fish species inhabit coral reefs. These fishes can feed on coral, the **cryptofauna** (invertebrates found within the calcium carbonate structures of the coral reefs), or the seaweed and algae that are associated with the coral. These species include predators, herbivores, or planktivores. Predators are animal species that hunt and are carnivores or "flesh eaters." Herbivores eat plant material, and **planktivores** eat plankton.

Figure 20.29 Coral reefs are formed by the calcium carbonate skeletons of coral organisms, which are marine invertebrates in the phylum Cnidaria. (credit: Terry Hughes)

Global Decline of Coral Reefs

It takes a long time to build a coral reef. The animals that create coral reefs do so over thousands of years, continuing to slowly deposit the calcium carbonate that forms their characteristic ocean homes. Bathed in warm tropical waters, the coral animals and their symbiotic protist partners evolved to survive at the upper limit of ocean water temperature.

Together, climate change and human activity pose dual threats to the long-term survival of the world's coral reefs. The main cause of killing of coral reefs is warmer-than-usual surface water. As global warming raises ocean temperatures, coral reefs are suffering. The excessive warmth causes the coral organisms to expel their endosymbiotic, food-producing protists, resulting in a phenomenon known as bleaching. The colors of corals are a result of the particular protist endosymbiont, and when the protists leave, the corals lose their color and turn white, hence the term "bleaching."

Rising levels of atmospheric carbon dioxide further threaten the corals in other ways; as carbon dioxide dissolves in ocean waters, it lowers pH, thus increasing ocean acidity. As acidity increases, it interferes with the calcification that normally occurs as coral animals build their calcium carbonate homes.

When a coral reef begins to die, species diversity plummets as animals lose food and shelter. Coral reefs are also economically important tourist destinations, so the decline of coral reefs poses a serious threat to coastal economies.

Human population growth has damaged corals in other ways, too. As human coastal populations increase, the runoff of sediment and agricultural chemicals has increased, causing some of the once-clear tropical waters to become cloudy. At the same time, overfishing of popular fish species has allowed the predator species that eat corals to go unchecked.

Although a rise in global temperatures of 1°C–2°C (a conservative scientific projection) in the coming decades may not seem large, it is very significant to this biome. When change occurs rapidly, species can become extinct before evolution leads to newly adapted species. Many scientists believe that global warming, with its rapid (in terms of evolutionary time) and inexorable increases in temperature, is tipping the balance beyond the point at which many of the world's coral reefs can recover.

Estuaries: Where the Ocean Meets Fresh Water

Estuaries are biomes that occur where a river, a source of fresh water, meets the ocean. Therefore, both fresh water and salt water are found in the same vicinity; mixing results in a diluted (brackish) salt water. Estuaries form protected areas where many of the offspring of crustaceans, mollusks, and fish begin their lives. Salinity is an important factor that influences the organisms and the adaptations of the organisms found in estuaries. The salinity of estuaries varies and is based on the rate of flow of its freshwater sources. Once or twice a day, high tides bring salt water into the estuary. Low tides occurring at the same frequency reverse the current of salt water (Figure 20.30).

Figure 20.30 As estuary is where fresh water and salt water meet, such as the mouth of the Klamath River in California, shown here. (credit: U.S. Army Corps of Engineers)

The daily mixing of fresh water and salt water is a physiological challenge for the plants and animals that inhabit estuaries. Many estuarine plant species are halophytes, plants that can tolerate salty conditions. Halophytic plants are adapted to deal with salt water spray and salt water on their roots. In some halophytes, filters in the roots remove the salt from the water that the plant absorbs. Animals, such as mussels and clams (phylum Mollusca), have developed behavioral adaptations that expend a lot of energy to function in this rapidly changing environment. When these animals are exposed to low salinity, they stop feeding, close their shells, and switch from aerobic respiration (in which they use gills) to anaerobic respiration (a process that does not require oxygen). When high tide returns to the estuary, the salinity and oxygen content of the water increases, and these animals open their shells, begin feeding, and return to aerobic respiration.

Freshwater Biomes

Freshwater biomes include lakes, ponds, and wetlands (standing water) as well as rivers and streams (flowing water). Humans rely on freshwater biomes to provide aquatic resources for drinking water, crop irrigation, sanitation, recreation, and industry. These various roles and human benefits are referred to as **ecosystem services**. Lakes and ponds are found in terrestrial landscapes and are therefore connected with abiotic and biotic factors influencing these terrestrial biomes.

Lakes and Ponds

Lakes and ponds can range in area from a few square meters to thousands of square kilometers. Temperature is an important abiotic factor affecting living things found in lakes and ponds. During the summer in temperate regions, thermal stratification of deep lakes occurs when the upper layer of water is warmed by the Sun and does not mix with deeper, cooler water. The process produces a sharp transition between the warm water above and cold water beneath. The two layers do not mix until cooling temperatures and winds break down the stratification and the water in the lake mixes from top to bottom. During the period of stratification, most of the productivity occurs in the warm, well-illuminated, upper layer, while dead organisms slowly rain down into the cold, dark layer below where decomposing bacteria and cold-adapted species such as lake trout exist. Like the ocean, lakes and ponds have a photic layer in which photosynthesis can occur. Phytoplankton (algae and cyanobacteria) are found here and provide the base of the food web of lakes and ponds. Zooplankton, such as rotifers and small crustaceans, consume these phytoplankton. At the bottom of lakes and ponds, bacteria in the aphotic zone break down dead organisms that sink to the bottom.

Nitrogen and particularly phosphorus are important limiting nutrients in lakes and ponds. Therefore, they are determining factors in the amount of phytoplankton growth in lakes and ponds. When there is a large input of nitrogen and phosphorus (e.g., from sewage and runoff from fertilized lawns and farms), the growth of algae skyrockets, resulting in a large accumulation of algae called an **algal bloom**. Algal blooms (Figure 20.31) can become so extensive that they reduce light penetration in water. As a result, the lake or pond becomes aphotic and photosynthetic plants cannot survive. When the algae die and decompose, severe oxygen depletion of the water occurs. Fishes and other organisms that require oxygen are then more likely to die.

Figure 20.31 The uncontrolled growth of algae in this waterway has resulted in an algal bloom.

Rivers and Streams

Rivers and the narrower streams that feed into the rivers are continuously moving bodies of water that carry water from the source or headwater to the mouth at a lake or ocean. The largest rivers include the Nile River in Africa, the Amazon River in South America, and the Mississippi River in North America (Figure 20.32).

(a)   (b)

Figure 20.32 Rivers range from (a) narrow and shallow to (b) wide and slow moving. (credit a: modification of work by Cory Zanker; credit b: modification of work by David DeHetre)

Abiotic features of rivers and streams vary along the length of the river or stream. Streams begin at a point of origin referred to as **source water**. The source water is usually cold, low in nutrients, and clear. The **channel** (the width of the river or stream) is narrower here than at any other place along the length of the river or stream. Headwater streams are of necessity at a higher elevation than the mouth of the river and often originate in regions with steep grades leading to higher flow rates than lower elevation stretches of the river.

Faster-moving water and the short distance from its origin results in minimal silt levels in headwater streams; therefore, the water is clear. Photosynthesis here is mostly attributed to algae that are growing on rocks; the swift current inhibits the growth of phytoplankton. Photosynthesis may be further reduced by tree cover reaching over the narrow stream. This shading also keeps temperatures lower. An additional input of energy can come from leaves or other organic material that falls into a river or stream from the trees and other plants that border the water. When the leaves decompose, the organic material and nutrients in the leaves are returned to the water. The leaves also support a food chain of invertebrates that eat them and are in turn eaten by predatory invertebrates and fish. Plants and animals have adapted to this fast-moving water. For instance, leeches (phylum Annelida) have elongated bodies and suckers on both ends. These suckers attach to the substrate, keeping the leech anchored in place. In temperate regions, freshwater trout species (phylum Chordata) may be an important predator in these fast-moving and colder river and streams.

As the river or stream flows away from the source, the width of the channel gradually widens, the current slows, and the temperature characteristically increases. The increasing width results from the increased volume of water from more and more tributaries. Gradients are typically lower farther along the river, which accounts for the slowing flow. With increasing volume can come increased silt, and as the flow rate slows, the silt may settle, thus increasing the deposition of sediment. Phytoplankton can also be suspended in slow-moving water. Therefore, the water will not be as clear as it is near the source. The water is also warmer as a result of longer exposure to sunlight and the absence of tree cover over wider expanses between banks. Worms (phylum Annelida) and insects (phylum Arthropoda) can be found burrowing into the mud. Predatory vertebrates (phylum Chordata) include waterfowl, frogs, and fishes. In heavily silt-laden rivers, these predators must find food in the murky waters, and, unlike the trout in the clear waters at the source, these vertebrates cannot use vision as their primary sense to find food. Instead, they are more likely to use taste or chemical cues to find prey.

When a river reaches the ocean or a large lake, the water typically slows dramatically and any silt in the river water will settle. Rivers with high silt content discharging into oceans with minimal currents and wave action will build deltas, low-elevation areas of sand and mud, as the silt settles onto the ocean bottom. Rivers with low silt content or in areas where ocean currents or wave action are high create estuarine areas where the fresh water and salt water mix.

Wetlands

Wetlands are environments in which the soil is either permanently or periodically saturated with water. Wetlands are different from lakes and ponds because wetlands exhibit a near continuous cover of emergent vegetation. **Emergent vegetation** consists of wetland plants that are rooted in the soil but have portions of leaves, stems, and flowers extending above the water's surface. There are several types of wetlands including marshes, swamps, bogs, mudflats, and salt marshes (Figure 20.33).

Figure 20.33 Located in southern Florida, Everglades National Park is vast array of wetland environments, including sawgrass marshes, cypress swamps, and estuarine mangrove forests. Here, a great egret walks among cypress trees. (credit: NPS)

Freshwater marshes and swamps are characterized by slow and steady water flow. Bogs develop in depressions where water flow is low or nonexistent. Bogs usually occur in areas where there is a clay bottom with poor percolation. Percolation is the movement of water through the pores in the soil or rocks. The water found in a bog is stagnant and oxygen depleted because the oxygen that is used during the decomposition of organic matter is not replaced. As the oxygen in the water is depleted, decomposition slows. This leads to organic acids and other acids building up and lowering the pH of the water. At a lower pH, nitrogen becomes unavailable to plants. This creates a challenge for plants because nitrogen is an important limiting resource. Some types of bog plants (such as sundews, pitcher plants, and Venus flytraps) capture insects and extract the nitrogen from their bodies. Bogs have low net primary productivity because the water found in bogs has low levels of nitrogen and oxygen.

KEY TERMS

abyssal zone the deepest part of the ocean at depths of 4000 m or greater

acid rain a corrosive rain caused by rainwater mixing with sulfur dioxide gas as it fall through the atmosphere, turning it into weak sulfuric acid, causing damage to aquatic ecosystems

algal bloom a rapid increase of algae in an aquatic system

apex consumer an organism at the top of the food chain

aphotic zone the part of the ocean where photosynthesis cannot occur

arctic tundra a biome characterized by low average temperatures, brief growing seasons, the presence of permafrost, and limited precipitation largely in the form of snow in which the dominant vegetation are low shrubs, lichens, mosses, and small herbaceous plants

autotroph an organism capable of synthesizing its own food molecules from smaller inorganic molecules

benthic realm (also, benthic zone) the part of the ocean that extends along the ocean bottom from the shoreline to the deepest parts of the ocean floor

biogeochemical cycle the cycling of minerals and nutrients through the biotic and abiotic world

biomagnification an increasing concentration of persistent, toxic substances in organisms at each trophic level, from the producers to the apex consumers

biome a large-scale community of organisms, primarily defined on land by the dominant plant types that exist in geographic regions of the planet with similar climatic conditions

boreal forest a biome found in temperate and subarctic regions characterized by short growing seasons and dominated structurally by coniferous trees

canopy the branches and foliage of trees that form a layer of overhead coverage in a forest

channel the bed and banks of a river or stream

chaparral a biome found in temperate coastal regions characterized by low trees and dry-adapted shrubs and forbs

chemoautotroph an organism capable of synthesizing its own food using energy from inorganic molecules

coral reef an ocean ridge formed by marine invertebrates living in warm shallow waters within the photic zone

cryptofauna the invertebrates found within the calcium carbonate substrate of coral reefs

dead zone an area in a lake and ocean near the mouths of rivers where large areas are depleted of their normal flora and fauna; these zones can be caused by eutrophication, oil spills, dumping of toxic chemicals, and other human activities

detrital food web a type of food web that is supported by dead or decaying organisms rather than by living autotrophs; these are often associated with grazing food webs within the same ecosystem

ecosystem a community of living organisms and their interactions with their abiotic environment

ecosystem services the human benefits provided by natural ecosystems

emergent vegetation the plants living in bodies of water that are rooted in the soil but have portions of leaves, stems, and flowers extending above the water's surface

equilibrium the steady state of a system in which the relationships between elements of the system do not change

estuary a region where fresh water and salt water mix where a river discharges into an ocean or sea

eutrophication the process whereby nutrient runoff causes the excess growth of microorganisms and plants in aquatic systems

fallout the direct deposition of solid minerals on land or in the ocean from the atmosphere

food chain a linear sequence of trophic (feeding) relationships of producers, primary consumers, and higher level consumers

food web a web of trophic (feeding) relationships among producers, primary consumers, and higher level consumers in an ecosystem

grazing food web a type of food web in which the producers are either plants on land or phytoplankton in the water; often associated with a detrital food web within the same ecosystem

gross primary productivity the rate at which photosynthetic producers incorporate energy from the Sun

hydrosphere the region of the planet in which water exists, including the atmosphere that contains water vapor and the region beneath the ground that contains groundwater

intertidal zone the part of the ocean that is closest to land; parts extend above the water at low tide

neritic zone the part of the ocean that extends from low tide to the edge of the continental shelf

net primary productivity the energy that remains in the producers after accounting for the organisms' respiration and heat loss

non-renewable resource a resource, such as a fossil fuel, that is either regenerated very slowly or not at all

oceanic zone the part of the ocean that begins offshore where the water measures 200 m deep or deeper

pelagic realm (also, pelagic zone) the open ocean waters that are not close to the bottom or near the shore

permafrost a perennially frozen portion of the Arctic tundra soil

photic zone the upper layer of ocean water in which photosynthesis is able to take place

photoautotroph an organism that uses sunlight as an energy source to synthesize its own food molecules

planktivore an animal that eats plankton

primary consumer the trophic level that obtains its energy from the producers of an ecosystem

producer the trophic level that obtains its energy from sunlight, inorganic chemicals, or dead or decaying organic material

resilience (ecological) the speed at which an ecosystem recovers equilibrium after being disturbed

resistance (ecological) the ability of an ecosystem to remain at equilibrium in spite of disturbances

savanna a biome located in the tropics with an extended dry season and characterized by a grassland with sparsely distributed trees

secondary consumer a trophic level in an ecosystem, usually a carnivore that eats a primary consumer

source water the point of origin of a river or stream

subduction the movement of one tectonic plate beneath another

subtropical desert a biome found in the subtropics with hot daily temperatures, very low and unpredictable precipitation, and characterized by a limited dry-adapted vegetation

temperate forest a biome found in temperate regions with moderate rainfall and dominated structurally by deciduous trees

temperate grassland a biome dominated by grasses and herbaceous plants due to low precipitation, periodic fires, and grazing

tertiary consumer a trophic level in an ecosystem, usually carnivores that eat other carnivores

trophic level the position of a species or group of species in a food chain or a food web

tropical rainforest a biome found near the equator characterized by stable temperatures with abundant and seasonal rainfall in which trees form the structurally important vegetation

wetland environment in which the soil is either permanently or periodically saturated with water

CHAPTER SUMMARY

20.1 Energy Flow through Ecosystems

Ecosystems exist underground, on land, at sea, and in the air. Organisms in an ecosystem acquire energy in a variety of ways, which is transferred between trophic levels as the energy flows from the base to the top of the food web, with energy being lost at each transfer. There is energy lost at each trophic level, so the lengths of food chains are limited because there is a point where not enough energy remains to support a population of consumers. Fat soluble compounds biomagnify up a food chain causing damage to top consumers. even when environmental concentrations of a toxin are low.

20.2 Biogeochemical Cycles

Mineral nutrients are cycled through ecosystems and their environment. Of particular importance are water, carbon, nitrogen, phosphorus, and sulfur. All of these cycles have major impacts on ecosystem structure and function. As human activities have caused major disturbances to these cycles, their study and modeling is especially important. Ecosystems have been damaged by a variety of human activities that alter the natural biogeochemical cycles due to pollution, oil spills, and events causing global climate change. The health of the biosphere depends on understanding these cycles and how to protect the environment from irreversible damage.

20.3 Terrestrial Biomes

Earth has terrestrial and aquatic biomes. Aquatic biomes include both freshwater and marine environments. There are eight major terrestrial biomes: tropical rainforests, savannas, subtropical deserts, chaparral, temperate grasslands, temperate forests, boreal forests, and Arctic tundra. The same biome can occur in different geographic locations with similar climates. Temperature and precipitation, and variations in both, are key abiotic factors that shape the composition of animal and plant communities in terrestrial biomes. Some biomes, such as temperate grasslands and temperate forests, have distinct seasons with cold and hot weather alternating throughout the year. In warm, moist biomes, such as the tropical rainforest, net primary productivity is high as warm temperatures, abundant water, and a year-round growing season fuel plant growth. Other biomes, such as deserts and tundra, have low primary productivity due to extreme temperatures and a shortage of water.

20.4 Aquatic and Marine Biomes

Aquatic biomes include both saltwater and freshwater biomes. The abiotic factors important for the structuring of aquatic biomes can be different than those seen in terrestrial biomes. Sunlight is an important factor in bodies of water, especially those that are very deep, because of the role of photosynthesis in sustaining certain organisms. Other important factors include temperature, water movement, and salt content. Oceans may be thought of as consisting of different zones based on water depth, distance from the shoreline, and light penetrance. Different kinds of organisms are adapted to the conditions found in each zone. Coral reefs are unique marine ecosystems that are home to a wide variety of species. Estuaries are found where rivers meet the ocean; their shallow waters provide nourishment and shelter for young crustaceans, mollusks, fishes, and many other species. Freshwater biomes include lakes, ponds, rivers, streams, and wetlands. Bogs are an interesting type of wetland characterized by standing water, a lower pH, and a lack of nitrogen.

ART CONNECTION QUESTIONS

1. Figure 20.12 Which of the following statements about the nitrogen cycle is false?

a. Ammonification converts organic nitrogenous matter from living organisms into ammonium (NH_4^+).

b. Denitrification by bacteria converts nitrates (NO_3^-) to nitrogen gas (N_2).

c. Nitrification by bacteria converts nitrates (NO_3^-) to nitrites (NO_2^-).

d. Nitrogen fixing bacteria convert nitrogen gas (N_2) into organic compounds.

2. Figure 20.28 In which of the following regions would you expect to find photosynthetic organisms?

a. The aphotic zone, the neritic zone, the oceanic zone, and the benthic realm.

b. The photic zone, the intertidal zone, the neritic zone, and the oceanic zone.

c. The photic zone, the abyssal zone, the neritic zone, and the oceanic zone.

d. The pelagic realm, the aphotic zone, the neritic zone, and the oceanic zone.

REVIEW QUESTIONS

3. Decomposers are associated with which class of food web?

a. grazing
b. detrital
c. inverted
d. aquatic

4. The producer in an ocean grazing food web is usually a _____.

a. plant
b. animal
c. fungi
d. plankton

5. Which term describes the process whereby toxic substances increase along trophic levels of an ecosystem?

a. biomassification
b. biomagnification
c. bioentropy
d. heterotrophy

6. The majority of the water found on Earth is:

a. ice
b. water vapor
c. fresh water
d. salt water

7. The process whereby oxygen is depleted by the growth of microorganisms due to excess nutrients in aquatic systems is called _____.

a. dead zoning
b. eutrophication
c. retrophication

d. depletion

8. Which of the following biomes is characterized by abundant water resources?

a. deserts
b. boreal forests
c. savanna
d. tropical wet forests

9. Which of the following biomes is characterized by short growing seasons?

a. deserts
b. tropical wet forests
c. Arctic tundra
d. savanna

10. Why is the tundra treeless?

a. lack of sufficient water
b. permanently frozen ground
c. winters too harsh
d. too many fires

11. Where would you expect to find the most photosynthesis in an ocean biome?

a. aphotic zone
b. abyssal zone
c. benthic realm
d. intertidal zone

12. A key feature of estuaries is

a. low light conditions and high productivity
b. salt water and fresh water
c. frequent algal blooms
d. little or no vegetation

CRITICAL THINKING QUESTIONS

13. Compare grazing and detrital food webs. Why would they both be present in the same ecosystem?

14. Why are drinking water supplies still a major concern for many countries?

15. The extremely low precipitation of subtropical desert biomes might lead one to expect fire to be a major disturbance factor; however, fire is more common in the temperate grassland biome than in the subtropical desert biome. Why is this?

16. In what ways are the subtropical desert and the Arctic tundra similar?

17. Describe the conditions and challenges facing organisms living in the intertidal zone.

21 | CONSERVATION AND BIODIVERSITY

Figure 21.1 Habitat destruction through deforestation, especially of tropical rainforests as seen in this satellite view of Amazon rainforests in Brazil, is a major cause of the current decline in biodiversity. (credit: modification of work by Jesse Allen and Robert Simmon, NASA Earth Observatory)

Chapter Outline
21.1: Importance of Biodiversity
21.2: Threats to Biodiversity
21.3: Preserving Biodiversity

Introduction

Biologists estimate that species extinctions are currently 500–1000 times the rate seen previously in Earth's history when there were no unusual geological or climatic events occurring. Biologists call the previous rate the "background" rate of extinction. The current high rates will cause a precipitous decline in the biodiversity (the diversity of species) of the planet in the next century or two. The losses will include many species we know today. Although it is sometimes difficult to predict which species will become extinct, many are listed as endangered (at great risk of extinction). However, the majority of extinctions will be of species that science has not yet even described.

Most of these "invisible" species that will become extinct currently live in tropical rainforests like those of the Amazon basin. These rainforests are the most diverse ecosystems on the planet and are being destroyed rapidly by deforestation, which biologists believe is driving many rare species with limited distributions extinct. Between 1970 and 2011, almost 20 percent of the Amazon rainforest was lost. Rates are higher in other tropical rainforests. What we are likely to notice on a day-to-day basis as a result of biodiversity loss is that food will be more difficult to produce, clean water will be more difficult to find, and the rate of development of new medicines will become slower, as we depend upon other species for much of these services. This increased loss of biodiversity is almost entirely a result of human activities as we destroy species' habitats, introduce disruptive species into ecosystems, hunt some species to extinction, continue to warm the planet with greenhouse gases, and influence nature in other ways. Slowing the loss of biodiversity is within our abilities if we make dramatic changes in our consumptive behavior and identify and protect the elements of our ecosystems that we depend on for our lives and welfare.

21.1 | Importance of Biodiversity

By the end of this section, you will be able to:

- Describe biodiversity as the equilibrium of naturally fluctuating rates of extinction and speciation
- Identify benefits of biodiversity to humans

Figure 21.2 This tropical lowland rainforest in Madagascar is an example of a high biodiversity habitat. This particular location is protected within a national forest, yet only 10 percent of the original coastal lowland forest remains, and research suggests half the original biodiversity has been lost. (credit: Frank Vassen)

Biodiversity is a broad term for biological variety, and it can be measured at a number of organizational levels. Traditionally, ecologists have measured **biodiversity** by taking into account both the number of species and the number of individuals in each of those species. However, biologists are using measures of biodiversity at several levels of biological organization (including genes, populations, and ecosystems) to help focus efforts to preserve the biologically and technologically important elements of biodiversity.

When biodiversity loss through extinction is thought of as the loss of the passenger pigeon, the dodo, or, even, the woolly mammoth there seems to be no reason to care about it because these events happened long ago. How is the loss practically important for the welfare of the human species? Would these species have made our lives any better? From the perspective of evolution and ecology, the loss of a particular individual species, with some exceptions, may seem unimportant, but the current accelerated extinction rate means the loss of tens of thousands of species within our lifetimes. Much of this loss is occurring in tropical rainforests like the one pictured in Figure 21.2, which are especially high-diversity ecosystems that are being cleared for timber and agriculture. This is likely to have dramatic effects on human welfare through the collapse of ecosystems and in added costs to maintain food production, clean air and water, and improve human health.

Biologists recognize that human populations are embedded in ecosystems and are dependent on them, just as is every other species on the planet. Agriculture began after early hunter-gatherer societies first settled in one place and heavily modified their immediate environment: the ecosystem in which they existed. This cultural transition has made it difficult for humans to recognize their dependence on living things other than crops and domesticated animals on the planet. Today our technology smoothes out the extremes of existence and allows many of us to live longer, more comfortable lives, but ultimately the human species cannot exist without its surrounding ecosystems. Our ecosystems provide our food. This includes living plants that grow in soil ecosystems and the animals that eat these plants (or other animals) as well as photosynthetic organisms in the oceans and the other organisms that eat them. Our ecosystems have provided and will provide many of the medications that maintain our health, which are commonly made from compounds found in living organisms. Ecosystems provide our clean water, which is held in lake and river ecosystems or passes through terrestrial ecosystems on its way into groundwater.

Types of Biodiversity

A common meaning of biodiversity is simply the number of species in a location or on Earth; for example, the American Ornithologists' Union lists 2078 species of birds in North and Central America. This is one measure of the bird biodiversity on the continent. More sophisticated measures of diversity take into account the relative abundances of species. For example, a forest with 10 equally common species of trees is more diverse than a forest that has 10 species of trees wherein just one of those species makes up 95 percent of the trees rather than them being equally distributed. Biologists have also identified alternate measures of biodiversity, some of which are important in planning how to preserve biodiversity.

Genetic and Chemical Biodiversity

Genetic diversity is one alternate concept of biodiversity. **Genetic diversity** (or variation) is the raw material for adaptation in a species. A species' future potential for adaptation depends on the genetic diversity held in the genomes of the individuals in populations that make up the species. The same is true for higher taxonomic categories. A genus with very different types of species will have more genetic diversity than a genus with species that look alike and have similar ecologies. The genus with the greatest potential for subsequent evolution is the most genetically diverse one.

Most genes code for proteins, which in turn carry out the metabolic processes that keep organisms alive and reproducing. Genetic diversity can also be conceived of as **chemical diversity** in that species with different genetic makeups produce different assortments of chemicals in their cells (proteins as well as the products and byproducts of metabolism). This chemical diversity is important for humans because of the potential uses for these chemicals, such as medications. For example, the drug eptifibatide is derived from rattlesnake venom and is used to prevent heart attacks in individuals with certain heart conditions.

At present, it is far cheaper to discover compounds made by an organism than to imagine them and then synthesize them in a laboratory. Chemical diversity is one way to measure diversity that is important to human health and welfare. Through selective breeding, humans have domesticated animals, plants, and fungi, but even this diversity is suffering losses because of market forces and increasing globalism in human agriculture and migration. For example, international seed companies produce only a very few varieties of a given crop and provide incentives around the world for farmers to buy these few varieties while abandoning their traditional varieties, which are far more diverse. The human population depends on crop diversity directly as a stable food source and its decline is troubling to biologists and agricultural scientists.

Ecosystems Diversity

It is also useful to define **ecosystem diversity**: the number of different ecosystems on Earth or in a geographical area. Whole ecosystems can disappear even if some of the species might survive by adapting to other ecosystems. The loss of an ecosystem means the loss of the interactions between species, the loss of unique features of coadaptation, and the loss of biological productivity that an ecosystem is able to create. An example of a largely extinct ecosystem in North America is the prairie ecosystem (Figure 21.3). Prairies once spanned central North America from the boreal forest in northern Canada down into Mexico. They are now all but gone, replaced by crop fields, pasture lands, and suburban sprawl. Many of the species survive, but the hugely productive ecosystem that was responsible for creating our most productive agricultural soils is now gone. As a consequence, their soils are now being depleted unless they are maintained artificially at greater expense. The decline in soil productivity occurs because the interactions in the original ecosystem have been lost; this was a far more important loss than the relatively few species that were driven extinct when the prairie ecosystem was destroyed.

Figure 21.3 The variety of ecosystems on Earth—from coral reef to prairie—enables a great diversity of species to exist. (credit "coral reef": modification of work by Jim Maragos, USFWS; credit: "prairie": modification of work by Jim Minnerath, USFWS)

Current Species Diversity

Despite considerable effort, knowledge of the species that inhabit the planet is limited. A recent estimate suggests that the eukaryote species for which science has names, about 1.5 million species, account for less than 20 percent of the total number of eukaryote species present on the planet (8.7 million species, by one estimate). Estimates of numbers of prokaryotic species are largely guesses, but biologists agree that science has only just begun to catalog their diversity. Even with what is known, there is no centralized repository of names or samples of the described species; therefore, there is no way to be sure that the 1.5 million descriptions is an accurate number. It is a best guess based on the opinions of experts on different taxonomic groups. Given that Earth is losing species at an accelerating pace, science knows little about what is being lost. Table 21.1 presents recent estimates of biodiversity in different groups.

Estimated Numbers of Described and Predicted species

	Source: Mora et al 2011		Source: Chapman 2009		Source: Groombridge and Jenkins 2002	
	Described	Predicted	Described	Predicted	Described	Predicted
Animals	1,124,516	9,920,000	1,424,153	6,836,330	1,225,500	10,820,000
Photosynthetic protists	17,892	34,900	25,044	200,500	—	—
Fungi	44,368	616,320	98,998	1,500,000	72,000	1,500,000
Plants	224,244	314,600	310,129	390,800	270,000	320,000
Non-photosynthetic protists	16,236	72,800	28,871	1,000,000	80,000	600,000
Prokaryotes	—	—	10,307	1,000,000	10,175	—
Total	1,438,769	10,960,000	1,897,502	10,897,630	1,657,675	13,240,000

Table 21.1 This table shows the estimated number of species by taxonomic group—including both described (named and studied) and predicted (yet to be named) species.

There are various initiatives to catalog described species in accessible and more organized ways, and the internet is facilitating that effort. Nevertheless, at the current rate of species description, which according to the State of Observed Species[1] reports is 17,000–20,000 new species a year, it would take close to 500 years to describe all of the species currently in existence. The task, however, is becoming increasingly impossible over time as **extinction** removes species from Earth faster than they can be described.

Naming and counting species may seem an unimportant pursuit given the other needs of humanity, but it is not simply an accounting. Describing species is a complex process by which biologists determine an organism's unique characteristics and whether or not that organism belongs to any other described species. It allows biologists to find and recognize the species

---

1. International Institute for Species Exploration (IISE), *2011 State of Observed Species (SOS)*. Tempe, AZ: IISE, 2011. Accessed May, 20, 2012. http://species.asu.edu/SOS.

after the initial discovery to follow up on questions about its biology. That subsequent research will produce the discoveries that make the species valuable to humans and to our ecosystems. Without a name and description, a species cannot be studied in depth and in a coordinated way by multiple scientists.

Patterns of Biodiversity

Biodiversity is not evenly distributed on the planet. Lake Victoria contained almost 500 species of cichlids (only one family of fishes present in the lake) before the introduction of an exotic species in the 1980s and 1990s caused a mass extinction. All of these species were found only in Lake Victoria, which is to say they were endemic. **Endemic species** are found in only one location. For example, the blue jay is endemic to North America, while the Barton Springs salamander is endemic to the mouth of one spring in Austin, Texas. Endemics with highly restricted distributions, like the Barton Springs salamander, are particularly vulnerable to extinction. Higher taxonomic levels, such as genera and families, can also be endemic.

Lake Huron contains about 79 species of fish, all of which are found in many other lakes in North America. What accounts for the difference in diversity between Lake Victoria and Lake Huron? Lake Victoria is a tropical lake, while Lake Huron is a temperate lake. Lake Huron in its present form is only about 7,000 years old, while Lake Victoria in its present form is about 15,000 years old. These two factors, latitude and age, are two of several hypotheses biogeographers have suggested to explain biodiversity patterns on Earth.

Biogeography

Biogeography is the study of the distribution of the world's species both in the past and in the present. The work of biogeographers is critical to understanding our physical environment, how the environment affects species, and how changes in environment impact the distribution of a species.

There are three main fields of study under the heading of biogeography: ecological biogeography, historical biogeography (called paleobiogeography), and conservation biogeography. Ecological biogeography studies the current factors affecting the distribution of plants and animals. Historical biogeography, as the name implies, studies the past distribution of species. Conservation biogeography, on the other hand, is focused on the protection and restoration of species based upon the known historical and current ecological information. Each of these fields considers both zoogeography and phytogeography—the past and present distribution of animals and plants.

One of the oldest observed patterns in ecology is that biodiversity in almost every taxonomic group of organism increases as latitude declines. In other words, biodiversity increases closer to the equator (Figure 21.4).

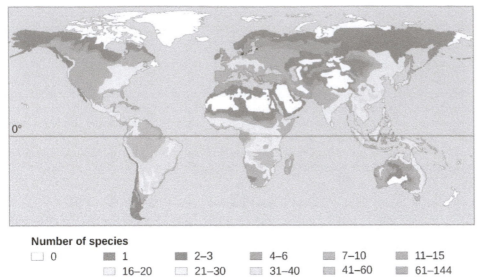

Number of species

☐ 0	▉ 1	▉ 2–3	▉ 4–6	▉ 7–10	▉ 11–15
	▨ 16–20	☐ 21–30	▨ 31–40	▨ 41–60	▨ 61–144

Figure 21.4 This map illustrates the number of amphibian species across the globe and shows the trend toward higher biodiversity at lower latitudes. A similar pattern is observed for most taxonomic groups.

It is not yet clear why biodiversity increases closer to the equator, but hypotheses include the greater age of the ecosystems in the tropics versus temperate regions, which were largely devoid of life or drastically impoverished during the last ice age. The greater age provides more time for speciation. Another possible explanation is the greater energy the tropics receive from the sun versus the lesser energy input in temperate and polar regions. But scientists have not been able to explain how greater energy input could translate into more species. The complexity of tropical ecosystems may promote speciation by increasing the **habitat heterogeneity**, or number of ecological niches, in the tropics relative to higher latitudes. The greater heterogeneity provides more opportunities for coevolution, specialization, and perhaps greater selection pressures leading to population differentiation. However, this hypothesis suffers from some circularity—ecosystems with more species encourage speciation, but how did they get more species to begin with? The tropics have been perceived as being more stable than temperate regions, which have a pronounced climate and day-length seasonality. The tropics have their own forms of seasonality, such as rainfall, but they are generally assumed to be more stable environments and this stability might promote speciation.

Regardless of the mechanisms, it is certainly true that biodiversity is greatest in the tropics. The number of endemic species is higher in the tropics. The tropics also contain more biodiversity hotspots. At the same time, our knowledge of the species living in the tropics is lowest and because of recent, heavy human activity the potential for biodiversity loss is greatest.

Importance of Biodiversity

Loss of biodiversity eventually threatens other species we do not impact directly because of their interconnectedness; as species disappear from an ecosystem other species are threatened by the changes in available resources. Biodiversity is important to the survival and welfare of human populations because it has impacts on our health and our ability to feed ourselves through agriculture and harvesting populations of wild animals.

Human Health

Many medications are derived from natural chemicals made by a diverse group of organisms. For example, many plants produce **secondary plant compounds**, which are toxins used to protect the plant from insects and other animals that eat them. Some of these secondary plant compounds also work as human medicines. Contemporary societies that live close to the land often have a broad knowledge of the medicinal uses of plants growing in their area. For centuries in Europe, older knowledge about the medical uses of plants was compiled in herbals—books that identified the plants and their uses. Humans are not the only animals to use plants for medicinal reasons. The other great apes, orangutans, chimpanzees, bonobos, and gorillas have all been observed self-medicating with plants.

Modern pharmaceutical science also recognizes the importance of these plant compounds. Examples of significant medicines derived from plant compounds include aspirin, codeine, digoxin, atropine, and vincristine (Figure 21.5). Many medications were once derived from plant extracts but are now synthesized. It is estimated that, at one time, 25 percent of modern drugs contained at least one plant extract. That number has probably decreased to about 10 percent as natural plant ingredients are replaced by synthetic versions of the plant compounds. Antibiotics, which are responsible for extraordinary improvements in health and lifespans in developed countries, are compounds largely derived from fungi and bacteria.

Figure 21.5 *Catharanthus roseus*, the Madagascar periwinkle, has various medicinal properties. Among other uses, it is a source of vincristine, a drug used in the treatment of lymphomas. (credit: Forest and Kim Starr)

In recent years, animal venoms and poisons have excited intense research for their medicinal potential. By 2007, the FDA had approved five drugs based on animal toxins to treat diseases such as hypertension, chronic pain, and diabetes. Another five drugs are undergoing clinical trials and at least six drugs are being used in other countries. Other toxins under investigation come from mammals, snakes, lizards, various amphibians, fish, snails, octopuses, and scorpions.

Aside from representing billions of dollars in profits, these medications improve people's lives. Pharmaceutical companies are actively looking for new natural compounds that can function as medicines. It is estimated that one third of pharmaceutical research and development is spent on natural compounds and that about 35 percent of new drugs brought to market between 1981 and 2002 were from natural compounds.

Finally, it has been argued that humans benefit psychologically from living in a biodiverse world. The chief proponent of this idea is entomologist E. O. Wilson. He argues that human evolutionary history has adapted us to living in a natural environment and that built environments generate stresses that affect human health and well-being. There is considerable research into the psychologically regenerative benefits of natural landscapes that suggest the hypothesis may hold some truth.

Agricultural

Since the beginning of human agriculture more than 10,000 years ago, human groups have been breeding and selecting crop varieties. This crop diversity matched the cultural diversity of highly subdivided populations of humans. For example, potatoes were domesticated beginning around 7,000 years ago in the central Andes of Peru and Bolivia. The people in this region traditionally lived in relatively isolated settlements separated by mountains. The potatoes grown in that region belong to seven species and the number of varieties likely is in the thousands. Each variety has been bred to thrive at particular elevations and soil and climate conditions. The diversity is driven by the diverse demands of the dramatic elevation changes, the limited movement of people, and the demands created by crop rotation for different varieties that will do well in different fields.

Potatoes are only one example of agricultural diversity. Every plant, animal, and fungus that has been cultivated by humans has been bred from original wild ancestor species into diverse varieties arising from the demands for food value, adaptation to growing conditions, and resistance to pests. The potato demonstrates a well-known example of the risks of low crop diversity: during the tragic Irish potato famine (1845–1852 AD), the single potato variety grown in Ireland became susceptible to a potato blight—wiping out the crop. The loss of the crop led to famine, death, and mass emigration. Resistance to disease is a chief benefit to maintaining crop biodiversity and lack of diversity in contemporary crop species carries similar risks. Seed companies, which are the source of most crop varieties in developed countries, must continually breed new varieties to keep up with evolving pest organisms. These same seed companies, however, have participated in the decline of the number of varieties available as they focus on selling fewer varieties in more areas of the world replacing traditional local varieties.

The ability to create new crop varieties relies on the diversity of varieties available and the availability of wild forms related to the crop plant. These wild forms are often the source of new gene variants that can be bred with existing varieties to create varieties with new attributes. Loss of wild species related to a crop will mean the loss of potential in crop improvement. Maintaining the genetic diversity of wild species related to domesticated species ensures our continued supply of food.

Since the 1920s, government agriculture departments have maintained seed banks of crop varieties as a way to maintain crop diversity. This system has flaws because over time seed varieties are lost through accidents and there is no way to replace them. In 2008, the Svalbard Global seed Vault, located on Spitsbergen island, Norway, (Figure 21.6) began storing seeds from around the world as a backup system to the regional seed banks. If a regional seed bank stores varieties in Svalbard, losses can be replaced from Svalbard should something happen to the regional seeds. The Svalbard seed vault is deep into the rock of the arctic island. Conditions within the vault are maintained at ideal temperature and humidity for seed survival, but the deep underground location of the vault in the arctic means that failure of the vault's systems will not compromise the climatic conditions inside the vault.

art CONNECTION

Figure 21.6 The Svalbard Global Seed Vault is a storage facility for seeds of Earth's diverse crops. (credit: Mari Tefre, Svalbard Global Seed Vault)

The Svalbard seed vault is located on Spitsbergen island in Norway, which has an arctic climate. Why might an arctic climate be good for seed storage?

Although crops are largely under our control, our ability to grow them is dependent on the biodiversity of the ecosystems in which they are grown. That biodiversity creates the conditions under which crops are able to grow through what are known as ecosystem services—valuable conditions or processes that are carried out by an ecosystem. Crops are not grown, for the most part, in built environments. They are grown in soil. Although some agricultural soils are rendered sterile using controversial pesticide treatments, most contain a huge diversity of organisms that maintain nutrient cycles—breaking down organic matter into nutrient compounds that crops need for growth. These organisms also maintain soil texture that affects water and oxygen dynamics in the soil that are necessary for plant growth. Replacing the work of these organisms in forming arable soil is not practically possible. These kinds of processes are called ecosystem services. They occur within ecosystems, such as soil ecosystems, as a result of the diverse metabolic activities of the organisms living there, but they provide benefits to human food production, drinking water availability, and breathable air.

Other key ecosystem services related to food production are plant pollination and crop pest control. It is estimated that honeybee pollination within the United States brings in $1.6 billion per year; other pollinators contribute up to $6.7 billion. Over 150 crops in the United States require pollination to produce. Many honeybee populations are managed by beekeepers who rent out their hives' services to farmers. Honeybee populations in North America have been suffering large losses caused by a syndrome known as colony collapse disorder, a new phenomenon with an unclear cause. Other pollinators include a diverse array of other bee species and various insects and birds. Loss of these species would make growing crops requiring pollination impossible, increasing dependence on other crops.

Finally, humans compete for their food with crop pests, most of which are insects. Pesticides control these competitors, but these are costly and lose their effectiveness over time as pest populations adapt. They also lead to collateral damage by killing non-pest species as well as beneficial insects like honeybees, and risking the health of agricultural workers and consumers. Moreover, these pesticides may migrate from the fields where they are applied and do damage to other ecosystems like streams, lakes, and even the ocean. Ecologists believe that the bulk of the work in removing pests is actually done by predators and parasites of those pests, but the impact has not been well studied. A review found that in 74 percent of studies that looked for an effect of landscape complexity (forests and fallow fields near to crop fields) on natural enemies of pests, the greater the complexity, the greater the effect of pest-suppressing organisms. Another experimental study found that introducing multiple enemies of pea aphids (an important alfalfa pest) increased the yield of alfalfa significantly. This study shows that a diversity of pests is more effective at control than one single pest. Loss of diversity in pest enemies will inevitably make it more difficult and costly to grow food. The world's growing human population faces significant challenges in the increasing costs and other difficulties associated with producing food.

Wild Food Sources

In addition to growing crops and raising food animals, humans obtain food resources from wild populations, primarily wild fish populations. For about one billion people, aquatic resources provide the main source of animal protein. But since 1990, production from global fisheries has declined. Despite considerable effort, few fisheries on Earth are managed sustainability.

Fishery extinctions rarely lead to complete extinction of the harvested species, but rather to a radical restructuring of the marine ecosystem in which a dominant species is so over-harvested that it becomes a minor player, ecologically. In addition to humans losing the food source, these alterations affect many other species in ways that are difficult or impossible to predict. The collapse of fisheries has dramatic and long-lasting effects on local human populations that work in the fishery. In addition, the loss of an inexpensive protein source to populations that cannot afford to replace it will increase the cost of living and limit societies in other ways. In general, the fish taken from fisheries have shifted to smaller species and the larger species are overfished. The ultimate outcome could clearly be the loss of aquatic systems as food sources.

Visit this website (http://openstaxcollege.org/l/decliningfish2) to view a brief video discussing a study of declining fisheries.

21.2 | Threats to Biodiversity

By the end of this section, you will be able to:

- Identify significant threats to biodiversity
- Explain the effects of habitat loss, exotic species, and hunting on biodiversity
- Identify the early and predicted effects of climate change on biodiversity

The core threat to biodiversity on the planet, and therefore a threat to human welfare, is the combination of human population growth and the resources used by that population. The human population requires resources to survive and grow, and those resources are being removed unsustainably from the environment. The three greatest proximate threats to biodiversity are habitat loss, overharvesting, and introduction of exotic species. The first two of these are a direct result of human population growth and resource use. The third results from increased mobility and trade. A fourth major cause of extinction, anthropogenic (human-caused) climate change, has not yet had a large impact, but it is predicted to become significant during this century. Global climate change is also a consequence of human population needs for energy and the use of fossil fuels to meet those needs (Figure 21.7). Environmental issues, such as toxic pollution, have specific targeted effects on species, but are not generally seen as threats at the magnitude of the others.

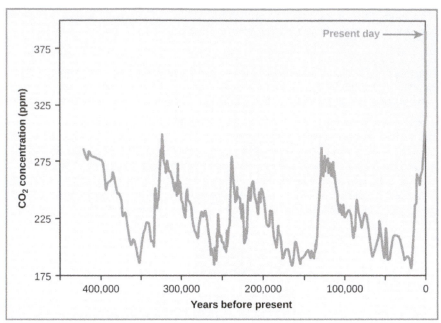

Figure 21.7 Atmospheric carbon dioxide levels fluctuate in a cyclical manner. However, the burning of fossil fuels in recent history has caused a dramatic increase in the levels of carbon dioxide in the Earth's atmosphere, which have now reached levels never before seen on Earth. Scientists predict that the addition of this "greenhouse gas" to the atmosphere is resulting in climate change that will significantly impact biodiversity in the coming century.

Habitat Loss

Humans rely on technology to modify their environment and replace certain functions that were once performed by the natural ecosystem. Other species cannot do this. Elimination of their habitat—whether it is a forest, coral reef, grassland, or flowing river—will kill the individuals in the species. Remove the entire habitat within the range of a species and, unless they are one of the few species that do well in human-built environments, the species will become extinct. Human destruction of habitats (habitats generally refer to the part of the ecosystem required by a particular species) accelerated in the latter half of the twentieth century. Consider the exceptional biodiversity of Sumatra: it is home to one species of orangutan, a species of critically endangered elephant, and the Sumatran tiger, but half of Sumatra's forest is now gone. The neighboring island of Borneo, home to the other species of orangutan, has lost a similar area of forest. Forest loss continues in protected areas of Borneo. The orangutan in Borneo is listed as endangered by the International Union for Conservation of Nature (IUCN), but it is simply the most visible of thousands of species that will not survive the disappearance of the forests of Borneo. The forests are removed for timber and to plant palm oil plantations (Figure 21.8). Palm oil is used in many products including food products, cosmetics, and biodiesel in Europe. A 5-year estimate of global forest cover loss for the years from 2000 to 2005 was 3.1 percent. Much loss (2.4 percent) occurred in the humid tropics where forest loss is primarily from timber extraction. These losses certainly also represent the extinction of species unique to those areas.

Figure 21.8 An oil palm plantation in Sabah province Borneo, Malaysia, replaces native forest habitat that a variety of species depended on to live. (credit: Lian Pin Koh)

biology IN ACTION

Preventing Habitat Destruction with Wise Wood Choices

Most consumers do not imagine that the home improvement products they buy might be contributing to habitat loss and species extinctions. Yet the market for illegally harvested tropical timber is huge, and the wood products often find themselves in building supply stores in the United States. One estimate is that 10 percent of the imported timber stream in the United States, which is the world's largest consumer of wood products, is potentially illegally logged. In 2006, this amounted to $3.6 billion in wood products. Most of the illegal products are imported from countries that act as intermediaries and are not the originators of the wood.

How is it possible to determine if a wood product, such as flooring, was harvested sustainably or even legally? The Forest Stewardship Council (FSC) certifies sustainably harvested forest products; therefore, looking for their certification on flooring and other hardwood products is one way to ensure that the wood has not been taken illegally from a tropical forest. Certification applies to specific products, not to a producer; some producers' products may not have certification while other products are certified. There are certifications other than the FSC, but these are run by timber companies creating a conflict of interest. Another approach is to buy domestic wood species. While it would be great if there was a list of legal versus illegal woods, it is not that simple. Logging and forest management laws vary from country to country; what is illegal in one country may be legal in another. Where and how a product is harvested and whether the forest from which it comes is being sustainably maintained all factor into whether a wood product will be certified by the FSC. It is always a good idea to ask questions about where a wood product came from and how the supplier knows that it was harvested legally.

Habitat destruction can affect ecosystems other than forests. Rivers and streams are important ecosystems and are frequently the target of habitat modification through building and from damming or water removal. Damming of rivers affects flows and access to all parts of a river. Altering a flow regime can reduce or eliminate populations that are adapted to seasonal changes in flow. For example, an estimated 91 percent of river lengths in the United States have been modified with damming or bank modifications. Many fish species in the United States, especially rare species or species with restricted distributions, have seen declines caused by river damming and habitat loss. Research has confirmed that species of amphibians that must carry out parts of their life cycles in both aquatic and terrestrial habitats are at greater risk of population declines and extinction because of the increased likelihood that one of their habitats or access between them will be lost. This is of particular concern because amphibians have been declining in numbers and going extinct more rapidly than many other groups for a variety of possible reasons.

Overharvesting

Overharvesting is a serious threat to many species, but particularly to aquatic species. There are many examples of regulated fisheries (including hunting of marine mammals and harvesting of crustaceans and other species) monitored by fisheries scientists that have nevertheless collapsed. The western Atlantic cod fishery is the most spectacular recent collapse. While it was a hugely productive fishery for 400 years, the introduction of modern factory trawlers in the 1980s and the pressure on the fishery led to it becoming unsustainable. The causes of fishery collapse are both economic and political in nature. Most fisheries are managed as a common resource, available to anyone willing to fish, even when the fishing territory lies within a country's territorial waters. Common resources are subject to an economic pressure known as the **tragedy of the commons**, in which fishers have little motivation to exercise restraint in harvesting a fishery when they do not own the fishery. The general outcome of harvests of resources held in common is their overexploitation. While large fisheries are regulated to attempt to avoid this pressure, it still exists in the background. This overexploitation is exacerbated when access to the fishery is open and unregulated and when technology gives fishers the ability to overfish. In a few fisheries, the biological growth of the resource is less than the potential growth of the profits made from fishing if that time and money were invested elsewhere. In these cases—whales are an example—economic forces will drive toward fishing the population to extinction.

Explore a U.S. Fish & Wildlife Service interactive map (http://openstaxcollege.org/l/habitat_map2) of critical habitat for endangered and threatened species in the United States. To begin, select "Visit the online mapper."

For the most part, fishery extinction is not equivalent to biological extinction—the last fish of a species is rarely fished out of the ocean. But there are some instances in which true extinction is a possibility. Whales have slow-growing populations and are at risk of complete extinction through hunting. Also, there are some species of sharks with restricted distributions that are at risk of extinction. The groupers are another population of generally slow-growing fishes that, in the Caribbean, includes a number of species that are at risk of extinction from overfishing.

Coral reefs are extremely diverse marine ecosystems that face peril from several processes. Reefs are home to 1/3 of the world's marine fish species—about 4000 species—despite making up only one percent of marine habitat. Most home marine aquaria house coral reef species that are wild-caught organisms—not cultured organisms. Although no marine species is known to have been driven extinct by the pet trade, there are studies showing that populations of some species have declined in response to harvesting, indicating that the harvest is not sustainable at those levels. There are also concerns about the effect of the pet trade on some terrestrial species such as turtles, amphibians, birds, plants, and even the orangutans.

View a brief video (http://openstaxcollege.org/l/ocean_matters2) discussing the role of marine ecosystems in supporting human welfare and the decline of ocean ecosystems.

Bush meat is the generic term used for wild animals killed for food. Hunting is practiced throughout the world, but hunting practices, particularly in equatorial Africa and parts of Asia, are believed to threaten several species with extinction. Traditionally, bush meat in Africa was hunted to feed families directly; however, recent commercialization of the practice now has bush meat available in grocery stores, which has increased harvest rates to the level of unsustainability. Additionally, human population growth has increased the need for protein foods that are not being met from agriculture. Species threatened by the bush meat trade are mostly mammals including many monkeys and the great apes living in the Congo basin.

Download for free at https://openstax.org/details/books/concepts-biology

Exotic Species

Exotic species are species that have been intentionally or unintentionally introduced by humans into an ecosystem in which they did not evolve. Human transportation of people and goods, including the intentional transport of organisms for trade, has dramatically increased the introduction of species into new ecosystems. These new introductions are sometimes at distances that are well beyond the capacity of the species to ever travel itself and outside the range of the species' natural predators.

Most exotic species introductions probably fail because of the low number of individuals introduced or poor adaptation to the ecosystem they enter. Some species, however, have characteristics that can make them especially successful in a new ecosystem. These exotic species often undergo dramatic population increases in their new habitat and reset the ecological conditions in the new environment, threatening the species that exist there. When this happens, the exotic species also becomes an invasive species. Invasive species can threaten other species through competition for resources, predation, or disease.

Explore this interactive global database (http://openstaxcollege.org/l/exotic_invasiv2) of exotic or invasive species.

Lakes and islands are particularly vulnerable to extinction threats from introduced species. In Lake Victoria, the intentional introduction of the Nile perch was largely responsible for the extinction of about 200 species of cichlids. The accidental introduction of the brown tree snake via aircraft (Figure 21.9) from the Solomon Islands to Guam in 1950 has led to the extinction of three species of birds and three to five species of reptiles endemic to the island. Several other species are still threatened. The brown tree snake is adept at exploiting human transportation as a means to migrate; one was even found on an aircraft arriving in Corpus Christi, Texas. Constant vigilance on the part of airport, military, and commercial aircraft personnel is required to prevent the snake from moving from Guam to other islands in the Pacific, especially Hawaii. Islands do not make up a large area of land on the globe, but they do contain a disproportionate number of endemic species because of their isolation from mainland ancestors.

Figure 21.9 The brown tree snake, *Boiga irregularis*, is an exotic species that has caused numerous extinctions on the island of Guam since its accidental introduction in 1950. (credit: NPS)

Many introductions of aquatic species, both marine and freshwater, have occurred when ships have dumped ballast water taken on at a port of origin into waters at a destination port. Water from the port of origin is pumped into tanks on a ship empty of cargo to increase stability. The water is drawn from the ocean or estuary of the port and typically contains living organisms such as plant parts, microorganisms, eggs, larvae, or aquatic animals. The water is then pumped out before the ship takes on cargo at the destination port, which may be on a different continent. The zebra mussel was introduced to the Great Lakes from Europe prior to 1988 in ship ballast. The zebra mussels in the Great Lakes have cost the industry millions of dollars in clean up costs to maintain water intakes and other facilities. The mussels have also altered the ecology of the lakes dramatically. They threaten native mollusk populations, but have also benefited some species, such as smallmouth bass. The mussels are filter feeders and have dramatically improved water clarity, which in turn has allowed aquatic plants to grow along shorelines, providing shelter for young fish where it did not exist before. The European green crab, *Carcinus maenas*, was introduced to San Francisco Bay in the late 1990s, likely in ship ballast water, and has spread north along the coast to Washington. The crabs have been found to dramatically reduce the abundance of native clams and crabs with resulting increases in the prey of native crabs.

Invading exotic species can also be disease organisms. It now appears that the global decline in amphibian species recognized in the 1990s is, in some part, caused by the fungus *Batrachochytrium dendrobatidis*, which causes the disease **chytridiomycosis** (Figure 21.10). There is evidence that the fungus is native to Africa and may have been spread throughout the world by transport of a commonly used laboratory and pet species: the African clawed frog, *Xenopus laevis*. It may well be that biologists themselves are responsible for spreading this disease worldwide. The North American bullfrog, *Rana catesbeiana*, which has also been widely introduced as a food animal but which easily escapes captivity, survives most infections of *B. dendrobatidis* and can act as a reservoir for the disease.

Figure 21.10 This Limosa harlequin frog (*Atelopus limosus*), an endangered species from Panama, died from a fungal disease called chytridiomycosis. The red lesions are symptomatic of the disease. (credit: Brian Gratwicke)

Early evidence suggests that another fungal pathogen, *Geomyces destructans*, introduced from Europe is responsible for **white-nose syndrome**, which infects cave-hibernating bats in eastern North America and has spread from a point of origin in western New York State (Figure 21.11). The disease has decimated bat populations and threatens extinction of species already listed as endangered: the Indiana bat, *Myotis sodalis*, and potentially the Virginia big-eared bat, *Corynorhinus townsendii virginianus*. How the fungus was introduced is unknown, but one logical presumption would be that recreational cavers unintentionally brought the fungus on clothes or equipment from Europe.

Figure 21.11 This little brown bat in Greeley Mine, Vermont, March 26, 2009, was found to have white-nose syndrome. (credit: modification of work by Marvin Moriarty, USFWS)

Climate Change

Climate change, and specifically the anthropogenic warming trend presently underway, is recognized as a major extinction threat, particularly when combined with other threats such as habitat loss. Anthropogenic warming of the planet has been observed and is hypothesized to continue due to past and continuing emission of greenhouse gases, primarily carbon dioxide and methane, into the atmosphere caused by the burning of fossil fuels and deforestation. These gases decrease the degree to which Earth is able to radiate heat energy created by the sunlight that enters the atmosphere. The changes in climate and energy balance caused by increasing greenhouse gases are complex and our understanding of them depends on predictions generated from detailed computer models. Scientists generally agree the present warming trend is caused by humans and some of the likely effects include dramatic and dangerous climate changes in the coming decades. However, there is still debate and a lack of understanding about specific outcomes. Scientists disagree about the likely magnitude of the effects on extinction rates, with estimates ranging from 15 to 40 percent of species committed to extinction by 2050. Scientists do agree that climate change will alter regional climates, including rainfall and snowfall patterns, making habitats less hospitable to the species living in them. The warming trend will shift colder climates toward the north and south poles, forcing species to move with their adapted climate norms, but also to face habitat gaps along the way. The shifting ranges will impose new competitive regimes on species as they find themselves in contact with other species not present in their historic range. One such unexpected species contact is between polar bears and grizzly bears. Previously, these two species had separate ranges. Now, their ranges are overlapping and there are documented cases of these two species mating and producing viable offspring. Changing climates also throw off the delicate timing adaptations that species have to seasonal food resources and breeding times. Scientists have already documented many contemporary mismatches to shifts in resource availability and timing.

Range shifts are already being observed: for example, on average, European bird species ranges have moved 91 km (56.5 mi) northward. The same study suggested that the optimal shift based on warming trends was double that distance, suggesting that the populations are not moving quickly enough. Range shifts have also been observed in plants, butterflies, other insects, freshwater fishes, reptiles, amphibians, and mammals.

Climate gradients will also move up mountains, eventually crowding species higher in altitude and eliminating the habitat for those species adapted to the highest elevations. Some climates will completely disappear. The rate of warming appears to be accelerated in the arctic, which is recognized as a serious threat to polar bear populations that require sea ice to hunt seals during the winter months: seals are the only source of protein available to polar bears. A trend to decreasing sea ice coverage has occurred since observations began in the mid-twentieth century. The rate of decline observed in recent years is far greater than previously predicted by climate models (Figure 21.12).

Figure 21.12 The effect of global warming can be seen in the continuing retreat of Grinnell Glacier. The mean annual temperature in Glacier National Park has increased 1.33°C since 1900. The loss of a glacier results in the loss of summer meltwaters, sharply reducing seasonal water supplies and severely affecting local ecosystems. (credit: USGS, GNP Archives)

Finally, global warming will raise ocean levels due to meltwater from glaciers and the greater volume occupied by warmer water. Shorelines will be inundated, reducing island size, which will have an effect on some species, and a number of islands will disappear entirely. Additionally, the gradual melting and subsequent refreezing of the poles, glaciers, and higher elevation mountains—a cycle that has provided freshwater to environments for centuries—will be altered. This could result in an overabundance of salt water and a shortage of fresh water.

21.3 | Preserving Biodiversity

By the end of this section, you will be able to:

- Describe biodiversity as the equilibrium of naturally fluctuating rates of extinction and speciation
- Explain the legislative framework for conservation
- Identify the factors important in conservation preserve design
- Identify examples of the effects of habitat restoration
- Identify the role of zoos in biodiversity conservation

Preserving biodiversity is an extraordinary challenge that must be met by greater understanding of biodiversity itself, changes in human behavior and beliefs, and various preservation strategies.

Change in Biodiversity through Time

The number of species on the planet, or in any geographical area, is the result of an equilibrium of two evolutionary processes that are ongoing: speciation and extinction. Both are natural "birth" and "death" processes of macroevolution. When speciation rates begin to outstrip extinction rates, the number of species will increase; likewise, the reverse is true when extinction rates begin to overtake speciation rates. Throughout the history of life on Earth, as reflected in the fossil record, these two processes have fluctuated to a greater or lesser extent, sometimes leading to dramatic changes in the number of species on the planet as reflected in the fossil record (Figure 21.13).

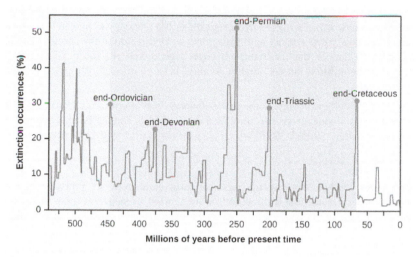

Figure 21.13 Extinction intensity as reflected in the fossil record has fluctuated throughout Earth's history. Sudden and dramatic losses of biodiversity, called mass extinctions, have occurred five times.

Paleontologists have identified five strata in the fossil record that appear to show sudden and dramatic (greater than half of all extant species disappearing from the fossil record) losses in biodiversity. These are called mass extinctions. There are many lesser, yet still dramatic, extinction events, but the five mass extinctions have attracted the most research into their causes. An argument can be made that the five mass extinctions are only the five most extreme events in a continuous series of large extinction events throughout the fossil record (since 542 million years ago). In most cases, the hypothesized causes are still controversial; in one, the most recent, the cause seems clear. The most recent extinction in geological time, about 65 million years ago, saw the disappearance of the dinosaurs and many other species. Most scientists now agree the cause of this extinction was the impact of a large asteroid in the present-day Yucatán Peninsula and the subsequent energy release and global climate changes caused by dust ejected into the atmosphere.

Recent and Current Extinction Rates

A sixth, or Holocene, mass extinction has mostly to do with the activities of *Homo sapiens*. There are numerous recent extinctions of individual species that are recorded in human writings. Most of these are coincident with the expansion of the European colonies since the 1500s.

One of the earlier and popularly known examples is the dodo bird. The dodo bird lived in the forests of Mauritius, an island in the Indian Ocean. The dodo bird became extinct around 1662. It was hunted for its meat by sailors and was easy prey because the dodo, which did not evolve with humans, would approach people without fear. Introduced pigs, rats, and dogs brought to the island by European ships also killed dodo young and eggs (Figure 21.14).

Figure 21.14 The dodo bird was hunted to extinction around 1662. (credit: Ed Uthman, taken in Natural History Museum, London, England)

Steller's sea cow became extinct in 1768; it was related to the manatee and probably once lived along the northwest coast of North America. Steller's sea cow was discovered by Europeans in 1741, and it was hunted for meat and oil. A total of 27 years elapsed between the sea cow's first contact with Europeans and extinction of the species. The last Steller's sea cow was killed in 1768. In another example, the last living passenger pigeon died in a zoo in Cincinnati, Ohio, in 1914. This species had once migrated in the millions but declined in numbers because of overhunting and loss of habitat through the clearing of forests for farmland.

These are only a few of the recorded extinctions in the past 500 years. The International Union for Conservation of Nature (IUCN) keeps a list of extinct and endangered species called the Red List. The list is not complete, but it describes 380 vertebrates that became extinct after 1500 AD, 86 of which were driven extinct by overhunting or overfishing.

Estimates of Present-day Extinction Rates

Estimates of **extinction rates** are hampered by the fact that most extinctions are probably happening without being observed. The extinction of a bird or mammal is often noticed by humans, especially if it has been hunted or used in some other way. But there are many organisms that are less noticeable to humans (not necessarily of less value) and many that are undescribed.

The background extinction rate is estimated to be about 1 per million species years (E/MSY). One "species year" is one species in existence for one year. One million species years could be one species persisting for one million years, or a million species persisting for one year. If it is the latter, then one extinction per million species years would be one of those million species becoming extinct in that year. For example, if there are 10 million species in existence, then we would expect 10 of those species to become extinct in a year. This is the background rate.

One contemporary extinction-rate estimate uses the extinctions in the written record since the year 1500. For birds alone, this method yields an estimate of 26 E/MSY, almost three times the background rate. However, this value may be underestimated for three reasons. First, many existing species would not have been described until much later in the time period and so their loss would have gone unnoticed. Second, we know the number is higher than the written record suggests because now extinct species are being described from skeletal remains that were never mentioned in written history. And third, some species are probably already extinct even though conservationists are reluctant to name them as such. Taking these factors into account raises the estimated extinction rate to nearer 100 E/MSY. The predicted rate by the end of the century is 1500 E/MSY.

A second approach to estimating present-time extinction rates is to correlate species loss with habitat loss, and it is based on measuring forest-area loss and understanding species–area relationships. The **species-area relationship** is the rate at which new species are seen when the area surveyed is increased (Figure 21.15). Likewise, if the habitat area is reduced, the number of species seen will also decline. This kind of relationship is also seen in the relationship between an island's area and the number of species present on the island: as one increases, so does the other, though not in a straight line. Estimates of extinction rates based on habitat loss and species–area relationships have suggested that with about 90 percent of habitat loss an expected 50 percent of species would become extinct. Figure 21.15 shows that reducing forest area from 100 km^2 to 10 km^2, a decline of 90 percent, reduces the number of species by about 50 percent. Species–area estimates have led to estimates of present-day species extinction rates of about 1000 E/MSY and higher. In general, actual observations do not show this amount of loss and one explanation put forward is that there is a delay in extinction. According to this explanation, it takes some time for species to fully suffer the effects of habitat loss and they linger on for some time after their habitat is destroyed, but eventually they will become extinct. Recent work has also called into question the applicability of the species-area relationship when estimating the loss of species. This work argues that the species–area relationship leads to an overestimate of extinction rates. Using an alternate method would bring estimates down to around 500 E/MSY in the coming century. Note that this value is still 500 times the background rate.

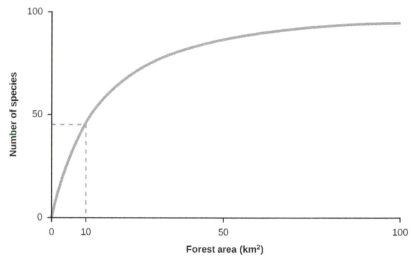

Figure 21.15 A typical species-area curve shows the cumulative number of species found as more and more area is sampled. The curve has also been interpreted to show the effect on species numbers of destroying habitat; a reduction in habitat of 90 percent from 100 km^2 to 10 km^2 reduces the number of species supported by about 50 percent.

Go to this website (http://openstaxcollege.org/l/whats_missing2) for an interactive exploration of endangered and extinct species, their ecosystems, and the causes of their endangerment or extinction.

Conservation of Biodiversity

The threats to biodiversity at the genetic, species, and ecosystem levels have been recognized for some time. In the United States, the first national park with land set aside to remain in a wilderness state was Yellowstone Park in 1890. However, attempts to preserve nature for various reasons have occurred for centuries. Today, the main efforts to preserve biodiversity involve legislative approaches to regulate human and corporate behavior, setting aside protected areas, and habitat restoration.

Changing Human Behavior

Legislation has been enacted to protect species throughout the world. The legislation includes international treaties as well as national and state laws. The Convention on International Trade in Endangered Species of Wild Fauna and Flora (CITES) treaty came into force in 1975. The treaty, and the national legislation that supports it, provides a legal framework for preventing "listed" species from being transported across nations' borders, thus protecting them from being caught or killed in the first place when the purpose involves international trade. The listed species that are protected to one degree or another by the treaty number some 33,000. The treaty is limited in its reach because it only deals with international movement of organisms or their parts. It is also limited by various countries' ability or willingness to enforce the treaty and supporting legislation. The illegal trade in organisms and their parts is probably a market in the hundreds of millions of dollars.

Within many countries there are laws that protect endangered species and that regulate hunting and fishing. In the United States, the Endangered Species Act was enacted in 1973. When an at-risk species is listed by the Act, the U.S. Fish & Wildlife Service is required by law to develop a management plan to protect the species and bring it back to sustainable numbers. The Act, and others like it in other countries, is a useful tool, but it suffers because it is often difficult to get a species listed, or to get an effective management plan in place once a species is listed. Additionally, species may be controversially taken off the list without necessarily having had a change in their situation. More fundamentally, the approach to protecting individual species rather than entire ecosystems (although the management plans commonly involve protection of the individual species' habitat) is both inefficient and focuses efforts on a few highly visible and often charismatic species, perhaps at the expense of other species that go unprotected.

The Migratory Bird Treaty Act (MBTA) is an agreement between the United States and Canada that was signed into law in 1918 in response to declines in North American bird species caused by hunting. The Act now lists over 800 protected species. It makes it illegal to disturb or kill the protected species or distribute their parts (much of the hunting of birds in the past was for their feathers). Examples of protected species include northern cardinals, the red-tailed hawk, and the American black vulture.

Global warming is expected to be a major driver of biodiversity loss. Many governments are concerned about the effects of anthropogenic global warming, primarily on their economies and food resources. Since greenhouse gas emissions do not respect national boundaries, the effort to curb them is an international one. The international response to global warming has been mixed. The Kyoto Protocol, an international agreement that came out of the United Nations Framework Convention on Climate Change that committed countries to reducing greenhouse gas emissions by 2012, was ratified by some countries, but spurned by others. Two countries that were especially important in terms of their potential impact that did not ratify the Kyoto protocol were the United States and China. Some goals for reduction in greenhouse gasses were met and exceeded by individual countries, but, worldwide, the effort to limit greenhouse gas production is not succeeding. The intended replacement for the Kyoto Protocol has not materialized because governments cannot agree on timelines and benchmarks. Meanwhile, the resulting costs to human societies and biodiversity predicted by a majority of climate scientists will be high.

As already mentioned, the non-profit, non-governmental sector plays a large role in conservation effort both in North America and around the world. The approaches range from species-specific organizations to the broadly focused IUCN and Trade Records Analysis of Flora and Fauna in Commerce (TRAFFIC). The Nature Conservancy takes a novel approach. It purchases land and protects it in an attempt to set up preserves for ecosystems. Ultimately, human behavior will change when human values change. At present, the growing urbanization of the human population is a force that mitigates against valuing biodiversity, because many people no longer come in contact with natural environments and the species that inhabit them.

Conservation in Preserves

Establishment of wildlife and ecosystem preserves is one of the key tools in conservation efforts (Figure 21.16). A preserve is an area of land set aside with varying degrees of protection for the organisms that exist within the boundaries of the preserve. Preserves can be effective for protecting both species and ecosystems, but they have some serious drawbacks.

Figure 21.16 National parks, such as Grand Teton National Park in Wyoming, help conserve biodiversity. (credit: Don DeBold)

A simple measure of success in setting aside preserves for biodiversity protection is to set a target percentage of land or marine habitat to protect. However, a more detailed preserve design and choice of location is usually necessary because of the way protected lands are allocated and how biodiversity is distributed: protected lands tend to contain less economically valuable resources rather than being set aside specifically for the species or ecosystems at risk. In 2003, the IUCN World Parks Congress estimated that 11.5 percent of Earth's land surface was covered by preserves of various kinds. This area is greater than previous goals; however, it only represents 9 out of 14 recognized major biomes and research has shown that 12 percent of all species live outside preserves; these percentages are much higher when threatened species are considered and when only high quality preserves are considered. For example, high quality preserves include only about 50 percent of threatened amphibian species. The conclusion must be that either the percentage of area protected must be increased, the percentage of high quality preserves must be increased, or preserves must be targeted with greater attention to biodiversity protection. Researchers argue that more attention to the latter solution is required.

A **biodiversity hotspot** is a conservation concept developed by Norman Myers in 1988. Hotspots are geographical areas that contain high numbers of endemic species. The purpose of the concept was to identify important locations on the planet for conservation efforts, a kind of conservation triage. By protecting hotspots, governments are able to protect a larger number of species. The original criteria for a hotspot included the presence of 1500 or more species of endemic plants and 70 percent of the area disturbed by human activity. There are now 34 biodiversity hotspots (Figure 21.17) that contain large numbers of endemic species, which include half of Earth's endemic plants.

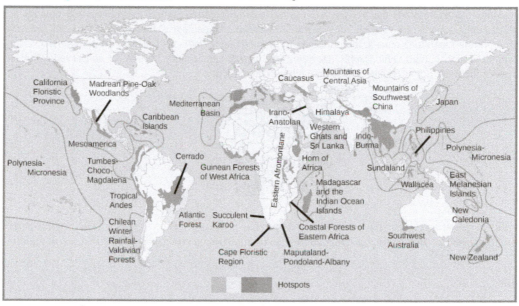

Figure 21.17 Conservation International has identified 34 biodiversity hotspots. Although these cover only 2.3 percent of the Earth's surface, 42 percent of the terrestrial vertebrate species and 50 percent of the world's plants are endemic to those hotspots.

There has been extensive research into optimal preserve designs for maintaining biodiversity. The fundamental principles behind much of the research have come from the seminal theoretical work of Robert H. MacArthur and Edward O. Wilson published in 1967 on island biogeography.[2] This work sought to understand the factors affecting biodiversity on islands. Conservation preserves can be seen as "islands" of habitat within "an ocean" of non-habitat. In general, large preserves are better because they support more species, including species with large home ranges; they have more core area of optimal habitat for individual species; they have more niches to support more species; and they attract more species because they can be found and reached more easily.

Preserves perform better when there are partially protected buffer zones around them of suboptimal habitat. The buffer allows organisms to exit the boundaries of the preserve without immediate negative consequences from hunting or lack of resources. One large preserve is better than the same area of several smaller preserves because there is more core habitat unaffected by less hospitable ecosystems outside the preserve boundary. For this same reason, preserves in the shape of a square or circle will be better than a preserve with many thin "arms." If preserves must be smaller, then providing wildlife corridors between them so that species and their genes can move between the preserves; for example, preserves along rivers and streams will make the smaller preserves behave more like a large one. All of these factors are taken into consideration when planning the nature of a preserve before the land is set aside.

In addition to the physical specifications of a preserve, there are a variety of regulations related to the use of a preserve. These can include anything from timber extraction, mineral extraction, regulated hunting, human habitation, and nondestructive human recreation. Many of the decisions to include these other uses are made based on political pressures rather than conservation considerations. On the other hand, in some cases, wildlife protection policies have been so strict that subsistence-living indigenous populations have been forced from ancestral lands that fell within a preserve. In other cases, even if a preserve is designed to protect wildlife, if the protections are not or cannot be enforced, the preserve status will have little meaning in the face of illegal poaching and timber extraction. This is a widespread problem with preserves in the tropics.

Some of the limitations on preserves as conservation tools are evident from the discussion of preserve design. Political and economic pressures typically make preserves smaller, never larger, so setting aside areas that are large enough is difficult. Enforcement of protections is also a significant issue in countries without the resources or political will to prevent poaching and illegal resource extraction.

---

2. Robert H. MacArthur and Edward O. Wilson, E. O., *The Theory of Island Biogeography* (Princeton, N.J.: Princeton University Press, 1967).

Climate change will create inevitable problems with the location of preserves as the species within them migrate to higher latitudes as the habitat of the preserve becomes less favorable. Planning for the effects of global warming on future preserves, or adding new preserves to accommodate the changes expected from global warming is in progress, but will only be as effective as the accuracy of the predictions of the effects of global warming on future habitats.

Finally, an argument can be made that conservation preserves reinforce the cultural perception that humans are separate from nature, can exist outside of it, and can only operate in ways that do damage to biodiversity. Creating preserves reduces the pressure on human activities outside the preserves to be sustainable and non-damaging to biodiversity. Ultimately, the political, economic, and human demographic pressures will degrade and reduce the size of conservation preserves if the activities outside them are not altered to be less damaging to biodiversity.

Check out this interactive global data system (http://openstaxcollege.org/l/protected_area2) of protected areas. Review data about specific protected areas by location or study statistics on protected areas by country or region.

Habitat Restoration

Habitat restoration holds considerable promise as a mechanism for maintaining or restoring biodiversity. Of course once a species has become extinct, its restoration is impossible. However, restoration can improve the biodiversity of degraded ecosystems. Reintroducing wolves, a top predator, to Yellowstone National Park in 1995 led to dramatic changes in the ecosystem that increased biodiversity. The wolves (Figure 21.18) function to suppress elk and coyote populations and provide more abundant resources to the guild of carrion eaters. Reducing elk populations has allowed revegetation of riparian (the areas along the banks of a stream or river) areas, which has increased the diversity of species in that habitat. Suppression of coyotes has increased the species previously suppressed by this predator. The number of species of carrion eaters has increased because of the predatory activities of the wolves. In this habitat, the wolf is a keystone species, meaning a species that is instrumental in maintaining diversity within an ecosystem. Removing a keystone species from an ecological community causes a collapse in diversity. The results from the Yellowstone experiment suggest that restoring a keystone species effectively can have the effect of restoring biodiversity in the community. Ecologists have argued for the identification of keystone species where possible and for focusing protection efforts on these species. It makes sense to return the keystone species to the ecosystems where they have been removed.

Figure 21.18 This photograph shows the Gibbon wolf pack in Yellowstone National Park, March 1, 2007. Wolves have been identified as a keystone species. (credit: Doug Smith, NPS)

Other large-scale restoration experiments underway involve dam removal. In the United States, since the mid-1980s, many aging dams are being considered for removal rather than replacement because of shifting beliefs about the ecological value of free-flowing rivers. The measured benefits of dam removal include restoration of naturally fluctuating water levels (often the purpose of dams is to reduce variation in river flows), which leads to increased fish diversity and improved water

Download for free at https://openstax.org/details/books/concepts-biology

quality. In the Pacific Northwest, dam removal projects are expected to increase populations of salmon, which is considered a keystone species because it transports nutrients to inland ecosystems during its annual spawning migrations. In other regions, such as the Atlantic coast, dam removal has allowed the return of other spawning anadromous fish species (species that are born in fresh water, live most of their lives in salt water, and return to fresh water to spawn). Some of the largest dam removal projects have yet to occur or have happened too recently for the consequences to be measured. The large-scale ecological experiments that these removal projects constitute will provide valuable data for other dam projects slated either for removal or construction.

The Role of Zoos and Captive Breeding

Zoos have sought to play a role in conservation efforts both through captive breeding programs and education (Figure 21.19). The transformation of the missions of zoos from collection and exhibition facilities to organizations that are dedicated to conservation is ongoing. In general, it has been recognized that, except in some specific targeted cases, captive breeding programs for endangered species are inefficient and often prone to failure when the species are reintroduced to the wild. Zoo facilities are far too limited to contemplate captive breeding programs for the numbers of species that are now at risk. Education, on the other hand, is a potential positive impact of zoos on conservation efforts, particularly given the global trend to urbanization and the consequent reduction in contacts between people and wildlife. A number of studies have been performed to look at the effectiveness of zoos on people's attitudes and actions regarding conservation; at present, the results tend to be mixed.

Figure 21.19 Zoos and captive breeding programs help preserve many endangered species, such as this golden lion tamarin. (credit: Garrett Ziegler)

KEY TERMS

biodiversity the variety of a biological system, typically conceived as the number of species, but also applying to genes, biochemistry, and ecosystems

biodiversity hotspot a concept originated by Norman Myers to describe a geographical region with a large number of endemic species and a large percentage of degraded habitat

bush meat a wild-caught animal used as food (typically mammals, birds, and reptiles); usually referring to hunting in the tropics of sub-Saharan Africa, Asia, and the Americas

chemical diversity the variety of metabolic compounds in an ecosystem

chytridiomycosis a disease of amphibians caused by the fungus *Batrachochytrium dendrobatidis;* thought to be a major cause of the global amphibian decline

ecosystem diversity the variety of ecosystems

endemic species a species native to one place

exotic species (also, invasive species) a species that has been introduced to an ecosystem in which it did not evolve

extinction the disappearance of a species from Earth; local extinction is the disappearance of a species from a region

extinction rate the number of species becoming extinct over time, sometimes defined as extinctions per million species–years to make numbers manageable (E/MSY)

genetic diversity the variety of genes and alleles in a species or other taxonomic group or ecosystem; the term can refer to allelic diversity or genome-wide diversity

habitat heterogeneity the number of ecological niches

secondary plant compound a compound produced as a byproduct of plant metabolic processes that is typically toxic, but is sequestered by the plant to defend against herbivores

species-area relationship the relationship between area surveyed and number of species encountered; typically measured by incrementally increasing the area of a survey and determining the cumulative numbers of species

tragedy of the commons an economic principle that resources held in common will inevitably be over-exploited

white-nose syndrome a disease of cave-hibernating bats in the eastern United States and Canada associated with the fungus *Geomyces destructans*

CHAPTER SUMMARY

21.1 Importance of Biodiversity

Biodiversity exists at multiple levels of organization, and is measured in different ways depending on the goals of those taking the measurements. These include numbers of species, genetic diversity, chemical diversity, and ecosystem diversity. The number of described species is estimated to be 1.5 million with about 17,000 new species being described each year. Estimates for the total number of eukaryotic species on Earth vary but are on the order of 10 million. Biodiversity is negatively correlated with latitude for most taxa, meaning that biodiversity is higher in the tropics. The mechanism for this pattern is not known with certainty, but several plausible hypotheses have been advanced.

Humans use many compounds that were first discovered or derived from living organisms as medicines: secondary plant compounds, animal toxins, and antibiotics produced by bacteria and fungi. More medicines are expected to be discovered in nature. Loss of biodiversity will impact the number of pharmaceuticals available to humans. Biodiversity may provide important psychological benefits to humans.

Crop diversity is a requirement for food security, and it is being lost. The loss of wild relatives to crops also threatens breeders' abilities to create new varieties. Ecosystems provide ecosystem services that support human agriculture: pollination, nutrient cycling, pest control, and soil development and maintenance. Loss of biodiversity threatens these ecosystem services and risks making food production more expensive or impossible. Wild food sources are mainly aquatic, but few are being managed for sustainability. Fisheries' ability to provide protein to human populations is threatened when extinction occurs.

21.2 Threats to Biodiversity

The core threats to biodiversity are human population growth and unsustainable resource use. To date, the most significant causes of extinction are habitat loss, introduction of exotic species, and overharvesting. Climate change is predicted to be a significant cause of extinction in the coming century. Habitat loss occurs through deforestation, damming of rivers, and other activities. Overharvesting is a threat particularly to aquatic species, but the taking of bush meat in the humid tropics threatens many species in Asia, Africa, and the Americas. Exotic species have been the cause of a number of extinctions and are especially damaging to islands and lakes. Exotic species' introductions are increasing because of the increased mobility of human populations and growing global trade and transportation. Climate change is forcing range changes that may lead to extinction. It is also affecting adaptations to the timing of resource availability that negatively affects species in seasonal environments. The impacts of climate change are currently greatest in the arctic. Global warming will also raise sea levels, eliminating some islands and reducing the area of all others.

21.3 Preserving Biodiversity

Five mass extinctions with losses of more than 50 percent of extant species are observable in the fossil record. Recent extinctions are recorded in written history and are the basis for one method of estimating contemporary extinction rates. The other method uses measures of habitat loss and species-area relationships. Estimates of contemporary extinction rates vary but are as high as 500 times the background rate, as determined from the fossil record, and are predicted to rise.

There is a legislative framework for biodiversity protection. International treaties such as CITES regulate the transportation of endangered species across international borders. Legislation within individual countries protecting species and agreements on global warming have had limited success; there is at present no international agreement on targets for greenhouse gas emissions. In the United States, the Endangered Species Act protects listed species but is hampered by procedural difficulties and a focus on individual species. The Migratory Bird Act is an agreement between Canada and the United States to protect migratory birds. The non-profit sector is also very active in conservation efforts in a variety of ways.

Conservation preserves are a major tool in biodiversity protection. Presently, 11 percent of Earth's land surface is protected in some way. The science of island biogeography has informed the optimal design of preserves; however, preserves have limitations imposed by political and economic forces. In addition, climate change will limit the effectiveness of present preserves in the future. A downside of preserves is that they may lessen the pressure on human societies to function more sustainably outside the preserves.

Habitat restoration has the potential to restore ecosystems to previous biodiversity levels before species become extinct. Examples of restoration include reintroduction of keystone species and removal of dams on rivers. Zoos have attempted to take a more active role in conservation and can have a limited role in captive breeding programs. Zoos also have a useful role in education.

ART CONNECTION QUESTIONS

1. Figure 21.6 The Svalbard seed vault is located on Spitsbergen island in Norway, which has an arctic climate. Why might an arctic climate be good for seed storage?

REVIEW QUESTIONS

2. The number of currently described species on the planet is about _____.
    a. 17,000
    b. 150,000
    c. 1.5 million
    d. 10 million

3. A secondary plant compound might be used for which of the following?
    a. a new crop variety
    b. a new drug

    c. a soil nutrient
    d. a crop pest

4. Pollination is an example of _____.
    a. a possible source of new drugs
    b. chemical diversity
    c. an ecosystem service
    d. crop pest control

5. Converting a prairie to a farm field is an example of _____.
    a. overharvesting

b. habitat loss
c. exotic species
d. climate change

6. Which two extinction risks may be a direct result of the pet trade?
a. climate change and exotic species introduction
b. habitat loss and overharvesting
c. overharvesting and exotic species introduction
d. habitat loss and climate change

7. What kind of ecosystem are exotic species especially threatening to?
a. deserts
b. marine ecosystems
c. islands
d. tropical forests

8. Certain species of parrot cannot be brought to the United States to be sold as pets. What is the name of the legislation that makes this illegal?
a. Red List
b. Migratory Bird Act
c. CITES
d. Endangered Species Act (ESA)

9. What is the name of the first international agreement on climate change?
a. Red List
b. Montreal Protocol
c. International Union for the Conservation of Nature (IUCN)
d. Kyoto Protocol

CRITICAL THINKING QUESTIONS

10. Explain how biodiversity loss can impact crop diversity.

11. Describe two types of compounds from living things that are used as medications.

12. Describe the mechanisms by which human population growth and resource use causes increased extinction rates.

13. Explain what extinction threats a frog living on a mountainside in Costa Rica might face.

14. Describe two considerations in conservation preserve design.

15. Describe what happens to an ecosystem when a keystone species is removed.

APPENDIX A | THE PERIODIC TABLE OF ELEMENTS

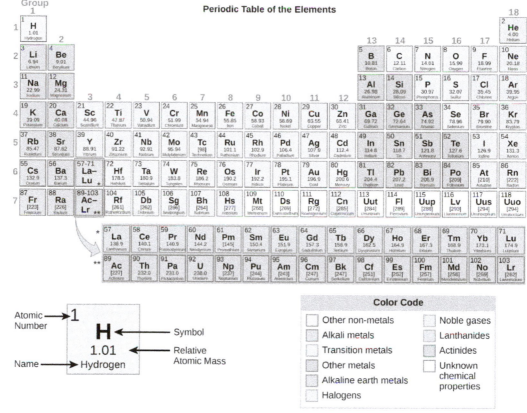

Figure A1

APPENDIX B | GEOLOGICAL TIME

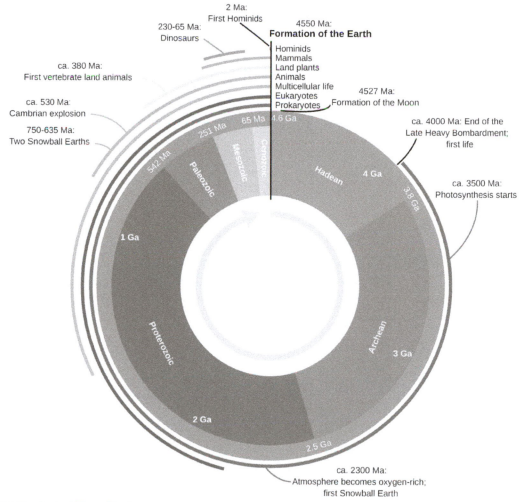

Figure B1 Geological Time Clock

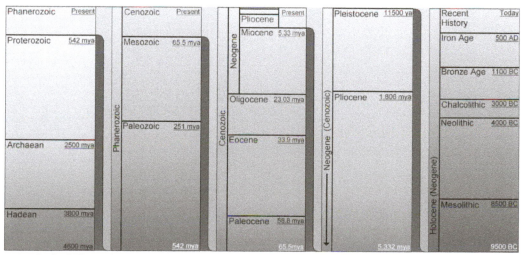

Figure B2 Geological Time Chart
(credit: Richard S. Murphy, Jr.)

APPENDIX C | MEASUREMENTS AND THE METRIC SYSTEM

C1 | Measurements and the Metric System

Measurements and the Metric System

Measurement	Unit	Abbreviation	Metric Equivalent	Approximate Standard Equivalent
Length	nanometer	nm	$1 \text{ nm} = 10^{-9} \text{ m}$	$1 \text{ mm} = 0.039 \text{ inch}$ $1 \text{ cm} = 0.394 \text{ inch}$ $1 \text{ m} = 39.37 \text{ inches}$ $1 \text{ m} = 3.28 \text{ feet}$ $1 \text{ m} = 1.093 \text{ yards}$ $1 \text{ km} = 0.621 \text{ miles}$
	micrometer	μm	$1 \text{ μm} = 10^{-6} \text{ m}$	
	millimeter	mm	$1 \text{ mm} = 0.001 \text{ m}$	
	centimeter	cm	$1 \text{ cm} = 0.01 \text{ m}$	
	meter	m	$1 \text{ m} = 100 \text{ cm}$ $1 \text{ m} = 1000 \text{ mm}$	
	kilometer	km	$1 \text{ km} = 1000 \text{ m}$	
Mass	microgram	μg	$1 \text{ μg} = 10^{-6} \text{ g}$	$1 \text{ g} = 0.035 \text{ ounce}$ $1 \text{ kg} = 2.205 \text{ pounds}$
	milligram	mg	$1 \text{ mg} = 10^{-3} \text{ g}$	
	gram	g	$1 \text{ g} = 1000 \text{ mg}$	
	kilogram	kg	$1 \text{ kg} = 1000 \text{ g}$	
Volume	microliter	μl	$1 \text{ μl} = 10^{-6} \text{ l}$	$1 \text{ ml} = 0.034 \text{ fluid ounce}$ $1 \text{ l} = 1.057 \text{ quarts}$ $1 \text{ kl} = 264.172 \text{ gallons}$
	milliliter	ml	$1 \text{ ml} = 10^{-3} \text{ l}$	
	liter	l	$1 \text{ l} = 1000 \text{ ml}$	
	kiloliter	kl	$1 \text{ kl} = 1000 \text{ l}$	
Area	square centimeter	cm^2	$1 \text{ cm}^2 = 100 \text{ mm}^2$	$1 \text{ cm}^2 = 0.155 \text{ square inch}$ $1 \text{ m}^2 = 10.764 \text{ square feet}$ $1 \text{ m}^2 = 1.196 \text{ square yards}$ $1 \text{ ha} = 2.471 \text{ acres}$
	square meter	m^2	$1 \text{ m}^2 = 10,000 \text{ cm}^2$	
	hectare	ha	$1 \text{ ha} = 10,000 \text{ m}^2$	
Temperature	Celsius	°C	—	$1 \text{ °C} = 5/9 \times (\text{°F} - 32)$

Table C1

ANSWER KEY

Chapter 1

1 Figure 1.8 B **3** C **5** A **7** Researchers can approach biology from the smallest to the largest, and everything in between. For instance, an ecologist may study a population of individuals, the population's community, the community's ecosystem, and the ecosystem's part in the biosphere. When studying an individual organism, a biologist could examine the cell and its organelles, the tissues that the cells make up, the organs and their respective organ systems, and the sum total—the organism itself.

Chapter 2

1 Figure 2.3 Potassium-39 has twenty neutrons. Potassium-40 has twenty one neutrons. **2** A **4** A **6** C **8** D **10** A **12** Hydrogen bonds and van der Waals interactions form weak associations between different molecules. They provide the structure and shape necessary for proteins and DNA within cells so that they function properly. Hydrogen bonds also give water its unique properties, which are necessary for life. **14** Water molecules are polar, meaning they have separated partial positive and negative charges. Because of these charges, water molecules are able to surround charged particles created when a substance dissociates. The surrounding layer of water molecules stabilizes the ion and keeps differently charged ions from reassociating, so the substance stays dissolved. **16** A change in gene sequence can lead to a different amino acid being added to a polypeptide chain instead of the normal one. This causes a change in protein structure and function. For example, in sickle cell anemia, the hemoglobin β chain has a single amino acid substitution. Because of this change, the disc-shaped red blood cells assume a crescent shape, which can result in serious health problems.

Chapter 3

1 Figure 3.7 Plant cells have plasmodesmata, a cell wall, a large central vacuole, chloroplasts, and plastids. Animal cells have lysosomes and centrosomes. **3** Figure 3.22 No, it must have been hypotonic, as a hypotonic solution would cause water to enter the cells, thereby making them burst. **4** C **6** D **8** D **10** A **12** C **15** The advantages of light microscopes are that they are easily obtained, and the light beam does not kill the cells. However, typical light microscopes are somewhat limited in the amount of detail that they can reveal. Electron microscopes are ideal because you can view intricate details, but they are bulky and costly, and preparation for the microscopic examination kills the specimen. Transmission electron microscopes are designed to examine the internal structures of a cell, whereas a scanning electron microscope only allows visualization of the surface of a structure. **17** "Form follows function" refers to the idea that the function of a body part dictates the form of that body part. As an example, organisms like birds or fish that fly or swim quickly through the air or water have streamlined bodies that reduce drag. At the level of the cell, in tissues involved in secretory functions, such as the salivary glands, the cells have abundant Golgi. **19** Water moves through a semipermeable membrane in osmosis because there is a concentration gradient across the membrane of solute and solvent. The solute cannot effectively move to balance the concentration on both sides of the membrane, so water moves to achieve this balance.

Chapter 4

1 Figure 4.6 A compost pile decomposing is an exergonic process. A baby developing from a fertilized egg is an endergonic process. Tea dissolving into water is an exergonic process. A ball rolling downhill is an exergonic process. **3** Figure 4.16 The illness is caused by lactic acid build-up. Lactic acid levels rise after exercise, making the symptoms worse. Milk sickness is rare today, but was common in the Midwestern United States in the early 1800s. **4** D **6** C **8** D **10** C **12** B **14** Physical exercise involves both anabolic and catabolic processes. Body cells break down sugars to provide ATP to do the work necessary for exercise, such as muscle contractions. This is catabolism. Muscle cells also must repair muscle tissue damaged by exercise by building new muscle. This is anabolism. **16** Most vitamins and minerals act as cofactors and coenzymes for enzyme action. Many enzymes require the binding of certain cofactors or coenzymes to be able to catalyze their reactions. Since enzymes catalyze many important reactions, it is critical to obtain sufficient vitamins and minerals from diet and supplements. Vitamin C (ascorbic acid) is a coenzyme necessary for the action of enzymes that build collagen. **18** The oxygen we inhale is the final electron acceptor in the electron transport chain and allows aerobic respiration to proceed, which is the most efficient pathway for harvesting energy in the form of ATP from food molecules. The carbon dioxide we breathe out is formed during the citric acid cycle when the bonds in carbon compounds are broken. **20** They are very economical. The substrates, intermediates, and products move between pathways and do so in response to finely tuned feedback inhibition loops that keep metabolism overall on an even keel. Intermediates in one pathway may occur in another, and they can move from one pathway to another fluidly in response to the needs of the cell.

Chapter 5

1 Figure 5.7 Levels of carbon dioxide (a reactant) will fall, and levels of oxygen (a product) will rise. As a result, the rate of photosynthesis will slow down. **2** C **4** C **6** C **8** B **10** A **12** To convert solar energy into chemical energy that cells can use to do work. **14** The energy is present initially as light. A photon of light hits chlorophyll, causing an electron to be energized.

The free electron travels through the electron transport chain, and the energy of the electron is used to pump hydrogen ions into the thylakoid space, transferring the energy into the electrochemical gradient. The energy of the electrochemical gradient is used to power ATP synthase, and the energy is transferred into a bond in the ATP molecule. In addition, energy from another photon can be used to create a high-energy bond in the molecule NADPH. **16** Photosynthesis takes the energy of sunlight and combines water and carbon dioxide to produce sugar and oxygen as a waste product. The reactions of respiration take sugar and consume oxygen to break it down into carbon dioxide and water, releasing energy. Thus, the reactants of photosynthesis are the products of respiration, and vice versa.

Chapter 6

1 Figure 6.4 D. The kinetochore becomes attached to the mitotic spindle. Sister chromatids line up at the metaphase plate. The kinetochore breaks apart and the sister chromatids separate. The nucleus reforms and the cell divides. **2** C **4** B **6** A **8** C **10** C **12** Human somatic cells have 46 chromosomes, including 22 homologous pairs and one pair of nonhomologous sex chromosomes. This is the 2n, or diploid, condition. Human gametes have 23 chromosomes, one each of 23 unique chromosomes. This is the n, or haploid, condition. **14** If one of the genes that produce regulator proteins becomes mutated, it produces a malformed, possibly non-functional, cell-cycle regulator. This increases the chance that more mutations will be left unrepaired in the cell. Each subsequent generation of cells sustains more damage. The cell cycle can speed up as a result of loss of functional checkpoint proteins. The cells can lose the ability to self-destruct. **16** The common components of eukaryotic cell division and binary fission are DNA duplication, segregation of duplicated chromosomes, and the division of the cytoplasmic contents.

Chapter 7

1 Figure 7.2 Yes, it will be able to reproduce asexually. **2** C **4** B **6** D **8** B **10** D **12** The offspring of sexually reproducing organisms are all genetically unique. Because of this, sexually reproducing organisms may have more successful survival of offspring in environments that change than asexually reproducing organisms, whose offspring are all genetically identical. In addition, the rate of adaptation of sexually reproducing organisms is higher, because of their increased variation. This may allow sexually reproducing organisms to adapt more quickly to competitors and parasites, who are evolving new ways to exploit or outcompete them. **14** Random alignment leads to new combinations of traits. The chromosomes that were originally inherited by the gamete-producing individual came equally from the egg and the sperm. In metaphase I, the duplicated copies of these maternal and paternal homologous chromosomes line up across the center of the cell to form a tetrad. The orientation of each tetrad is random. There is an equal chance that the maternally derived chromosomes will be facing either pole. The same is true of the paternally derived chromosomes. The alignment should occur differently in almost every meiosis. As the homologous chromosomes are pulled apart in anaphase I, any combination of maternal and paternal chromosomes will move toward each pole. The gametes formed from these two groups of chromosomes will have a mixture of traits from the individual's parents. Each gamete is unique. **16** The problems caused by trisomies arise because the genes on the chromosome that is present in three copies produce more product than genes on chromosomes with only two copies. The cell does not have a way to adjust the amount of product, and the lack of balance causes problems in development and the maintenance of the individual. Each chromosome is different, and the differences in survivability could have to do with the numbers of genes on the two chromosomes. Chromosome 21 may be a smaller chromosome, so there are fewer unbalanced gene products. It is also possible that chromosome 21 carries genes whose products are less sensitive to differences in dosage than chromosome 18. The genes may be less involved in critical pathways, or the differences in dosage may make less of a difference to those pathways.

Chapter 8

1 Figure 8.9 You cannot be sure if the plant is homozygous or heterozygous as the data set is too small: by random chance, all three plants might have acquired only the dominant gene even if the recessive one is present. **3** Figure 8.16 Half of the female offspring would be heterozygous ($X^W X^w$) with red eyes, and half would be homozygous recessive ($X^w X^w$) with white eyes. Half of the male offspring would be hemizygous dominant ($X^W Y$) with red eyes, and half would be hemizygous recessive ($X^w Y$) with white eyes. **4** B **6** A **8** C **10** D **12** C **14** The garden pea has flowers that close tightly during self-pollination. This helps to prevent accidental or unintentional fertilizations that could have diminished the accuracy of Mendel's data. **16** The Punnett square will be 2 × 2 and will have T and t along the top and T and t along the left side. Clockwise from the top left, the genotypes listed within the boxes will be TT, Tt, Tt, and tt. The genotypic ratio will be 1TT:2Tt:1tt. **18** Yes this child could have come from these parents. The child would have inherited an i allele from each parent and for this to happen the type A parent had to have genotype $I^A i$ and the type b parent had to have genotype $I^B i$.

Chapter 9

1 Figure 9.10 Ligase, as this enzyme joins together Okazaki fragments. **2** A **4** B **6** A **8** C **10** D **12** The DNA is wound around proteins called histones. The histones then stack together in a compact form that creates a fiber that is 30-nm thick. The fiber is further coiled for greater compactness. During metaphase of mitosis, the chromosome is at its most compact to facilitate chromosome movement. During interphase, there are denser areas of chromatin, called heterochromatin, that contain DNA that is not expressed, and less dense euchromatin that contains DNA that is expressed. **14** Telomerase has an inbuilt RNA template that extends the 3' end, so a primer is synthesized and extended. Thus, the ends are protected. **16** The cell controls which protein is expressed, and to what level that protein is expressed, in the cell. Prokaryotic cells alter the transcription rate to turn genes on or off. This method will increase or decrease protein levels in response to what is needed by the cell. Eukaryotic cells change the accessibility (epigenetic), transcription, or translation of a gene. This will alter the amount of RNA, and the lifespan of the RNA, to alter the amount of protein that exists. Eukaryotic cells also change the protein's translation to increase or decrease its overall levels. Eukaryotic organisms are much more complex and can manipulate protein levels by changing many stages in the process.

Chapter 10

1 Figure 10.7 Because even though the original cell came from a Scottish Blackface sheep and the surrogate mother was a Scottish Blackface, the DNA came from a Finn-Dorset. **2** B **4** A **6** C **8** D **10** The polymerase chain reaction is used to quickly produce many copies of a specific segment of DNA when only one or a very few copies are originally present. The benefit of PCR is that there are many instances in which we would like to know something about a sample of DNA when only very small amounts are available. PCR allows us to increase the number of DNA molecules so that other tests, such as sequencing, can be performed with it. **12** Genome mapping helps researchers to study disease-causing genes in humans. It also helps to identify traits of organisms that can be used in applications such as cleaning up pollution.

Chapter 11

1 Figure 11.7 Genetic drift is likely to occur more rapidly on an island, where smaller populations are expected to occur. **2** B **4** C **6** C **8** C **10** A **12** B **14** B **15** The plants that can best use the resources of the area, including competing with other individuals for those resources, will produce more seeds themselves and those traits that allowed them to better use the resources will increase in the population of the next generation. **17** The theory of natural selection stems from the observation that some individuals in a population survive longer and have more offspring than others, thus passing on more of their genes to the next generation. For example, a big, powerful male gorilla is much more likely than a smaller, weaker gorilla to become the population's silverback, the pack's leader who mates far more than the other males of the group. The pack leader will, therefore, father more offspring, who share half of his genes, and are thus likely to also grow bigger and stronger like their father. Over time, the genes for bigger size will increase in frequency in the population, and the population will, as a result, grow larger on average. **19** Organisms of one species can arrive to an island together and then disperse throughout the chain, each settling into different niches, exploiting different food resources and, evolving independently with little gene flow between different islands. **21** In science, a theory is a thoroughly tested and verified set of explanations for a body of observations of nature. It is the strongest form of knowledge in science. In contrast, a theory in common usage can mean a guess or speculation about something, meaning that the knowledge implied by the theory may be very weak.

Chapter 12

1 Figure 12.3 Cats and dogs are part of the same group at five levels: both are in the domain Eukarya, the kingdom Animalia, the phylum Chordata, the class Mammalia, and the order Carnivora. **3** C **5** D **7** B **9** A **11** B **13** The phylogenetic tree shows the order in which evolutionary events took place and in what order certain characteristics and organisms evolved in relation to others. It does not generally indicate time durations. **15** Dolphins are mammals and fish are not, which means that their evolutionary paths (phylogenies) are quite separate. Dolphins probably adapted to have a similar body plan after returning to an aquatic lifestyle, and therefore this trait is probably analogous. **17** The biologist looks at the state of the character in an outgroup, an organism that is outside the clade for which the phylogeny is being developed. The polarity of the character change is from the state of the character in the outgroup to the second state.

Chapter 13

1 Figure 13.6 A **2** B **4** D **6** C **8** D **10** C **12** C **14** Antibiotics kill bacteria that are sensitive to them; thus, only the resistant ones will survive. These resistant bacteria will reproduce, and therefore, after a while, there will be only resistant bacteria, making it more difficult to treat the diseases they may cause in humans. **16** Eukaryote cells arose through endosymbiotic events that gave rise to energy-producing organelles within the eukaryotic cells, such as mitochondria and plastids. The nuclear genome of eukaryotes is related most closely to the Archaea, so it may have been an early archaean that engulfed a bacterial cell that evolved into a mitochondrion. Mitochondria appear to have originated from an alpha-proteobacterium, whereas chloroplasts originated from a cyanobacterium. There is also evidence of secondary endosymbiotic events. Other cell components may have resulted from endosymbiotic events. **18** The trypanosomes that cause this disease are capable of expressing a glycoprotein coat with a different molecular structure with each generation. Because the immune system must respond to specific antigens to raise a meaningful defense, the changing nature of trypanosome antigens prevents the immune system from ever clearing this infection. Massive trypanosome infection eventually leads to host organ failure and death.

Chapter 14

1 Figure 14.19 B. The diploid zygote forms after the pollen tube has finished forming so that the male generative nucleus (sperm) can fuse with the female egg. **3** A **5** A **7** D **9** A **11** A **13** The sporangium of plants protects the spores from drying out. Apical meristems ensure that a plant is able to grow in the two directions required to acquire water and nutrients: up toward sunlight and down into the soil. The multicellular embryo is an important adaptation that improves survival of the developing plant in dry environments. The development of molecules that gave plants structural strength allowed them to grow higher on land and obtain more sunlight. A waxy cuticle prevents water loss from aerial surfaces. **15** It became possible to transport water and nutrients through the plant and not be limited by rates of diffusion. Vascularization allowed the development of leaves, which increased efficiency of photosynthesis and provided more energy for plant growth. **17** The resemblance between cycads and palm trees is only superficial. Cycads are gymnosperms and do not bear flowers or fruit. Unlike palms, cycads produce cones; large, female cones that produce naked seeds, and smaller male cones on separate plants.

Chapter 15

1 Figure 15.3 B   **3** Figure 15.33 A   **4** B   **6** D   **8** B   **10** A   **12** B   **14** C   **16** C   **18** A   **20** Specialized tissues allow more efficient functioning because differentiated tissue types can perform unique functions and work together in tandem to allow the animal to perform more functions. For example, specialized muscle tissue allows directed and efficient movement, and specialized nervous tissue allows for multiple sensory modalities as well as the ability to respond to various sensory information; these functions are not necessarily available to other non-animal organisms. **22** The sponges draw water carrying food particles into the spongocoel using the beating of flagella in the choanocytes. The food particles are caught by the collar of the choanocyte and brought into the cell by phagocytosis. Digestion of the food particle takes place inside the cell. The difference between this and the mechanisms of other animals is that digestion takes place within cells rather than outside of cells. It means that the organism can feed only on particles smaller than the cells themselves. **24** In a complete digestive system, food material is not mixed with waste material, so the digestion and uptake of nutrients can be more efficient. In addition, the complete digestive system allows for an orderly progression of digestion of food matter and the specialization of different zones of the digestive tract. **26** Mollusks have a large muscular foot that may be modified in various ways, such as into tentacles, but it functions in locomotion. They have a mantle, a structure of tissue that covers and encloses the dorsal portion of the animal and secretes the shell when it is present. The mantle encloses the mantle cavity, which houses the gills (when present), excretory pores, anus, and gonadopores. The coelom of mollusks is restricted to the region around the systemic heart. The main body cavity is a hemocoel. Many mollusks have a radula near the mouth that is used for scraping food. **28** During embryonic development, we also have a notochord, a dorsal hollow nerve tube, pharyngeal slits, and a post-anal tail. **30** A moist environment is required as frog eggs lack a shell and dehydrate quickly in dry environments.

Chapter 16

1 Figure 16.2 Pyrogens increase body temperature by causing the blood vessels to constrict, inducing shivering, and stopping sweat glands from secreting fluid. **3** Figure 16.9 B   **5** Figure 16.14 A   **6** C   **8** B   **10** C   **12** A   **14** C   **16** A   **18** A   **20** A   **22** B   **24** C   **26** A   **27** The body has a sensor that detects a deviation of the state of the cells or the body from the set point. The information is relayed to a control center, usually the brain, where signals go to effectors. Those effectors cause a negative feedback response that moves the state of the body in a direction back toward the set point. **29** Accessory organs play an important role in producing and delivering digestive juices to the intestine during digestion and absorption. Specifically, the salivary glands, liver, pancreas, and gallbladder play important roles. Malfunction of any of these organs can lead to disease states. **31** In the United States, obesity, particularly childhood obesity, is a growing concern. Some of the contributors to this situation include sedentary lifestyles and consuming more processed foods and less fruits and vegetables. As a result, even young children who are obese can face health concerns. **33** The sac-like structure of the alveoli increases their surface area. In addition, the alveoli are made of thin-walled cells. These features allows gases to easily diffuse across the cells. **35** The cells of both exocrine and endocrine glands produce a product that will be secreted by the gland. An exocrine gland has a duct and secretes its product to the outside of the gland, not into the bloodstream. An endocrine gland secretes its product into the bloodstream and does not use a duct. **37** Blood-glucose levels are regulated by hormones produced by the pancreas: insulin and glucagon. When blood-glucose levels are increasing, the pancreas releases insulin, which stimulates uptake of glucose by cells. When blood-glucose levels are decreasing, the pancreas releases glucagon, which stimulates the release of stored glucose by the liver to the bloodstream. **39** Neurons contain organelles common to all cells, such as a nucleus and mitochondria. They are unique because they contain dendrites, which can receive signals from other neurons, and axons that can send these signals to other cells. **41** The sympathetic nervous system prepares the body for "fight or flight," whereas the parasympathetic nervous system allows the body to "rest and digest." Sympathetic neurons release norepinephrine onto target organs; parasympathetic neurons release acetylcholine. Sympathetic neuron cell bodies are located in sympathetic ganglia. Parasympathetic neuron cell bodies are located in the brainstem and sacral spinal cord. Activation of the sympathetic nervous system increases heart rate and blood pressure and decreases digestion and blood flow to the skin. Activation of the parasympathetic nervous system decreases heart rate and blood pressure and increases digestion and blood flow to the skin.

Chapter 17

1 Figure 17.5 D   **3** Figure 17.17 If the blood of the mother and fetus mixes, memory cells that recognize the Rh antigen of the fetus can form in the mother late in the first pregnancy. During subsequent pregnancies, these memory cells launch an immune attack on the fetal blood cells of an Rh-positive fetus. Injection of anti-Rh antibody during the first pregnancy prevents the immune response from occurring. **4** B   **6** B   **8** B   **10** C   **12** A   **14** B   **16** The virus cannot attach to dog cells because dog cells do not express the receptors for the virus or there is no cell within the dog that is permissive for viral replication. **18** If the MHC class

I molecules expressed on donor cells differ from the MHC class I molecules expressed on recipient cells, NK cells may identify the donor cells as not normal and produce enzymes to induce the donor cells to undergo apoptosis, which would destroy the transplanted organ. **20** T cells bind antigens that have been digested and embedded in MHC molecules by APCs. In contrast, B cells function as APCs to bind intact, unprocessed antigens. **22** This is probably a delayed sensitivity reaction to one or more chemicals in the developer. An initial exposure would have sensitized the individual to the chemical and then subsequent exposures will induce a delayed inflammation reaction a day or two after exposure.

Chapter 18

1 Figure 18.12 D **3** B **5** A **7** A **9** D **11** B **13** Temperatures can vary from year to year and an unusually cold or hot year might produce offspring all of one sex, making it hard for individuals to find mates. **15** If multiple sperm fused with one egg, a zygote with a multiple ploidy level (multiple copies of the chromosomes) would form, and then would die. **17** Low levels of progesterone allow the hypothalamus to send GnRH to the anterior pituitary and cause the release of FSH and LH. FSH stimulates follicles on the ovary to grow and prepare the eggs for ovulation. As the follicles increase in size, they begin to release estrogen and a low level of progesterone into the blood. The level of estrogen rises to a peak, causing a spike in the concentration of LH. This causes the most mature follicle to rupture and ovulation occurs.

Chapter 19

1 Figure 19.2 Smaller animals require less food and others resources, so the environment can support more of them per unit area. **3** Figure 19.11 Stage 4 represents a population that is decreasing. **4** C **6** A **8** C **10** A **12** C **14** B **16** C **18** The researcher would mark a certain number of penguins with a tag, release them back into the population, and, at a later time, recapture penguins to see what percentage was tagged. This percentage would allow an estimation of the size of the penguin population. **20** If a natural disaster such as a fire happened in the winter, when populations are low, it would have a greater effect on the overall population and its recovery than if the same disaster occurred during the summer, when population levels are high. **22** The competitive exclusion principles states that no two species competing for the same resources at the same time and place can co-exist over time. Thus, one of the competing species will eventually dominate. On the other hand, if the species evolve such that they use resources from different parts of the habitat or at different times of day, the two species can exist together indefinitely.

Chapter 20

1 Figure 20.12 C: Nitrification by bacteria converts nitrates (NO_3^-) to nitrites (NO_3^-). **3** B **5** B **7** B **9** C **11** D **13** Grazing food webs have a producer at their base, which is either a plant for terrestrial ecosystems or a phytoplankton for aquatic ecosystems. The producers pass their energy to the various trophic levels of consumers. At the base of detrital food webs are the decomposers, which pass their energy to a variety of other consumers. Detrital food webs are important for the health of many grazing food webs because they eliminate dead and decaying organic material, thus clearing space for new organisms and removing potential causes of disease. **15** Fire is less common in desert biomes than in temperate grasslands because deserts have low net primary productivity, thus very little plant biomass to fuel a fire. **17** Organisms living in the intertidal zone must tolerate periodic exposure to air and sunlight and must be able to be periodically dry. They also must be able to endure the pounding waves; for this reason, some shoreline organisms have hard exoskeletons that provide protection while also reducing the likelihood of drying out.

Chapter 21

1 Figure 21.6 The ground is permanently frozen so the seeds will keep, even if the electricity fails. **2** C **4** C **6** C **8** C **10** Crop plants are derived from wild plants, and genes from wild relatives are frequently brought into crop varieties by plant breeders to add valued characteristics to the crops. If the wild species are lost, then this genetic variation would no longer be available. **12** Human population growth leads to unsustainable resource use, which causes habitat destruction to build new human settlements, create agricultural fields, and so on. Larger human populations have also led to unsustainable fishing and hunting of wild animal populations. Excessive use of fossil fuels also leads to global warming. **14** Larger preserves will contain more species. Preserves should have a buffer around them to protect species from edge effects. Preserves that are round or square are better than preserves with many thin arms.

INDEX

A

absorption spectrum, 124, 132
abyssal zone, 556, 563
acellular, 450, 472
acetyl CoA, 104, 113
acid, 51
Acid rain, 547
acid rain, 563
Acids, 38
acoelomate, 395
acoelomates, 360
Actinopterygii, 387, 395
action potential, 432, 440
activation energy, 97, 113
active immunity, 461, 472
active site, 98, 113
Active transport, 81
active transport, 85
adaptation, 253, 270
Adaptive immunity, 460
adaptive immunity, 472
adaptive radiation, 264, 270
adhesion, 37, 51
adrenal gland, 440
adrenal glands, 423
Age structure, 512
age structure, 525
algal bloom, 560, 563
allele, 194
alleles, 178
allergy, 469, 472
Allopatric speciation, 262
allopatric speciation, 270
allosteric inhibition, 100, 113
alternation of generations, 155, 170
alternative RNA splicing, 219, 220
alveoli, 415
alveolus, 440
amino acid, 51
Amino acids, 46
amniote, 395
amniotes, 389
amoebocyte, 395
Amoebocytes, 362
Amoebozoa, 306, 319
Amphibia, 388, 395
ampulla of Lorenzini, 395
ampullae of Lorenzini, 387
amygdala, 437, 440
amylase, 409, 440
anabolic, 93, 113

anaerobic, 292, 319
anaerobic cellular respiration, 113
analogous structure, 270, 283, 288
analogous structures, 253
anaphase, 140, 149
aneuploid, 165, 170
anion, 51
anions, 31
anneal, 245
annealing, 229
Annelida, 378, 395
anoxic, 292, 319
anther, 344, 351
Anthophyta, 347, 351
Anthropoids, 393
anthropoids, 395
antibody, 461, 472
antigen, 460, 472
antigen-presenting cell (APC), 462, 472
Anura, 388, 395
anus, 411, 440
aorta, 417, 440
apex consumer, 563
apex consumers, 531
aphotic zone, 555, 563
apical meristem, 329, 351
Apoda, 388, 395
apoptosis, 453, 472
appendicular skeleton, 428, 440
applied science, 22, 24
Archaeplastida, 306, 319
Arctic tundra, 553
arctic tundra, 563
Arteries, 419
artery, 440
Arthropoda, 371, 395
Ascomycota, 314, 319
Asexual reproduction, 478
asexual reproduction, 495
Asymmetrical, 358
asymmetrical, 395
atom, 9, 24
atomic number, 28, 51
ATP, 102, 113
ATP synthase, 107, 113
atrium, 417, 440
attenuation, 455, 472
auditory ossicles, 427, 440
autoantibody, 470, 472
Autoimmunity, 470
autoimmunity, 472
autonomic nervous system, 437, 440

autosome, 170
autosomes, 165
autotroph, 118, 132, 563
autotrophs, 535
axial skeleton, 426, 440
axon, 433, 440

B

B cell, 472
B cells, 460
Basal angiosperms, 348
basal angiosperms, 351
basal ganglia, 436, 440
base, 51
bases, 38
Basic science, 22
basic science, 24
Basidiomycota, 314
basidiomycota, 319
benthic realm, 555, 563
bicuspid valve, 417, 440
Bilateral symmetry, 359
bilateral symmetry, 395
Bile, 410
bile, 440
binary fission, 145, 149
binomial nomenclature, 276, 288
biodiversity, 568, 590
biodiversity hotspot, 586, 590
bioenergetics, 92, 113
biofilm, 294, 319
biogeochemical cycle, 537, 563
Biology, 5
biology, 24
Biomagnification, 536
biomagnification, 563
biomarker, 243, 245
biome, 531, 563
bioremediation, 301, 319
biosphere, 12, 24
Biotechnology, 225
biotechnology, 245
birth rate, 505, 525
Black Death, 297, 319
blastocyst, 483, 495
body plan, 356, 395
bolus, 409, 440
bones, 391
boreal forest, 552, 563
bottleneck effect, 256, 270
botulism, 299, 319
brachiation, 393, 395
brainstem, 437, 440
branch point, 279, 288

CPSIA information can be obtained
at www.ICGtesting.com
Printed in the USA
LVHW011330230122
709154LV00011B/1260

9 781680 921021